SCHAUM'S OUTLINE OF

THEORY AND PROBLEMS

OF

COLLEGE CHEMISTRY

SEVENTH EDITION

•

JEROME L. ROSENBERG, Ph.D.
Professor of Biological Sciences and Chemistry
University of Pittsburgh

and

LAWRENCE M. EPSTEIN, Ph.D.
Associate Professor of Chemistry, Emeritus
University of Pittsburgh

SCHAUM'S OUTLINE SERIES
McGRAW-HILL, INC.

New York   St. Louis   San Francisco   Auckland   Bogotá   Caracas
Hamburg   Lisbon   London   Madrid   Mexico   Milan   Montreal
New Delhi   Paris   San Juan   São Paulo   Singapore
Sydney   Tokyo   Toronto

JEROME L. ROSENBERG did his graduate work at Columbia University in physical chemistry, receiving his M.A. in 1944 and his Ph.D. in 1948. His research activities at several institutions resulted in many papers in photosynthesis, general photochemistry, and molecular electronic structure. Formerly Dean of the Faculty of Arts and Sciences and Vice Provost at the University of Pittsburgh he is currently Chairman of the Department of Biological Sciences in addition to being professor of biological sciences and chemistry there. Dr. Rosenberg has been associated with the preparation of *Schaum's Outline of College Chemistry* since the third edition, published in 1949.

LAWRENCE M. EPSTEIN started his career as a chemical engineer, then earned his M.S. in 1952 and Ph.D. in 1955 from Polytechnic University in the field of physical chemistry. He did research in the fields of radiation chemistry and in Mössbauer effect spectroscopy at Westinghouse Research Laboratories, and then served as Associate Professor and supervisor of the General Chemistry program at the University of Pittsburgh until he retired in 1986.

Schaum's Outline of Theory and Problems of
COLLEGE CHEMISTRY

2 3 4 5 6 7 8 9 10 11 12 13 14 15 16 17 18 19 20 SHP SHP   9 2 1 0

ISBN 0-07-053707-0

Sponsoring Editor, David Beckwith
Production Supervisor, Leroy Young
Editing Supervisor, Meg Tobin

Library of Congress Cataloging-in-Publication Data

Rosenberg, Jerome Laib
    Schaum's outline of theory and problems of college chemistry /
  Jerome L. Rosenberg and Lawrence M. Epstein.—7th ed.
        p.   cm.—(Schaum's outline series)
    Includes index.
    ISBN 0-07-053707-0
    1. Chemistry—Outlines, syllabi, etc.   I. Epstein, Lawrence M.
  II. Title.   III. Title: Outline of theory and problems of college
  chemistry.
  QD41.R67   1990   89-34888
  540—dc20   CIP

Cover design by Amy E. Becker.

Preface

This book is designed to help the student of college chemistry by summarizing the chemical principles of each topic and relating the solution of quantitative problems to those fundamentals. Although the book is not intended to replace a textbook, its solved problems, with complete and detailed solutions, do cover most of the subject matter of a first course in college chemistry. Both the solved and the supplementary problems are arranged to allow a progression in difficulty within each topic.

Several important features had been introduced into the sixth edition, notably the kinetic theory of gases, a more formal treatment of thermochemistry, a modern treatment of atomic properties and chemical bonding, and a chapter on chemical kinetics.

In this seventh edition the early chapters were revised to conform more closely to the methods used in current textbooks to introduce calculational skills to the beginning student. Some changes in notation were made, and there is now a more consistent usage of SI units. Every problem in the book was checked for accuracy, and the latest values of various physical and chemical constants were used. Some problems were removed and new ones written to replace them. An attempt was made to increase the variety of stoichiometry problems, especially in the chapters on gases and solutions, while eliminating some of the very complex problems that arise in gaseous and aqueous equilibria.

In the treatment of chemical bonding the subject of molecular orbitals has been deemphasized in favor of VSEPR theory. An elementary treatment of bonding in metals has been added, and more attention is given to intermolecular weak forces in liquids and solids.

A new chapter on Organic and Biochemistry has been added, conforming to the trend in current texts.

The use of SI units is still not universal; liter, atmosphere, and calorie are still used where appropriate, but the joule is favored, and the reader is made aware of the conversion to the bar as the standard-state pressure.

The authors acknowledge with thanks the careful word processing of Masha Epstein and the perceptive and helpful editing of David Beckwith.

Jerome L. Rosenberg
Lawrence M. Epstein

Contents

Chapter 1

Quantities and Units

INTRODUCTION

Most of the measurements and calculations in chemistry and physics are concerned with different kinds of *quantities*, e.g., length, velocity, volume, mass, energy. Every measurement includes both a number and a unit. The *unit* simultaneously identifies the kind of dimension and the magnitude of the reference quantity used as a basis for comparison. Many units are commonly used for the dimension of length, e.g., inch, yard, mile, centimeter, meter, kilometer. The *number* obviously indicates how many of the reference units are contained in the quantity being measured. Thus the statement that the length of a room is 20 feet means that the length of the room is 20 times the length of the foot, which in this case is the unit of length chosen for comparison. Although *20 feet* has the dimension of length, *20* is a pure number and is dimensionless, being the ratio of two lengths, that of the room and that of the reference foot.

It will be assumed that readers are familiar with the use of exponents, particularly powers-of-ten notation, and with the rules for significant figures. If not, Appendices A and B should be studied in conjunction with Chapter 1.

SYSTEMS OF MEASUREMENT

Dimensional calculations are greatly simplified if the unit for each kind of measure is expressed in terms of the units of selected reference dimensions. The three independent reference dimensions for mechanics are *length*, *mass*, and *time*. As examples of relating other quantities to the reference dimensions, the unit of speed is defined as unit length per unit time, the unit of volume is the cube of the unit of length, etc. Other reference dimensions, such as those used to express electrical and thermal phenomena, will be introduced later. There are several systems of units still in use in the English-speaking nations, so that one must occasionally make calculations to convert values from one system to another, e.g., inches to centimeters, or pounds to kilograms.

INTERNATIONAL SYSTEM OF UNITS

Considerable progress is being made in the acceptance of a common international system of reference units within the world scientific community. This system, known as SI from the French name, Système International d'Unités, has been adopted by many international bodies, including the International Union of Pure and Applied Chemistry. In SI, the reference units for *length*, *mass*, and *time* are the *meter*, *kilogram*, and *second*, with the symbols m, kg, and s, respectively.

To express quantities much larger or smaller than the standard units, use may be made of multiples or submultiples of these units, defined by applying as multipliers of these units certain recommended powers of ten, listed in Table 1-1. The multiplier abbreviation is to precede the symbol of the base unit without any space or punctuation. Thus, picosecond (10^{-12} s) is ps, and kilometer (10^3 m) is km. Since for historical reasons the SI unit for mass, kilogram, already has a prefix, multiples for mass should be derived by applying the multiplier to the *gram* rather than to the *kilogram*. Thus, microgram (10^{-6} g), abbreviated μg, is 10^{-9} kg.

Compound units can be derived by applying algebraic operations to the simple units.

EXAMPLE 1   The unit for volume in SI is the cubic meter (m^3), since

$$\text{Volume} = \text{length} \times \text{length} \times \text{length} = m \times m \times m = m^3$$

1

Table 1-1. Multiples and Submultiples for Units

Prefix	Abbreviation	Multiplier	Prefix	Abbreviation	Multiplier
deci	d	10^{-1}	deka	da	10
centi	c	10^{-2}	hecto	h	10^2
milli	m	10^{-3}	kilo	k	10^3
micro	μ	10^{-6}	mega	M	10^6
nano	n	10^{-9}	giga	G	10^9
pico	p	10^{-12}	tera	T	10^{12}
femto	f	10^{-15}			
atto	a	10^{-18}			

The unit for speed is the unit for length (distance) divided by the unit for time, or the meter per second (m/s), since

$$\text{Speed} = \frac{\text{distance}}{\text{time}} = \frac{\text{m}}{\text{s}}$$

The unit for density is the unit for mass divided by the unit for volume, or the kilogram per cubic meter (kg/m^3), since

$$\text{Density} = \frac{\text{mass}}{\text{volume}} = \frac{\text{kg}}{\text{m}^3}$$

Table 1-2. Some SI and Non-SI Units

Physical Quantity	Unit Name	Unit Symbol	Definition
Length	angstrom	Å	10^{-10} m
	inch	in	2.54×10^{-2} m
Area	square meter (SI)	m^2	
Volume	cubic meter (SI)	m^3	
	liter	L	dm^3
	cubic centimeter	cm^3	
Mass	atomic mass unit (dalton)	u	$1.660\,57 \times 10^{-27}$ kg
	pound	lb	0.453 592 37 kg
Density	kilogram per cubic meter (SI)	kg/m^3	
Force	newton (SI)	N	kg·m/s^2
Pressure	pascal (SI)	Pa	N/m^2
	bar	bar	10^5 Pa
	atmosphere	atm	101 325 Pa
	torr (millimeter of mercury)	torr (mm Hg)	atm/760 *or* 133.32 Pa

Symbols for compound units may be expressed in any of the following formats.

1. Multiplication of units. Example: *kilogram second*.
   (a) Dot   kg·s
   (b) Spacing   kg s
       (*not used in this book*)
2. Division of units. Example: *meter per second*.
   (a) Division sign $\dfrac{m}{s}$ or m/s
   (b) Negative power $m \cdot s^{-1}$ (or m s^{-1})

Note that the term *per* in a word definition is equivalent to *divide by* in the mathematical notation. Note also that symbols are not followed by a period, except at the end of a sentence.

In recognition of long-standing traditions, the various international commissions have acknowledged that a few non-SI units will remain in use in certain fields of science and in certain countries during a transitional period. Table 1-2 lists some units, both SI and non-SI, which will be used in this book, involving the quantities length, mass, and time. Other units will be introduced in subsequent chapters.

TEMPERATURE

Temperature may be defined as that property of a body which determines the flow of heat. Two bodies are at the same temperature if there is no transfer of heat when they are placed together. Temperature is an independent dimension which cannot be defined in terms of mass, length, and time. The SI unit of temperature is the *kelvin*, and 1 kelvin (K) is defined as 1/273.16 times the *triple point* temperature. The *triple point* is the temperature at which water coexists in equilibrium with ice at the pressure exerted by water vapor only. The *triple point* is 0.01 K above the *normal freezing point* of water, at which water and ice coexist in equilibrium with air at standard atmospheric pressure. The SI unit of temperature is so defined that 0 K is the absolute zero of temperature; the SI or Kelvin scale is often called the *absolute temperature scale*. Although absolute zero is never actually attainable, it has been approached to within less than 10^{-4} K.

OTHER TEMPERATURE SCALES

Other temperature scales have been employed at various times. The most useful ones are those which are linearly related to the Kelvin scale; that is, the ratio, r, of the difference between any two temperatures on such a scale to the number of kelvins separating the same two temperatures is a fixed number. One such scale, for which $r = 1$, is the *Celsius* (sometimes called the *centigrade*) scale. On the Celsius scale, the freezing point of water is 0 °C, the boiling point is 100 °C at one atmosphere pressure, and absolute zero is −273.15 °C. The *Fahrenheit scale* is one for which $r = \frac{9}{5}$. The absolute zero and the freezing point and boiling point of water on the Fahrenheit scale are −459.67 °F, 32 °F, and 212 °F, respectively. The relationships among these three scales, illustrated in Fig. 1-1, are described by the following linear equations, in which temperature on the SI scale is designated by T, and on the other scales by t.

$$\frac{t}{^\circ\text{C}} = \frac{T}{\text{K}} - 273.15 \qquad \text{or} \qquad t = \left(\frac{T}{\text{K}} - 273.15\right)\,^\circ\text{C}$$

$$\frac{t}{^\circ\text{F}} = \frac{9}{5}\left(\frac{t}{^\circ\text{C}}\right) + 32 \qquad \text{or} \qquad t = \left[\frac{9}{5}\left(\frac{t}{^\circ\text{C}}\right) + 32\right]^\circ\text{F}$$

$$\frac{t}{^\circ\text{C}} = \frac{5}{9}\left(\frac{t}{^\circ\text{F}} - 32\right) \qquad \text{or} \qquad t = \frac{5}{9}\left(\frac{t}{^\circ\text{F}} - 32\right)^\circ\text{C}$$

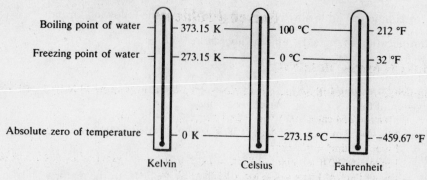

Fig. 1-1.

Here the ratios T/K, $t/°C$, $t/°F$ are the dimensionless numerical measures of the temperature on the Kelvin, Celsius, and Fahrenheit scales, respectively. Read the upper left equation as follows: The temperature in degrees *Celsius* equals the temperature in kelvins minus 273.15.

USE AND MISUSE OF UNITS

The units (e.g., cm, kg, g/mL, ft/s) must be regarded as a necessary part of the complete specification of a physical quantity. It is as foolish to separate the number of a measure from its unit as it is to separate a laboratory reagent bottle from its label. When physical quantities are subjected to mathematical operations, the units must be carried along with the numbers and must undergo the same operations as the numbers. Quantities cannot be added or subtracted directly unless they have not only the same dimensions but also the same units.

EXAMPLE 2   It is obvious that we cannot add 5 hours (time) to 20 miles/hour (speed) since *time* and *speed* have different physical significance. If we are to add 2 lb (mass) and 4 kg (mass), we must first convert lb to kg or kg to lb. Quantities of various types, however, can be combined in multiplication or division, in which *the units as well as the numbers* obey the algebraic laws of squaring, cancellation, etc. Thus:

1. $6 L + 2 L = 8 L$
2. $(5 \text{ cm})(2 \text{ cm}^2) = 10 \text{ cm}^3$
3. $(3 \text{ ft}^3)(200 \text{ lb/ft}^3) = 600 \text{ lb}$
4. $(2 \text{ s})(3 \text{ ft/s}^2) = 6 \text{ ft/s}$
5. $\dfrac{15 \text{ g}}{3 \text{ g/cm}^3} = 5 \text{ cm}^3$

ESTIMATION OF NUMERICAL ANSWERS

If one's calculator is working correctly and is accurately used, the answer will be correct. But if not, will an incorrect answer be recognized? A very important skill is to determine first, by visual inspection, an approximate answer. Especially important is the correct order of magnitude, represented by the location of the decimal point or the power of 10, which may go astray even though the digits are correct.

EXAMPLE 3   Consider the multiplication: $122 \text{ g} \times 0.051\,8 = 6.32 \text{ g}$. Visual inspection shows that 0.051 8 is a little more than $\frac{1}{20}$, and $\frac{1}{20}$ of 122 is a little more than 6. Hence the answer should be a little more than 6 g, which it is. If the answer were given incorrectly as 63.2 g or 0.632 g, visual inspection or mental checking of the result would indicate that the decimal point had been misplaced.

Solved Problems

UNITS BASED ON MASS OR LENGTH

1.1. The following examples illustrate conversions among various units of length, volume, or mass.

$$1 \text{ inch} = 2.54 \text{ cm} = 0.025\ 4 \text{ m} = 25.4 \text{ mm} = 2.54 \times 10^7 \text{ nm}$$
$$1 \text{ foot} = 12 \text{ in} = 12 \times 2.54 \text{ cm} = 30.48 \text{ cm} = 0.304\ 8 \text{ m} = 304.8 \text{ mm}$$
$$1 \text{ liter} = 1 \text{ dm}^3 = 10^{-3} \text{ m}^3$$
$$1 \text{ mile} = 5\ 280 \text{ ft} = 1.609 \times 10^5 \text{ cm} = 1.609 \times 10^3 \text{ m} = 1.609 \text{ km} = 1.609 \times 10^6 \text{ mm}$$
$$1 \text{ pound} = 0.453\ 6 \text{ kg} = 453.6 \text{ g} = 4.536 \times 10^5 \text{ mg}$$
$$1 \text{ metric ton} = 1\ 000 \text{ kg} = 10^6 \text{ g}$$

1.2. Convert 5.00 inches to (a) centimeters, (b) millimeters, (c) meters.

(a)
$$5.00 \text{ in} = (5.00 \text{ in})(2.54 \text{ cm/in}) = 12.7 \frac{\text{in} \cdot \text{cm}}{\text{in}} = 12.7 \text{ cm}$$

The procedure can be understood easily in terms of the word definition of the conversion factor: 2.54 is the number of cm *per* in; that is, the number of cm in 1 in. Thus the number of cm in 5 in is 5×2.54.

More formally, the conversion factor may be considered to be a statement of equality between 2.54 cm and 1 in. Since 2.54 cm = 1 in,

$$2.54 \text{ cm per in} = 2.54 \text{ cm/in} = \frac{2.54 \text{ cm}}{1 \text{ in}} = \frac{2.54 \text{ cm}}{2.54 \text{ cm}} = 1$$

Thus the conversion factor, 2.54 cm/in, is mathematically equal to 1, so that any quantity may be multiplied or divided by the conversion factor without changing the essential value of the quantity.

The inverse of the above conversion factor, i.e., 1 in/2.54 cm, must likewise be equal to 1 and is also a conversion factor.

In this problem the multiplication of 5.00 in by 2.54 cm/in leads to the useful result 12.7 cm, a statement of the same quantity of length in different units. The multiplication of 5.00 in by 1 in/2.54 cm, although mathematically allowed, does not lead to a useful result because the result, 1.97 in²/cm, is not expressed in a simple unit which has a physical interpretation. In general, any proper conversion factor is equal to 1 and either it or its inverse may be used as a multiplier of any physical quantity. The choice depends on the particular problem.

(b)
$$12.7 \text{ cm} = (12.7 \text{ cm})(10 \text{ mm/cm}) = 127 \frac{\text{cm} \cdot \text{mm}}{\text{cm}} = 127 \text{ mm}$$

(c)
$$12.7 \text{ cm} = (12.7 \text{ cm})(1 \text{ m}/100 \text{ cm}) = 0.127 \frac{\text{cm} \cdot \text{m}}{\text{cm}} = 0.127 \text{ m}$$

Obviously multiplication by a conversion factor is equivalent to division by its inverse. Thus:

$$12.7 \text{ cm} = \frac{12.7 \text{ cm}}{100 \text{ cm/m}} = 0.127 \frac{\text{cm} \cdot \text{m}}{\text{cm}} = 0.127 \text{ m}$$

Here it was appropriate to divide by the conversion factor 100 cm/m. If by mistake we had multiplied 12.7 cm by 100 cm/m, the answer would have been expressed in cm²/m, since

$$\text{cm} \times \frac{\text{cm}}{\text{m}} = \frac{\text{cm}^2}{\text{m}}$$

We would immediately realize our error, seeing that the answer is not expressed in meters.

1.3.   Convert (*a*) 14.0 cm and (*b*) 7.00 m to inches.

(*a*) $14.0 \text{ cm} = (14.0 \text{ cm})\left(\dfrac{1 \text{ in}}{2.54 \text{ cm}}\right) = 5.51 \text{ in}$ or $14.0 \text{ cm} = \dfrac{14.0 \text{ cm}}{2.54 \text{ cm/in}} = 5.51 \text{ in}$

Factors such as $\dfrac{1 \text{ in}}{2.54 \text{ cm}}$ will frequently appear in the form 1 in/2.54 cm, especially in computer displays, because it allows the entire calculation to appear on one line of type.

(*b*) $7.00 \text{ m} = (7.00 \text{ m})\left(\dfrac{100 \text{ cm}}{1 \text{ m}}\right)\left(\dfrac{1 \text{ in}}{2.54 \text{ cm}}\right) = 276 \text{ in}$

Note that the use of two successive conversion factors was necessary. Both the units, m and cm, canceled out leaving the desired unit, in.

1.4.   How many square inches in one square meter?

$$1 \text{ m} = (1 \text{ m})(100 \text{ cm/1 m})(1 \text{ in/2.54 cm}) = 39.37 \text{ in}$$

$$1 \text{ m}^2 = (1 \text{ m})^2 = (39.37 \text{ in})^2 = 1\,550 \text{ in}^2$$

alternatively,

$$1 \text{ m}^2 = (1 \text{ m})^2(100 \text{ cm/1 m})^2(1 \text{ in/2.54 cm})^2$$

$$= [(100)^2/(2.54)^2] \text{ in}^2 = 1\,550 \text{ in}^2$$

Note that since a conversion factor is equal to 1, it may be squared (or raised to any power) without changing its value.

1.5.   (*a*) How many cubic centimeters in one cubic meter?   (*b*) How many liters in one cubic meter? (*c*) How many cubic centimeters in one liter?

(*a*) $1 \text{ m}^3 = (1 \text{ m})^3\left(\dfrac{100 \text{ cm}}{1 \text{ m}}\right)^3 = (10^2 \text{ cm})^3 = 10^6 \text{ cm}^3$

(*b*) $1 \text{ m}^3 = (1 \text{ m})^3\left(\dfrac{10 \text{ dm}}{1 \text{ m}}\right)^3\left(\dfrac{1 \text{ L}}{1 \text{ dm}^3}\right) = 10^3 \text{ L}$

(*c*) $1 \text{ L} = 1 \text{ dm}^3 = (1 \text{ dm})^3\left(\dfrac{10 \text{ cm}}{1 \text{ dm}}\right)^3 = 10^3 \text{ cm}^3$

1.6.   Find the capacity in liters of a tank 0.6 m long, 10 cm wide, and 50 mm deep.

Convert to decimeters, since $1 \text{ L} = 1 \text{ dm}^3$.

$$\text{Volume} = (0.6 \text{ m})(10 \text{ cm})(50 \text{ mm})$$

$$= (0.6 \text{ m})\left(\dfrac{10 \text{ dm}}{1 \text{ m}}\right) \times (10 \text{ cm})\left(\dfrac{1 \text{ dm}}{10 \text{ cm}}\right) \times (50 \text{ mm})\left(\dfrac{1 \text{ dm}}{100 \text{ mm}}\right)$$

$$= (6 \text{ dm})(1 \text{ dm})(0.5 \text{ dm}) = 3 \text{ dm}^3 = 3 \text{ L}$$

1.7.   Determine the mass of 66 lb of sulfur in (*a*) kilograms and (*b*) grams. (*c*) Find the mass of 3.4 kg of copper in pounds.

(*a*) $66 \text{ lb} = (66 \text{ lb})(0.453\,6 \text{ kg/lb}) = 30 \text{ kg}$ or $66 \text{ lb} = (66 \text{ lb})(1 \text{ kg/2.2 lb}) = 30 \text{ kg}$

(b) $66\,lb = (66\,lb)(453.6\,g/lb) = 30\,000\,g$, or $3.0 \times 10^4\,g$

(c) $3.4\,kg = (3.4\,kg)(2.2\,lb/kg) = 7.5\,lb$

COMPOUND UNITS

1.8. Fatty acids spread spontaneously on water to form a monomolecular film. A benzene solution containing $0.1\,mm^3$ of stearic acid is dropped into a tray full of water. The acid is insoluble in water but spreads on the surface to form a continuous film area of $400\,cm^2$ after all of the benzene has evaporated. What is the average film thickness in angstroms?

$$1\,mm^3 = (10^{-3}\,m)^3 = 10^{-9}\,m^3 \qquad 1\,cm^2 = (10^{-2}\,m)^2 = 10^{-4}\,m^2$$

$$\text{Film thickness} = \frac{\text{volume}}{\text{area}} = \frac{(0.1\,mm^3)(10^{-9}\,m^3/mm^3)}{(400\,cm^2)(10^{-4}\,m^2/cm^2)} = 2.5 \times 10^{-9}\,m$$

$$= 2.5 \times 10^{-9}\,m \times 10^{10}\,\text{Å}/m = 25\,\text{Å}$$

1.9. A pressure of one atmosphere is equal to $101.3\,kPa$. Express this pressure in pounds force (lbf) per square inch.

The pound force (lbf) is equal to $4.448\,N$.

$$1\,atm = 101.3\,kPa = (101.3 \times 10^3\,N/m^2)\left(\frac{1\,lbf}{4.448\,N}\right)\left(\frac{2.54 \times 10^{-2}\,m}{1\,in}\right)^2 = 14.69\,lbf/in^2$$

Notice that the conversion factor between m and in is squared to give the conversion factor between m^2 and in^2.

1.10. A table is to be prepared showing the dependence of the braking distance of an automobile upon speed of travel. The braking distance is to be expressed in feet and the speed in miles per hour. If the entries in the table are to be pure (dimensionless) numbers, how should the columns be headed for the two physical quantities?

If d is to be the symbol for braking distance, the column showing braking distances should be headed, d/ft. An entry under that heading, such as 200, would then be a pure number, the ratio of the braking distance to the standard foot. If $d/\text{ft} = 200$, then $d = 200\,\text{ft}$. This symbolic heading is more acceptable than the headings

$$d(\text{ft}) \qquad d,\,\text{ft}$$

Similarly, if speed is to be designated as s, the column should be headed, $s/\text{mi} \cdot h^{-1}$ or $s/(\text{mi}/h)$. An entry in this column, such as 62, would be the ratio between the car speed and the standard mile per hour. (The heading $s \cdot h/\text{mi}$ might also be used, but the ratio form is clearer.)

1.11. New York City's 7.9 million people in 1978 had a daily per capita consumption of 173 gallons of water. How many tons of sodium fluoride (45% fluorine by weight) would be required per year to give this water a tooth-strengthening dose of 1 part (by weight) fluorine per million parts water? One U.S. gallon of water at normal room temperature weighs $8.34\,lbf$ (i.e., has a mass of $8.34\,lb$). One ton is $2\,000\,lb$.

Mass of water, in tons, required per year

$$= (7.9 \times 10^6\,\text{persons})\left(\frac{173\,\text{gal water}}{\text{person} \cdot \text{day}}\right)\left(\frac{365\,\text{days}}{\text{yr}}\right)\left(\frac{8.34\,\text{lb water}}{1\,\text{gal water}}\right)\left(\frac{1\,\text{ton}}{2\,000\,\text{lb}}\right)$$

$$= 2.08 \times 10^9\,\frac{\text{tons water}}{\text{yr}}$$

Note that all units cancel out except tons water/yr, which appears in the result.

Mass of sodium fluoride, in tons, required per year

$$= \left(2.08 \times 10^9 \; \frac{\text{tons water}}{\text{yr}}\right)\left(\frac{1 \text{ ton fluorine}}{10^6 \text{ tons water}}\right)\left(\frac{1 \text{ ton sodium fluoride}}{0.45 \text{ ton fluorine}}\right)$$

$$= 4.6 \times 10^3 \; \frac{\text{tons sodium fluoride}}{\text{yr}}$$

1.12. A tennis ball was observed to travel at a speed of 95 miles per hour. Express this in meters per second.

$$95 \text{ mi/h} = \left(95 \; \frac{\text{mi}}{\text{h}}\right)\left(1.609 \times 10^3 \; \frac{\text{m}}{\text{mi}}\right)\left(\frac{1 \text{ h}}{3.6 \times 10^3 \text{ s}}\right) = 42.5 \; \frac{\text{m}}{\text{s}}$$

The conversion factor relating the meter to the mile was taken from Problem 1.1.

1.13. Calculate the density, in g/cm³, of a body that weighs 420 g (i.e., has a mass of 420 g) and has a volume of 52 cm³.

$$\text{Density} = \frac{\text{mass}}{\text{volume}} = \frac{420 \text{ g}}{52 \text{ cm}^3} = 8.1 \text{ g/cm}^3$$

1.14. Express the density of the above body in the standard SI unit, kg/m³.

$$(8.1 \text{ g/cm}^3)\left(\frac{1 \text{ kg}}{1\,000 \text{ g}}\right)\left(\frac{100 \text{ cm}}{1 \text{ m}}\right)^3 = 8.1 \times 10^3 \text{ kg/m}^3$$

1.15. What volume will 300 g of mercury occupy? Density of mercury is 13.6 g/cm³.

$$\text{Volume} = \frac{\text{mass}}{\text{density}} = \frac{300 \text{ g}}{13.6 \text{ g/cm}^3} = 22.1 \text{ cm}^3$$

1.16. The density of cast iron is 7 200 kg/m³. Calculate its density in pounds per cubic foot.

$$\text{Density} = \left(7\,200 \; \frac{\text{kg}}{\text{m}^3}\right)\left(\frac{1 \text{ lb}}{0.453\,6 \text{ kg}}\right)\left(\frac{0.304\,8 \text{ m}}{1 \text{ ft}}\right)^3 = 449 \text{ lb/ft}^3$$

The two conversion factors were taken from Problem 1.1.

1.17. A casting of an alloy in the form of a disk weighed 50.0 g. The disk was 0.250 inch thick and had a circular cross section of diameter 1.380 in. What is the density of the alloy, in g/cm³?

$$\text{Volume} = \left(\frac{\pi d^2}{4}\right)h = \left(\frac{\pi(1.380 \text{ in})^2(0.250 \text{ in})}{4}\right)\left(\frac{2.54 \text{ cm}}{1 \text{ in}}\right)^3 = 6.13 \text{ cm}^3$$

$$\text{Density of the alloy} = \frac{\text{mass}}{\text{volume}} = \frac{50.0 \text{ g}}{6.13 \text{ cm}^3} = 8.15 \text{ g/cm}^3$$

1.18. The density of zinc is 455 lb/ft³. Find the mass in grams of 9.00 cm³ of zinc.

First express the density in g/cm³.

$$\left(455 \; \frac{\text{lb}}{\text{ft}^3}\right)\left(\frac{1 \text{ ft}}{30.48 \text{ cm}}\right)^3\left(\frac{453.6 \text{ g}}{1 \text{ lb}}\right) = 7.29 \; \frac{\text{g}}{\text{cm}^3}$$

$$(9.00 \text{ cm}^3)(7.29 \text{ g/cm}^3) = 65.6 \text{ g}$$

1.19. Battery acid has a density of 1.285 g/cm^3 and contains 38.0% by weight H_2SO_4. How many grams of pure H_2SO_4 are contained in a liter of battery acid?

1 cm^3 of acid has a mass of 1.285 g. Then 1 L of acid (1 000 cm^3) has a mass of 1 285 g. Since 38.0% by weight (or by mass) of the acid is pure H_2SO_4, the number of grams of H_2SO_4 in 1 L of battery acid is

$$0.380 \times 1\ 285\ g = 488\ g$$

Formally, the above solution can be written as follows:

$$\text{Mass of } H_2SO_4 = (1\ 285\ g\ acid)\left(\frac{38\ g\ H_2SO_4}{100\ g\ acid}\right) = 488\ g\ H_2SO_4$$

Here the conversion factor

$$\frac{38\ g\ H_2SO_4}{100\ g\ acid}$$

is taken to be equal to 1. Although the condition 38 g H_2SO_4 = 100 g acid is not a universal truth in the same sense that 1 in always equals 2.54 cm, the condition is a rigid one of association of 38 g H_2SO_4 with every 100 g acid for this particular acid preparation. Mathematically, these two quantities may be considered to be equal for this problem, since one of the quantities implies the other. Liberal use will be made in subsequent chapters of conversion factors that are valid only for particular cases, in addition to the conversion factors that are universally valid.

1.20. (a) Calculate the mass of pure HNO_3 per cm^3 of the concentrated acid which assays 69.8% by weight HNO_3 and has a density of 1.42 g/cm^3. (b) Calculate the mass of pure HNO_3 in 60.0 cm^3 of concentrated acid. (c) What volume of the concentrated acid contains 63.0 g of pure HNO_3?

(a) 1 cm^3 of acid has a mass of 1.42 g. Since 69.8% of the total mass of the acid is pure HNO_3, then the number of grams of HNO_3 in 1 cm^3 of acid is

$$0.698 \times 1.42\ g = 0.991\ g$$

(b)   Mass of HNO_3 in 60.0 cm^3 of acid = (60.0 cm^3)(0.991 g/cm^3) = 59.5 g HNO_3

(c) 63.0g HNO_3 is contained in

$$\frac{63.0\ g}{0.991\ g/cm^3} = 63.6\ cm^3\ acid$$

TEMPERATURE

1.21. Ethyl alcohol (a) boils at 78.5 °C and (b) freezes at −117 °C, at one atmosphere pressure. Convert these temperatures to the Fahrenheit scale.

Use

$$t = \left[\frac{9}{5}\left(\frac{t}{°C}\right) + 32\right]°F$$

(a) $t = [\frac{9}{5}(78.5) + 32]\,°F = (141 + 32)\,°F = 173\,°F$

(b) $t = [\frac{9}{5}(-117) + 32]\,°F = (-211 + 32)\,°F = -179\,°F$

1.22. Mercury (a) boils at 675 °F and (b) solidifies at −38.0 °F, at one atmosphere pressure. Express these temperatures in degrees Celsius.

Use

$$t = \frac{5}{9}\left(\frac{t}{°F} - 32\right)°C$$

(a) $t = \frac{5}{9}(675 - 32)\,°C = \frac{5}{9}(643)\,°C = 357\,°C$

(b) $t = \frac{5}{9}(-38.0 - 32.0)\,°C = \frac{5}{9}(-70.0)\,°C = -38.9\,°C$

1.23. Change (a) 40 °C and (b) −5 °C to the Kelvin scale.

Use
$$T = \left(\frac{t}{°C} + 273\right) K$$

(a)
$$T = (40 + 273) K = 313 K$$

(b)
$$T = (-5 + 273) K = 268 K$$

1.24. Convert (a) 220 K and (b) 498 K to the Celsius scale.

Use
$$t = \left(\frac{T}{K} - 273\right) °C$$

(a)
$$t = (220 - 273) °C = -53 °C$$

(b)
$$t = (498 - 273) °C = 225 °C$$

1.25. During the course of an experiment, laboratory temperature rose 0.8 °C. Express this rise in degrees Fahrenheit.

Temperature *intervals* are converted differently than temperature *readings*. For intervals, it is seen from Fig. 1-1 that

$$100 °C = 180 °F \qquad or \qquad 5 °C = 9 °F$$

Hence
$$0.8 °C = (0.8 °C)\left(\frac{9 °F}{5 °C}\right) = 1.4 °F$$

Supplementary Problems

UNITS BASED ON MASS OR LENGTH

1.26. (a) Express 3.69 m in kilometers, in centimeters, and in millimeters. (b) Express 36.24 mm in centimeters and in meters.

Ans.   (a) 0.003 69 km, 369 cm, 3 690 mm; (b) 3.624 cm, 0.036 24 m

1.27. Determine the number of (a) millimeters in 10 in, (b) feet in 5 m, (c) centimeters in 4 ft 3 in.

Ans.   (a) 254 mm; (b) 16.4 ft; (c) 130 cm

1.28. Convert the molar volume, 22.4 liters, to cubic centimeters, to cubic meters, and to cubic feet.

Ans.   22 400 cm^3, 0.022 4 m^3, 0.791 ft^3

1.29. Express the weight (mass) of 32 g of oxygen in milligrams, in kilograms, and in pounds.

Ans.   32 000 mg, 0.032 kg, 0.070 5 lb

1.30. How many grams in 5.00 lb of copper sulfate? How many pounds in 4.00 kg of mercury? How many milligrams in 1 lb 2 oz of sugar?

Ans.   2 270 g, 8.82 lb, 510 000 mg

1.31. Convert the weight (mass) of 500 lb of coal to (a) kilograms, (b) metric tons, (c) U.S. tons (1 ton = 2 000 lb).

Ans.   (a) 227 kg; (b) 0.227 metric ton; (c) 0.250 ton

1.32. The color of light depends on its wavelength. The longest visible rays, at the red end of the visible spectrum, are 7.8×10^{-7} m in length. Express this length in micrometers, in nanometers, and in angstroms.

Ans.   0.78 μm, 780 nm, 7 800 Å

1.33. Determine the number of   (*a*) cubic centimeters in a cubic inch,   (*b*) cubic inches in a liter,   (*c*) cubic feet in a cubic meter.

Ans.   (*a*) 16.4 cm^3;   (*b*) 61.0 cm^3;   (*c*) 35.3 ft^3

1.34. In a crystal of platinum, centers of individual atoms are 2.8 Å apart along the direction of closest packing. How many atoms would lie on a one-centimeter length of a line in this direction?

Ans. 3.5×10^7

1.35. The blue iridescence of butterfly wings is due to striations that are 0.15 μm apart, as measured by the electron microscope. What is this distance in centimeters? How does this spacing compare with the wavelength of blue light, about 4 500 Å?

Ans. 1.5×10^{-5} cm, $\frac{1}{3}$ wavelength of blue light

1.36. The thickness of a soap bubble film at its thinnest (bimolecular) stage is about 60 Å.   (*a*) What is this thickness in centimeters?   (*b*) How does this thickness compare with the wavelength of yellow sodium light, which is 0.589 0 μm?

Ans.   (*a*) 6.0×10^{-7} cm;   (*b*) about 0.01 of wavelength of yellow light

1.37. An average man requires about 2.00 mg of riboflavin (vitamin B_2) per day. How many pounds of cheese would a man have to eat per day if this were his only source of riboflavin and if the cheese contained 5.5 μg riboflavin per gram?

Ans.   0.80 lb/day

1.38. When a sample of healthy human blood is diluted to 200 times its initial volume and microscopically examined in a layer 0.10 mm thick, an average of 30 red corpuscles are found in each 100×100 micrometer square.   (*a*) How many red cells are in a cubic millimeter of blood?   (*b*) The red blood cells have an average life of 1 month, and the adult blood volume is about 5 L. How many red cells are generated every second in the bone marrow of the adult?

Ans.   (*a*) 6×10^6 cells/min^3;   (*b*) 1×10^7 cells/s

1.39. A porous catalyst for chemical reactions has an internal surface area of 800 m^2 per cm^3 of bulk material. Fifty percent of the bulk volume consists of the pores (holes), while the other 50 percent of the volume is made up of the solid substance. Assume that the pores are all cylindrical tubules of uniform diameter d and length l, and that the measured internal surface area is the total area of the curved surfaces of the tubules. What is the diameter of each pore? (*Hint*: Find the number of tubules per bulk cm^3, n, in terms of l and d, by using the formula for the volume of a cylinder. $V = \frac{1}{4}\pi d^2 l$. Then apply the surface-area formula, $S = \pi dl$, to the cylindrical surfaces of n tubules.)

Ans.   25 Å

COMPOUND UNITS

1.40. The density of water is 1.000 g/cm^2 at 4 °C. Calculate the density of water in pounds per cubic foot at the same temperature.

Ans.   62.4 lb/ft^3

1.41. What is the average speed, in miles per hour, of a sprinter in doing the 100-m dash in 10.1 s?

    Ans.   22 mi/h

1.42. The silica gel which is used to protect sealed overseas shipments from moisture seepage has a surface area of 6.0×10^5 m^2 per kilogram. What is this surface area in square feet per gram?

    Ans. 6.5×10^3 ft^2/g

1.43. There is reason to think that the length of the day, determined from the earth's period of rotation, is increasing uniformly by about 0.001 s every century. What is this variation in parts per billion?

    Ans. 3×10^{-4} s per 10^9 s

1.44. The bromine content of average ocean water is 65 parts by weight per million. Assuming 100 percent recovery, how many cubic meters of ocean water must be processed to produce one pound of bromine? Assume that the density of seawater is 1.0×10^3 kg/m^3.

    Ans.   7.0 m^3

1.45. An important physical quantity has the value 8.314 joules or 0.082 06 liter·atmosphere. What is the conversion factor from liter·atmospheres to joules?

    Ans.   101.3 J/L·atm

1.46. Find the density of ethyl alcohol if 80.0 cm^3 weighs 63.3 g.

    Ans.   0.791 g/cm^3

1.47. Find the volume in liters of 40 kg of carbon tetrachloride, whose density is 1.60 g/cm^3.

    Ans.   25 L

1.48. Determine the mass in kilograms of 20.0 cubic feet of aluminum (density = 2.70 g/cm^3).

    Ans. 1.53×10^3 kg

1.49. Air weighs about 8 lbf per 100 cubic feet. Find its density in (*a*) grams per cubic foot, (*b*) grams per liter, (*c*) kilograms per cubic meter.

    Ans.   (*a*) 36 g/ft^3; (*b*) 1.3 g/L; (*c*) 1.3 kg/m^3

1.50. What is the density in SI units of a steel ball which has a diameter of 7.50 mm and a mass of 1.765 g? (Volume of a sphere of radius r is $\frac{4}{3}\pi r^3$.)

    Ans.   7 990 kg/m^3

1.51. A wood block, 10 in × 6.0 in × 2.0 in, weighs 3 lb 10 oz. What is the density of the wood in SI units?

    Ans.   840 kg/m^3

1.52. An alloy was machined into a flat disk, 31.5 mm in diameter and 4.5 mm thick, with a hole 7.5 mm in diameter drilled through the center. The disk weighed 20.2 g. What is the density of the alloy in SI units?

    Ans.   6 100 kg/m^3

1.53. A glass vessel weighed 20.237 6 g when empty and 20.310 2 g when filled to an etched mark with water at 4 °C. The same vessel was then dried and filled to the same mark with a solution at 4 °C. The vessel was now found to weigh 20.330 0 g. What is the density of the solution?

    Ans.   1.273 g/cm^3

1.54. What should the column headings be for a table of dimensionless numerical entries showing the critical densities (d_c), critical temperatures (T_c), and critical pressures (P_c) of several gases, where the quantities are measured in SI units?

Ans. $\dfrac{d_c}{\text{kg/m}^3}, \dfrac{T_c}{\text{K}}, \dfrac{P_c}{\text{Pa}}$ or $\dfrac{P_c}{\text{N/m}^2}$ or $d_c/(\text{kg/m}^3), T_c/\text{K}, P_c \, / \, Pa$ or $P_c/(\text{N}/\text{m}^2)$

1.55. A sample of concentrated sulfuric acid is 95.7% H_2SO_4 by weight and its density is 1.84 g/cm^3.   (*a*) How many grams of pure H_2SO_4 are contained in one liter of the acid?   (*b*) How many cubic centimeters of acid contain 100 g of pure H_2SO_4?

Ans.   (*a*) $1\,760$ g;   (*b*) 56.8 cm^3

1.56. Analysis shows that 20.0 cm^3 of concentrated hydrochloric acid of density 1.18 g/cm^3 contains 8.36 g HCl. (*a*) Find the mass of HCl per cubic centimeter of acid solution.   (*b*) Find the percent by weight (mass) of HCl in the concentrated acid.

Ans.   (*a*) 0.418 g HCl/cm^3 acid;   (*b*) 35.4% HCl

1.57. An electrolytic tin-plating process gives a coating 30 millionths of an inch thick. How many square meters can be coated with one kilogram of tin, density $7\,300$ kg/m^3?

Ans. 180 m^2

1.58. A piece of gold leaf (density 19.3 g/cm^3) weighing 1.93 mg can be beaten further into a transparent film covering an area of 14.5 cm^2.   (*a*) What is the volume of 1.93 mg of gold?   (*b*) What is the thickness of the transparent film, in angstroms?

Ans.   (*a*) 1.00×10^{-4} cm^3;   (*b*) 690 Å

1.59. A piece of capillary tubing was calibrated in the following manner. A clean sample of the tubing weighed 3.247 g. A thread of mercury, drawn into the tube, occupied a length of 23.75 mm, as observed under a microscope. The weight of the tube with the mercury was 3.489 g. The density of mercury is 13.60 g/cm^3. Assuming that the capillary bore is a uniform cylinder, find the diameter of the bore.

Ans. 0.98 mm

1.60. The General Sherman tree, located in Sequoia National Park, is believed to be the most massive of living things. If the overall density of the tree trunk is assumed to be 850 kg/m^3, calculate the mass of the trunk by assuming that it may be approximated by two right conical frusta having lower and upper diameters of 11.2 and 5.6 m, and 5.6 and 3.3 m, respectively, and respective heights of 2.4 and 80.6 m. A frustum is a portion of a cone bounded by two planes, both perpendicular to the axis of the cone. The volume of a frustum is given by

$$\tfrac{1}{3}\pi h(r_1^2 + r_2^2 + r_1\,r_2)$$

where h is the height and r_1 and r_2 are the radii of the circular ends of the frusta.

Ans. 1.20×10^6 kg $= 1\,200$ metric tons

TEMPERATURE

1.61. (*a*) Convert $68\,°$F to $°$C; $5\,°$F to $°$C; $176\,°$F to $°$C.   (*b*) Convert $30\,°$C to $°$F; $5\,°$C to $°$F; $-20\,°$C to $°$F.

Ans.   (*a*) $20\,°$C, $-15\,°$C, $80\,°$C;   (*b*) $86\,°$F, $41\,°$F, $-4\,°$F

1.62. Convert the following temperatures: $-195.5\,°$C to $°$F; $-430\,°$F to $°$C; $1\,705\,°$C to $°$F.

Ans. $-319.9\,°$F, $-256.7\,°$C, $3\,101\,°$F

1.63. The temperature of dry ice (sublimation temperature at normal pressure) is $-109\,°F$. Is this higher or lower than the temperature of boiling ethane (a component of bottled gas), which is $-88\,°C$?

   Ans.   higher

1.64. Gabriel Fahrenheit in 1714 suggested for the zero point on his scale the lowest temperature then obtainable from a mixture of salts and ice, and for his 100° point he suggested the highest known normal animal temperature. Express these "extremes" in degrees Celsius.

   Ans. $-17.8\,°C,\ 37.8\,°C$

1.65. Sodium metal has a very wide liquid range, melting at $98\,°C$ and boiling at $892\,°C$. Express the liquid range in degrees Celsius, kelvins, and degrees Fahrenheit.

   Ans. $794\,°C,\ 794\,K,\ 1\,429\,°F$

1.66. Convert 300 K, 760 K, 180 K, to degrees Celsius.

   Ans. $27\,°C,\ 487\,°C,\ -93\,°C$

1.67. Express 8 K, 273 K, in degrees Fahrenheit.

   Ans. $-445\,°F,\ 32\,°F$

1.68. Convert $14\,°F$ to degrees Celsius and kelvins.

   Ans. $-10\,°C,\ 263\,K$

1.69. At what temperature have the Celsius and Fahrenheit readings the same numerical value?

   Ans. $-40°$

1.70. A water-stabilized electric arc was reported to have reached a temperature of $25\,600\,°F$. On the absolute scale, what is the ratio of this temperature to that of an oxyacetylene flame, $3\,500\,°C$?

   Ans.   3.84

1.71. Construct a temperature scale on which the freezing and boiling points of water are 100° and 400°, respectively, and the degree interval is a constant multiple of the Celsius degree interval. What is the absolute zero on this scale, and what is the melting point of sulfur, which is $444.6\,°C$?

   Ans. $-719°,\ 1\,433.8°$

Chapter 2

Atomic Weights, Molecular Weights, and Moles

ATOMS

In the form of the atomic theory proposed by John Dalton in 1805, all atoms of a given element were thought to be identical. Chemists in the following decades set themselves the task of finding the relative masses of the atoms of the different elements by precise quantitative chemical analysis. Over a hundred years after Dalton's proposal was made, investigations with radioactive substances showed that not all atoms of a given element are identical.

Modern atomic weight tables must recognize this fact because one of the properties that distinguishes these different kinds of atoms for a given element is the atomic mass. Each element can exist in several *isotopic* forms, and all atoms of the same isotope are identical.

NUCLEI

Every atom has a positively charged nucleus which contains over 99.9 percent of the total mass of the atom. Although the structure of the nucleus is not thoroughly understood, every nucleus may be described as being made up of two different kinds of particles. These two kinds of particles, called *nucleons*, are the *proton* and the *neutron*, and they have nearly the same mass. Of these two, only the proton is charged and the sign of its charge is positive. The total charge of any nucleus is equal to the number of protons times the charge of one proton. The magnitude of the protonic charge may indeed be considered *the* fundamental unit of charge for atomic and nuclear phenomena, since no smaller charge than this has been discovered in any free particle. The nuclear charge expressed in this unit is then equal to the number of protons in the nucleus.

The atoms of all isotopes of any one element have the same number of protons. This number is called the *atomic number*, Z, and is a characteristic of the element. The nuclei of the different isotopes differ, however, in the number of neutrons and therefore in the total number of nucleons per nucleus. The total number of nucleons, A, is called the *mass number*, and the number of neutrons is therefore $A - Z$. Atoms of the different isotopic forms of an element are called *nuclides* and are distinguished by using the mass number as a supserscript to the elementary symbol on the left. Thus ^{15}N refers to the nitrogen of mass number 15.

RELATIVE ATOMIC WEIGHTS

The masses of the individual atoms are very small. Even the heaviest atom has a mass less than 5×10^{-25} kg. It has been convenient, therefore, to define a special unit in which the masses of the atoms can be expressed without having to use exponents. This unit is called the *atomic mass unit* and is designated by the symbol u. (Biochemists sometimes refer to this unit as the *dalton*.) It is defined as exactly $\frac{1}{12}$ the mass of a ^{12}C atom. Thus the mass of the ^{12}C atom is exactly 12 u; the mass of the ^{23}Na atom is 22.989 8 u. Table 2-1 lists the masses of some nuclides to which reference will be made in this chapter and in Chapter 21.

Most chemical reactions do not discriminate significantly among the various isotopes. For example, the percentages of iron atoms which are ^{54}Fe, ^{56}Fe, ^{57}Fe, and ^{58}Fe are 5.8, 91.8, 2.1, and 0.3, respectively, in all iron ores, meteorites, and iron compounds prepared synthetically. For chemical purposes it is of interest to know the *average mass* of an iron atom in this natural isotopic mixture. These average masses are also tabulated in terms of the unit u and are given a different name to distinguish them from the nuclidic masses. The average masses are called the *relative atomic weights*, or simply the *atomic weights*, and are designated by $A_r(E)$, where E is the symbol for the particular element. It should be emphasized

Table 2-1. Some Nuclidic Masses

^{1}H	1.007 83 u	^{12}C	12.000 00 u	^{17}O	16.999 13 u	^{35}Cl	34.968 85 u
^{2}H	2.014 10	^{13}C	13.003 35	^{18}O	17.999 16	^{37}Cl	36.965 90
^{3}H	3.016 05	^{14}C	14.003 24	^{18}F	18.000 94	^{36}Ar	35.967 55
^{4}He	4.002 60	^{16}C	16.014 70	^{18}Ne	18.005 71	^{38}Ar	37.962 73
^{6}He	6.018 89	^{14}N	14.003 07	^{28}Si	27.976 93	^{40}Ar	39.962 38
^{6}Li	6.015 12	^{15}N	15.000 11	^{29}Si	28.976 49	^{87}Rb	86.909 19
^{7}Li	7.016 00	^{16}N	16.006 10	^{30}Si	29.973 77		
^{7}Be	7.016 93	^{16}O	15.994 91	^{32}S	31.972 07		

that they are the averaged masses for the atoms of the elements as they exist in nature. These values, **which are listed facing the inside back cover of this book**, form the basis for practically all chemical weight calculations. The atomic weights are determined either by carrying out precise chemical analyses on a chemical compound of known formula in which all the atomic weights are known but one, or by making a physical determination of the masses of all the stable nuclides of an element and averaging them according to their relative proportions in natural sources of the element.

MOLE

An ordinary chemical experiment involves the reaction of enormous numbers of atoms or molecules. It has been convenient to define a new term, the *mole*, as denoting a collection of a large fixed number of fundamental chemical entities, comparable to the quantity that might be involved in an actual experiment. In fact, the mole is recognized in SI as the unit for one of the dimensionally independent quantities, the *amount of substance*. The abbreviation of the unit is *mol*. A mole of atoms of any element is defined as that amount of substance containing the same number of atoms as there are C atoms in exactly 12 g of pure ^{12}C. This number is called *Avogadro's constant*, N_A; its value may be related to the value of the u, listed in Table 1-2, as follows.

$$\text{Mass of 1 mol of } {}^{12}C \text{ atoms} = N_A \times (\text{mass of one } {}^{12}C \text{ atom})$$

$$12 \text{ g/mol} = N_A \times 12 \text{ u}$$

$$N_A = \frac{12 \text{ g/mol}}{12 \text{ u}} = \frac{1 \text{ g/mol}}{1 \text{ u}} = \frac{1 \text{ g/mol}}{(1.660\ 57 \times 10^{-27} \text{ kg})(10^3 \text{ g/kg})}$$

$$= 6.022\ 0 \times 10^{23}/\text{mol}$$

All units in the expression for N_A cancelled except for mol, which remained in the denominator, and thus appears as mol^{-1}.

Let us now consider a mole of atoms of some other element, of atomic weight A_r. The average mass of an atom of this element is A_ru and the mass of a mole of such atoms is $N_A \times A_r$u, or simply A_rg/mol. In other words, the mass in grams of a mole of atoms of an element is equal to the atomic weight, and A_r may be considered to have the units g/mol.

As will be seen below, the mole may be used to refer to N_A elementary entities of types other than atoms, such as formula units, molecules, and charged particles.

SYMBOLS, FORMULAS, FORMULA WEIGHTS

The symbol of an element is used to designate that element as distinct from other elements. In a formula the symbol designates an atom of the element. In addition, the symbol is often used to designate a particular amount of that element, 1 mole. Thus Au sometimes means simply gold. In other contexts it refers to 1 mol or 197.0 g of gold.

A *formula* indicates the relative number of atoms of each element in a substance. Thus Fe_3O_4 refers to a compound in which 3 atoms of iron are present for every 4 atoms of oxygen. If the subscripts are the smallest possible integral numbers that express the relative numbers of atoms, the formula is called an *empirical formula*. Fe_3O_4 is an example; it is a descriptive notation for this particular oxide of iron. Fe_3O_4 is also used in some contexts to mean a particular amount of the compound, the amount that contains 3 moles of iron atoms and 4 atoms of oxygen atoms. This amount of compound contains N_A simplest formula units (where the formula unit contains 3 iron atoms and 4 oxygen atoms) and is, therefore, called a *mole* of Fe_3O_4. It is composed of

$$(3)(55.85) = 167.55 \text{ g iron} \quad \text{and} \quad (4)(16.00) = 64.00 \text{ g oxygen}$$

combined to make $167.55 + 64.00 = 231.55$ grams of compound. 231.55 is called the *formula weight* of the compound. In general, the formula weight is equal to the sum of numbers, each number being the product of the atomic weight of one of the elements appearing in the compound multiplied by the number of atoms of that element indicated in the formula.

MOLECULAR FORMULAS, MOLECULAR WEIGHTS

A formula that expresses not only the relative number of atoms of each element but also the actual number of atoms of each element in one molecule of the compound is called a *molecular formula*. The formula weight is then called the *molecular weight*.

EXAMPLE 1   The molecular formula for benzene is C_6H_6. This indicates that the benzene molecule is composed of 6 atoms of carbon and 6 atoms of hydrogen, and the molecular weight of benzene is

$$(6)(12.011) + (6)(1.008) = 78.114$$

The empirical or simplest formula for benzene is CH; it indicates only that benzene is composed of the elements carbon and hydrogen in the ratio of 1 atom of C to 1 atom of H. The empirical formula does not represent the actual numbers of C and H in one molecule of benzene, and thus it does not represent the molecular weight of benzene.

To determine the molecular formula of a compound, it is necessary to know the molecular weight of the compound. The molecular formula is an integral multiple (1, 2, 3, etc.) of the empirical formula.

The molecular formula C_6H_6 may be used to mean 78.114 g of benzene. This amount of compound contains $6N_A$ atoms (6 moles) of carbon and $6N_A$ hydrogen atoms (6 moles of hydrogen atoms). It contains N_A molecules (C_6H_6 units) and is, therefore, called a *mole* of benzene.

In short, the mass in grams of one mole of any chemical entity is equal to the formula weight of that entity: atom, molecule, ion, or simplest collection of atoms in the fixed proportions of a chemical compound. A mole always contains N_A of the individual elementary entities.

Solved Problems

ATOMIC WEIGHTS

2.1.   It has been found by mass spectrometric analysis that in nature the relative abundances of the various isotopic atoms of silicon are 92.23% ^{28}Si, 4.67% ^{29}Si, and 3.10% ^{30}Si. Calculate the atomic weight of silicon from this information and from the nuclidic masses.

The atomic weight is the average mass of the three nuclides, each weighted according to its own relative abundance. The nuclidic masses are given in Table 2-1.

$$A_r = (0.922\,3)(27.977) + (0.046\,7)(28.976) + (0.031\,0)(29.974)$$

$$= 25.803 + 1.353 + 0.929 = 28.085$$

2.2. Naturally occurring carbon consists of two isotopes, ^{12}C and ^{13}C. What are the percentage abundances of the two isotopes in a sample of carbon whose atomic weight is 12.011 12?

Let $x = \%$ abundance of ^{13}C; then $100 - x$ is $\%$ of ^{12}C.

$$A_r = 12.011\,12 = \frac{(12.000\,00)(100 - x) + (13.003\,35)x}{100}$$

$$= 12.000\,00 + \frac{(13.003\,35 - 12.000\,00)x}{100} = 12.000\,00 + (0.010\,033\,5)x$$

Thus

$$x = \frac{12.011\,12 - 12.000\,00}{0.010\,033\,5} = \frac{0.011\,12}{0.010\,033\,5} = 1.108\%\ ^{13}C$$

and

$$100 - x = 98.892\%\ ^{12}C$$

2.3. Before 1961, a physical atomic weight scale was used whose basis was an assignment of the value 16.000 00 to ^{16}O. What would have been the physical atomic weight of ^{12}C on the old scale?

The ratio of the masses of any two nuclides must be independent of the establishment of the reference point.

$$\left(\frac{A_r \text{ of } ^{12}C}{A_r \text{ of } ^{16}O}\right)_{old\ scale} = \left(\frac{A_r \text{ of } ^{12}C}{A_r \text{ of } ^{16}O}\right)_{present\ scale} = \frac{12.000\,00}{15.994\,91}$$

or

$$(A_r \text{ of } ^{12}C)_{old\ scale} = (16.000\,0)\left(\frac{12.000\,00}{15.994\,91}\right) = 12.003\,82$$

2.4. A 1.527 6 g sample of $CdCl_2$ was converted to metallic cadmium and cadmium-free products by an electrolytic process. The weight of the metallic cadmium was 0.936 7 g. If the atomic weight of chlorine is taken as 35.453, what must be the atomic weight of Cd from this experiment?

Throughout this book we will specify the amount of a substance in terms of the chemist's unit, the mole. We will use the symbol $n(X)$ to refer to the number of moles of substance or chemical constituent X. Moreover, since atomic masses are called atomic weights, we shall not bother to distinguish between "mass" and "weight." In this case, we calculate first the number of moles of Cl atoms in the weighed sample.

$$\begin{aligned}
\text{Weight of } CdCl_2 &= 1.527\,6 \text{ g} \\
\text{Weight of Cd in } CdCl_2 &= \underline{0.936\,7 \text{ g}} \\
\text{Weight of Cl in } CdCl_2 &= 0.590\,9 \text{ g}
\end{aligned}$$

$$n(Cl) = 0.590\,9 \text{ g} \times \frac{1 \text{ mol}}{35.453 \text{ g}} = 0.016\,667 \text{ mol}$$

From the formula $CdCl_2$ we see that the number of moles of Cd is exactly half the number of moles of Cl.

$$n(Cd) = \tfrac{1}{2}n(Cl) = \tfrac{1}{2}(0.016\,667) = 0.008\,333 \text{ mol}$$

The atomic weight is the weight per mole.

$$A_r(Cd) = \frac{0.936\,7 \text{ g}}{0.008\,333 \text{ mol}} = 112.41 \text{ g/mol}$$

2.5. In a chemical determination of the atomic weight of vanadium, 2.893 4 g of pure $VOCl_3$ was allowed to undergo a set of reactions as a result of which all the chlorine contained in this compound was converted to AgCl. The weight of the AgCl was 7.180 1 g. Assuming the atomic weights of Ag and Cl are 107.868 and 35.453, what is the experimental value for the atomic weight of vanadium?

This problem is similar to Problem 2.4, except that $n(Cl)$ must be obtained by way of $n(AgCl)$. The three Cl atoms of $VOCl_3$ are converted to 3 formula units of AgCl, the formula weight of which is 143.321 (the sum of 107.868 and 35.453).

$$n(AgCl) = 7.180\,1 \text{ g} \times \frac{1 \text{ mol}}{143.321 \text{ g}} = 0.050\,098 \text{ mol}$$

From the formula AgCl,

$$n(Cl) = n(AgCl) = 0.050\,098 \text{ mol Cl}$$

Also, from the formula $VOCl_3$,

$$n(V) = \tfrac{1}{3}n(Cl) = \tfrac{1}{3}(0.050\,098) = 0.016\,699 \text{ mol V}$$

To find the weight of vanadium in the weighed sample of $VOCl_3$, we must subtract the weights of chlorine and oxygen contained. If we designate the mass of any substance or chemical constituent X by $m(X)$, then

$$m(X) = n(X) \times FW(X)$$

where $FW(X)$ is the formula weight of X.

$$m(Cl) = n(Cl) \times A_r(Cl) = (0.050\,098 \text{ mol})(35.453 \text{ g/mol}) = 1.776\,1 \text{ g Cl}$$

Noting from the formula $VOCl_3$ that $n(O) = n(V)$,

$$m(O) = n(O) \times A_r(O) = (0.016\,699 \text{ mol})(15.999 \text{ g/mol}) = 0.267\,2 \text{ g O}$$

By difference,
$$m(V) = m(VOCl_3) - m(O) - m(Cl)$$
$$= (2.893\,4 - 0.267\,2 - 1.776\,1) \text{ g} = 0.850\,1 \text{ g}$$

Then
$$A_r(V) = \frac{m(V)}{n(V)} = \frac{0.850\,1 \text{ g}}{0.016\,699 \text{ mol}} = 50.91 \text{ g/mol}$$

Note that this result differs slightly from the accepted value. The difference can be ascribed to experimental error in this determination.

FORMULA WEIGHTS AND MOLECULAR WEIGHTS

2.6. Determine (a) the formula weight of potassium hexachloroiridate(IV), K_2IrCl_6 and (b) the molecular weight of trifluorosilane, $SiHF_3$.

Potassium hexachloroiridate(IV) does not exist as discrete molecules represented by the empirical formula, but trifluorosilane does. The term *formula weight* may be used in every case to describe the relative weight of the indicated formula unit; further chemical experience is needed to define the applicability of the term *molecular weight*. Many chemists prefer to use only one term, *molecular weight*, even when referring to a nonmolecular substance. To avoid this contradiction the term *molar mass* has come into use for either *formula weight* or *molecular weight*.

(a) 2 K = 2(39.098) = 78.20   (b) 1 Si = 1(28.086) = 28.086

     1 Ir = 1(192.22) = 192.22   1 H = 1(1.008) = 1.008

     6 Cl = 6(35.453) = 212.72   3 F = 3(18.998\,4) = 56.995

     Formula weight = 483.14   Molecular weight = 86.089

Note that the atomic weights are not all known to the same number of significant figures or to the same number of decimal places in u units. In general, the rules for significant figures discussed in Appendix B apply. The value of u for Ir is known to only 0.01 u. Note that in order to express 6 times the atomic weight of Cl to 0.01 u, it was necessary to use the atomic weight to 0.001 u. Similarly, an extra figure was used in the atomic weight for fluorine to give the maximum significance to the last digit in the sum column.

2.7. How many (a) grams of H_2S, (b) moles of H and of S, (c) grams of H and of S, (d) molecules of H_2S, (e) atoms of H and of S, are contained in 0.400 mol H_2S?

Atomic weight of H is 1.008; of S, 32.066. Molecular weight of H_2S is $2(1.008) + 32.066 = 34.08$.

Note that it was not necessary to express the molecular weight to 0.001 u, even though the atomic weights are known to this significance. Since the limiting factor in this problem is $n(H_2S)$, known to one part in 400, the value 34.08 (expressed to one part in over 3 000) for the molecular weight is more than adequate.

(a)   Number of grams of compound = (number of moles) × (weight of 1 mole)

Number of grams of H_2S = (0.400 mol)(34.08 g/mol) = 13.63 g H_2S

(b) One mole of H_2S contains 2 moles of H and 1 mole of S. Then 0.400 mol H_2S contains

$$(0.400 \text{ mol } H_2S)\left(\frac{2 \text{ mol H}}{1 \text{ mol } H_2S}\right) = 0.800 \text{ mol H}$$

and 0.400 mol S.

(c)   Number of grams of element = (number of moles) × (weight of 1 mole)

Number of grams of H = (0.800 mol)(1.008 g/mol) = 0.806 g H

Number of grams of S = (0.400 mol)(32.066 g/mol) = 12.83 g S

(d)   Number of molecules = (number of moles) × (number of molcules in 1 mole)
$$= (0.400 \text{ mol})(6.02 \times 10^{23} \text{ molecules/mol}) = 2.41 \times 10^{23} \text{ molecules}$$

(e)   Number of atoms of element = (number of moles) × (number of atoms per mole)

Number of atoms of H = (0.800 mol)(6.02 × 10^{23} atoms/mol) = 4.82 × 10^{23} atoms H

Number of atoms of S = (0.400 mol)(6.02 × 10^{23} atoms/mol) = 2.41 × 10^{23} atoms S

2.8. How many moles of atoms are contained in (a) 10.02 g calcium, (b) 92.91 g phosphorus? (c) How many moles of molecular phosphorus are contained in 92.91 g phosphorus if the formula of the molecule is P_4? (d) How many atoms are contained in 92.91 g phosphorus? (e) How many molecules are contained in 92.91 g phosphorus?

Atomic weights of Ca and P are 40.08 and 30.974. Hence

$$1 \text{ mol Ca} = 40.08 \text{ g Ca} \qquad 1 \text{ mol P} = 30.974 \text{ g P}$$

(a) $$n(Ca) = \frac{\text{weight of Ca}}{\text{atomic weight of Ca}} = \frac{10.02 \text{ g}}{40.08 \text{ g/mol}} = 0.250 \text{ mol Ca atoms}$$

Observe that the above calculation has been presented in a slightly different style up to this point, i.e:

$$n(C_a) = 10.02 \text{ g} \times \frac{1 \text{ mol}}{40.08 \text{ g}} = 0.250 \text{ mol } C_a \text{ atoms}$$

(b) $$n(P) = \frac{\text{weight of P}}{\text{atomic weight of P}} = \frac{92.91 \text{ g}}{30.974 \text{ g/mol}} = 3.000 \text{ mol P atoms}$$

(c) Molecular weight of P_4 is $(4)(30.974) = 123.90$. Then

$$n(P_4) = \frac{\text{weight of P}}{\text{molecular weight of } P_4} = \frac{92.91 \text{ g}}{123.90 \text{ g/mol}} = 0.750\,0 \text{ mol } P_4 \text{ molecules}$$

(d)   Number of atoms of P = (3.000 mol)(6.023 × 10^{23} atoms/mol) = 1.807 × 10^{24} atoms P

(e)   Number of molecules of P_4 = (0.750 0 mol)(6.023 × 10^{23} molecules/mol)
$$= 4.517 \times 10^{23} \text{ molecules } P_4$$

2.9. How many moles are represented by (a) 9.54 g SO_2, (b) 85.16 g NH_3, (c) 25.02 g $TiS_{1.85}$?

Atomic weight of S, 32.07; of O, 16.00; of N, 14.007; of H, 1.008; of Ti, 47.88. From these,

$$\text{Molecular weight of SO}_2 = 32.07 + 2(16.00) = 64.07$$
$$\text{Molecular weight of NH}_3 = 14.007 + 3(1.008) = 17.031$$
$$\text{Formula weight of TiS}_{1.85} = 47.88 + 1.85\,(32.07) = 107.21$$

Hence 1 mol $SO_2 = 64.07$ g SO_2, 1 mol $NH_3 = 17.031$ g NH_3, 1 mol $TiS_{1.85} = 107.21$ g $TiS_{1.85}$.

(a) $$\text{Amount of SO}_2 = \frac{\text{mass of SO}_2}{\text{mass per mole of SO}_2} = \frac{9.54 \text{ g}}{64.07 \text{ g/mol}} = 0.148\,9 \text{ mol SO}_2$$

(b) $$\text{Amount of NH}_3 = \frac{\text{mass of NH}_3}{\text{mass per mole of NH}_3} = \frac{85.16 \text{ g}}{17.031 \text{ g/mol}} = 5.000 \text{ mol NH}_3$$

(c) Titanium sulfides belong to a relatively small class of nonstoichiometric solid compounds, whose structures allow a limited variability in composition. The actual atomic ratios of titanium and sulfur may vary by as much as 10 percent, depending on the details of preparation. The formula given here describes a particular preparation. The mole concept may be applied just as well to entities with nonintegral formulas as to those with integral formulas.

$$\text{Amount of TiS}_{1.85} = \frac{\text{mass of TiS}_{1.85}}{\text{mass per mole of TiS}_{1.85}} = \frac{25.02 \text{ g}}{107.21 \text{ g/mol}} = 0.233\,4 \text{ mol TiS}_{1.85}$$

MULTIPLE PROPORTIONS

2.10. Three common gaseous compounds of nitrogen and oxygen of different elementary composition are known, (A) laughing gas containing 63.65% nitrogen, (B) a colorless gas containing 46.68% nitrogen, and (C) a brown gas containing 30.45% nitrogen. Show how these data illustrate the *law of multiple proportions*.

According to the law of multiple proportions, the relative amounts of an element combining with some fixed amount of a second element in a series of compounds are the ratios of small whole numbers.

On the basis of 100 g of each compound, we tabulate below the mass of N, the mass of O (obtained by difference), and the mass of N per gram of O.

	Compound A	Compound B	Compound C
g of N	63.65	46.68	30.45
g of O	36.35	53.32	69.55
(g of N)/(g of O)	1.751 0	0.875 5	0.437 8

The relative amounts of nitrogen in the three cases are not affected if all three amounts are divided by the smallest of them.

$$1.751\,0 : 0.875\,5 : 0.437\,8 = \frac{1.751\,0}{0.437\,8} : \frac{0.875\,5}{0.437\,8} : \frac{0.437\,8}{0.437\,8} = 4.000 : 2.000 : 1.000$$

These relative amounts are indeed the ratios of small whole numbers, 4, 2, and 1, within the precision of the analyses.

Supplementary Problems

ATOMIC WEIGHTS

2.11. Naturally occurring argon consists of three isotopes, the atoms of which occur in the following abundances: 0.34% ^{36}Ar, 0.07% ^{38}Ar, and 99.59% ^{40}Ar. Calculate the atomic weight of argon from these data.

Ans. 39.948

2.12. Naturally occurring boron consists of 80.22% ^{11}B (nuclidic mass = 11.009) and 19.78% another isotope. To account for the atomic weight, 10.810, what must be the nuclidic mass of the other isotope?

Ans. 10.01

2.13. ^{35}Cl and ^{37}Cl are the only naturally occurring chlorine isotopes. What percentage distribution accounts for the atomic weight, 35.4527?

Ans. 24.23% ^{37}Cl

2.14. To account for nitrogen's atomic weight of 14.00674, what must be the ratio of ^{15}N to ^{14}N atoms in natural nitrogen? Ignore the small amount of ^{16}N.

Ans. 0.00369

2.15. At one time there was a chemical atomic weight scale based on the assignment of the value 16.0000 to naturally occurring oxygen. What would have been the atomic weight, on such a table, of silver, if current information had been available? The atomic weights of oxygen and silver on the present table are 15.9994 and 107.8682.

Ans. 107.872

2.16. The nuclidic mass of ^{90}Sr had been determined on the old physical scale ($^{16}O = 16.0000$) as 89.936. Recompute this to the present atomic weight scale, on which ^{16}O is 15.9949.

Ans. 89.907

2.17. In a chemical atomic weight determination, the tin content of 3.7692 g of $SnCl_4$ was found to be 1.7170 g. If the atomic weight of chlorine is taken as 35.453, what is the value for the atomic weight of tin determined from this experiment?

Ans. 118.65

2.18. A 12.5843-g sample of $ZrBr_4$ was dissolved and, after several chemical steps, all of the combined bromine was precipitated as AgBr. The silver content of the AgBr was found to be 13.2160 g. Assume the atomic weights of silver and bromine to be 107.868 and 79.904. What value was obtained for the atomic weight of Zr from this experiment?

Ans. 91.23

2.19. The atomic weight of sulfur was determined by decomposing 6.2984 g of Na_2CO_3 with sulfuric acid and weighing the resultant Na_2SO_4 formed. The weight was found to be 8.4380 g. Taking the atomic weights of C, O, and Na as 12.011, 15.999, and 22.990, respectively, what value is computed for the atomic weight of sulfur?

Ans. 32.017

FORMULA WEIGHTS AND MOLECULAR WEIGHTS

2.20. Determine the molecular weights (or formula weights) to 0.01 u for NaOH, HNO_3, F_2, S_8, $Ca_3(PO_4)_2$, $Fe_4[Fe(CN)_6]_3$, $TiO_{1.12}$.

Ans. 40.00, 63.02, 38.00, 256.53, 310.18, 859.28, 65.80

2.21. How many grams of each of the constituent elements are contained in one mole of   (a) CH_4,   (b) Fe_2O_3, (c) Ca_3P_2? How many atoms of each element are contained in the same amount of compound?

Ans. (a) 12.01 g C, 4.032 g H; 6.02×10^{23} atoms C, 2.41×10^{24} atoms H
        (b) 111.69 g Fe, 48.00 g O; 1.204×10^{24} atoms Fe, 1.81×10^{24} atoms O
        (c) 120.23 g Ca, 61.95 g P; 1.81×10^{24} atoms Ca, 1.204×10^{24} atoms P

2.22. Calculate the number of grams in a mole of each of the following common substances: (a) calcite, $CaCO_3$; (b) quartz, SiO_2; (c) cane sugar, $C_{12}H_{22}O_{11}$; (d) gypsum, $CaSO_4 \cdot 2H_2O$; (e) white lead, $Pb(OH)_2 \cdot 2PbCO_3$.

Ans.   (a) 100.09 g; (b) 60.09 g; (c) 342.3 g; (d) 172.2 g; (e) 775.7 g

2.23. What is the average weight in kilograms of (a) a hydrogen atom, (b) an oxygen atom, (c) a uranium atom?

Ans.   (a) 1.67×10^{-27} kg; (b) 2.66×10^{-26} kg; (c) 3.95×10^{-25} kg

2.24. What is the weight of one molecule of (a) CH_3OH, (b) $C_{60}H_{122}$, (c) $C_{1\,200}H_{2\,000}O_{1\,000}$?

Ans.   (a) 5.32×10^{-26} kg; (b) 1.40×10^{-24} kg; (c) 5.38×10^{-23} kg

2.25. How many moles of atoms are contained in (a) 32.7 g Zn, (b) 7.09 g Cl, (c) 95.4 g Cu, (d) 4.31 g Fe, (e) 0.378 g S?

Ans.   (a) 0.500 mol; (b) 0.200 mol; (c) 1.50 mol; (d) 0.077 2 mol; (e) 0.011 8 mol

2.26. How many moles are represented by (a) 24.5 g H_2SO_4, (b) 4.00 g O_2?

Ans.   (a) 0.250 mol; (b) 0.125 mol

2.27. (a) How many moles of Ba and of Cl are contained in 107.0 g of $Ba(ClO_3)_2 \cdot H_2O$? (b) How many molecules of water of hydration are in this same amount?

Ans.   (a) 0.332 mol Ba, 0.664 mol Cl; (b) 2.00×10^{23} molecules H_2O

2.28. How many moles of Fe and of S are contained in (a) 1 mol of FeS_2 (pyrite), (b) 1 kg of FeS_2? (c) How many kilograms of S are contained in exactly 1 kg of FeS_2?

Ans.   (a) 1 mol Fe, 2 mol S; (b) 8.33 mol Fe, 16.7 mol S; (c) 0.535 kg S

2.29. A certain public water supply contained 0.10 ppb (part per billion) of chloroform, $CHCl_3$. How many molecules of $CHCl_3$ would be contained in a 0.05-mL drop of this water?

Ans. 2.5×10^{10}

MULTIPLE PROPORTIONS

2.30. Verify the law of multiple proportions for an element, X, which forms oxides having percentages of X equal to 77.4%, 63.2%, 69.6%, and 72.0%.

Ans.   The relative amounts of X combining with a fixed amount of oxygen are 2, 1, $\frac{4}{3}$, and $\frac{3}{2}$. The relative amounts of oxygen combining with a fixed amount of X are, 1, 2, $\frac{3}{2}$, and $\frac{4}{3}$.

Chapter 3

Formulas and Composition Calculations

EMPIRICAL FORMULA FROM COMPOSITION

As defined in Chapter 2, an empirical formula expresses the relative numbers of atoms of the different elements in a compound using the smallest integers possible. These integers may be found by converting analytical weight composition data to the amount of each element, expressed in moles of atoms, contained in some fixed weight of the compound. Consider a compound that analyzes 17.09% magnesium, 37.93% aluminum, and 44.98% oxygen. (Unless stated to the contrary, percentage is a *weight* percentage, i.e., number of grams of the element per 100 g of the compound.) Table 3-1 gives a systematic scheme for handling the data.

Table 3-1

(1) Element, E	(2) Mass of E per Fixed Amount of Compound (in this case, 100 g), $m(E)$	(3) Atomic Weight of E, $A_r(E)$	(4) Amount of E in Moles of Atoms, $n(E) = \dfrac{m(E)}{A_r(E)}$	(5) $\dfrac{n(E)}{\text{Smallest } n(E)}$
Mg	17.09 g	24.31 g/mol	0.703 mol	1.00
Al	37.93 g	26.98 g/mol	1.406 mol	2.00
O	44.98 g	16.00 g/mol	2.812 mol	4.00

The numbers in column (4) are the numbers of moles of atoms of the component elements in the fixed weight of compound, 100 g, chosen as the basis. Any set of numbers obtained by multiplying or dividing each of the numbers in column (4) by the same factor will be in the same ratio to each other as the numbers in (4). The set in column (5) is such a set, obtained by dividing each $n(E)$ value in (4) by the *smallest* entry in (4), 0.703. Column (5) shows that the relative numbers of moles of atoms, and therefore of atoms themselves, of Mg, Al, and O in this compound are 1:2:4. Therefore the empirical formula is $MgAl_2O_4$.

Problems will not be given in this chapter involving empirical formula determinations for nonstoichiometric compounds in which the relative numbers of atoms cannot be given as the ratio of small whole numbers, unless specific mention is made of this fact.

COMPOSITION FROM FORMULA

The existence of a formula for a compound implies that fixed relationships exist between the weights of any two elements in the compound or between the weight of any element and the weight of the compound as a whole. These relationships can best be seen by writing the formula in a vertical form, as illustrated in Table 3-2 for the compound Al_2O_3.

The sum of the entries in column (4) for the elements equals the formula weight (FW) of the compound. The entries in column (5) represent the *fractional* content of the various elements in the compound. These numbers are really dimensionless (g/g) and are the same in any unit of mass. Thus, 1 gram (pound, ton, etc.) of Al_2O_3 contains 0.529 gram (pound, ton, etc.) of Al and 0.471 gram (pound, ton, etc.) of O. It is obvious that the sum of the constituent fractions of any compound must equal 1.000.

The *percentage* of aluminum in Al_2O_3 is the number of parts by weight of Al in 100 parts by weight of Al_2O_3. It follows that the percentage is expressed by a number 100 times as great as the fraction. Thus, the

Table 3-2

(1)	(2) $n(E)$ per Mole of Compound	(3) $A_r(E)$ (atomic weight of element)	(4) $m(E)$ per Mole of Compound = $n(E) \times A_r(E)$	(5) $m(E)$ per g of Compound
Al_2	2 mol	27.0 g/mol	54.0 g	$\dfrac{54.0 \text{ g Al}}{102.0 \text{ g Al}_2O_3} = 0.529$ g Al/g Al_2O_3
O_3	3 mol	16.0 g/mol	48.0 g	$\dfrac{48.0 \text{ g O}}{102.0 \text{ g Al}_2O_3} = 0.471$ g O/g Al_2O_3
Al_2O_3	1 mol		FW = 102.0	**Check:** 1.000

percentages of aluminum and oxygen are 52.9% and 47.1%, respectively. The sum of the constituent percentages of any compound must equal 100.0%.

Sometimes the composition of a substance with respect to a particular element is expressed in terms of a simple compound that can be prepared from that element. For example, the aluminum content of a glass may be expressed as 1.3% Al_2O_3. This means that if all the aluminum contained in 100 g of the glass were converted to Al_2O_3, the weight of the Al_2O_3 would be 1.3 g. This convention is not meant to imply that the aluminum in the glass is in the chemical form Al_2O_3. In many cases oxide notations are the result of historical errors in the assignment of chemical structures to complex substances. Whatever the origin, it is a straightforward procedure to convert the data in such a form to direct elementary composition, or vice versa, by the use of a *quantitative factor* such as that found in column (5) of Table 3-2. The ratio

$$\frac{54.0 \text{ g Al}}{102.0 \text{ g Al}_2O_3} \quad \text{or} \quad \frac{102.0 \text{ g Al}_2O_3}{54.0 \text{ g Al}}$$

is called a quantitative factor and may be used as a special conversion factor in numerical problems. (See the solution to Problem 1.19.)

NONSTOICHIOMETRIC FACTORS

A similar use of conversion factors limited to particular cases is common even when the relative proportions are not fixed by a chemical formula. Consider a silver alloy used for jewelry containing 86% silver. Factors based on this composition, such as

$$\frac{0.86 \text{ g Ag}}{1 \text{ g alloy}} \quad \text{or} \quad \frac{100 \text{ g alloy}}{86 \text{ g Ag}}$$

may be used as conversion factors in all problems involving alloys of this particular composition.

MOLECULAR MASS NUMBERS AND CHEMICAL FORMULAS

The molecular weight of a compound, computed from the atomic weights of the constituent elements, is the average mass (in u) of a molecule, weighted among the various isotopic forms of the different component elements. A nuclidic molecular weight may be defined for a molecule made up of particular nuclides by adding nuclidic atomic masses in the same way that the usual molecular weight is computed from the atomic masses.

A *mass spectrometer* is an instrument that separates particles of different isotopic composition and measures their individual relative masses. If a nuclidic mass of an unknown compound is known with great precision from mass spectrometry, the exact molecular formula can often be deduced directly from this information without resort to a quantitative chemical composition analysis.

EXAMPLE 1   Consider the three gases CO, N_2, and C_2H_4. Since ^{12}C, ^{16}O, ^{14}N, and ^1H dominate over all other isotopes, the mass spectrometer will reveal the presence of a particle of approximate mass 28 in all three cases. If the measurements are made with great precision, the three gases can easily be distinguished on the basis of their nuclidic masses, which are calculated below.

$$^{12}C^{16}O \qquad \begin{array}{l} 12.000\,0 \\ \underline{15.994\,9} \\ 27.994\,9 \text{ u} \end{array} \qquad\qquad ^{14}N_2 \qquad 2(14.003\,07) = 28.006\,1 \text{ u}$$

$$^{12}C_2{}^1H_4 \qquad \begin{array}{l} 2(12.000\,0) = 24.000\,0 \\ 4(1.007\,83) = \underline{\ \ 4.031\,3} \\ \qquad\qquad\quad\ 28.031\,3 \text{ u} \end{array}$$

EXAMPLE 2   Find the formula of an organic compound whose dominant nuclidic species was found to have a precise molecular weight of 44.025. It is known that no elements other than C, H, O, and N are present.

The number of carbon atoms in the molecule, $n(C)$, must be at least 1, otherwise the compound would not be organic. $n(C)$ cannot be greater than 3, because 4 carbon atoms would contribute 48 to the total mass number of the molecule, 44. Similar constraints limit the number of O or N atoms per molecule. The possible combinations of C, O, and N consistent with the limiting mass are listed in column (1) of Table 3-3.

Column (2) lists the mass numbers of the (C, O, N) skeletons. Column (3) lists the number of hydrogen atoms needed to bring the mass number of the molecule to 44. Column (4) lists the maximum number of H atoms consistent with rules for molecular structure as discussed in Chapters 9 and 15. One such rule is that $n(H, max)$ is equal to twice the number of carbon atoms plus the number of nitrogen atoms plus 2. Column (5) lists the allowed formulas consistent with the total mass number and with all the assumptions and rules. Note that all skeletons for which the number in column (3) (the mass shortage to be made up by hydrogen) exceeds the number in column (4) (the amount of hydrogen allowable for the skeleton by the rules of valence) are rejected. Column (6) tabulates the nuclidic molecular weights for the allowed formulas, computed from the nuclidic masses in Table 2-1. When the computed molecular weights are compared with the experimental value, 44.025, it is seen that C_2OH_4 is the only allowable formula that fits the data within the claimed precision; therefore this must be the formula of the substance.

Not all the allowable molecular formulas in column (5) represent stable chemical substances. Nevertheless, they might possibly show up in a mass spectrometer reading, which details not only stable compounds but also ions and

Table 3-3

(1) (C, O, N) Skeleton	(2) Mass Number of Skeleton	(3) 44 minus Mass Number of Skeleton	(4) $n(H, max)$	(5) Molecular Formula	(6) Nuclidic Molecular Weight
C	12	32	4		
C_2	24	20	6		
C_3	36	8	8	C_3H_8	44.063
CO	28	16	4		
CO_2	44	0	4	CO_2	43.990
C_2O	40	4	6	C_2OH_4	44.026
CN	26	18	5		
CN_2	40	4	6	CN_2H_4	44.037
C_2N	38	6	7	C_2NH_6	44.050
CON	42	2	5	$CONH_2$	44.014

decomposition fragments. For those who use this method frequently, compilations exist of molecular formulas consistent with any mass number.

Solved Problems

CALCULATION OF FORMULAS

3.1. Derive the empirical formula of a hydrocarbon that on analysis gave the following percentage composition: C = 85.63%, H = 14.37%.

The tabular solution, based on 100 g of compound, is as follows:

E	$m(E)$	$A_r(E)$	$n(E) = \dfrac{m(E)}{A_r(E)}$	$\dfrac{n(E)}{7.129 \text{ mol}}$
C	85.63 g	12.011 g/mol	7.129 mol	1.000
H	14.37 g	1.008 g/mol	14.26 mol	2.000

where E = element; $m(E)$ = mass of element per 100 g of compound; $A_r(E)$ = atomic weight of element; $n(E)$ = amount of element per 100 g of compound, expressed in moles of atoms.

The procedure of dividing $n(E)$ by $n(C)$ is equivalent to finding the number of atoms of each element for every atom of carbon. The ratio of H to C atoms is 2:1. Hence, the empirical formula is CH_2. (The formulas C_2H_4, C_3H_6, C_4H_8, etc., imply the same percentage composition as does CH_2, but for the empirical formula the smallest possible integers are chosen.)

There is no stable substance with the formula CH_2. It is necessary to determine the molecular weight to determine the molecular formula. If this hydrocarbon were a gas or an easily volatilized liquid, its molecular weight could be estimated from the density of the gas, as shown in Chapter 5. Supposing such a determination yields a molecular weight of about 55 g/mol, what is the molecular formula?

The formula weight of CH_2 is 14. Since 55 is very close to 4×14 (but far from 3×14 or 5×14), the compound must be C_4H_8.

3.2. A compound gave on analysis the following percentage composition: K = 26.57%, Cr = 35.36%, O = 38.07%. Derive the empirical formula of the compound.

The standard tabular solution, as applied to 100 g of compound, follows.

(1) E	(2) $m(E)$	(3) $A_r(E)$	(4) $n(E) = \dfrac{m(E)}{A_r(E)}$	(5) $\dfrac{n(E)}{0.680\,0 \text{ mol}}$	(6) $\dfrac{n(E)}{0.680\,0 \text{ mol}} \times 2$
K	26.57 g	39.10 g/mol	0.680 0 mol	1.000	2
Cr	35.36 g	52.00 g/mol	0.680 0 mol	1.000	2
O	38.07 g	16.00 g/mol	2.379 mol	3.499	7

In contrast to previous examples, the numbers in column (5) are not all integers. The ratio of the numbers of atoms of two elements in a compound must be the ratio of small whole numbers, in order to satisfy one of the postulates of Dalton's atomic theory. Allowing for experimental and calculational uncertainty, we see that the entry of oxygen in column (5), 3.499, is, to within the allowed error, 3.500 or $\frac{7}{2}$, indeed the ratio of small whole numbers. By rounding off in this way and multiplying each of the entries in column (5) by 2, we arrive at the set of smallest integers that correctly represent the relative numbers of atoms in the compound, as tabulated in column (6). The formula is thus $K_2Cr_2O_7$.

3.3.   A 15.00-g sample of an unstable hydrated salt, $Na_2SO_4 \cdot xH_2O$, was found to contain 7.05 g of water. Determine the empirical formula of the salt.

Hydrates are compounds containing water molecules loosely bound to the other components. H_2O may usually be removed intact by heating such compounds and may then be replaced by wetting. The Na_2SO_4 and H_2O groups may thus be considered as the units of which the compound is made, and their formula weights are used in place of atomic weights. It is clearly more convenient to base the tabular solution in this case on 15 g of compound (which contains $15.00 - 7.05 = 7.95$ g Na_2SO_4).

X	$m(X)$	FW(X)	$n(X) = \dfrac{m(X)}{FW(X)}$	$\dfrac{n(X)}{0.055\,9\,mol}$
Na_2SO_4	7.95 g	142.1 g/mol	0.055 9 mol	1.00
H_2O	7.05 g	18.02 g/mol	0.391 mol	6.99

The mole ratio of H_2O to Na_2SO_4 is, to within the allowed error, 7 to 1, and the empirical formula is $Na_2SO_4 \cdot 7H_2O$.

3.4.   A 2.500-g sample of uranium was heated in the air. The resulting oxide weighed 2.949 g. Determine the empirical formula of the oxide.

2.949 g of the oxide contains 2.500 g U and 0.449 g O (obtained by subtracting the initial mass from the final mass). Calculation shows that 2.949 g of the oxide contains 0.028 06 mol oxygen atoms and 0.010 50 mol uranium atoms, or 2.672 mol oxygen atoms per mol uranium atoms. The smallest multiplying integer that will give whole numbers is 3.

$$\frac{n(O)}{n(U)} = \frac{2.672 \text{ mol O}}{1.000 \text{ mol U}} = \frac{3(2.672 \text{ mol O})}{3(1.000 \text{ mol U})} = \frac{8.02 \text{ mol O}}{3.00 \text{ mol U}}$$

The empirical formula is U_3O_8.

Emphasis must be placed on the importance of carrying out the computations to as many significant figures as the analytical precision requires. If numbers in the ratio 2.67:1 had been multiplied by 2 to give 5.34:2 and these numbers had been rounded off to 5:2, the wrong formula would have been obtained. This would have been unjustified because it would have assumed an error of 34 parts in 500 in the analysis of oxygen. The weight of oxygen, 0.449 g, indicates a possible error of only a few parts in 500. When the multiplying factor 3 was used, the rounding off was from 8.02 to 8.00, the assumption being made that the analysis of oxygen may have been in error by 2 parts in 800; this amount of error is more reasonable.

3.5.   A 1.367-g sample of an organic compound was combusted in a stream of air to yield 3.002 g CO_2 and 1.640 g H_2O. If the original compound contained only C, H, and O, what is its empirical formula?

It is necessary to use quantitative factors for CO_2 and H_2O to find how much C and H are present in the combustion products and, thus, in the original sample.

$$m(C) = \left(\frac{1 \text{ mol C}}{1 \text{ mol CO}_2} \right)(3.002 \text{ g CO}_2) = \left(\frac{12.01 \text{ g C}}{44.01 \text{ g CO}_2} \right)(3.002 \text{ g CO}_2) = 0.819 \text{ g C}$$

$$m(H) = \left(\frac{2 \text{ mol H}}{1 \text{ mol H}_2O} \right)(1.640 \text{ g H}_2O) = \left[\frac{2(1.008 \text{ g H})}{18.02 \text{ g H}_2O} \right](1.640 \text{ g H}_2O) = 0.183\,5 \text{ g H}$$

The amount of oxygen in the original sample cannot be obtained from the weight of combustion products, since the CO_2 and H_2O contain oxygen that came partly from the combined oxygen in the compound and partly from the air stream used in the combustion process. The oxygen content of the sample can be obtained, however, by difference.

$$m(O) = m(\text{compound}) - m(C) - n(H) = 1.367 - 0.819 - 0.184 = 0.364 \text{ g}$$

The problem can now be solved by the usual procedures. The numbers of the moles of the elementary atoms in 1.367 g of compound are found to be C, 0.068 2; H, 0.182 0; O, 0.022 8. These numbers are in the ratio 3:8:1, and the empirical formula is C_3H_8O.

COMPOSITION PROBLEMS

3.6. A strip of electrolytically pure copper weighing 3.178 g is strongly heated in a stream of oxygen until it is all converted to 3.978 g of the black oxide. What is the percentage composition of this oxide?

$$\text{Total weight of black oxide} = 3.978 \text{ g}$$
$$\text{Weight of copper in oxide} = \underline{3.178 \text{ g}}$$
$$\text{Weight of oxygen in oxide} = 0.800 \text{ g}$$

$$\text{Fraction of coppper} = \frac{\text{weight of copper in oxide}}{\text{total weight of oxide}} = \frac{3.178 \text{ g}}{3.978 \text{ g}} = 0.799 = 79.9\%$$

$$\text{Fraction of oxygen} = \frac{\text{weight of oxygen in oxide}}{\text{total weight of oxide}} = \frac{0.800 \text{ g}}{3.978 \text{ g}} = 0.201 = 20.1\%$$

$$\textbf{Check:}\quad \underline{100.0\%}$$

3.7. (a) Determine the percentages of iron in $FeCO_3$, Fe_2O_3, and Fe_3O_4.   (b) How many kilograms of iron could be obtained from 2.000 kg of Fe_2O_3?

(a)   Formula weight of $FeCO_3$ is 115.86; of Fe_2O_3, 159.69; of Fe_3O_4, 231.54. Consider 1 mol of each compound.

$$\text{Fraction of Fe in FeCo}_3 = \frac{m(1 \text{ mol Fe})}{m(1 \text{ mol FeCO}_3)} = \frac{55.847 \text{ g}}{115.86 \text{ g}} = 0.482\,0 = 48.20\%$$

$$\text{Fraction of Fe in Fe}_2\text{O}_3 = \frac{m(2 \text{ mol Fe})}{m(1 \text{ mol Fe}_2\text{O}_3)} = \frac{2(55.847 \text{ g})}{159.69 \text{ g}} = 0.699\,4 = 69.94\%$$

$$\text{Fraction of Fe in Fe}_3\text{O}_4 = \frac{m(3 \text{ mol Fe})}{m(1 \text{ mol Fe}_3\text{O}_4)} = \frac{3(55.847 \text{ g})}{231.54 \text{ g}} = 0.723\,6 = 72.36\%$$

(b)   From (a), the weight of Fe in 2.000 kg Fe_2O_3 is $0.699\,4 \times 2.000$ kg $= 1.399$ kg Fe.

3.8. Given the formula K_2CO_3, determine the percentage composition of potassium carbonate.

One formula weight of K_2CO_3 contains

$$2 \text{ atomic weights K} = 2(39.098\,3) = \;\; 78.197 \text{ parts K by weight}$$
$$1 \text{ atomic weight C} = 1(12.011) = \;\; 12.011 \text{ parts C by weight}$$
$$3 \text{ atomic weights O} = 3(15.999\,4) = \;\; \underline{47.998} \text{ parts O by weight}$$
$$\text{Formula weight of K}_2\text{CO}_3 = 138.206 \text{ parts by weight}$$

$$\text{Fraction of K in K}_2\text{CO}_3 = \frac{78.197}{138.206} = 0.565\,8 = 56.58\%$$

$$\text{Fraction of C in K}_2\text{CO}_3 = \frac{12.011}{138.206} = 0.086\,9 = \;\; 8.69\%$$

$$\text{Fraction of O in K}_2\text{CO}_3 = \frac{47.998}{138.206} = 0.347\,3 = 34.73\%$$

$$\textbf{Check:}\quad \underline{100.00\%}$$

3.9.   (a) Calculate the percentage of CaO in $CaCO_3$.   (b) How many pounds of CaO can be obtained from 1 ton of limestone that is 97.0% $CaCO_3$? One ton is 2000 lb.

(a)   The quantitative factor can be written by considering the conservation of Ca atoms. 1 mol $CaCO_3$ contains 1 mol Ca, which is the same amount of Ca as is contained in 1 mol CaO. Therefore,

$$\text{Fraction of CaO in } CaCO_3 = \frac{\text{formula weight CaO}}{\text{formula weight } CaCO_3} = \frac{56.1}{100.1} = 0.560 = 56.0\%$$

(b)   Weight of $CaCO_3$ in 1 ton limestone $= 0.970 \, (2000 \text{ lb}) = 1940 \text{ lb } CaCO_3$

$$\text{Weight of CaO} = (\text{fraction of CaO in } CaCO_3)(\text{weight of } CaCO_3)$$

$$= 0.560(1940 \text{ lb}) = 1090 \text{ lb CaO in 1 ton limestone}$$

3.10.   How much 58.0% sulfuric acid solution is needed to provide 150 g of H_2SO_4?

Let $x =$ mass of sulfuric acid solution

$$\left(\frac{58.0 \text{ g } H_2SO_4}{100 \text{ g soln}}\right)x = 150 \text{ g } H_2SO_4$$

$$x = 259 \text{ g soln}$$

Using the concept of an inverse conversion factor avoids the use of algebra:

$$(150 \text{ g } H_2SO_4)\left(\frac{100 \text{ g soln}}{58.0 \text{ g } H_2SO_4}\right) = 259 \text{ g soln}$$

3.11.   How much calcium is in the amount of $Ca(NO_3)_2$ that contains 20.0 g of nitrogen?

It is not necessary to find the weight of $Ca(NO_3)_2$ containing 20.0 g of nitrogen. The relationship between two component elements of a compound may be found directly from the formula.

$$\text{Weight of Ca} = (20.0 \text{ g N})\left(\frac{1 \text{ mol Ca}}{2 \text{ mol N}}\right) = (20.0 \text{ g N})\left[\frac{40.08 \text{ g Ca}}{2(14.01 \text{ g N})}\right] = 28.6 \text{ g Ca}$$

3.12.   (a) How much H_2SO_4 could be produced from 500 kg of sulfur?   (b) How many kilograms of Glauber's salt, $Na_2SO_4 \cdot 10H_2O$, could be obtained from 1.000 kg H_2SO_4?

(a)   The formula H_2SO_4 indicates that 1 mol S (32.07 g S) will give 1 mol H_2SO_4 (98.08 g H_2SO_4). Then, since the *ratio* of any two constituents is independent of the mass units,

$$\text{Weight of } H_2SO_4 = (500 \cancel{\text{ kg S}})\left(\frac{98.08 \text{ kg } H_2SO_4}{32.07 \text{ kg S}}\right) = 1529 \text{ kg } H_2SO_4$$

(b)   1 mol H_2SO_4 (98.08 g H_2SO_4) will give 1 mol $Na_2SO_4 \cdot 10H_2O$ (322.2 g $Na_2SO_4 \cdot 10H_2O$), since each substance contains one sulfate (SO_4) group. Then

$$\text{Weight of } Na_2SO_4 \cdot 10H_2O = (1.000 \cancel{\text{ kg } H_2SO_4})\left(\frac{322.2 \text{ kg } Na_2SO_4 \cdot 10H_2O}{98.08 \cancel{\text{ kg } H_2SO_4}}\right)$$

$$= 3.285 \text{ kg } Na_2SO_4 \cdot 10H_2O$$

3.13.   How many tons of $Ca_3(PO_4)_2$ must be treated with carbon and sand in an electric furnace to make 1 ton of phosphorus?

The formula $Ca_3(PO_4)_2$ indicates that 2 mol P ($2 \times 30.974 \text{ g} = 61.95 \text{ g P}$) is contained in 1 mol $Ca_3(PO_4)_2$ [310.2 g $Ca_3(PO_4)_2$]. Then, changing grams to tons in the weight ratio, we obtain

$$\text{Weight of } Ca_3(PO_4)_2 = (1 \cancel{\text{ ton P}})\left(\frac{310.2 \text{ tons } Ca_3(PO_4)_2}{61.95 \cancel{\text{ tons P}}}\right) = 5.01 \text{ tons } Ca_3(PO_4)_2$$

3.14. A 5.82-g silver coin is dissolved in nitric acid. When sodium chloride is added to the solution, all the silver is precipitated as AgCl. The AgCl precipitate weighs 7.20 g. Determine the percentage of silver in the coin.

$$\text{Fraction of Ag in AgCl} = \frac{\text{atomic weight Ag}}{\text{formula weight AgCl}} = \frac{107.9}{143.3} = 0.753$$

$$\text{Weight of Ag in 7.20 g AgCl} = (0.753)(7.20 \text{ g}) = 5.42 \text{ g Ag}$$

Hence the 5.82-g coin contains 5.42 g Ag.

$$\text{Fraction of Ag in coin} = \frac{5.42 \text{ g}}{5.82 \text{ g}} = 0.931 = 93.1\% \text{ Ag}$$

3.15. A sample of impure sulfide ore contains 42.34% Zn. Find the percentage of pure ZnS in the sample.

The formula ZnS shows that 1 formula weight ZnS contains 1 atomic weight Zn; hence a suitable conversion factor is

$$\frac{\text{Formula weight ZnS}}{\text{Atomic weight Zn}} = \frac{97.46 \text{ g ZnS}}{65.39 \text{ g Zn}}$$

Consider 100.0 g of sample; it contains 42.34 g Zn. Then

$$(42.34 \text{ g Zn})\left(\frac{97.46 \text{ g ZnS}}{65.39 \text{ g Zn}}\right) = 63.11 \text{ g ZnS in 100 g of sample, or } 63.11\% \text{ pure ZnS}$$

3.16. A granulated sample of aircraft alloy (Al, Mg, Cu) weighing 8.72 g was first treated with alkali to dissolve the aluminum, then with very dilute HCl to dissolve the magnesium, leaving a residue of copper. The residue after the alkali treatment weighed 2.10 g, and the residue after the HCl treatment weighed 0.69 g. What is the composition of the alloy?

$$\text{Weight of Al} = 8.72 \text{ g} - 2.10 \text{ g} = 6.62 \text{ g}$$
$$\text{Weight of Mg} = 2.10 \text{ g} - 0.69 \text{ g} = 1.41 \text{ g}$$
$$\text{Weight of Cu} = 0.69 \text{ g}$$

$$\text{Fraction of Al} = \frac{6.62 \text{ g}}{8.72 \text{ g}} = 0.759 = 75.9\%$$

$$\text{Fraction of Mg} = \frac{1.41 \text{ g}}{8.72 \text{ g}} = 0.162 = 16.2\%$$

$$\text{Fraction of Cu} = \frac{0.69 \text{ g}}{8.72 \text{ g}} = 0.079 = 7.9\%$$

Check: 100.0%

3.17. A Pennsylvania bituminous coal is analyzed as follows: Exactly 2.500 g is weighed into a fused silica crucible. After drying for 1 hour at 110 °C, the moisture-free residue weighs 2.415 g. The crucible, next, is covered with a vented lid and strongly heated until no volatile matter remains. The residual coke button weighs 1.528 g. The crucible is then heated without the cover until all specks of carbon have disappeared, and the final ash weighs 0.245 g. What is the *proximate analysis* of this coal, i.e., the percents of moisture, volatile combustible matter (VCM), fixed carbon (FC), and ash?

$$\text{Moisture} = 2.500 \text{ g} - 2.415 \text{ g} = 0.085 \text{ g}$$
$$\text{VCM} = 2.415 \text{ g} - 1.528 \text{ g} = 0.887 \text{ g}$$
$$\text{FC} = 1.528 \text{ g} - 0.245 \text{ g} = 1.283 \text{ g}$$
$$\text{Ash} = \underline{0.245 \text{ g}}$$
$$\text{Total} \quad 2.500 \text{ g coal}$$

$$\text{Fraction of moisture} = \frac{0.085 \text{ g}}{2.500 \text{ g}} = 0.034 = 3.4\%$$

Similarly, the other percentages are calculated to be 35.5% VCM, 51.3% FC, 9.8% ash.

3.18. On the "dry basis" a sample of coal analyzes as follows: VCM, 21.06%; fixed carbon, 71.80%; ash, 7.14%. If the moisture present in the coal is 2.49%, what is the analysis on the "wet basis"?

Consider as a basis 100.0 g of wet coal. Of this, 100.0 g − 2.49 g = 97.5 g is dry coal. The weights of VCM, FC, and ash then are

VCM	$(0.210\,6)(97.5) = 20.5$ g or 20.5% of the 100 g of wet coal
FC	$(0.718\,0)(97.5) = 70.0$ g or 70.0%
Ash	$(0.071\,4)(97.5) = 7.0$ g or 7.0%

The sum of the above percentages plus the 2.5% moisture comes to 100%.

3.19. When the *Bayer process* is used for recovering aluminum from siliceous ores, some aluminum is always lost because of the formation of an unworkable "mud" having the following average formula: $3\,Na_2O \cdot 3\,Al_2O_3 \cdot 5\,SiO_2 \cdot 5\,H_2O$. Since aluminum and sodium ions are always in excess in the solution from which this precipitate is formed, the precipitation of the silicon in the "mud" is complete. A certain ore contained 13% (by weight) kaolin ($Al_2O_3 \cdot 2\,SiO_2 \cdot 2\,H_2O$) and 87% gibbsite ($Al_2O_3 \cdot 3\,H_2O$). What percent of the total aluminum in this ore is recoverable in the Bayer process?

Consider 100 g ore, which contains 13 g kaolin and 87 g gibbsite.

$$\text{Weight of Al in 13 g kaolin} = 13 \text{ g kaolin} \times \frac{2 \text{ mol Al}}{1 \text{ mol kaolin}} = 13 \times \frac{54.0}{258} = 2.7 \text{ g Al}$$

$$\text{Weight of Al in 87 g gibbsite} = 87 \text{ g gibbsite} \times \frac{2 \text{ mol Al}}{1 \text{ mol gibbsite}} = 87 \times \frac{54.0}{156} = 30.1 \text{ g Al}$$

$$\text{Total weight of Al in 100 g ore} = 2.7 \text{ g} + 30.1 \text{ g} = 32.8 \text{ g}$$

Kaolin has equal numbers of Al and Si atoms, and 13 g kaolin contains 2.7 g Al. The mud takes 6 Al atoms for 5 Si atoms or 6 Al atoms lost for every 5 Si atoms in the kaolin. Hence the precipitation of all the Si from 13 g kaolin involves the loss of $\frac{6}{5}(2.7 \text{ g}) = 3.2 \text{ g Al}$.

$$\text{Fraction of Al recoverable} = \frac{\text{recoverable Al}}{\text{total Al}} = \frac{(32.8 - 3.2) \text{ g}}{32.8 \text{ g}} = 0.90 = 90\%$$

3.20. A clay was partially dried and then contained 50% silica and 7% water. The original clay contained 12% water. What is the percentage of silica in the original sample?

Assuming that the only substance lost in the drying process was water, the original and dried clays have the compositions

	% Water	% Silica	% Other
Original	12	x	$88 - x$
Dried	7	50	43

The ratio of silica to the other dry constituents must be the same in both clays; hence

$$\frac{x}{88 - x} = \frac{50}{43}$$

Solving, $x = 47$; i.e, there is 47% silica in the original clay.

3.21. Two unblended manganese ores contain 40% and 25% manganese, respectively. How many pounds of each ore must be mixed to give 100 lb of blended ore containing 35% manganese?

Let x = pounds of 40% ore required; then 100 lb − x = pounds of 25% ore required.

Mn from 40% ore + Mn from 25% ore = total Mn in 100 lb of mixture

$$(0.40)x + (0.25)(100 \text{ lb} - x) = (0.35)(100 \text{ lb})$$

Solving, x = 67 lb of 40% ore. Then 100 lb − x = 33 lb of 25% ore.

3.22. A nugget of gold and quartz weighs 100 g and has a density of 6.4 g/cm³. The densities of gold and quartz are 19.3 and 2.65 g/cm³, respectively. Determine the weight of gold in the nugget.

Let x = grams of gold in nugget; then 100 g − x = grams of quartz in nugget.

Volume of nugget = (volume of gold in nugget) + (volume of quartz in nugget)

$$\frac{100 \text{ g}}{6.4 \text{ g/cm}^3} = \frac{x}{19.3 \text{ g/cm}^3} + \frac{100 \text{ g} - x}{2.65 \text{ g/cm}^3}$$

from which x = 68 g gold.

MOLECULAR WEIGHTS

3.23. A purified cytochrome protein isolated from a bacterial preparation was found to contain 0.376% iron. What can be deduced about the molecular weight of the protein?

The protein must contain at least one atom of iron per molecule. If it contains only one, of weight 55.8 u = 55.8 daltons, then the molecular weight $\mathcal{M}$ is given by

$$0.003\,76\,\mathcal{M} = 55.8 \text{ daltons}$$

$$\mathcal{M} = 14\,800 \text{ daltons}$$

If the protein molecule contained x atoms of Fe, the molecular weight would be 14 800x daltons.

The method given here is useful for determining the *minimum* molecular weight of a macromolecular substance when an analysis can be done for one of the minor components. Frequently the approximate molecular weight can be determined by a physical method such as osmotic pressure or sedimentation rate, thus fixing the value of the integer x.

3.24. A purified pepsin isolated from a bovine preparation was subject to an amino acid analysis of its hydrolytic product. The amino acid present in smallest amount was lysine, $C_6H_{14}N_2O_2$, and the amount of lysine recovered was found to be 0.43 g per 100 g protein. What is the minimum molecular weight of the protein?

Proteins do not contain free amino acids, but they do contain chemically linked forms of amino acids, which on degradative hydrolysis can be reconverted to the free amino acid form. The molecular weight of free lysine is 146, and we let $\mathcal{M}$ be the minimum molecular weight of the protein. As in Problem 3.23, the protein molecule must be at least heavy enough to contain one lysine residue.

No. of moles lysine = No. of moles protein

$$0.43 \text{ g} \times \frac{1 \text{ mol}}{146 \text{ g}} = 100 \text{ g} \times \frac{1 \text{ mol}}{\mathcal{M} \text{ g}}$$

from which $\mathcal{M}$ = 34 000.

3.25. A sample of potato starch was ground in a ball mill to give a starchlike substance of lower molecular weight. The product analyzed 0.086% phosphorus. If each molecule of product is assumed to contain one atom of phosphorus, what is the average molecular weight of the material?

By assumption, 1 mol of phosphorus atoms (31.0 g P) is contained in 1 mol of the material. Since 0.086 g P is contained in 100 g of material, then 31.0 g P is contained in

$$31.0 \text{ g of P} \times \frac{100 \text{ g of material}}{0.086 \text{ g of P}} = 36\,000 \text{ g of material}$$

Hence the average molecular weight of the material is 36 000.

The computational procedures in Problems 3.23, 3.24, and 3.25 are obviously equivalent; that is, the numbers are combined in the same way eventually, even though the reasoning may seem to be somewhat different.

FORMULA FROM PRECISE MOLECULAR WEIGHT

3.26 An organic compound was prepared containing at least one and no more than two sulfur atoms per molecule. The compound had no nitrogen, but oxygen could have been present. The mass-spectrometrically determined molecular weight of the predominant nuclidic species was 110.020. (*a*) What are the allowable molecular formulas consistent with the mass number 110 and with the facts about the elementary composition?   (*b*) What is the molecular formula of the compound?

(*a*)   The nonhydrogen skeleton would be made up of the elements C, O, and S. The number of possible skeletons can be reduced by the following considerations: (i) The maximum number of carbon atoms is 6, since the mass number of 7 carbons plus 1 sulfur would be 116, in excess of the given value.   (ii) The maximum number of hydrogen atoms is $2n(C) + 2 = 14$.   (iii) The (C, O, S) skeleton must therefore contribute between 96 and 110, inclusive, to the mass number. A rather short list, Table 3-4, will now suffice.

(*b*)   Of the six formulas consistent with the known mass number, only C_6SH_6 is consistent with the precise molecular weight.

Table 3-4

(1) (C, O, S) Skeleton	(2) Mass Number of Skeleton	(3) 110 − (2)	(4) $n(H, \text{max})$	(5) Molecular Formula	(6) Nuclidic Molecular Weight
CO_4S	108	2	4	CO_4SH_2	109.967
CO_2S_2	108	2	4	$CO_2S_2H_2$	109.949
C_2O_3S	104	6	6	$C_2O_3SH_6$	110.004
C_2OS_2	104	6	6	$C_2OS_2H_6$	109.986
C_3O_2S	100	10	8		
C_3OS_2	100	10	8		
C_4OS	96	14	10		
C_5OS	108	2	12	C_5OSH_2	109.983
C_6S	104	6	14	C_6SH_6	110.019

Supplementary Problems

CALCULATION OF FORMULAS

3.27. Derive the empirical formulas of the substances having the following percentage compositions: (a) Fe = 63.53%, S = 36.47%;   (b) Fe = 46.55%, S = 53.45%;   (c) Fe = 53.73%, S = 46.27%.

Ans.   (a) FeS;   (b) FeS_2;   (c) Fe_2S_3

3.28. A compound contains 21.6% sodium, 33.3% chlorine, 45.1% oxygen. Derive its empirical formula. Take Na = 23.0, Cl = 35.5, 0 = 16.0.

Ans. $NaClO_3$

3.29. When 1.010 g of zinc vapor is burned in air, 1.257 g of the oxide is produced. What is the empirical formula of the oxide?

Ans.   ZnO

3.30. A compound has the following percentage composition: H = 2.24%, C = 26.69%, O = 71.07%. Its molecular weight is 90. Derive its molecular formula.

Ans. $H_2C_2O_4$

3.31. Determine the simplest formula of a compound that has the following composition: Cr = 26.52%, S = 24.52%, O = 48.96%.

Ans. $Cr_2S_3O_{12}$ or $Cr_2(SO_4)_3$

3.32. A 3.245-g sample of a titanium chloride was reduced with sodium to metallic titanium. After the resultant sodium chloride was washed out, the residual titanium metal was dried, and weighed 0.819 g. What is the empirical formula of this titanium chloride?

Ans. $TiCl_4$

3.33. A compound contains 63.1% carbon, 11.92% hydrogen, and 24.97% fluorine. Derive its empirical formula.

Ans. C_4H_9F

3.34. An organic compound was found on analysis to contain 47.37% carbon and 10.59% hydrogen. The balance was presumed to be oxygen. What is the empirical formula of the compound?

Ans. $C_3H_8O_2$

3.35. Derive the empirical formulas of the minerals that have the following compositions: (a) $ZnSO_4$ = 56.14%, H_2O = 43.86%;   (b) MgO = 27.16%, SiO_2 = 60.70%, H_2O = 12.14%;   (c) Na = 12.10%, Al = 14.19%, Si = 22.14%, O = 42.09%, H_2O = 9.48%.

Ans.   (a) $ZnSO_4 \cdot 7H_2O$;   (b) $2MgO \cdot 3SiO_2 \cdot 2H_2O$;   (c) $Na_2Al_2Si_3O_{10} \cdot 2H_2O$

3.36. A *borane* (a compound containing only boron and hydrogen) analyzed 88.45% boron. What is its empirical formula?

Ans. B_5H_7

3.37. An experimental catalyst used in the polymerization of butadiene is 23.3% Co, 25.3% Mo, and 51.4% Cl. What is its empirical formula?

Ans. $Co_3Mo_2Cl_{11}$

3.38. A 1.500-g sample of a compound containing only C, H, and O was burned completely. The only combustion products were 1.738 g CO_2 and 0.711 g H_2O. What is the empirical formula of the compound?

Ans. $C_2H_4O_3$

3.39. Elementary analysis showed that an organic compound contained C, H, N, and O as its only elementary constituents. A 1.279-g sample was burned completely, as a result of which 1.60 g of CO_2 and 0.77 g of H_2O were obtained. A separately weighed 1.625-g sample contained 0.216 g nitrogen. What is the empirical formula of the compound?

Ans. $C_3H_7O_3N$

3.40. Manganese forms nonstoichiometric oxides having the general formula MnO_x. Find the value of x for a compound that analyzed 63.70% Mn.

Ans. 1.957

3.41. A hydrocarbon containing 92.3% C and 7.74% H was found (by measuring its gas density) to have a molecular weight of approximately 79. What is the molecular formula?

Ans. C_6H_6

COMPOSITION

3.42. A fusible alloy is made by melting together 10.6 lb bismuth, 6.4 lb lead, and 3.0 lb tin. (*a*) What is the percentage composition of the alloy? (*b*) How much of each metal is required to make 70.0 g of alloy? (*c*) What weight of alloy can be made from 4.2 lb of tin?

Ans. (*a*) 53% Bi, 32% Pb, 15% Sn; (*b*) 37.1 g Bi, 22.4 g Pb, 10.5 g Sn; (*c*) 28 lb alloy

3.43. Calculate the percentage of copper in each of the following minerals: cuprite, Cu_2O, copper pyrites, $CuFeS_2$; malachite, $CuCO_3 \cdot Cu(OH)_2$. How many kilograms of cuprite will give 500 kg of copper?

Ans. 88.82%, 34.63%, 57.48%; 563 kg

3.44. What is the nitrogen content (fertilizer rating) of NH_4NO_3? Of $(NH_4)_2SO_4$?

Ans. 35.0% N, 21.2% N

3.45. Determine the percentage composition of (*a*) silver chromate, Ag_2CrO_4; (*b*) calcium pyrophosphate, $Ca_2P_2O_7$.

Ans. (*a*) 65.03% Ag, 15.67% Cr, 19.29% O; (*b*) 31.54% Ca, 24.38% P, 44.08% O

3.46. Find the percentage of arsenic in a polymer having the empirical formula C_2H_8AsB.

Ans. 63.6% As

3.47. The specifications for a transistor material called for one boron atom in 10^{10} silicon atoms. What would be the boron content of 1 kg of such material?

Ans. 4×10^{-11} kg B

3.48. The purest form of carbon is prepared by decomposing pure sugar, $C_{12}H_{22}O_{11}$ (driving off the contained H_2O). What is the maximum number of grams of carbon that could be obtained from 500 g of sugar?

Ans. 211 g C

3.49. What weight of silver is present in 3.45 g Ag_2S?

   Ans.   3.00 g Ag

3.50. What weight of CuO will be required to furnish 200 kg copper?

   Ans.   250 kg CuO

3.51. Ordinary salt, NaCl, can be electrolyzed in the molten state to produce sodium and chlorine. Electrolysis of an aqueous solution produces sodium hydroxide (NaOH), hydrogen, and chlorine. The latter two products may be combined to form hydrogen chloride (HCl). How many pounds of metallic sodium and of liquid chlorine can be obtained from 1 ton of salt? Alternatively, how many pounds of NaOH and how many pounds of hydrogen chloride?

   Ans.   787 lb Na, 1 213 lb liquid Cl_2, 1 370 lb NaOH, 1 248 lb HCl

3.52. Compute the amount of zinc in a ton of ore containing 60.0% zincite, ZnO.

   Ans.   964 lb Zn

3.53. How much phosphorus is contained in 5.00 g of the compound $CaCO_3 \cdot 3Ca_3(PO_4)_2$? How much P_2O_5?

   Ans.   0.902 g P, 2.07 g P_2O_5

3.54. A 10.00-g sample of crude ore contains 2.80 g of HgS. What is the percentage of mercury in the ore?

   Ans.   24.1% Hg

3.55. A procedure for analyzing the oxalic acid content of a solution involves the formation of the insoluble complex $Mo_4O_3(C_2O_4)_3 \cdot 12H_2O$.   (a) How many grams of this complex would form per gram of oxalic acid, $H_2C_2O_4$, if 1 mol of the complex results from the reaction with 3 mol of oxalic acid?   (b) How many grams of molybdenum are contained in the complex formed by reaction with 1 g of oxalic acid?

   Ans.   (a) 3.38 g complex;   (b) 1.42 g Mo

3.56. The arsenic content of an agricultural insecticide was reported as 28% As_2O_5. What is the % arsenic in this preparation?

   Ans.   18% As

3.57. Express the potassium content of a fertilizer in % K_2O if its elementary potassium content is 4.5%.

   Ans.   5.4% K_2O

3.58. A typical analysis of a Pyrex glass showed 12.9% B_2O_3, 2.2% Al_2O_3, 3.8% Na_2O, 0.4% K_2O, and the balance SiO_2. Assume that the oxide percentages add up to 100%. What is the ratio of silicon to boron atoms in the glass?

   Ans.   3.6

3.59. A piece of plumber's solder weighing 3.00 g was dissolved in dilute nitric acid, then treated with dilute H_2SO_4. This precipitated the lead as $PbSO_4$, which after washing and drying weighed 2.93 g. The solution was then neutralized to precipitate stannic acid, which was decomposed by heating, yielding 1.27 g SnO_2. What is the analysis of the solder as % Pb and % Sn?

   Ans.   66.7% Pb, 33.3% Sn

3.60. Determine the weight of sulfur required to make 1 metric ton of H_2SO_4.

   Ans.   327 kg S

3.61. A sample of impure cuprite Cu_2O contains 66.6% copper. What is the percentage of pure Cu_2O in the sample?

   Ans. 75.0% Cu_2O

3.62. A cold cream sample weighing 8.41 g lost 5.83 g of moisture on heating to $110\,°C$. The residue on extracting with water and drying lost 1.27 g of water-soluble glycerol. The balance was oil. Calculate the composition of this cream.

   Ans. 69.3% moisture, 15.1% glycerol, 15.6% oil

3.63. A household cement gave the following analytical data: A 28.5-g sample, on dilution with acetone, yielded a residue of 4.6 g of aluminum powder. The filtrate, on evaporation of the acetone and solvent, yielded 3.2 g of plasticized nitrocellulose, which contained 0.8 g of benzene-soluble plasticizer. Determine the composition of this cement.

   Ans. 16.2% Al, 72.6% solvent, 2.8% plasticizer, 8.4% nitrocellulose

3.64. A coal contains 2.4% water. After drying, the moisture-free residue contains 71.0% carbon. Determine the percentage of carbon on the "wet basis."

   Ans. 69.3% C

3.65. A clay contains 45% silica and 10% water. What is the percentage of silica in the clay on a dry (water-free) basis?

   Ans. 50% silica

3.66. A liter flask contains a mixture of two liquids (A and B) of specific gravity 1.4. (Specific gravity means the density relative to that of water.) The specific gravity of liquid A is 0.8 and of liquid B, 1.8. What volume of each must have been put into the flask? Assume no change of volume on mixing.

   Ans.   400 mL of A, 600 mL of B

3.67. There is available 10.0 tons of a coal containing 2.50% sulfur, and also supplies of two coals containing 0.80% and 1.10% sulfur. How many tons of each of the latter should be mixed with the original 10.0 tons to give 20.0 tons containing 1.70% sulfur?

   Ans.   6.7 tons of 0.80%, 3.3 tons of 1.10%

3.68. A taconite ore consisted of 35.0% Fe_3O_4 and the balance siliceous impurities. How many tons of the ore must be processed in order to recover a ton of metallic iron   (a) if there is 100% recovery,   (b) if there is only 75% recovery?

   Ans.   (a) 3.94 tons;   (b) 5.25 tons

3.69. A typical formulation for a cationic asphalt emulsion calls for 0.5% tallow amine emulsifier and 70% asphalt, the rest consisting of water and water-soluble ingredients. How much asphalt can be emulsified per pound of the emulsifier?

   Ans.   140 lb

3.70. Uranium hexafluoride, UF_6, is used in the gaseous diffusion process for separating uranium isotopes. How many kilograms of elementary uranium can be converted to UF_6 per kilogram of combined fluorine?

   Ans.   2.09 kg

MOLECULAR WEIGHTS

3.71. One of the earliest methods for determining the molecular weight of proteins was based on chemical analysis. A hemoglobin preparation was found to contain 0.335% iron. (a) If the hemoglobin molecule contains 1 atom of iron, what is its molecular weight? (b) If it contains 4 atoms of iron, what is its molecular weight?

Ans. (a) 16 700; (b) 66 700

3.72. A polymeric substance, tetrafluoroethylene, can be represented by the formula $(C_2F_4)_x$, where x is a large number. The material was prepared by polymerizing C_2F_4 in the presence of a sulfur-bearing catalyst that served as a nucleus upon which the polymer grew. The final product was found to contain 0.012% S. What is the value of x if each polymeric molecule contains (a) 1 sulfur atom, (b) 2 sulfur atoms? In either case, assume that the catalyst contributes a negligible amount to the total mass of the polymer.

Ans. (a) 2 700; (b) 5 300

3.73. A peroxidase enzyme isolated from human red blood cells was found to contain 0.29% selenium. What is the minimum molecular weight of the enzyme?

Ans. 27 000 daltons

3.74. A sample of polystyrene prepared by heating styrene with tribromobenzoyl peroxide in the absence of air has the formula $Br_3C_6H_3(C_8H_8)_n$. The number n varies with the conditions of preparation. One sample of polystyrene prepared in this manner was found to contain 10.46% bromine. What is the value of n?

Ans. 19

FORMULA FROM PRECISE MOLECULAR WEIGHT

In all the following problems the nuclidic molecular weight reported is that of the species containing the most prevalent nuclide of each of its elements.

3.75. An alkaloid was extracted from the seed of a plant and purified. The molecule was known to contain 1 atom of nitrogen, no more than 4 atoms of oxygen, and no other elements besides C and H. The mass-spectrometrically determined nuclidic molecular weight was found to be 297.138 (a) How many molecular formulas are consistent with mass number 297 and with the other known facts except the precise molecular weight? (b) What is the probable molecular formula?
Ans. (a) 17; (b) $C_{18}O_3NH_{19}$

3.76. An organic ester was decomposed inside a mass spectrometer. An ionic decomposition product had the molecular weight 117.090. What is the molecular formula of this product, if it is known in advance that the only possible constituent elements are C, O, and H, and that no more than 4 oxygen atoms are present in the molecule?

Ans. $C_6O_2H_{13}$

3.77. An intermediate in the synthesis of a naturally occurring alkaloid had a mass-spectrometrically determined molecular weight of 205.147. The compound is known to have no more than 1 nitrogen atom and no more than 2 oxygen atoms per molecule. (a) What is the most probable molecular formula of the compound? (b) What must the precision of the measurement be to exclude the next to most probable formula?

Ans. (a) $C_{13}ONH_{19}$ (nuclidic molecular weight is 205.147). (b) The closest molecular weight is 205.159 for $C_{14}OH_{21}$. The range of uncertainty in the experimental value should not exceed half the difference between 205.147 and 205.159, i.e., it should be less than 0.006, or about 1 part in 35 000.

Chapter 4

Calculations from Chemical Equations

INTRODUCTION

Calculations based on chemical equations are among the most important in general chemistry because of the large amount of descriptive and quantitative knowledge which is condensed into these equations. Knowledge about a chemical change is represented by an equation of formulas, just as each formula represents the composition of a substance in terms of the constituent atoms.

The balanced chemical equation is an algebraic equation with all the reactants on the left side and all the products on the right side; hence the equals sign usually is replaced by an arrow showing the rightward course of the reaction. If the reverse reaction also takes place, the double-arrow of equilibrium equations is used.

MOLECULAR RELATIONS FROM EQUATIONS

The *relative numbers of reacting and resulting molecules* are indicated by the coefficients of the formulas representing these molecules. For example, the combustion of ammonia in oxygen is described by the balanced chemical equation

$$4NH_3 + 3O_2 \rightarrow 2N_2 + 6H_2O$$
$$\text{(4 molecules)} \quad \text{(3 molecules)} \quad \text{(2 molecules)} \quad \text{(6 molecules)}$$

in which the algebraic coefficients 4, 3, 2, and 6 indicate that 4 molecules of NH_3 react with 3 molecules of O_2 to form 2 molecules of N_2 and 6 molecules of H_2O. The balanced equation does not necessarily mean that if 4 molecules of NH_3 are mixed with 3 molecules of O_2 the reaction as indicated will go to completion. Some reactions between chemical substances occur almost instantaneously upon mixing, some occur completely after sufficient time has elapsed, and some reactions go to only a partial extent even after an infinite time. The common interpretation of the balanced equation for all categories is as follows: If a large number of NH_3 and O_2 molecules are mixed, a certain number of N_2 and H_2O molecules will be formed. At a given instant it is not necessary that either the NH_3 or O_2 is all consumed, but whatever reaction does occur takes place in the molecular ratio prescribed by the equation.

In the above reaction, the atoms in seven indicated molecules ($4NH_3$, $3O_2$) rearrange to form eight molecules ($2N_2$, $6H_2O$); there is no algebraic rule governing these numbers of molecules, but *the number of atoms on each side of the equation must balance for each element*, since the reaction obeys the laws of conservation of matter and of nontransmutability of the elements. Thus the equation is balanced and checked by counting the atoms of each kind (4N, 12H, 6O), not the molecules.

The number of atoms of any element occurring in a given substance is found by multiplying the subscript of that element in the formula by the coefficient of the formula. Thus $4NH_3$ represents 12 atoms of H, because there are 3 atoms of H in each of 4 molecules of NH_3. In some more complex formulas, several subscripts must be multiplied together before multiplying by the coefficient of the entire formula. Thus $3(NH_4)_2SO_4$ would represent 24 atoms of H, because each of the 3 formula units of $(NH_4)_2SO_4$ contains 2 NH_4 groups, each of which in turn contains 4H atoms.

MASS RELATIONS FROM EQUATIONS

Because 1 mol of any substance is N_A molecules (Chapter 2), *the relative numbers of moles undergoing reaction are the same as the relative numbers of molecules*. In terms of the molecular weights $NH_3 = 17$, $O_2 = 32$, $N_2 = 28$, and $H_2O = 18$, the above combustion equation,

$$4NH_3 + 3O_2 \rightarrow 2N_2 + 6H_2O$$
$$\text{(4 mol = 68 g)} \quad \text{(3 mol = 96 g)} \quad \text{(2 mol = 56 g)} \quad \text{(6 mol = 108 g)}$$

shows that 4 mol of NH_3 (4×17 g NH_3) reacts with 3 mol of O_2 (3×32 g O_2) to form 2 mol of N_2 (2×28 g N_2) and 6 mol of H_2O (6×18 g H_2O). More generally, the equation shows that the masses of NH_3, O_2, N_2, and H_2O consumed or formed in the reaction—expressed in any mass unit whatever—are in the ratio 68:96:56:108 (or 17:24:14:27).

In all cases, the law of conservation of mass requires that the sum of the reactant masses ($68 + 96$) be equal to the sum of the resultant masses ($56 + 108$).

The importance of the above mass relations may be summarized as follows.

1. Mass relations are as exacting as the law of conservation of mass.
2. Mass relations do not require any knowledge about the variable conditions; for example, whether the H_2O is water or steam.
3. Mass relations do not require any knowledge of the true molecular formulas. In the above example, the masses or the numbers of atoms would be unchanged if the oxygen were assumed to be ozone, $2O_3$, instead of $3O_2$. In either case, the equation would be balanced with 6 oxygen atoms on each side. Similarly, if the water molecules were polymerized, mass relations would be the same whether the equation contained $6H_2O$, $3H_4O_2$, or $2H_6O_3$. This principle is very important in cases where the true molecular formulas are not known. Mass relations are valid for the many equations involving molecules that may dissociate (S_8, P_4, H_6F_6, N_2O_4, I_2, etc.) or those that associate to form complex polymers, such as the many industrially important derivatives of formaldehyde, starch, cellulose, nylon, synthetic rubbers, silicones, etc., regardless of whether empirical or molecular formulas are used.

TYPES OF CHEMICAL REACTIONS

Skill in balancing equations will increase rapidly with practice, especially as one learns to recognize common types of reactions. Also recognition of the reaction type often helps in the prediction of the products of the reaction, if they are not stated. A few examples of the more predictable types are given below.

1. *Combustion reactions*: Oxygen (usually from the air) combines with organic compounds (carbon, hydrogen, and oxygen) to produce carbon dioxide (CO_2) and water (H_2O) as the sole products.

$$C_3H_8 + 5O_2 \rightarrow 3CO_2 + 4H_2O$$

2. *Replacement (displacement) reactions*: A more active element replaces a less active one in a compound.

$$Zn + CuSO_4 \rightarrow ZnSO_4 + Cu$$

$$2Mg + TiCl_4 \rightarrow 2MgCl_2 + Ti$$

3. *Double displacement reactions (metathesis)*: Particularly common for ionic reactions in solution; atoms or groups of atoms will "exchange partners" if an insoluble salt results.

$$AgNO_3 + NaCl \rightarrow NaNO_3 + AgCl \text{ (insoluble)}$$

$$BaCl_2 + Na_2SO_4 \rightarrow 2NaCl + BaSO_4 \text{ (insoluble)}$$

4. *Acid-base reactions (neutralization)*: An acid, which contributes H, and a base, which contributes OH, undergo metathesis to produce water (HOH, or H_2O) and a salt.

$$HCl + NaOH \rightarrow NaCl + HOH$$

$$H_2SO_4 + Mg(OH)_2 \rightarrow MgSO_4 + 2HOH$$

Two very simple reaction types, combination and decomposition, unfortunately afford no simple ways to predict the products.

5. *Combination reactions*:   Elements or compounds simply combine into one product.

$$2SO_2 + O_2 \rightarrow 2SO_3$$

$$P_4 + 6Cl_2 \rightarrow 4PCl_3 \quad \text{or} \quad P_4 + 10Cl_2 \rightarrow 4PCl_5$$

depending upon ratio of reactions and conditions of temperature and pressure.

6. *Decomposition*:   A single reactant is transformed by heat or electricity into two or more products.

$$2H_2O \xrightarrow{\text{(elec)}} 2H_2 + O_2$$

$$2HgO \xrightarrow{\text{(heat)}} 2Hg + O_2 \quad \text{also} \quad 4HgO \xrightarrow{\text{(heat)}} 2Hg_2O + O_2$$

depending on the temperature and oxygen pressure.

Solved Problems

4.1   Balance the following skeleton equations:

(*a*) $FeS_2 + O_2 \rightarrow Fe_2O_3 + SO_2$ (*b*) $C_7H_6O_2 + O_2 \rightarrow CO_2 + H_2O$

(*a*)   There are no fixed rules for balancing equations. Often a trial-and-error procedure must be used. It is commonly helpful to start with the most complex formula. Fe_2O_3 has two different elements and a greater total number of atoms than any of the other substances, so we might start with it. We note that oxygen atoms occur in pairs in the molecules O_2 and SO_2 but not in the formula unit Fe_2O_3. If we write the equation with symbols representing the *integral* coefficients,

$$wFeS_2 + xO_2 \rightarrow yFe_2O_3 + zSO_2$$

then the total number of oxygen atoms on the left, $2x$, is even for any integral value of x. The total number on the right, $3y + 2z$, can be even or odd, depending on whether y is even or odd. We conclude from the required equality of $2x$ with $3y + 2z$ that y must be even. We can now try the smallest even number, 2, and proceed from there.

$$wFeS_2 + xO_2 \rightarrow 2Fe_2O_3 + zSO_2$$

To balance iron atoms, w must equal 4.

$$4FeS_2 + xO_2 \rightarrow 2Fe_2O_3 + zSO_2$$

To balance sulfur, z must equal 8.

$$4FeS_2 + xO_2 \rightarrow 2Fe_2O_3 + 8SO_2$$

Finally, to balance oxygen, $2x = 6 + 16$,   or $x = 11$.

$$4FeS_2 + 11O_2 \rightarrow 2Fe_2O_3 + 8SO_2$$

Note that the coefficient of the simplest substance, elementary oxygen in this case, was evaluated last. This is the usual consequence of beginning the balancing procedure with the most complex substance.

(*b*)   The most complex substance in this equation is $C_7H_6O_2$. We may assume 1 molecule of this substance and immediately write the coefficients for CO_2 and H_2O that will lead to balance of C and H, respectively.

$$C_7H_6O_2 + xO_2 \rightarrow 7CO_2 + 3H_2O$$

The balance of oxygen atoms is saved for last because an adjustment of x would not interfere with the balance of any other element. An arithmetic balance now demands that x equal $\frac{15}{2}$, giving an equation

$$C_7H_6O_2 + \tfrac{15}{2}O_2 \rightarrow 7CO_2 + 3H_2O$$

that, while balanced, violates the rule of integral coefficients. The correct ratio is preserved and the fraction eliminated by multiplying each coefficient by 2.

$$2C_7H_6O_2 + 15O_2 \rightarrow 14CO_2 + 6H_2O$$

This is the correct form. This can be recognized as a combustion reaction. The balancing procedure is the same for countless other combustion reactions.

4.2 Complete and balance the following equations. [Note: Barium phosphate, $Ba_3(PO_4)_2$, is very insoluble; tin is a more active metal than silver.]

(a) $Ba(NO_3)_2 + Na_3PO_4 \rightarrow$ (b) $Sn + AgNO_3 \rightarrow$ (c) $HC_2H_3O_2 + Ba(OH)_2 \rightarrow$

(a) Write the products of this metathesis rection by exchanging partners to get the skeleton equation. Note that the left-hand partner in each compound is a single metallic element but the right-hand partner is a group of atoms. Chapter 9 addresses the question of how to write the proper formulas of the product compounds.

$$Ba(NO_3)_2 + Na_3PO_4 \rightarrow NaNO_3 + Ba_3(PO_4)_2$$

Consider 1 formula unit of the most complex compound $Ba_3(PO_4)_2$ and insert the coefficients 3 and 2 in the reactants to balance the Ba atoms and the PO_4 groups. Then the coefficient 6 for $NaNO_3$ simultaneously balances both the Na atoms and the NO_3 groups. The final result is

$$3Ba(NO_3)_2 + 2Na_3PO_4 \rightarrow 6NaNO_3 + Ba_3(PO_4)_2$$

(b) Write the products of this replacement reaction by exchanging Ag and Sn to get the skeleton equation

$$Sn + AgNO_3 \rightarrow Ag + Sn(NO_3)_2$$

The most complex formula $Sn(NO_3)_2$ demands 2 of the $AgNO_3$, which in turn requires 2Ag among the products.

$$Sn + 2AgNO_3 \rightarrow 2Ag + Sn(NO_3)_2$$

(c) Recognizing that this is an acid-base reaction you can write it in the balanced form at once by supplying enough acid molecules to furnish the same number of H atoms as OH groups, which in turn is the number of water molecules.

$$2HC_2H_3O_2 + Ba(OH)_2 \rightarrow Ba(C_2H_3O_2)_2 + 2H_2O$$

4.3 Caustic soda, NaOH, is often prepared commercially by the reaction of Na_2CO_3 with slaked lime. $Ca(OH)_2$. How many grams of NaOH can be obtained by treating 1 kg of Na_2CO_3 with $Ca(OH)_2$?

First write the balanced equation for the reaction.

$$\underset{\text{(1 mol}=106.0\text{ g)}}{Na_2CO_3} + Ca(OH_2) \rightarrow \underset{\text{[2 mol}=2(40.0)=80.0\text{ g]}}{2NaOH} + CaCO_3$$

The pertinent mass ratio, 106/80, is all that is needed to solve the problem. We show four conceptual methods, all equivalent arithmetically.

First Method.

$$106.0 \text{ g } Na_2CO_3 \quad \text{gives} \quad 80.0 \text{ g NaOH}$$

hence $$1 \text{ g } Na_2CO_3 \quad \text{gives} \quad \frac{80.0}{106.0} \text{ g NaOH}$$

and $$1\,000 \text{ g } Na_2CO_3 \quad \text{gives} \quad 1\,000 \times \frac{80.0}{106.0} = 775 \text{ g NaOH}$$

Mole Method.

As in Chapter 2, the symbol $n(X)$ will be used to refer to the number of moles of a substance whose formula is X, and $m(X)$ will denote the mass of substance X. Consider $1\,000$ g Na_2CO_3.

$$n(Na_2CO_3) = \frac{1\,000\ g}{106.0\ g/mol} = 9.434\ \text{mol } Na_2CO_3$$

From the coefficients in the balanced equation, $n(NaOH) = 2n(Na_2CO_3) = 2(9.434) = 18.87$ mol NaOH.

$$m(NaOH) = (18.87\ \text{mol NaOH})(40.0\ \text{g NaOH/mol NaOH}) = 755\ \text{g NaOH}$$

Proportion Method.

Let x = number of grams of NaOH obtained from $1\,000$ g Na_2CO_3. It is known that 106.0 g Na_2CO_3 gives 80.0 g NaOH; then, by proportion,

$$\frac{106.0\ g\ Na_2CO_3}{80.0\ g\ NaOH} = \frac{1\,000\ g\ Na_2CO_3}{x}$$

$$x = (1\,000\ g\ Na_2CO_3)\left(\frac{80.0\ g\ NaOH}{106.0\ g\ Na_2CO_3}\right) = 755\ \text{g NaOH}$$

Note: It should be evident that $1\,000$ pounds Na_2CO_3 will give 755 *pounds* NaOH, and that $1\,000$ *tons* Na_2CO_3 will give 755 *tons* NaOH.

Factor-Label Method.

The amount of NaOH is obtained by multiplying the $1\,000$ g Na_2CO_3 by successive conversion factors until it has the desired units g NaOH.

$$\text{Amt. of NaOH} = (1\,000\ g\ Na_2CO_3)\left(\frac{1\ mol\ Na_2CO_3}{106.0\ g\ Na_2CO_3}\right)\left(\frac{2\ mol\ NaOH}{1\ mol\ Na_2CO_3}\right)\left(\frac{40.0\ g\ NaOH}{1\ mol\ NaOH}\right) = 755\ \text{g NaOH}$$

The functions of the above conversion factors are: convert starting information from grams to moles; using the coefficients in the balanced equation convert to moles of desired substance; convert from moles to grams of desired substance. Be sure to label each factor completely with its units. Then if canceling of units does not result in the correct units for the answer you will know that there has been an error in your reasoning.

4.4. The equation for the preparation of phosphorus in an electric furnace is

$$2Ca_3(PO_4)_2 + 6SiO_2 + 10C \rightarrow 6CaSiO_3 + 10CO + P_4$$

Determine (*a*) the number of moles of phosphorus formed for each mole of $Ca_3(PO_4)_2$ used, (*b*) the number of grams of phosphorus formed for each mole of $Ca_3(PO_4)_2$ used, (*c*) the number of grams of phosphorus formed for each gram of $Ca_3(PO_4)_2$ used, (*d*) the number of pounds of phosphorus formed for each pound of $Ca_3(PO_4)_2$ used, (*e*) the number of tons of phosphorus formed for each ton of $Ca_3(PO_4)_2$ used, (*f*) the number of moles each of SiO_2 and C required for each mole $Ca_3(PO_4)_2$ used.

(*a*)   From the equation, 1 mol P_4 is obtained for 2 mol $Ca_3(PO_4)_2$ used, or, $\frac{1}{2}$ mol P_4 per mole $Ca_3(PO_4)_2$.
(*b*)   Molecular weight of P_4 is 124. Then $\frac{1}{2}$ mol $P_4 = \frac{1}{2} \times 124 = 62$ g P_4.
(*c*)   One mol of $Ca_3(PO_4)_2$ (319 g) yields $\frac{1}{2}$ mol P_4 (62 g). Then 1 g $Ca_3(PO_4)_2$ gives $\frac{62}{310} = 0.20$ g P_4.
(*d*)   0.20 lb; the relative amounts are the same regardless of units.
(*e*)   0.20 ton; as above.
(*f*)   From the coefficients in the balanced equation, 1 mol $Ca_3(PO_4)_2$ required 3 mol SiO_2 and 5 mol C.

4.5. Much of the commercial hydrochloric acid is prepared by heating NaCl with concentrated H_2SO_4. How much sulfuric acid containing 90.0% H_2SO_4 by weight is needed for the production of 1 000 kg of concentrated hydrochloric acid containing 42.0% HCl by weight?

(1) Amount of pure HCl in 1 000 kg of 42.0% acid is (0.420) (1 000 kg) = 420 kg.
(2) Determine the amount of H_2SO_4 required to produce 420 kg HCl (H_2SO_4 = 98.1, HCl = 36.46).

$$2NaCl + \underset{(1\,mol\,=\,98.1\,g)}{H_2SO_4} \rightarrow Na_2SO_4 + \underset{[2\,mol\,=\,2(36.46)\,=\,72.92\,g]}{2HCl}$$

From the equation

$$72.92 \text{ g HCl}\quad \text{requires}\quad 98.1 \text{ g } H_2SO_4$$

$$1 \text{ g HCl}\quad \text{requires}\quad \frac{98.1}{72.92} \text{ g } H_2SO_4$$

$$1 \text{ kg HCl}\quad \text{requires}\quad \frac{98.1}{72.92} \text{ kg } H_2SO_4$$

and

$$420 \text{ kg HCl}\quad \text{requires}\quad (420)\left(\frac{98.1}{72.92} \text{ kg}\right) = 565 \text{ kg } H_2SO_4$$

(3) Finally, determine the amount of sulfuric acid solution containing 90.0% H_2SO_4 that can be made from 565 kg of pure H_2SO_4.
    Since 0.900 kg of pure H_2SO_4 makes 1 kg of 90.0% solution, then

$$565 \text{ kg } H_2SO_4 \times \frac{1 \text{ kg soln}}{0.900 \text{ kg } H_2SO_4} = 628 \text{ kg soln}$$

Factor-Label Method.

This method has the advantage of allowing one to write down the solution in one step. Carefully note the function of each factor, and check to confirm that all units cancel except the desired one.

$$\text{Amt. of 90.0\% } H_2SO_4 = (1\,000 \text{ kg } 42.0\% \text{ HCl})\left(\frac{42.0 \text{ kg HCl}}{100 \text{ kg } 42.0\% \text{ HCl}}\right)\left(\frac{1\,000 \text{ g}}{1 \text{ kg}}\right)\left(\frac{1 \text{ mol HCl}}{36.46 \text{ g HCl}}\right)$$

$$\times \left(\frac{1 \text{ mol } H_2SO_4}{2 \text{ mol HCl}}\right)\left(\frac{98.1 \text{ g } H_2SO_4}{1 \text{ mol } H_2SO_4}\right)\left(\frac{100 \text{ g } 90.0\% \text{ } H_2SO_4}{90.0 \text{ g } H_2SO_4}\right)\left(\frac{1 \text{ kg}}{1\,000 \text{ g}}\right)$$

$$= 628 \text{ kg } 90.0\% \text{ } H_2SO_4$$

4.6. A particular 100-octane aviation gasoline used 1.00 cm^3 of tetraethyl lead, $(C_2H_5)_4Pb$, of density 1.66 g/cm^3, per liter of product. This compound is made as follows:

$$4C_2H_5Cl + 4NaPb \rightarrow (C_2H_5)_4Pb + 4NaCl + 3Pb$$

How many grams of ethyl chloride, C_2H_5Cl, are needed to make enough tetraethyl lead for 1 L of gasoline?

The mass of 1.00 cm^3 $(C_2H_5)_4Pb$ is (1.00 cm^3)(1.66 g/cm^3) = 1.66 g; this is the amount needed per liter. In terms of moles,

$$\text{Number of moles of } (C_2H_5)_4Pb \text{ needed} = \frac{1.66 \text{ g}}{323 \text{ g/mol}} = 0.005\,1 \text{ mol}$$

The chemical equation shows that 1 mol $(C_2H_5)_4Pb$ requires 4 mol C_2H_5Cl. Hence 4(0.005 1) = 0.020 4 mol C_2H_5Cl is needed.

$$m(C_2H_5Cl) = 0.020\,4 \text{ mol} \times 64.5 \text{ g/mol} = 1.32 \text{ g } C_2H_5Cl$$

Factor-Label Method.

$$\text{Amt. of } C_2H_5Cl = (1 \text{ L gasoline})\left(\frac{1.00 \text{ cm}^3 \text{ tet lead}}{1 \text{ L gasoline}}\right)\left(\frac{1.66 \text{ g}}{1.00 \text{ cm}^3}\right)\left(\frac{1 \text{ mol tet lead}}{323 \text{ g tet lead}}\right)$$

$$\times \left(\frac{4 \text{ mol } C_2H_5Cl}{1 \text{ mol tet lead}}\right)\left(\frac{64.5 \text{ g } C_2H_5Cl}{1 \text{ mol } C_2H_5Cl}\right)$$

$$= 1.33 \text{ g } C_2H_5Cl$$

4.7. A solution containing 2.00 g of $Hg(NO_3)_2$ was added to a solution containing 2.00 g of Na_2S. Calculate the mass of HgS that was formed according to the reaction

$$Hg(NO_3)_2 + Na_2S \rightarrow HgS + NaNO_3$$

The special feature of this problem is that the starting quantities of *two* reactants are specified. It is necessary first to determine which, if any, substance is in excess. The simplest approach is to calculate the number of moles of each.

$$n[Hg(NO_3)_2] = \frac{2.00 \text{ g}}{324.6 \text{ g/mol}} = 6.16 \times 10^{-3} \text{ mol } Hg(NO_3)_2$$

$$n(Na_2S) = \frac{2.00 \text{ g}}{78.00 \text{ g/mol}} = 2.56 \times 10^{-2} \text{ mol } Na_2S$$

The equation indicates that equimolar quantities of reactants are required. Hence the given amount of $Hg(NO_3)_2$ requires only 6.16×10^{-3} mol of Na_2S. The given amount of Na_2S is far in excess of this. The $Hg(NO_3)_2$ is the *limiting* reagent in this case and determines the quantity of product.

$$\text{Amt. of HgS} = [6.16 \times 10^{-3} \text{ mol } Hg(NO_3)_2]\left[\frac{1 \text{ mol HgS}}{1 \text{ mol } Hg(NO_3)_2}\right]\left(\frac{232.6 \text{ g HgS}}{1 \text{ mol HgS}}\right) = 1.43 \text{ g HgS}$$

The amount of excess Na_2S is 2.56×10^{-2} mol minus the 6.16×10^{-3} mol consumed, or 1.94×10^{-2} mol, which amounts to 1.94×10^{-2} mol $\times$ 78.0 g/mol = 1.51 g.
The amount of $NaNO_3$ is

$$(6.16 \times 10^{-3} \text{ mol } Na_2S)\left(\frac{2 \text{ mol } NaNO_3}{1 \text{ mol } Na_2S}\right)\left(\frac{85.0 \text{ g } NaNO_3}{1 \text{ mol } NaNO_3}\right) = 1.05 \text{ g } NaNO_3$$

In summary, the original 4.00 g of reagents has been transformed to 1.43 g of HgS product, 1.05 g of by-product $NaNO_3$, with 1.51 g of excess reagent Na_2S remaining. (Note that the sum of the final masses is 3.99 g, within round-off error of the original 4.00 g.)

4.8. How many grams of $Ca_3(PO_4)_2$ can be made according to the reaction

$$3CaCl_2 + 2K_3PO_4 \rightarrow Ca_3(PO_4)_2 + 6KCl$$

by mixing a solution containing 5.00 g of $CaCl_2$ with another containing 8.00 g of K_3PO_4?

$$n(CaCl_2) = \frac{5.00 \text{ g}}{111.0 \text{ g/mol}} = 4.50 \times 10^{-2} \text{mol} \qquad n(K_3PO_4) = \frac{8.00 \text{ g}}{212.3 \text{ g/mol}} = 3.77 \times 10^{-2} \text{ mol}$$

$$(4.50 \times 10^{-2} \text{ mol } CaCl_2)\left(\frac{2 \text{ mol } K_3PO_4}{3 \text{ mol } CaCl_2}\right) = 3.00 \times 10^{-2} \text{ mol } K_3PO_4$$

required for the given $CaCl_2$. Thus the K_3PO_4 is in excess and $CaCl_2$ is the *limiting* reagent.

$$\text{Amt. of } Ca_3(PO_4)_2 = (4.50 \times 10^{-2} \text{ mol } CaCl_2)\left[\frac{1 \text{ mol } Ca_3(PO_4)_2}{3 \text{ mol } CaCl_2}\right]\left[\frac{310.2 \text{ g } Ca_3(PO_4)_2}{1 \text{ mol } Ca_3(PO_4)_2}\right]$$

$$= 4.65 \text{ g } Ca_3(PO_4)_2$$

If the calculation were begun with K_3PO_4 instead of $CaCl_2$:

$$(3.77 \times 10^{-2} \text{ mol } K_3PO_4)\left(\frac{3 \text{ mol } CaCl_2}{2 \text{ mol } K_3PO_4}\right) = 5.66 \text{ mol } CaCl_2$$

required for the given K_3PO_4. Since this is greater than the given amount of $CaCl_2$, again it is concluded that $CaCl_2$ is the *limiting* reagent.

Completing the mass balance will provide a check on the work.

$$\text{Excess } K_3PO_4 = [(3.77 - 3.00) \times 10^{-2} \text{ mol}] (212.3 \text{ g/mol}) = 1.63 \text{ g}$$

$$\text{Amt. of KCl} = (4.50 \times 10^{-2} \text{ mol } CaCl_2)\left(\frac{6 \text{ mol KCl}}{3 \text{ mol } CaCl_2}\right)\left(\frac{74.5 \text{ g KCl}}{1 \text{ mol KCl}}\right) = 6.71 \text{ g KCl}$$

The total of the product, by-product, and excess reagent is $4.65 + 6.71 + 1.63 = 12.99$ g, in agreement with the $8.00 + 5.00 = 13.00$ g of reactants.

4.9. In one process for waterproofing, a fabric is exposed to $(CH_3)_2SiCl_2$ vapor. The vapor reacts with hydroxyl groups on the surface of the fabric or with traces of water to form the waterproofing film $[(CH_3)_2SiO]_n$, by the reaction

$$n(CH_3)_2SiCl_2 + 2nOH^- \rightarrow 2nCl^- + nH_2O + [(CH_3)_2SiO]_n$$

where n stands for a large integer. The waterproofing film is deposited on the fabric layer upon layer. Each layer is 6 Å thick [the thickness of the $(CH_3)_2SiO$ group]. How much $(CH_3)_2SiCl_2$ is needed to waterproof one side of a piece of fabric, 1 m by 2 m, with a film 300 layers thick? The density of the film is 1.0 g/cm^3.

$$\text{Mass of film} = (\text{volume of film})(\text{density of film})$$
$$= (\text{area of film})(\text{thickness of film})(\text{density of film})$$
$$= (100 \text{ cm} \times 200 \text{ cm})(300 \times 6 \text{ Å} \times 10^{-8} \text{ cm/Å})(1.0 \text{ g/cm}^3) = 0.36 \text{ g}$$

$$\text{Amt. of } (CH_3)_2SiCl_2 = \{0.36 \text{ g } [(CH_3)_2 SiO]_n\}\left\{\frac{1 \text{ mol } [(CH_3)_2SiO]_n}{74n \text{ g } [(CH_3)_2SiO]_n}\right\}$$

$$\times \left\{\frac{n \text{ mol } (CH_3)_2SiCl_2}{1 \text{ mol } [(CH_3)_2SiO]_n}\right\}\left[\frac{129 \text{ g } (CH_3)_2SiCl_2}{1 \text{ mol } (CH_3)_2SiCl_2}\right]$$

$$= 0.63 \text{ g } (CH_3)_2SiCl_2$$

Note that the unknown integer n canceled in the above *factor-label* calculation along with the units (except for grams).

4.10. What is the % free SO_3 in an oleum (considered as a solution of SO_3 in H_2SO_4) that is labeled "109% H_2SO_4"? Such a designation refers to the total weight of pure H_2SO_4, 109 g, that would be present after dilution of 100 g of the oleum, when all free SO_3 would combine with water to form H_2SO_4.

Nine g H_2O will combine with all the free SO_3 in 100 g of the oleum to give a total of 109 g H_2SO_4. The equation $H_2O + SO_3 \rightarrow H_2SO_4$ indicates that 1 mol H_2O (18 g) combines with 1 mole SO_3 (80 g). Then

$$(9 \text{ g } H_2O)\left(\frac{80 \text{ g } SO_3}{18 \text{ g } H_2O}\right) = 40 \text{ g } SO_3$$

Thus 100 g of the oleum contains 40 g SO_3, or the % free SO_3 in the oleum is 40%.

4.11. $KClO_4$ may be made by the following series of reactions:

$$Cl_2 + 2KOH \rightarrow KCl + KClO + H_2O$$
$$3KClO \rightarrow 2KCl + KClO_3$$
$$4KClO_3 \rightarrow 3KClO_4 + KCl$$

How much Cl_2 is needed to prepare 100 g $KClO_4$ by the above sequence?

The mole method and the factor-label method are the simplest routes to the solution of this problem.

Mole Method.

$$n(KClO) = n(Cl_2)$$

$$n(KClO_3) = \tfrac{1}{3}n(KClO) = \tfrac{1}{3}n(Cl_2)$$

$$n(KClO_4) = \tfrac{3}{4}n(KClO_3) = (\tfrac{3}{4})(\tfrac{1}{3})n(Cl_2) = \tfrac{1}{4}n(Cl_2)$$

$$n(KClO_4) = \frac{100 \text{ g } KClO_4}{138.6 \text{ g } KClO_4/\text{mol } KClO_4} = 0.721\,5 \text{ mol } KClO_4$$

$$n(Cl_2) = 4(0.721\,5) = 2.886 \text{ mol } Cl_2$$

$$m(Cl_2) = (2.886 \text{ mol } Cl_2)(70.9 \text{ g } Cl_2/\text{mol } Cl_2) = 205 \text{ g } Cl_2$$

Factor-Label Method.

In using this method it is not necessary to obtain the masses of the intermediate products, but one must determine the molar ratios from every step in the process and combine them to obtain the correct ratio of initial reactant moles to product moles.

$$\text{Amt. of } Cl_2 = (100 \text{ g } KClO_4)\left(\frac{1 \text{ mol } KClO_4}{138.6 \text{ g } KClO_4}\right)\left(\frac{4 \text{ mol } KClO_3}{3 \text{ mol } KClO_4}\right)\left(\frac{3 \text{ mol } KClO}{1 \text{ mol } KClO_3}\right)$$

$$\times \left(\frac{1 \text{ mol } Cl_2}{1 \text{ mol } KClO}\right)\left(\frac{70.9 \text{ g } Cl_2}{1 \text{ mol } Cl_2}\right)$$

$$= 205 \text{ g } Cl_2$$

4.12. A 1.204 8-g impure sample of Na_2CO_3 is dissolved and allowed to react with a solution of $CaCl_2$. The resulting $CaCO_3$, after precipitation, filtration, and drying, was found to weigh 1.026 2 g. Assuming that the impurities do not contribute to the weight of the precipitate, calculate the percentage purity of the Na_2CO_3.

The equation for the reaction is as follows:

$$Na_2CO_3 + CaCl_2 \rightarrow CaCO_3 + 2NaCl$$

The amount of $CaCO_3$ should first be found.

$$n(CaCO_3) = \frac{1.026\,2 \text{ g } CaCO_3}{100.09 \text{ g } CaCO_3/\text{mol}} = 0.010\,253 \text{ mol}$$

From the coefficients in the balanced equation,

$$n(Na_2CO_3) = n(CaCO_3) = 0.010\,253 \text{ mol}$$

Now calculate the mass of pure Na_2CO_3 in the sample.

$$m(Na_2CO_3) = (0.010\,253 \text{ mol})(105.99 \text{ g } Na_2CO_3/\text{mol}) = 1.086\,7 \text{ g } Na_2CO_3$$

The percentage purity is obtained by dividing the mass of Na_2CO_3 by the mass of the sample, and multiplying by 100.

$$\% \text{ purity} = \frac{1.086\,7 \text{ g}}{1.204\,8 \text{ g}}(100\%) = 90.20\%$$

4.13. A mixture of NaCl and KCl weighed 5.489 2 g. The sample was dissolved in water and reacted with an excess of silver nitrate in solution. The resulting AgCl weighed 12.705 2 g. What was the percentage NaCl in the mixture?

The two parallel reactions are

$$NaCl + AgNO_3 \rightarrow AgCl + NaNO_3 \qquad KCl + AgNO_3 \rightarrow AgCl + KNO_3$$

Here the conservation of Cl atoms requires that the number of moles of AgCl formed equal the *sum* of the numbers of moles of NaCl and KCl.

$$n(AgCl) = \frac{12.705\,2 \text{ g AgCl}}{143.321 \text{ g AgCl/mol}} = 0.088\,649 \text{ mol} = n(NaCl) + n(KCl)$$

Let $x = $ mass of NaCl and $y = $ mass of KCl. Then

$$\frac{x}{58.443 \text{ g/mol}} + \frac{y}{74.551 \text{ g/mol}} = 0.088\,649 \text{ mol} \qquad (1)$$

A second equation for the unknown masses is provided by the data:

$$x + y = 5.489\,2 \text{ g} \qquad (2)$$

Eliminating y between (1) and (2), and solving for x, we obtain $x = m(NaCl) = 4.062\,4$ g. Then

$$\% \text{ NaCl} = \frac{4.062\,4 \text{ g}}{5.489\,2 \text{ g}}(100\%) = 74.01\% \text{ NaCl}$$

Supplementary Problems

BALANCING EQUATIONS

Balance the following equations.

4.14. $C_6H_6 + O_2 \rightarrow CO_2 + H_2O$

4.15. $Na + H_2O \rightarrow NaOH + H_2$

4.16. $Fe + FeCl_3 \rightarrow FeCl_2$

4.17. $Ba(OH)_2 + AlCl_3 \rightarrow Al(OH)_3 + BaCl_2$

4.18. $H_2C_2O_4 + KOH \rightarrow K_2C_2O_4 + H_2O$

4.19. $C_2H_2Cl_4 + Ca(OH)_2 \rightarrow C_2HCl_3 + CaCl_2 + H_2O$

4.20. $(NH_4)_2Cr_2O_7 \rightarrow N_2 + Cr_2O_3 + H_2O$

4.21. $Zn_3Sb_2 + H_2O \rightarrow Zn(OH)_2 + SbH_3$

4.22. $HClO_4 + P_4O_{10} \rightarrow H_3PO_4 + Cl_2O_7$

4.23. $C_6H_5Cl + SiCl_4 + Na \rightarrow (C_6H_5)_4Si + NaCl$

4.24. $Sb_2S_3 + HCl \rightarrow H_3SbCl_6 + H_2S$

4.25. $IBr + NH_3 \rightarrow NI_3 + NH_4Br$

4.26. $KrF_2 + H_2O \rightarrow Kr + O_2 + HF$

4.27. $Na_2CO_3 + C + N_2 \rightarrow NaCN + CO$

4.28. $K_4Fe(CN)_6 + H_2SO_4 + H_2O \rightarrow K_2SO_4 + FeSO_4 + (NH_4)_2SO_4 + CO$

4.29. $Fe(CO)_5 + NaOH \rightarrow Na_2Fe(CO)_4 + Na_2CO_3 + H_2O$

4.30. $H_3PO_4 + (NH_4)_2MoO_4 + HNO_3 \rightarrow (NH_4)_3PO_4 \cdot 12MoO_3 + NH_4NO_3 + H_2O$

4.31. Identify each of the following types of reactions, write the products, and balance the equation.
(a) $HCl + Mg(OH)_2 \rightarrow$
(b) $PbCl_2 + K_2SO_4 \rightarrow$
(c) $CH_3CH_2OH + O_2(excess) \rightarrow$
(d) $NaOH + H_2C_6H_6O_6 \rightarrow$
(e) $Fe + AgNO_3 \rightarrow$

Partial answers   (a) acid-base (neutralization), products are H_2O and $MgCl_2$;   (b) double displacement (metathesis), products are KCl and $PbSO_4$;   (c) combustion, products are CO_2 and H_2O;   (d) acid-base (neutralization), products are $Na_2C_6H_6O_6$ and H_2O;   (e) replacement (displacement), products are $Fe(NO_3)_2$ and Ag

MASS RELATIONS

4.32. Consider the combustion of amyl alcohol, $C_5H_{11}OH$.

$$2C_5H_{11}OH + 15O_2 \rightarrow 10CO_2 + 12H_2O$$

(a) How many moles of O_2 are needed for the combustion of 1 mol of amyl alcohol?   (b) How many moles of H_2O are formed for each mole of O_2 consumed?   (c) How many grams of CO_2 are produced for each mole of amyl alcohol burned?   (d) How many grams of CO_2 are produced for each gram of amyl alcohol burned?   (e) How many tons of CO_2 are produced for each ton of amyl alcohol burned?

Ans.   (a) 7.5 mol O_2;   (b) 0.80 mol H_2O;   (c) 220 g CO_2;   (d) 2.49 g CO_2;   (e) 2.49 tons CO_2

4.33. A portable hydrogen generator utilizes the reaction $CaH_2 + 2H_2O \rightarrow Ca(OH)_2 + 2H_2$. How many grams of H_2 can be produced by a 50 g cartridge of CaH_2?

Ans.   4.8 g H_2

4.34. Iodine can be made by the reaction $2NaIO_3 + 5NaHSO_3 \rightarrow 3NaHSO_4 + 2Na_2SO_4 + H_2O + I_2$. To produce each kilogram of iodine, how much $NaIO_3$ and how much $NaHSO_3$ must be used?

Ans.   1.56 kg $NaIO_3$, 2.05 kg $NaHSO_3$

4.35. Calculate how much $KClO_3$ must be heated to obtain 3.50 g of oxygen, assuming that the only other product is KCl.

Ans.   8.94 g $KClO_3$

4.36. How much ferric oxide will be produced by the complete oxidation of 100 g of iron? The reaction is $4Fe + 3O_2 \rightarrow 2Fe_2O_3$.

Ans.   143 g

4.37. (a) How many pounds of ZnO will be formed when 1 lb of zinc blende, ZnS, is strongly heated in air?

$$2ZnS + 3O_2 \rightarrow 2ZnO + 2SO_2$$

(b) How many tons of ZnO will be formed from 1 ton ZnS?   (c) How many kilograms of ZnO will be formed from 1 kg ZnS?

Ans.   (a) 0.835 lb;   (b) 0.835 ton;   (c) 0.835 kg

4.38. In a rocket motor fueled with butane, C_4H_{10}, how many kilograms of liquid oxygen should be provided with each kilogram of butane to provide for complete combustion?

$$2C_4H_{10} + 13O_2 \rightarrow 8CO_2 + 10H_2O$$

Ans.   3.58 kg

4.39. Chloropicrin, CCl_3NO_2, can be made cheaply for use as an insecticide by a process which utilizes the reaction

$$CH_3NO_2 + 3Cl_2 \rightarrow CCl_3NO_2 + 3HCl$$

How much nitromethane, CH_3NO_2, is needed to form 500 g of chloropicrin?

Ans.   186 g

4.40. Ethyl alcohol (C_2H_5OH) is made by the fermentation of glucose ($C_6H_{12}O_6$), as indicated by the equation

$$C_6H_{12}O_6 \rightarrow 2C_2H_5OH + 2CO_2$$

How many metric tons of alcohol can be made from 2.00 metric tons of glucose?

Ans.   1.02 metric tons

4.41. How many kilograms of H_2SO_4 can be prepared from 1 kg of cuprite, Cu_2S, if each atom of S in Cu_2S is converted into 1 molecule of H_2SO_4?

Ans.   0.616 kg

4.42. (a) How much bismuth nitrate, $Bi(NO_3)_3 \cdot 5H_2O$, would be formed from a solution of 10.4 g of bismuth in nitric acid?

$$Bi + 4HNO_3 + 3H_2O \rightarrow Bi(NO_3)_3 \cdot 5H_2O +\ NO$$

(b) How much 30.0% nitric acid (containing 30.0% HNO_3 by mass) is required to react with this amount of bismuth?

Ans.   (a) 24.1 g;   (b) 41.8 g

4.43 One of the reactions used in the petroleum industry for improving the octane rating of fuels is

$$C_7H_{14} \rightarrow C_7H_8 + 3H_2$$

The two hydrocarbons appearing in this equation are liquids; the hydrogen formed is a gas. What is the percentage reduction in liquid weight accompanying the completion of the above reaction?

Ans.   6.2%

4.44. In the *Mond process* for purifying nickel, the volatile nickel carbonyl, $Ni(CO)_4$, is produced by the following reaction:

$$Ni + 4CO \rightarrow Ni(CO)_4$$

How much CO is used up in volatilizing each kilogram of nickel?

Ans.   1.91 kg CO

4.45. When copper is heated with an excess of sulfur, Cu_2S is formed. How many grams of Cu_2S could be produced if 100 g of copper is heated with 50 g of sulfur?

Ans.   125 g Cu_2S

4.46. The reduction of Cr_2O_3 by Al proceeds quantitatively on ignition of a suitable fuse.

$$2Al + Cr_2O_3 \rightarrow Al_2O_3 + 2Cr$$

(a) How much metallic chromium can be made by bringing to reaction temperature a mixture of 5.0 kg Al and 20.0 kg Cr_2O_3?   (b) Which reactant remains at the completion of the reaction, and how much?

Ans.   (a) 9.6 kg Cr;   (b) 5.9 kg Cr_2O_3

4.47. A mixture of 1 ton of CS_2 and 2 tons of Cl_2 is passed through a hot reaction tube, where the following reaction takes place:

$$CS_2 + 3Cl_2 \rightarrow CCl_4 + S_2Cl_2$$

(a) How much CCl_4 can be made by complete reaction of the limiting starting material?   (b) Which starting material is in excess, and how much of it remains unreacted?

Ans.   (a) 1.45 tons CCl_4;   (b) 0.28 ton CS_2

4.48. A gram (dry weight) of green algae was able to absorb 4.7×10^{-3} mol CO_2 per hour by photosynthesis. If the fixed carbon atoms were all stored after photosynthesis as starch, $(C_6H_{10}O_5)_n$, how long would it take for the algae to double their own weight? Neglect the increase in photosynthetic rate due to the increasing amount of living matter.

Ans.   7.9 h

4.49. Carbon disulfide, CS_2, can be made from by-product SO_2. The overall reaction is

$$5C + 2SO_2 \rightarrow CS_2 + 4CO$$

How much CS_2 can be produced from 450 kg of waste SO_2 with excess coke, if the SO_2 conversion is 82%?

Ans.   219 kg

4.50. A 50.0 g sample of impure zinc reacts with exactly 129 cm³ of hydrochloric acid which has a density of 1.18 g/cm³ and contains 35.0% HCl by mass. What is the percent of metallic zinc in the sample? Assume that the impurity is inert to HCl.

Ans.   96% Zn

4.51. The chemical formula of the chelating agent Versene is $C_2H_4N_2(C_2H_2O_2Na)_4$. If each mole of this compound could bind 1 mole of Ca^{2+}, what would be the rating of pure Versene, expressed as mg $CaCO_3$ bound per gram of chelating agent? Here the Ca^{2+} is expressed in terms of the amount of $CaCO_3$ it could form.

Ans.   264 mg $CaCO_3$ per g

4.52. When CaC_2 is made in an electric furnace by the reaction

$$CaO + 3C \rightarrow CaC_2 + CO$$

the crude product is typically 85% CaC_2 and 15% unreacted CaO. How much CaO is to be added to the furnace charge for each 50 tons   (a) of CaC_2 produced?   (b) of crude product?

Ans.   (a) 53 tons CaO;   (b) 45 tons CaO

4.53.   The plastics industry uses large amounts of phthalic anhydride, $C_8H_4O_3$, made by the controlled oxidation of naphthalene.

$$2C_{10}H_8 + 9O_2 \rightarrow 2C_8H_4O_3 + 4CO_2 + 4H_2O$$

Since some of the naphthalene is oxidized to other products, only 70 percent of the maximum yield predicted by the above equation is actually obtained. How much phthalic anhydride would be produced in practice by the oxidation of 100 lb of $C_{10}H_8$?

Ans.   81 lb

4.54.   The empirical formula of a commercial ion-exchange resin is $C_8H_7SO_3Na$. The resin can be used to soften water according to the reaction

$$Ca^{2+} + 2C_8H_7SO_3Na \rightarrow (C_8H_7SO_3)_2Ca + 2Na^+$$

What would be the maximum uptake of Ca^{2+} by the resin, expressed in moles per gram of resin?

Ans.   0.002 4 mol Ca^{2+}/g resin

4.55.   The insecticide DDT is made by the reaction

$$CCl_3CHO \text{ (chloral)} + 2C_6H_5Cl \text{ (chlorobenzene)} \rightarrow (ClC_6H_4)_2CHCCl_3 (DDT) + H_2O$$

If 100 lb of chloral were reacted with 100 lb of chlorobenzene, how much DDT would be formed? Assume the reaction goes to completion without side reactions or losses.

Ans.   157 lb DDT

4.56.   Commercial sodium "hydrosulfite" is 90% pure $Na_2S_2O_4$. How much of the commercial product could be made by using 100 tons of zinc with a sufficient supply of the other reactants? The reactions are

$$Zn + 2SO_2 \rightarrow ZnS_2O_4$$
$$ZnS_2O_4 + Na_2CO_3 \rightarrow ZnCO_3 + Na_2S_2O_4$$

Ans.   296 tons

4.57.   Fluorocarbon polymers can be made by fluorinating polyethylene according to the reaction

$$(CH_2)_n + 4nCoF_3 \rightarrow (CF_2)_n + 2nHF + 4nCoF_2$$

where n is a large integer. The CoF_3 can be regenerated by the reaction

$$2CoF_2 + F_2 \rightarrow 2CoF_3$$

(a) If the HF formed in the first reaction cannot be reused, how many kilograms of fluorine are consumed per kilogram of fluorocarbon produced, $(CF_2)_n$?   (b) If the HF can be recovered and electrolyzed to hydrogen and fluorine, and if this fluorine is used for regenerating CoF_3, what is the net consumption of fluorine per kilogram of fluorocarbon?

Ans.   (a) 1.52 kg;   (b) 0.76 kg

4.58.   A process designed to remove organic sulfur from coal prior to combustion involves the following reactions:

$$X{-}S{-}Y + 2NaOH \rightarrow X{-}O{-}Y + Na_2S + H_2O$$
$$CaCO_3 \rightarrow CaO + CO_2$$
$$Na_2S + CO_2 + H_2O \rightarrow Na_2CO_3 + H_2S$$
$$CaO + H_2O \rightarrow Ca(OH)_2$$
$$Na_2CO_3 + Ca(OH)_2 \rightarrow CaCO_3 + 2NaOH$$

In the processing of 100 metric tons of a coal having a 1.0% sulfur content, how much limestone ($CaCO_3$) must be decomposed to provide enough $Ca(OH)_2$ to regenerate the NaOH used in the original leaching step?

Ans.   3.12 metric tons

4.59.   Silver may be removed from solutions of its salts by reaction with metallic zinc according to the reaction

$$Zn + 2Ag^+ \rightarrow Zn^{2+} + Ag$$

A 50-g piece of zinc was thrown into a 100-L vat containing 3.5 g Ag^+/L. (*a*) Which reactant was completely consumed?   (*b*) How much of the other substance remained unreacted?

Ans.   (*a*) zinc;   (*b*) 1.9 g Ag^+/L

4.60.   The following reaction proceeds until the limiting substance is all consumed:

$$2Al + 3MnO \rightarrow Al_2O_3 + 3Mn$$

A mixture containing 100 g Al and 200 g MnO was heated to initiate the reaction. Which initial substance remained in excess, and how much?

Ans.   Al, 49 g

4.61.   A mixture of $NaHCO_3$ and Na_2CO_3 weighed 1.023 5 g. The dissolved mixture was reacted with excess $Ba(OH)_2$ to form 2.102 8 g $BaCO_3$, by the reactions

$$Na_2CO_3 + Ba(OH)_2 \rightarrow BaCO_3 + 2NaOH$$
$$NaHCO_3 + Ba(OH)_2 \rightarrow BaCO_3 + NaOH + H_2O$$

What was the percentage $NaHCO_3$ in the original mixture?

Ans.   39.51% $NaHCO_3$

4.62.   A mixture of NaCl and NaBr, weighing 3.508 4 g, was dissolved and treated with enough $AgNO_3$ to precipitate all the chloride and bromide as AgCl and AgBr. The washed precipitate was treated with KCN to solubilize the silver and the resulting solution was electrolyzed. The equations are

$$NaCl + AgNO_3 \rightarrow AgCl + NaNO_3$$
$$NaBr + AgNO_3 \rightarrow AgBr + NaNO_3$$
$$AgCl + 2KCN \rightarrow KAg(CN)_2 + KCl$$
$$AgBr + 2KCN \rightarrow KAg(CN)_2 + KBr$$
$$4KAg(CN)_2 + 4KOH \rightarrow 4Ag + 8KCN + O_2 + 2H_2O$$

After the final step was complete, the deposit of metallic silver weighed 5.502 8 g. What was the composition of the initial mixture?

Ans.   65.23% NaCl, 34.77% NaBr

Chapter 5

Measurement of Gases

GAS VOLUMES

Gases, unlike liquids and solids, occupy volumes that depend very sensitively on pressure and temperature. Special attention must therefore be given to factors influencing the volume of gases.

PRESSURE

Pressure is defined as the force acting normally on a unit area of surface.

$$\text{Pressure} = \frac{\text{force acting perpendicular to an area}}{\text{area over which the force is distributed}}$$

$$\text{Pressure (in pascals)} = \frac{\text{force (in newtons)}}{\text{area (in square meters)}}$$

Thus the *pascal* is defined by $1\ \text{Pa} = 1\ \text{N/m}^2 = (1\ \text{kg} \cdot \text{m/s}^2)/\text{m}^2 = 1\ \text{kg/m} \cdot \text{s}^2$.

The pressure exerted by a column of fluid is

$$\text{Pressure} = \text{height} \times \text{density of fluid} \times \text{acceleration of gravity}$$

$$\text{Pressure (in Pa)} = \text{height (in m)} \times \text{density (in kg/m}^3) \times 9.81\ \text{m/s}^2$$

STANDARD ATMOSPHERIC PRESSURE

Air has weight and therefore exerts a pressure. The atmospheric pressure is due to the weight of the overlying air. A *standard atmosphere* (1 atm) is defined as exactly 101 325 Pa. It is approximately equal to the average pressure of the atmosphere at sea level. The standard atmosphere also is approximately equal (within several parts in 10^7) to the pressure exerted by a column of mercury 760 mm high, at 0 °C and at sea level. The *torr* is defined by: 760 torr = 1 atm. For problems in this book, the torr and the millimeter of mercury (mmHg) will be taken to be the same. The bar is defined by: 1 bar = exactly 10^5 Pa. One bar has recently replaced 1 atm as the standard pressure for reporting thermodynamic data.

STANDARD CONDITIONS (S.T.P.)

Standard conditions (S.T.P.) denotes a temperature of 0 °C (273.15 K, rounded off for most problems in this book to 273 K) and normal atmospheric pressure (1 atm = 760 torr). As both the volume and density of any gas are affected by changes of temperature and pressure, it is customary to reduce all gas volumes to standard conditions for purposes of comparison.

GAS LAWS

At sufficiently low pressures and sufficiently high temperatures, all gases have been found to obey three simple laws. These laws relate the volume of a gas to the pressure and temperature. When a gas obeys these laws, it is said to behave as an *ideal gas* or *perfect gas*. These laws, which are described below, may be applied only to gases which do not undergo a change in chemical complexity when the temperature or pressure is varied. A nonideal gas, for example, is NO_2, which undergoes a dimerization to N_2O_4 at increasing pressures or decreasing temperatures.

BOYLE'S LAW

When the temperature is kept constant, the volume of a given mass of an ideal gas varies inversely with the pressure to which the gas is subjected. In mathematical terms, the product pressure × volume of a given amount of gas remains constant. Thus, in comparing the properties of a given amount of an ideal gas under two conditions, which we may call the *initial* and *final* states, we may write the following equation, applicable at constant temperature:

$$(PV)_{\text{initial}} = (PV)_{\text{final}} \qquad \text{or} \qquad P_1 V_1 = P_2 V_2$$

A given subscript, 1 or 2, refers to a given state of the gas; 1 usually refers to an initial state and 2 to a final state. This law furnishes the most direct test of how well a real gas corresponds to ideal behavior.

CHARLES' LAW

At constant pressure, the volume of a given mass of gas varies directly with the *absolute temperature*. Then, at constant pressure,

$$\left(\frac{V}{T}\right)_{\text{initial}} = \left(\frac{V}{T}\right)_{\text{final}} \qquad \text{or} \qquad \frac{V_1}{T_1} = \frac{V_2}{T_2}$$

where T_1 and T_2 denote the *absolute* temperatures of the gas at the two states being compared.

GAY-LUSSAC'S LAW

At constant volume, the pressure of a given mass of gas varies directly with the *absolute temperature*. Then, at constant volume,

$$\frac{P_1}{T_1} = \frac{P_2}{T_2}$$

COMBINED GAS LAW

Any two of the above three gas laws can be employed to derive a law which applies to all possible combinations of changes:

$$\frac{P_1 V_1}{T_1} = \frac{P_2 V_2}{T_2} = \text{constant}$$

for a fixed mass of gas. (See Problem 5.6.)

Since many gas calculations are concerned with determining a new volume given an old volume, the combined gas law is often written as

$$V_2 = V_1\left(\frac{T_2}{T_1}\right)\left(\frac{P_1}{P_2}\right) \qquad \text{(fixed mass)}$$

This expression indicates that the volume of a given mass of gas varies directly with the absolute temperature and inversely with the pressure. Notice that, while T_1 and T_2 must be in kelvins, any convenient pressure unit may be used for P_1 and P_2, and any convenient volume unit may be used for V_1 and V_2.

DENSITY OF AN IDEAL GAS

As the volume of a given mass of gas increases, the mass per unit voume (i.e., the density) decreases proportionately. Therefore the density (d) of a gas varies inversely with its volume. For an ideal gas, the combined gas law then gives

$$d_2 = d_1 \frac{V_1}{V_2} = d_1\left(\frac{T_1}{T_2}\right)\left(\frac{P_2}{P_1}\right)$$

DALTON'S LAW OF PARTIAL PRESSURES

The *partial pressure* of a component of a gas mixture is the pressure which that component would exert if it alone occupied the entire volume. According to Dalton's law, the total pressure of a gaseous mixture is equal to the sum of the partial pressures of the components.

Dalton's law is rigidly accurate only for ideal gases. At pressures of only a few atmospheres or below, gas mixtures may be regarded as ideal gases, and this law may be applied in calculations in this book.

COLLECTING GASES OVER A LIQUID

If a gas is collected over a volatile liquid, such as water, a correction is made for the amount of water vapor present with the gas. A gas collected over water is saturated with water vapor, which occupies the total gas volume and exerts a partial pressure. The partial pressure of the water vapor is a constant for each temperature and is independent of the nature or pressure of the confined gas. This definite value of the vapor pressure of water may be found tabulated as a function of temperature in handbooks or in other reference books. If the total pressure (of the gas *plus* water vapor) is what is measured, the vapor pressure must be subtracted from the total pressure in order to obtain the partial pressure of the gas.

$$\text{Pressure of gas} = (\text{total pressure}) - (\text{vapor pressure of water})$$

When a gas is collected over mercury, it is not necessary to make a correction for the vapor pressure of mercury, which is negligible at ordinary temperatures.

DEVIATIONS FROM IDEAL BEHAVIOR

The laws discussed above are strictly valid only for ideal gases. The very fact that all gases can be liquefied if they are compressed and cooled sufficiently is an indication that all gases become nonideal at high pressures and low temperatures. The ideal properties are observed at low pressures and high temperatures, conditions far removed from those of the liquid state. At pressures below a few atmospheres practically all gases are sufficiently dilute for the application of the ideal gas laws with a reliability of a few percent or better.

Solved Problems

PRESSURE DEFINED

5.1. Calculate the difference in pressure between the top and bottom of a vessel exactly 76 cm deep when filled at 25 °C with   (a) water,   (b) mercury. Density of mercury at 25 °C is 13.53 g/cm^3; of water 0.997 g/cm^3.

Using the equivalent values in SI units for height and density,

(a)
$$\text{Pressure} = \text{height} \times \text{density} \times g$$
$$= (0.76 \text{ m})(997 \text{ kg/m}^3)(9.81 \text{ m/s}^2) = 7.43 \times 10^3 \text{ Pa}$$

or 7.43 kPa.

(b)
$$\text{Pressure} = (0.76 \text{ m})(13\,530 \text{ kg/m}^3)(9.81 \text{ m/s}^2) = 100.9 \times 10^3 \text{ Pa}$$

or 100.9 kPa.

5.2. How high a column of air would be necessary to cause the barometer to read 76 cm of mercury, if the atmosphere were of uniform density 1.2 kg/m³? The density of mercury is 13.53×10^3 kg/m³.

$$\text{Pressure of Hg} = \text{pressure of air}$$

$$\text{height of Hg} \times \text{density of Hg} \times \cancel{g} = \text{height of air} \times \text{density of air} \times \cancel{g}$$

$$(0.76 \text{ m})(13\,530 \text{ kg/m}^3) = h(1.2 \text{ kg/m}^3)$$

$$h = \frac{(0.76 \text{ m})(13\,530 \text{ kg/m}^3)}{1.2 \text{ kg/m}^3} = 8.6 \text{ km}$$

Actually the density of air decreases with increasing height so that the atmosphere extends much beyond 8.6 km. Modern aircraft routinely fly higher than 8.6 km (28 000 ft).

GAS LAWS

5.3. A mass of oxygen occupies 5.00 L under a pressure of 740 torr. Determine the volume of the same mass of gas at standard pressure, the temperature remaining constant.

Figure 5-1 illustrates the change. Standard pressure is 760 torr. Boyle's law gives

$$P_1 V_1 = P_2 V_2 \qquad \text{or} \qquad V_2 = \frac{P_1}{P_2} V_1 = \frac{740 \text{ torr}}{760 \text{ torr}}(5.00 \text{ L}) = 4.87 \text{ L}$$

Observe that any convenient pressure and volume units can be chosen (in this case, torr and L), since only ratios of like quantities are involved.

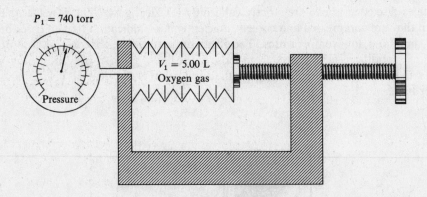

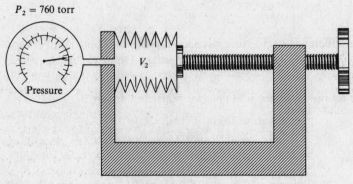

Fig. 5-1

5.4. A mass of neon occupies 200 cm³ at 100 °C. Find its volume at 0 °C, the pressure remaining contant.

See Figure 5-2. Charles law gives

$$\frac{V_1}{T_1} = \frac{V_2}{T_2} \quad \text{or} \quad V_2 = \frac{T_2}{T_1} V_1 = \frac{(0+273)\,\text{K}}{(100+273)\,\text{K}} (200\,\text{cm}^3) = 146\,\text{cm}^3$$

Absolute temperatures must be used in the gas laws.

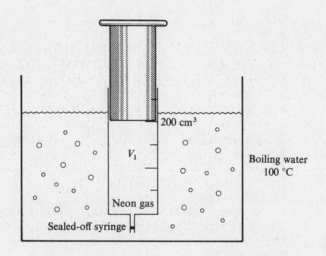

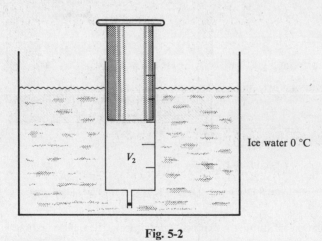

Fig. 5-2

5.5. A steel tank contains carbon dioxide at 27 °C and a pressure of 12.0 atm. Determine the internal gas pressure when the tank and its contents are heated to 100 °C.

See Figure 5-3. Gay-Lussac's law gives

$$\frac{P_1}{T_1} = \frac{P_2}{T_2} \quad \text{or} \quad P_2 = \frac{T_2}{T_1} P_1 = \frac{(100+273)\,\text{K}}{(27+273)\,\text{K}} (12.0\,\text{atm}) = 14.9\,\text{atm}$$

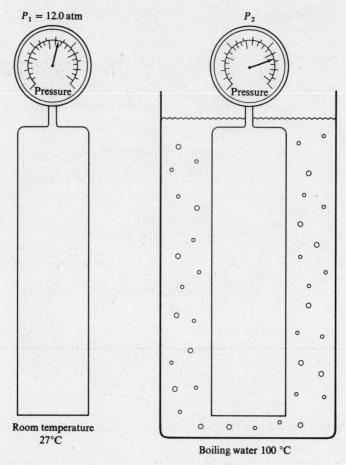

Fig. 5-3

5.6.   Obtain the combined gas law from Boyle's law and Charles' law.

Visualize the gas expanding in two stages as pictured below.

$$\boxed{P_1 V_1 T_1} \longrightarrow \boxed{P_2 V_i T_1} \longrightarrow \boxed{P_2 V_2 T_2}$$

T is constant P is constant

$$P_1 V_1 = P_2 V_i \qquad \frac{V_i}{T_1} = \frac{V_2}{T_2} \quad \text{whence} \quad V_i = \frac{T_1 V_2}{T_2}$$

In the left equation substitute the expression for V_i from the right equation.

$$P_1 V_1 = P_2 \frac{T_1 V_2}{T_2} \qquad \text{or} \qquad \frac{P_1 V_1}{T_1} = \frac{P_2 V_2}{T_2}$$

5.7.   Given 20.0 L of ammonia at 5 °C and 760 torr, determine its volume at 30 °C and 800 torr

See Figure 5-4. The combined gas law gives   (5 °C = 278 K, 30 °C = 303 K):

$$\frac{P_1 V_1}{T_1} = \frac{P_2 V_2}{T_2} \qquad \text{or} \qquad V_2 = V_1 \left(\frac{T_2}{T_1}\right)\left(\frac{P_1}{P_2}\right) = (20.0 \text{ L})\left(\frac{303 \text{ K}}{278 \text{ K}}\right)\left(\frac{760 \text{ torr}}{800 \text{ torr}}\right) = 20.7 \text{ L}$$

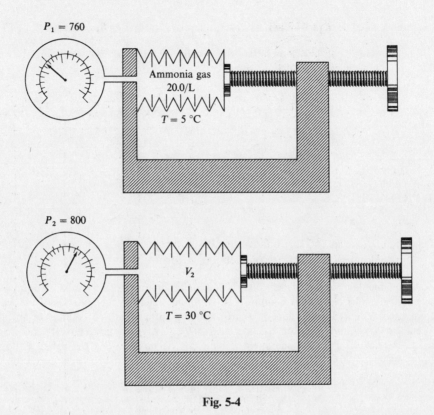

Fig. 5-4

5.8. The volume of a quantity of sulfur dioxide at 18 °C and 1 500 torr is 5.0 ft³. Calculate its volume at S.T.P.

S.T.P. means 0 °C and 760 torr. Converting to absolute temperatures and applying the combined gas law,

$$V_2 = V_1\left(\frac{T_2}{T_1}\right)\left(\frac{P_1}{P_2}\right) = (5.0 \text{ ft}^3)\left(\frac{273 \text{ K}}{291 \text{ K}}\right)\left(\frac{1\,500 \text{ torr}}{760 \text{ torr}}\right) = 9.3 \text{ ft}^3$$

5.9. To how many atmospheres pressure must 1 L of gas measured at 1 atm and −20 °C be subjected to be compressed to $\frac{1}{2}$ L when the temperature is 40 °C?

The combined gas law gives

$$\frac{P_1 V_1}{T_1} = \frac{P_2 V_2}{T_2} \quad \text{or} \quad P_2 = P_1\left(\frac{V_1}{V_2}\right)\left(\frac{T_2}{T_1}\right) = (1 \text{ atm})\left(\frac{1 \text{ L}}{\frac{1}{2} \text{ L}}\right)\left(\frac{313 \text{ K}}{253 \text{ K}}\right) = 2.47 \text{ atm}$$

DENSITY OF A GAS

5.10. The density of helium is 0.178 6 kg/m³ at S.T.P. If a given mass of helium at S.T.P. is allowed to expand to 1.500 times its initial volume by changing the temperature and pressure, compute its resultant density.

The density of a gas varies inversely with the volume.

$$\text{Resultant density} = (0.178\ 6 \text{ kg/m}^3)\left(\frac{1}{1.500}\right) = 0.119\ 1 \text{ kg/m}^3$$

Note that 1 kg/m³ = 1 g/L.

5.11. The density of oxygen is 1.43 g/L at S.T.P. Determine the density of oxygen at 17 °C and 700 torr.

The combined gas law shows that the density of an ideal gas varies inversely with the absolute temperature and directly with the pressure.

$$d_2 = d_1\left(\frac{T_1}{T_2}\right)\left(\frac{P_2}{P_1}\right) = (1.43 \text{ g/L})\left(\frac{273\text{K}}{290 \text{ K}}\right)\left(\frac{700 \text{ torr}}{760 \text{ torr}}\right) = 1.24 \text{ g/L}$$

PARTIAL PRESSURE

5.12. A mixture of gases at 760 torr contains 65.0% nitrogen, 15.0% oxygen, and 20.0% carbon dioxide by volume. What is the pressure of each gas in torr?

A fundamental property of ideal gases is that each component of a gas mixture occupies the entire volume of the mixture. The percentage composition by volume refers to the volumes of the separate gases *before* mixing, each measured at the same pressure and temperature. Thus 65 volumes of nitrogen, 15 volumes of oxygen, and 20 volumes of carbon dioxide, each at 760 torr, are mixed to give 100 volumes of mixture at 760 torr. As the volume of each component is increased by mixing, its pressure must (Boyle's law) decrease proportionately.

$$\text{Partial pressure of N}_2 = (760 \text{ torr})(\tfrac{65}{100}) = 494 \text{ torr}$$
$$\text{Partial pressure of O}_2 = (760 \text{ torr})(\tfrac{15}{100}) = 114 \text{ torr}$$
$$\text{Partial pressure of CO}_2 = (760 \text{ torr})(\tfrac{20}{100}) = \underline{152 \text{ torr}}$$
$$\textbf{Check:} \text{ Total pressure} = \text{sum of partial pressures} = 760 \text{ torr}$$

5.13. In a gaseous mixture at 20 °C the partial pressures of the components are: hydrogen, 200 torr; carbon dioxide, 150 torr; methane, 320 torr; ethylene, 105 torr. What are the total pressure of the mixture and the volume percent of hydrogen?

$$\text{Total pressure of mixture} = \text{sum of partial pressures} = 200 + 150 + 320 + 105 = 775 \text{ torr}$$

$$\text{Volume fraction of H}_2 = \frac{\text{partial pressure of H}_2}{\text{total pressure of mixture}} = \frac{200 \text{ torr}}{775 \text{ torr}} = 0.258 = 25.8\%$$

5.14. A 200-mL flask contained oxygen at 200 torr, and a 300 mL flask contained nitrogen at 100 torr, as shown in Fig. 5-5. The two flasks were then interconnected so that each gas filled its combined volumes. Assuming no change in temperature, what was the partial pressure of each gas in the final mixture and what was the total pressure?

The final total volume was 500 mL.

$$\text{Oxygen} \quad P_{\text{final}} = P_{\text{initial}}\left(\frac{V_{\text{initial}}}{V_{\text{final}}}\right) = (200 \text{ torr})\left(\frac{200}{500}\right) = 80 \text{ torr}$$

$$\text{Nitrogen} \quad P_f = P_i\left(\frac{V_i}{V_f}\right) = (100 \text{ torr})\left(\frac{300}{500}\right) = 60 \text{ torr}$$

$$\text{Total pressure} = (80 \text{ torr}) + (60 \text{ torr}) = 140 \text{ torr}$$

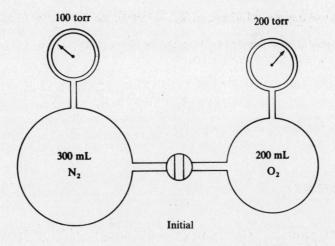

Initial

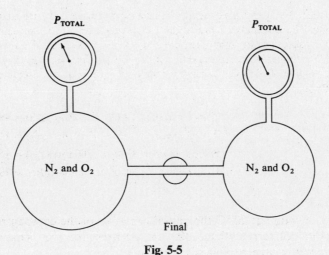

Final

Fig. 5-5

COLLECTING GASES OVER A LIQUID

5.15. Exactly 100 cm³ of oxygen is collected over water at 23 °C and 800 torr. Compute the standard volume of the dry oxygen. Vapor pressure of water at 23 °C is 21.1 torr.

The gas collected is actually a mixture of oxygen and water vapor. The partial pressure of water vapor in the mixture at 23 °C is 21.1 torr. Hence

$$\text{Pressure of dry oxygen} = (\text{total pressure}) - (\text{vapor pressure of water})$$
$$= (800 \text{ torr}) - (21 \text{ torr}) = 779 \text{ torr}$$

Thus, for the dry oxygen, $V_1 = 100 \text{ cm}^3, T_1 = 23 + 273 = 296 \text{ K}, P = 779$ torr. Converting to S.T.P.,

$$V_2 = V_1\left(\frac{T_2}{T_1}\right)\left(\frac{P_1}{P_2}\right) = (100 \text{ cm}^3)\left(\frac{273 \text{ K}}{296 \text{ K}}\right)\left(\frac{779 \text{ torr}}{760 \text{ torr}}\right) = 94.5 \text{ cm}^3$$

5.16. In a basal metabolism measurement timed at exactly 6 min, a patient exhaled 52.5 L of air, measured over water at 20 °C. The vapor pressure of water of 20 °C is 17.5 torr. The barometric pressure was 750 torr. The exhaled air analyzed 16.75 volume % oxygen, and the inhaled air 20.32

volume % oxygen, both on a dry basis. Neglecting any solubility of the gases in water and any difference in the total volumes of inhaled and exhaled air, calculate the rate of oxygen consumption by the patient in cm^3 (S.T.P.) per minute.

$$\text{Volume of dry air at S.T.P.} = (52.5 \text{ L})\left(\frac{273 \text{ K}}{293 \text{ K}}\right)\left[\frac{(750 - 17.5) \text{ torr}}{760 \text{ torr}}\right] = 47.1 \text{ L}$$

$$\text{Rate of oxygen consumption} = \frac{\text{volume of oxygen (S.T.P.) consumed}}{\text{time in which this volume was consumed}}$$

$$= \frac{(0.203\ 2 - 0.167\ 5)(47.1 \text{ L})}{6 \text{ min}} = 0.280 \text{ L/min} = 280 \text{ cm}^3/\text{min}$$

5.17. A quantity of gas is collected in a graduated tube over mercury. The volume of gas at 20 °C is 50.0 cm^3, and the level of the mercury in the tube is 200 mm above the outside mercury level, as shown in Fig. 5.6. The barometer reads 750 torr. Find the volume at S.T.P.

After a gas is collected over a liquid, the receiver is often adjusted so that the level of liquid inside and outside the receiver is the same. When this cannot be done conveniently, it is necessary to correct for the difference in levels.

Since the level of mercury inside the tube is 200 mm higher than outside, the pressure of the gas is 200 mmHg = 200 torr less than the atmospheric pressure of 750 torr.

$$\text{Volume at S.T.P.} = (50.0 \text{ cm}^3)\left(\frac{273 \text{ K}}{293 \text{ K}}\right)\left[\frac{(750 - 200) \text{ torr}}{760 \text{ torr}}\right] = 33.7 \text{ cm}^3$$

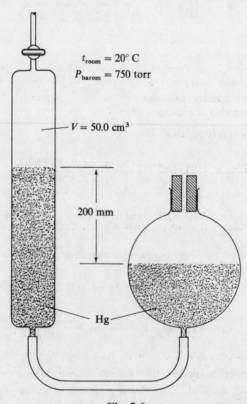

$t_{room} = 20° \text{ C}$

$P_{barom} = 750 \text{ torr}$

$V = 50.0 \text{ cm}^3$

200 mm

Hg

Fig. 5-6

Supplementary Problems

5.18. Express the standard atmosphere in   (a) bars,   (b) pounds force per square inch.

   Ans.   (a) 1.013 bars;   (b) 14.70 lbf/in^2

5.19. The vapor pressure of water at 25 °C is 23.8 torr. Express this in   (a) atmospheres,   (b) kilopascals.

   Ans.   (a) 0.031 3 atm;   (b) 3.17 kPa

5.20. Camphor has been found to undergo a crystalline modification at a temperature of 148 °C and a pressure of 3.09×10^9 N/m^2. What is the transition pressure in atmospheres?

   Ans. 3.05×10^4 atm

5.21. A mass of oxygen occupies 40.0 ft^3 at 758 torr. Compute its volume at 635 torr, temperature remaining constant.

   Ans.   47.7 ft^3

5.22. Ten liters of hydrogen under 1 atm pressure is contained in a cylinder which has a movable piston. The piston is moved in until the same mass of gas occupies 2 L at the same temperature. Find the pressure in the cylinder.

   Ans.   5 atm

5.23. A given mass of chlorine occupies 38.0 cm^3 at 20 °C. Determine its volume at 45 °C, pressure remaining constant.

   Ans.   41.2 cm^3

5.24. A quantity of hydrogen is confined in a platinum chamber of constant volume. When the chamber is immersed in a bath of melting ice, the pressure of the gas is 1 000 torr.   (a) What is the Celsius temperature when the pressure manometer indicates an absolute pressure of 100 torr?   (b) What pressure will be indicated when the chamber is brought to 100 °C?

   Ans.   (a) -246 °C;   (b) 1 366 torr

5.25. Given 1 000 ft^3 of helium at 15 °C and 763 torr. Calculate the volume at -6 °C and 420 torr.

   Ans.   1 685 ft^3

5.26. A mass of gas at 50 °C and 785 torr occupies 350 mL. What volume will the gas occupy at S.T.P.?

   Ans.   306 mL

5.27. If a gas occupies 15.7 ft^3 at 60 °F and 14.7 lbf/in^2, what volume would it occupy at 100 °F and 25 lbf/in^2?

   Ans.   9.9 ft^3

5.28. Exactly 500 cm^3 of nitrogen is collected over water at 25 °C and 755 torr. The gas is saturated with water vapor. Compute the volume of the nitrogen in the dry condition at S.T.P. Vapor pressure of water at 25 °C is 23.8 torr.

   Ans.   441 cm^3

5.29.   A dry gas occupied 127 cm^3 at S.T.P. If this same mass of gas were collected over water at 23 °C and a total gas pressure of 745 torr, what volume would it occupy? The vapor pressure of water at 23 °C is 21 torr.

Ans.   145 cm^3

5.30.   A mass of gas occupies 0.825 L at −30 °C and 556 Pa. What is the pressure if the volume becomes 1 L and the temperature 20 °C?

Ans.   553 Pa

5.31.   One mole of a gas occupies 22.4 L at S.T.P.   (*a*) What pressure would be required to compress 1 mol of oxygen into a 5-L container held at 100 °C?   (*b*) What maximum Celsius temperature would be permitted if this amount of oxygen were held in 5 L at a pressure not exceeding 3 atm?   (*c*) What capacity would be required to hold this same amount if the conditions were fixed at 100 °C and 3 atm?

Ans.   (*a*) 6.12 atm;   (*b*) −90 °C;   (*c*) 10.2 L

5.32.   If the density of a certain gas at 30 °C and 768 torr is 1.253 kg/m^3, find its density at S.T.P.

Ans.   1.376 kg/m^3

5.33.   A certain container holds 2.55 g of neon at S.T.P. What mass of neon will it hold at 100 °C and 10.0 atm?

Ans.   18.7 g

5.34.   At the top of a mountain the thermometer reads 10 °C and the barometer reads 700 mmHg. At the bottom of the mountain the temperature is 30 °C and the pressure is 760 mmHg. Compare the density of the air at the top with that at the bottom.

Ans.   0.986 (top) to 1.000 (bottom)

5.35.   A volume of 95 cm^3 of nitrous oxide at 27 °C is collected over mercury in a graduated tube, the level of mercury inside the tube being 60 mm above the outside mercury level when the barometer reads 750 torr. (*a*) Compute the volume of the same mass of gas at S.T.P.   (*b*) What volume would the same mass of gas occupy at 40 °C, the barometric pressure being 745 torr and the level of mercury inside the tube 25 mm below that outside?

Ans.   (*a*) 78 cm^3;   (*b*) 89 cm^3

5.36.   At a certain altitude in the upper atmosphere, the temperature is estimated to be −100 °C and the density just 10^{-9} that of the earth's atmosphere at S.T.P. Assuming a uniform atmospheric composition, what is the pressure, in torr, at this altitude?

Ans. 4.82×10^{-7} torr

5.37.   At 0 °C the density of nitrogen at 1 atm is 1.25 kg/m^3. The nitrogen which occupied 1 500 cm^3 at S.T.P. was compressed at 0 °C to 575 atm and the gas volume was observed to be 3.92 cm^3, in violation of Boyle's law. What was the final density of this nonideal gas?

Ans.   478 kg/m^3

5.38.   The respiration of a suspension of yeast cells was measured by observing the decrease in pressure of gas above the cell suspension. The apparatus, shown in Fig. 5-7, was arranged so that the gas was confined to a constant volume, 16.0 cm^3, and the entire pressure change was caused by uptake of oxygen by the cells. The pressure was measured in a manometer the fluid of which had a density of 1.034 g/cm^3. The supply of fluid was adjustable so that the level on the closed side was kept constant. The entire apparatus was immersed in a thermostat at 37 °C. In a 30-min observation period the fluid in the open side of the manometer dropped 37 mm. Neglecting the solubility of oxygen in the yeast suspension, compute the rate of oxygen consumption by the cells in cubic millimeters of O_2 (S.T.P.) per hour.

Ans.   105 mm^3/h

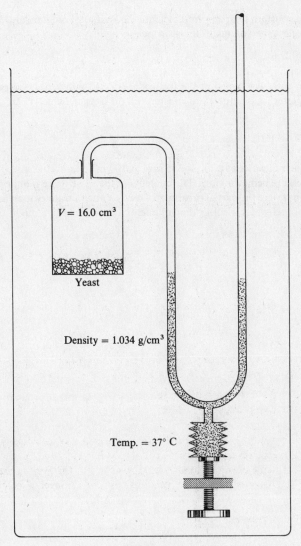

$V = 16.0 \text{ cm}^3$

Yeast

Density = 1.034 g/cm^3

Temp. = 37° C

Fig. 5-7

5.39. A mixture of N_2, NO, and NO_2 was analyzed by selective absorption of the oxides of nitrogen. The initial volume of the mixture was 2.74 cm^3. After treatment with water, which absorbed the NO_2, the volume was 2.02 cm^3. A ferrous sulfate solution was then shaken with the residual gas to absorb the NO, after which the volume was 0.25 cm^3. All volumes were measured at barometric pressure. Neglecting water vapor, what was the volume percentage of each gas in the original mixture?

Ans.   9.1 % N_2, 64.6 % NO, 26.3 % NO_2

5.40. A handball with an internal volume of 60 cm^3 was filled with air to a pressure of 1.35 atm. A dishonest player filled a syringe to the 25-cm^3 mark with air at 1.00 atm and injected it into the handball. Calculate the pressure inside the tampered handball, assuming its volume did not change.

Ans.   1.77 atm

5.41. A 250-mL flask contained krypton at 500 torr. A 450-mL flask contained helium at 950 torr. The contents of the two flasks were mixed by opening a stopcock connecting them. Assuming that all operations were carried out at a uniform constant temperature, calculate the final total pressure and the volume percent of each gas in the mixture. Neglect the volume of the stopcock.

Ans. 789 torr, 22.6 % Kr

5.42. A glass vacuum tube was sealed at the factory at 750 °C with a residual pressure of air of 4.5×10^{-7} torr. Then a metal "getter" was actuated to consume all the oxygen (which is 21 % by volume of air). What was the final pressure in the tube at 22 °C?

Ans. 1.03×10^{-7} torr

5.43. The vapor pressure of water at 80 °C is 355 torr. A 100-mL vessel contained water-saturated oxygen at 80 °C, the total gas pressure being 760 torr. The contents of the vessel were pumped into a 50 mL vessel at the same temperature. What were the partial pressures of oxygen and of water vapor, and what was the total pressure in the final equilibrated state? Neglect the volume of any water which might condense.

Ans. 810 torr, 355 torr, 1 165 torr

Chapter 6

Molecular Weights of Gases and Kinetic Theory

AVOGADRO'S HYPOTHESIS

Avogadro's hypothesis states that *equal volumes of all gases under the same conditions of temperature and pressure contain the same number of molecules.* Thus 1 L (or mL or m³ or other unit of volume) of oxygen contains the same number of molecules as 1 L (or mL, etc.) of hydrogen or of any other gas, provided the volumes are measured under the same conditions.

Avogadro's hypothesis enables us to determine the relative weights of the molecules (molecular weights) of gases. The logic is shown in Example 1.

EXAMPLE 1   At standard conditions, 1 L of oxygen weighs 1.43 g and 1 L of carbon monoxide weighs 1.25 g. By Avogadro's hypothesis, 1 L of carbon monoxide (S.T.P.) contains the same number of molecules as 1 L of oxygen (S.T.P.). Hence a molecule of carbon monoxide must weigh 1.25/1.43 times as much as a molecule of oxygen. Accordingly, if we take the molecular weight of oxygen as 32, then the molecular weight of carbon monoxide is

$$\frac{1.25}{1.43} \times 32 = 28$$

When applied precisely, the gas density method may be used for atomic weight determinations, particularly for the lighter elements. Even crude gas density experiments may be used, in conjunction with chemical composition data, to establish the molecular weight, and hence molecular formula of a gaseous compound.

EXAMPLE 2   A hydride of silicon that has the empirical formula SiH_3 was found to have an approximate gas density at S.T.P. of 2.9 g/L. By comparison with oxygen, whose molecular weight and density are known, the molecular weight of the hydride is then

$$\frac{2.9}{1.43} \times 32 = 65$$

Although this approximate molecular weight might be in error by as much as 10 percent, it is sufficiently accurate to allow assignment of the molecular formula Si_2H_6 (molecular weight 62.2) and to rule out SiH_3 (31.1), Si_3H_9 (93.3), and higher multiples of the empirical formula weight.

MOLAR VOLUME

If 1 mol of any gas has the same number of molecules, N_A, as 1 mol of any other gas (Chapter 2), and if equal numbers of molecules correspond to equal volumes at S.T.P. (Avogadro's hypothesis), then 1 mol of any gas has the same volume at S.T.P. as 1 mol of any other gas. This standard *molar volume* has the value 22.414 L.

Because of the deviations from ideal behavior (and thus from Avogadro's hypothesis) shown by real gases, the actual observed molar volume of a gas at S.T.P. may be slightly different from 22.414 L, usually lower. In the rest of this chapter, the rounded value 22.4 L will be used for all real gases.

IDEAL GAS LAW

Let us apply the combined gas law (Chapter 5) to 1 mol of ideal gas, using a subscript zero to denote standard conditions.

$$\frac{PV}{T} = \frac{P_0 V_0}{T_0} = \frac{(1 \text{ atm})(22.4 \text{ L/mol})}{273 \text{ K}} = 0.0821 \text{ L} \cdot \text{atm} \cdot \text{K}^{-1} \cdot \text{mol}^{-1}$$

The quantity $R = 0.0821$ L·atm·K^{-1}·mol^{-1} is known as the *universal gas constant*. For n moles of ideal gas at the same temperature and pressure, the volume would be n times as great. Thus $PV/T = nR$, or

$$PV = nRT$$

This is the *ideal gas law*; it should be memorized by the student, along with the numerical value of R. The above value of R is used when P is in atmospheres, V is in liters, T is in kelvins, and n is in moles. When SI units are used for P and V (pascals and cubic meters), then

$$R = 8.314 \text{ J·K}^{-1}\text{·mol}^{-1}$$

must be used.

The mass, in grams, of gas present is given by

$$w = n\mathcal{M}$$

where $\mathcal{M}$ is the molecular weight, in grams per mole; or by

$$w = dV$$

where d is the gas density, in grams per liter if V is in liters. Hence the alternative forms of the ideal gas law:

$$PV = \left(\frac{w}{\mathcal{M}}\right)RT$$

$$P = \left(\frac{d}{\mathcal{M}}\right)RT$$

For an ideal gas, V is proportional to n when P and T are fixed. Hence the awkward concept of volume percentage (or fraction) discussed in Problem 5.12 can now be replaced with mole percentage, or mole fraction, where each gas is assumed to occupy the entire volume of the mixture but is at its own partial pressure.

GAS VOLUME RELATIONS FROM EQUATIONS

A chemical equation representing the reaction or production of two or more gaseous substances directly indicates the *volumes* of the gases participating in the reaction when T and P are fixed. The volumes are related to the numbers of molecules indicated in the equation and may be evaluated without reference to the reacting weights of the gases. For example:

$$4\text{NH}_3(gas) + 3\text{O}_2(gas) \rightarrow 2\text{N}_2(gas) + 6\text{H}_2\text{O}(vapor)$$

4 molecules	3 molecules	2 molecules	6 molecules
4 mol	3 mol	2 mol	6 mol
4×22.4 L	3×22.4 L	2×22.4 L	6×22.4 L
4 L	3 L	2 L	6 L
4 ft^3	3 ft^3	2 ft^3	6 ft^3

The above interpretation is valid so long as all the water remains in the vapor state under the conditions of temperature and pressure specified. At S.T.P. the water would be condensed to a negligible volume of liquid, and so 7 L of reactants would produce only 2 L of products rather than 8 L.

GAS STOICHIOMETRY INVOLVING MASS

In chemical reactions involving gases, the amounts of some reactants or products may be specified in units of mass. These problems are solved by the same methods as in Chapter 4, except that the number of moles of gas may be determined from its volume, temperature, and pressure using the ideal gas law. As always we turn to the balanced equation for the relative numbers of moles of all reactants and products.

EXAMPLE 3   Carbon dioxide can be removed from the recirculated air aboard a spaceship by passing it over lithium hydroxide.

$$2\,LiOH(s) + CO_2(g) \rightarrow Li_2CO_3(s) + H_2O(g)$$

Calculate the number of grams of LiOH consumed in the above reaction when 100 L of air containing 1.20% CO_2, at 29 °C and 776 torr, is passed through.

$$n_{CO_2} = \frac{(0.012\,0)(100\ L)[776\ torr/(760\ torr/atm)]}{0.082\,1\ L\cdot atm\cdot K^{-1}\cdot mol^{-1}\cdot 302\ K} = 0.049\,4\ mol$$

$$n_{LiOH} = 2(n_{CO_2}) = 2(0.049\,4\ mol) = 0.098\,8\ mol$$

$$\text{Amount of LiOH} = (0.098\,8\ mol)(23.9\ g\ mol^{-1}) = 2.36\ g$$

BASIC ASSUMPTIONS OF THE KINETIC THEORY OF GASES

The ideal gas law can be derived solely from theoretical principles by making a few assumptions about the nature of gases and the meaning of temperature. Such a derivation can be found in any physical chemistry textbook. Herewith, though, is a listing of the basic assumptions.

1. A gas consists of a very large number of molecules which are in a state of continual random motion. A molecule has negligible size. Pressure is a consequence of the force of molecular collisions on a measuring surface, such as the container walls.
2. Collisions between molecules or between a molecule and an inert surface are perfectly elastic, with no change in the total kinetic energy of the gas molecules.
3. No forces are exerted between molecules except through collisions. Therefore, between collisions, a molecule travels in a straight line at constant speed.
4. The average kinetic energy per gas molecule, $(\frac{1}{2}mu^2)_{avg}$, is independent of the nature of the gas and is directly proportional to the temperature. Here, the kinetic energy is defined by the standard physical formula in terms of the molecular mass, m, and the molecular speed, u. This statement may be taken as a more exact definition of temperature than the qualitative one given in Chapter 1. A list of some common energy units is given in Table 6-1.

Table 6-1. Some Common Energy Units

Unit	Unit Symbol	Definition
joule (SI)	J	$m^2\cdot kg\cdot s^{-2} = N\cdot m$
calorie (thermochemical)	cal	4.184 J
kilocalorie (thermochemical)	kcal	10^3 cal
British thermal unit (thermochemical)	Btu	252 cal = 1 054 J*

* Approximate definition.

PREDICTIONS OF THE KINETIC THEORY

(a)   A mechanical argument based on Assumptions 1, 2, and 3 above shows that for a gas composed of N molecules

$$PV = \tfrac{2}{3}N[(\tfrac{1}{2}mu^2)_{avg}]$$

(b)   Using Assumption 4, the distribution of speeds over the assemblage of molecules is predicted in closed mathematical form, known as the *Maxwell-Boltzmann distribution*. Figure 6-1 shows, for

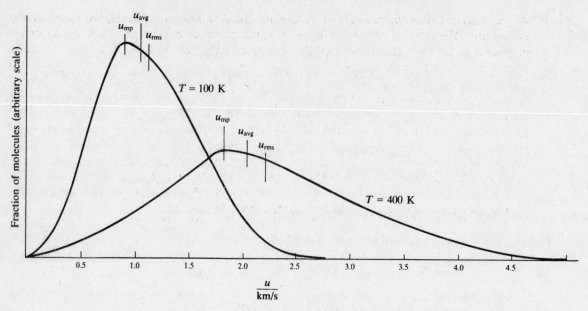

Fig. 6-1

hydrogen, some plots of the fraction of molecules having speeds close to a given value u, as a function of u, and at two different temperatures. The speed for which a distribution curve has its maximum is called the *most probable speed*, u_{mp}. It is slightly smaller than the average speed, u_{avg}. Still another speed, slightly larger than u_{avg}, is the *root-mean-square speed*, u_{rms}; it is defined as that speed for which a molecule's kinetic energy would be equal to the average kinetic energy over the whole sample, i.e.,

$$\tfrac{1}{2}m(u_{rms})^2 = (\tfrac{1}{2}mu^2)_{avg}$$

For the Maxwell-Boltzmann distribution:

$$u_{mp} = \left(\frac{2RT}{\mathcal{M}}\right)^{1/2} \qquad u_{avg} = \left(\frac{8RT}{\pi\mathcal{M}}\right)^{1/2} \qquad u_{rms} = \left(\frac{3RT}{\mathcal{M}}\right)^{1/2}$$

(c) From the results of (a) and (b), and using $N = nN_A$, and $N_A m = \mathcal{M}$, we can derive the ideal gas law.

$$PV = \frac{2N}{3}\left(\frac{mu_{rms}^2}{2}\right) = \frac{2nN_A}{3}\left(\frac{3mRT}{2\mathcal{M}}\right) = nRT$$

The agreement of the above with the empirically determined ideal gas law validates the important definition of temperature in assumption 4, above.

(d) The frequency of molecular collisions with a unit area of container wall is predicted as

$$Z = \frac{N_A P}{(2\pi \mathcal{M}RT)^{1/2}}$$

Now, the rate at which a gas *effuses* (flows out) through a small hole in the container into a vacuum is exactly the rate at which the molecules would collide with a wall area equal to the area of the hole. Hence, from the above equation, the effusion rates of two gases, both at the same pressure and temperature, are in the ratio

$$\frac{Z_1}{Z_2} = \left(\frac{\mathcal{M}_2}{\mathcal{M}_1}\right)^{1/2}$$

where $\mathcal{M}$ is the molecular weight. This explains *Graham's law of effusion*, the experimental finding that the rates of effusion of gases at equal pressures and temperatures are inversely proportional to the square roots of their densities. (By the ideal gas law, gas density is proportional to molecular weight.)

(e) The inverse dependence of average velocities and of effusion rates on the square roots of molecular weights shows up in the treatment of other transport phenomena, such as *diffusion*, *thermal conduction*, and *nonturbulent flow*, although the theory is not as exact in these phenomena, since it must be augmented by a detailed consideration of the nature of intermolecular collisions, a process involving some nonideal aspects of gases. Problem 6.60 treats an important case of diffusion, the process in which one gas penetrates another on the way to equalization of concentrations throughout a gas mixture. The assumption will be made in this book that the relative diffusion rates of two gases are inversely proportional to the square roots of their molecular weights, as is the case more exactly for effusion.

Solved Problems

VOLUMES AND MOLECULAR WEIGHTS OF GASES

6.1. Determine the approximate molecular weight of a gas if 560 cm^3 weighs 1.55 g at S.T.P.

$$PV = \left(\frac{w}{\mathcal{M}}\right)RT$$

$$\mathcal{M} = \frac{wRT}{PV} = \frac{(1.55 \text{ g})(0.082\,1 \text{ L} \cdot \text{atm} \cdot \text{K}^{-1} \cdot \text{mol}^{-1})(273 \text{ K})}{(1 \text{ atm})(0.560 \text{ L})} = 62.0 \text{ g} \cdot \text{mol}^{-1}$$

Another Method.

Molecular weight = weight of 1 L at S.T.P. × number of liters at S.T.P. per mole

$$= \left(\frac{1.55 \text{ g}}{0.560 \text{ L}}\right)(22.4 \text{ L/mol}) = 62.0 \text{ g/mol}$$

This second method is faster if the data happen to be given at S.T.P.

6.2. At 18 °C and 765 torr, 1.29 L of a gas weighs 2.71 g. Calculate the approximate molecular weight of the gas.

For variety in calculational procedure, first convert the given gas data to moles instead of relying on a formula for $\mathcal{M}$ as in Problem 6.1.

$$PV = nRT \qquad n = \frac{PV}{RT} \qquad P = \frac{765}{760} \text{ atm} \qquad T = (18 + 273) \text{ K} = 291 \text{ K}$$

$$n = \frac{(\frac{765}{760} \text{ atm})(1.29 \text{ L})}{(0.082\,1 \text{ L} \cdot \text{atm} \cdot \text{K}^{-1} \cdot \text{mol}^{-1})(291 \text{ K})} = 0.054\,4 \text{ mol}$$

Then $\mathcal{M} = 2.71 \text{ g}/0.054\,4 \text{ mol} = 49.8 \text{ g/mol}$.

6.3. Determine the volume occupied by 4.0 g of oxygen at S.T.P. Molecular weight of oxygen is 32.

Volume of 4.0 g O_2 at S.T.P. = number of moles in 4.0 g O_2 × standard molar volume

$$= \left(\frac{4.0 \text{ g}}{32 \text{ g/mol}}\right)(22.4 \text{ L/mol}) = 2.8 \text{ L}$$

6.4. What volume would 15.0 g of argon occupy at 90 °C and 735 torr?

$$V = \frac{wRT}{\mathscr{M}P} = \frac{(15.0 \text{ g})(0.082\,1 \text{ L} \cdot \text{atm} \cdot \text{K}^{-1} \cdot \text{mol}^{-1})(363 \text{ K})}{(39.9 \text{ g} \cdot \text{mol}^{-1})(\frac{735}{760} \text{ atm})} = 11.6 \text{ L}$$

Another Method.

At S.T.P. this much argon would occupy

$$V_1 = \left(\frac{15.0}{39.9} \text{ mol}\right)(22.4\text{L/mol}) = 8.4 \text{ L}$$

This problem can be completed by the usual procedure for converting from S.T.P to an arbitrary set of conditions.

$$V_2 = V_1\left(\frac{T_2}{T_1}\right)\left(\frac{P_1}{P_2}\right) = (8.4 \text{ L})\left(\frac{363 \text{ K}}{273 \text{ K}}\right)\left(\frac{760 \text{ torr}}{735 \text{ torr}}\right) = 11.6 \text{ L}$$

6.5. Compute the approximate density of methane, CH_4, at 20 °C and 5.00 atm. The molecular weight of methane is 16.0.

$$PV = \left(\frac{w}{\mathscr{M}}\right)RT \qquad P = \left(\frac{w}{V \cdot \mathscr{M}}\right)RT = \left(\frac{d}{\mathscr{M}}\right)RT$$

$$d = \frac{\mathscr{M}P}{RT} = \frac{(16.0 \text{ g} \cdot \text{mol}^{-1})(5.00 \text{ atm})}{(0.082\,1 \text{ L} \cdot \text{atm} \cdot \text{K}^{-1} \cdot \text{mol}^{-1})(293 \text{ K})} = 3.33 \text{ g/L}$$

Another Method.

$$\text{Density} = \frac{\text{mass of 1 mol}}{\text{volume of 1 mol}} = \frac{16.0 \text{ g}}{(22.4 \text{ L})\left(\dfrac{1.00 \text{ atm}}{5.00 \text{ atm}}\right)\left(\dfrac{293 \text{ K}}{273 \text{ K}}\right)} = 3.33 \text{ g/L}$$

6.6. Dry air consists of approximately 21 % O_2, 78 % N_2, and 1 % Ar by moles. Calculate the density of dry air at S.T.P. What is the average (i.e. "apparent") molecular weight of dry air?

Compute the mass of 22.4 L of air, i.e., 1 mol, which contains 0.21 mol O_2, 0.78 mol N_2, and 0.01 mol Ar.

$$(0.21 \text{ mol})(32.0 \text{ g/mol}) + (0.78 \text{ mol})(28.0 \text{ g/mol}) + (0.01 \text{ mol})(39.95 \text{ g/mol}) = 29 \text{ g}$$

The mass of 22.4 L of a gas at S.T.P. would be its molecular weight if it were a pure gaseous substance. Hence 29 g/mol appears to be the molecular weight of air. Its density is

$$\frac{29 \text{ g}}{22.4 \text{ L}} = 1.29 \text{ g/L}$$

6.7. An organic compound had the following analysis: C = 55.8 %, H = 7.03 %, O = 37.2 %. A 1.500-g sample was vaporized and was found to occupy 530 cm³ at 100 °C and 740 torr. What is the molecular formula of the compound?

The approximate molecular weight, calculated from the gas density data, is 89.0. The empirical formula, calculated from the percentage composition data, is C_2H_3O, and the empirical formula weight is 43.0. The exact molecular weight must therefore be $(2)(43.0) = 86.0$, since this is the only integral multiple of 43.0 which is reasonably close to the approximate weight of 89. The molecule must therefore contain twice the number of atoms in the empirical formula, and the molecular formula must be $C_4H_6O_2$.

Another Method.

Instead of calculating the empirical formula, we can use the composition data to calculate the number of moles of atoms of each element in 89 g of compound.

$$n(C) = \frac{(0.558)(89 \text{ g})}{12.0 \text{ g/mol}} = 4.1 \qquad n(H) = \frac{(0.070\,3)(89 \text{ g})}{1.01 \text{ g/mol}} = 6.2 \qquad n(O) = \frac{(0.372)(89 \text{ g})}{16.0 \text{ g/mol}} = 2.1$$

These numbers approximate the numbers of atoms in a molecule, the small deviations from integral values resulting from the approximate nature of the molecular weight. The molecular formula, $C_4H_6O_2$, is obtained without going through the intermediate evaluation of an empirical formula.

6.8. Mercury diffusion pumps may be used in the laboratory to produce a high vacuum. Cold traps are generally placed between the pump and the system to be evacuated. These cause the condensation of mercury vapor and prevent mercury from diffusing back into the system. The maximum pressure of mercury that can exist in the system is the vapor pressure of mercury at the temperature of the cold trap. Calculate the number of mercury vapor molecules per unit volume in a cold trap maintained at $-120\,°C$. The vapor pressure of mercury at this temperature is 10^{-16} torr.

$$\text{Moles per liter} = \frac{n}{V} = \frac{P}{RT} = \frac{(10^{-16}/760) \text{ atm}}{(0.082\,1 \text{ L} \cdot \text{atm} \cdot \text{K}^{-1} \cdot \text{mol}^{-1})(153 \text{ K})} = 1.0 \times 10^{-20} \text{ mol/L}$$

$$\text{Molecules per liter} = (1.0 \times 10^{-20} \text{ mol/L})(6.0 \times 10^{23} \text{ molecules/mol}) = 6\,000 \text{ molecules/L}$$

or 6 molecules/cm^3.

REACTIONS OF GASES

In each of the next four problems all gases are measured at the same temperature and pressure.

6.9. What volume of hydrogen will combine with 12 L of chlorine to form hydrogen chloride? What volume of hydrogen chloride will be formed?

The balanced equation for this reaction is

$$H_2(gas) + Cl_2(gas) \rightarrow 2\,HCl(gas)$$
$$\text{(1 molecule)} \qquad \text{(1 molecule)} \qquad \text{(2 molecules)}$$

The equation shows that 1 molecule H_2 reacts with 1 molecule Cl_2 to form 2 molecules HCl. But, by Avogadro's hypothesis, equal numbers of molecules of *gases* under the same conditions of temperature and pressure occupy equal volumes. Therefore the equation also indicates that 1 volume of H_2 reacts with 1 volume Cl_2 to form 2 volumes of HCl (*gas*).

Thus 12 L of H_2 combines with 12 L of Cl_2 to form $2 \times 12 \text{ L} = 24 \text{ L}$ of HCl.

6.10. What volume of hydrogen will unite with 6 ft^3 of nitrogen to form ammonia? What volume of ammonia will be produced?

$$N_2(gas) + 3\,H_3(gas) \rightarrow 2\,NH_3(gas)$$

The equation indicates that, since only gases are involved, 1 volume N_2 reacts with 3 volumes H_2 to form 2 volumes NH_3. Therefore 6 ft^3 N_2 reacts with $3 \times 6 = 18 \text{ ft}^3$ H_2 to form $2 \times 6 = 12 \text{ ft}^3$ NH_3.

6.11. Sixty-four liters of NO is mixed with 40 L of O_2 and the following reaction allowed to go to completion:

$$2NO(gas) + O_2(gas) \rightarrow 2NO_2(gas)$$

Calculate the total volume of gas after completion of the reaction.

Since the volume of oxygen required is only half that of the NO, $\frac{1}{2}(64) = 32$ L of O_2 will be consumed and $40 - 32 = 8$ L will be in excess. The volume of NO_2 formed is the same as that of the NO reacted. The final volume of gas then will be 64 L of NO_2 plus 8 L of excess O_2 or a total of 72 L.

6.12. What volume of O_2 at S.T.P. is required for the complete combustion of 1 mol of carbon disulfide, CS_2? What volumes of CO_2 and SO_2 at S.T.P. are produced?

$$CS_2(liquid) + 3O_2(gas) \rightarrow CO_2(gas) + 2SO_2(gas)$$
$$\text{(1 mol)} \qquad \text{(3 mol)} \qquad \text{(1 mol)} \qquad \text{(2 mol)}$$

The equation shows that 1 mol CS_2 reacts with 3 mol O_2 to form 1 mol CO_2 and 2 mol SO_2. One mole of a *gas* at S.T.P. occupies 22.4 L. Therefore

Volume of 3 mol O_2 at S.T.P. $= 3(22.4) = 67.2$ L O_2

Volume of 1 mol CO_2 at S.T.P. $= 1(22.4) = 22.4$ L CO_2

Volume of 2 mol SO_2 at S.T.P. $= 2(22.4) = 44.8$ L SO_2

6.13. How many liters of oxygen, at standard conditions, can be obtained from 100 g of potassium chlorate?

$$2KClO_3 \ (solid) \rightarrow 2KCl + 3O_2 \ (gas)$$
$$\text{(2 mol)} \qquad\qquad\qquad \text{(3 mol)}$$

Molar Method.

The equation shows that 2 mol $KClO_3$ gives 3 mol O_2. As in previous chapters, the symbol n will be used for the number of moles.

$$n(KClO_3) = \frac{100 \text{ g}}{122.6 \text{ g/mol}} = 0.816 \text{ mol } KClO_3$$

$$n(O_2) = \tfrac{3}{2}n(KClO_3) = \tfrac{3}{2}(0.816) = 1.224 \text{ mol } O_2$$

Volume of 1.224 mol O_2 at S.T.P. $= (1.224 \text{ mol } O_2)(22.4 \text{ L/mol}) = 27.4$ L O_2

Another Method.

The equation shows that 2 mol $KClO_3[2(122.6) = 245.2 \text{ g}]$ gives 3 molar volumes $O_2[3(22.4 \text{ L}) = 67.2 \text{ L}]$. Then

245.2 g $KClO_3$ gives   67.2 L O_2 at S.T.P.

1 g $KClO_3$ gives $\dfrac{67.2}{245.2}$ L O_2 at S.T.P.

and   100 g $KClO_3$ gives $100\left(\dfrac{67.2}{245.2} \text{ L}\right) = 27.4$ L O_2 at S.T.P.

Note: It was not necessary in either method of solution to calculate the *mass* of oxygen formed.

6.14. What volume of oxygen, at 18 °C and 750 torr, can be obtained from 100 g of $KClO_3$?

This problem is identical with Problem 6.13, except that the 27.4 L of O_2 at $0\,°C$ and 760 torr must be converted to liters of O_2 at $18\,°C$ and 750 torr.

$$\text{Volume at } 18\,°C \text{ and } 750 \text{ torr} = (27.4 \text{ L})\left[\frac{(273+18)\text{ K}}{273\text{ K}}\right]\left(\frac{760\text{ torr}}{750\text{ torr}}\right) = 29.6 \text{ L}$$

Another Method.

Since we are not concerned with S.T.P., it is more direct to calculate the volume from the number of moles.

$$V = \frac{nRT}{P} = \frac{(1.224 \text{ mol})(0.082\,1 \text{ L} \cdot \text{atm} \cdot \text{K}^{-1} \cdot \text{mol}^{-1})(291\text{ K})}{\frac{750}{760}\text{ atm}} = 29.6 \text{ L}$$

6.15. How many grams of zinc must be dissolved in sulfuric acid in order to obtain 500 cm^3 of hydrogen at $20\,°C$ and 770 torr?

$$\underset{(1 \text{ mol})}{\text{Zn }(\textit{solid})} + H_2SO_4 \rightarrow ZnSO_4 + \underset{(1 \text{ mol})}{H_2(\textit{gas})}$$

The number of moles of H_2 may be found by either of the following methods, of which the first is more direct.

$$n(H_2) = \frac{PV}{RT} = \frac{(\frac{770}{760}\text{ atm})(0.500\text{ L})}{(0.082\,1 \text{ L} \cdot \text{atm} \cdot \text{K}^{-1} \cdot \text{mol}^{-1})(293\text{ K})} = 0.021\,1 \text{ mol } H_2$$

or

$$V_{\text{S.T.P.}} = (500 \text{ cm}^3)\left[\frac{273\text{ K}}{(273+20)\text{ K}}\right]\left(\frac{770\text{ torr}}{760\text{ torr}}\right) = 472 \text{ cm}^3 = 0.472 \text{ L}$$

$$n(H_2) = \frac{0.472 \text{ L}}{22.4 \text{ L/mol}} = 0.021\,1 \text{ mol } H_2$$

The equation shows that 1 mol H_2 requires 1 mol Zn. Then 0.021 1 mol H_2 requires 0.021 1 mol Zn, the mass of which is

$$(0.021\,1 \text{ mol})(65.4 \text{ g/mol}) = 1.38 \text{ g Zn}$$

6.16. A natural gas sample contains 84% (by volume) CH_4, 10% C_2H_6, 3% C_3H_8, and 3% N_2. If a series of catalytic reactions could be used for converting all the carbon atoms of the gas into butadiene, C_4H_6, with 100% efficiency, how much butadiene could be prepared from 100 g of the natural gas?

Molecular weights: $CH_4 = 16$, $C_2H_6 = 30$, $C_3H_8 = 44$, $N_2 = 28$, $C_4H_6 = 54$. The volume percent of a gas mixture is the same as the mole percent.

$$100 \text{ mol mixture} = 84 \text{ mol } CH_4 + 10 \text{ mol } C_2H_6 + 3 \text{ mol } C_3H_8 + 3 \text{ mol } N_2$$
$$= 84(16 \text{ g } CH_4) + 10(30 \text{ g } C_2H_6) + 3(44 \text{ g } C_3H_8) + 3(28 \text{ g } N_2)$$
$$= 1\,860 \text{ g natural gas}$$

The number of moles of C in 100 mol mixture is $84(1) + 10(2) + 3(3) = 113$ mol C. Since 4 mol C gives 1 mol (54 g) C_4H_6, 113 mol C gives

$$(\tfrac{113}{4} \text{ mol})(54 \text{ g/mol}) = 1\,530 \text{ g } C_4H_6$$

Then   1 860 g natural gas yields   1 530 g C_4H_6

and   100 g natural gas yields $\frac{100}{1\,860}(1\,530 \text{ g}) = 82 \text{ g } C_4H_6$

6.17. A reaction mixture for the combustion of SO_2 was prepared by opening a stopcock connecting two separate chambers, one having a volume of 2.125 L filled at 0.750 atm with SO_2 and the other having a 1.500-L volume filled at 0.500 atm with O_2; both gases were at $80\,°C$.   (*a*) What were the mole fraction of SO_2 in the mixture, the total pressure, and the partial pressures? (b) If the mixture was passed over a catalyst that promoted the formation of SO_3 and then was returned to the original two connected vessels, what were the mole fractions in the final mixture, and what was the final total pressure? Assume that the final temperature is 80 °C, and that the conversion of SO_2 is complete to the extent of the availability of O_2.

(a)
$$n(SO_2) = \frac{PV}{RT} = \frac{(0.750 \text{ atm})(2.125 \text{ L})}{(0.082\,1 \text{ L} \cdot \text{atm} \cdot \text{K}^{-1} \cdot \text{mol}^{-1})(353 \text{ K})} = 0.055\,0 \text{ mol}$$

$$n(O_2) = \frac{PV}{RT} = \frac{(0.500 \text{ atm})(1.500 \text{ L})}{(0.082\,1 \text{ L} \cdot \text{atm} \cdot \text{K}^{-1} \cdot \text{mol}^{-1})(353 \text{ K})} = 0.025\,9 \text{ mol}$$

Each mole fraction is evaluated by dividing the number of moles of the component by the total number of moles in the mixture.

$$x(SO_2) = \frac{n(SO_2)}{n(SO_2) + n(O_2)} = \frac{0.055\,0 \text{ mol}}{(0.055\,0 + 0.025\,9) \text{ mol}} = \frac{0.055\,0}{0.080\,9} = 0.680$$

$$x(O_2) = \frac{0.025\,9}{0.080\,9} = 0.320$$

Note that the sum of the mole fractions is unity.

The total pressure can best be evaluated by using total volume (2.125 L + 1.500 L = 3.625 L) and total number of moles, n, (0.080 9 mol).

$$P(\text{tot}) = \frac{nRT}{V} = \frac{(0.080\,9 \text{ mol})(0.082\,1 \text{ L} \cdot \text{atm} \cdot \text{K}^{-1} \cdot \text{mol}^{-1})(353 \text{ K})}{(3.625 \text{ L})} = 0.647 \text{ atm}$$

Partial pressures in a mixture are easily computed by multiplying the respective mole fractions by the total pressure. This statement can be proved as follows:

$$\frac{P(X)}{P(\text{tot})} = \frac{n(X)RT/V}{nRT/V} = \frac{n(X)}{n} = x(X) \qquad \text{whence} \qquad P(X) = x(X)P(\text{tot})$$

Thus,
$$P(SO_2) = x(SO_2) \times P(\text{tot}) = (0.680)(0.647 \text{ atm}) = 0.440 \text{ atm}$$
$$P(O_2) = x(O_2) \times P(\text{tot}) = (0.320)(0.647 \text{ atm}) = 0.207 \text{ atm}$$

The sum of the partial pressures must of course equal the total pressure.

(b) The chemical equation for the reaction is

$$2SO_2 + O_2 \rightarrow 2SO_3$$

All substances are gases under the experimental conditions. The number of moles of O_2 required for stoichiometric conversion is half the number of moles of SO_2; but the number of moles of O_2 available, 0.025 9, was less than half of 0.055 0. Therefore all the O_2 was used up and an excess of SO_2 remained. For the quantities of gases participating or formed in the reaction,

$$n(O_2) = 0.025\,9 \text{ mol}$$
$$n(SO_2) = 2 \times n(O_2) = (2)(0.025\,9 \text{ mol}) = 0.051\,8 \text{ mol}$$

The product would be

$$n(SO_3) = n(SO_2) = 0.051\,8 \text{ mol}$$

so that the residue would be (0.055 0 − 0.051 8) mol = 0.003 2 mol SO_2. After the reaction is completed,

$$x(SO_3) = \frac{n(SO_3)}{n(SO_3) + n(SO_2)} = \frac{0.051\,8 \text{ mol}}{(0.051\,8 + 0.003\,2) \text{ mol}} = \frac{0.051\,8}{0.055\,0} = 0.942$$

$$x(SO_2) = \frac{n(SO_2)}{n(SO_3) + n(SO_2)} = \frac{0.003\,2 \text{ mol}}{0.055\,0 \text{ mol}} = 0.058$$

$$P(\text{tot}) = \frac{n(\text{tot})RT}{V} = \frac{(0.055\,0 \text{ mol})(0.082\,1 \text{ L} \cdot \text{atm} \cdot \text{K}^{-1} \cdot \text{mol}^{-1})(353 \text{ K})}{3.625 \text{ L}} = 0.440 \text{ atm}$$

Note that the final *total* pressure is the same as the initial *partial* pressure of SO_2. This happens because the only gases present at the end are SO_2 and SO_3. Because each mole of SO_3 was made at the expense of 1 mol of SO_2, the sum of the numbers of moles of the two gases had to equal the initial number of moles of SO_2. The final total pressure is less than the initial total pressure because the consumption of the oxygen caused a decrease in the total number of moles of gas.

KINETIC THEORY

6.18. (a) Show how the value of R in $L \cdot atm \cdot K^{-1} \cdot mol^{-1}$ can be derived from the value in SI units.
(b) Express R in calories.

(a) $$R = 8.314 \, J \cdot K^{-1} \cdot mol^{-1}$$

$$= (8.314 \, N \cdot m \cdot K^{-1} \cdot mol^{-1})\left(\frac{1 \, atm}{1.013 \times 10^5 \, N \cdot m^{-2}}\right)\left(\frac{10^3 \, dm^3}{1 \, m^3}\right)\left(\frac{1 \, L}{dm^3}\right)$$

$$= 0.0821 \, L \cdot atm \cdot K^{-1} \cdot mol^{-1}$$

(b) $$R = (8.314 \, J \cdot K^{-1} \cdot mol^{-1})\left(\frac{1 \, cal}{4.184 \, J}\right) = 1.987 \, cal \cdot K^{-1} \cdot mol^{-1}$$

6.19. Calculate the root-mean-square velocity of H_2 at $0\,°C$.

$$u_{rms} = \left(\frac{3RT}{\mathcal{M}}\right)^{1/2} = \left[\frac{3(8.314 \, J \cdot K^{-1} \cdot mol^{-1})(273 \, K)}{(2.016 \, g/mol)(1 \, kg/1\,000 \, g)}\right]^{1/2}$$

$$= 1\,837 \, (J/kg)^{1/2} = 1.84 \times 10^3 \left(\frac{kg \cdot m^2/s^2}{kg}\right)^{1/2} = 1.84 \times 10^3 \, m/s$$

or 1.84 km/s.

6.20. Compute the relative rates of effusion of H_2 and CO_2 through a fine pinhole.

$$\frac{r(H_2)}{r(CO_2)} = \sqrt{\frac{\mathcal{M}(CO_2)}{\mathcal{M}(H_2)}} = \sqrt{\frac{44}{2.0}} = 4.7$$

Supplementary Problems

VOLUMES, MOLECULAR WEIGHTS, AND REACTIONS OF GASES

6.21. If 200 cm^3 of a gas weighs 0.268 g at S.T.P., what is its molecular weight?

Ans.   30.0

6.22. Compute the volume of 11 g of nitrous oxide, N_2O, at S.T.P.

Ans.   5.6 L

6.23. What volume will 1.216 g of SO_2 gas occupy at 18 °C and 755 torr?

Ans.   456 cm^3

6.24. A 1.225 g mass of a volatile liquid is vaporized, giving 400 cm³ of vapor when measured over water at 30 °C and 770 torr. The vapor pressure of water at 30 °C is 32 torr. What is the molecular weight of the substance?

Ans.   78.4

6.25. Compute the weight of 1 L of ammonia gas, NH_3, at S.T.P.

Ans.   0.76 g/L

6.26. Determine the density of H_2S gas at 27 °C and 2.00 atm.

Ans.   2.77 g/L

6.27. Find the molecular weight of a gas whose density at 40 °C and 785 torr is 1.286 kg/m³.

Ans.   32.0

6.28. What weight of hydrogen at S.T.P. could be contained in a vessel that holds 4.0 g of oxygen at S.T.P.?

Ans.   0.25 g

6.29. One of the methods for estimating the temperature of the center of the sun is based on the ideal gas law. If the center is assumed to consist of gases whose average molecular weight is 2.0, and if the density and pressure are 1.4×10^3 kg/m³ and 1.3×10^9 atm, calculate the temperature.

Ans. 2.3×10^7 K

6.30. An electronic vacuum tube was sealed off during manufacture at a pressure of 1.2×10^{-5} torr at 27 °C. Its volume is 100 cm³. Compute the number of gas molecules remaining in the tube.

Ans. 3.9×10^{13} molecules

6.31. One of the important criteria of hydrogen as a vehicular fuel is its compactness. Compare the numbers of hydrogen atoms per cubic meter available in (a) hydrogen gas under a pressure of 14.0 MPa at 300 K; (b) liquid hydrogen at 20 K at a density of 70.0 kg/m³; (c) the solid compound $DyCo_3H_5$, having a density of 8 200 kg/m³ at 300 K, all of whose hydrogen may be made available for combustion.

Ans.   (a) 0.68×10^{28} atom/m³; (b) 4.2×10^{28} atoms/m³; (c) 7.2×10^{28} atoms/m³

6.32. An empty steel gas tank with valve weighed 125 lb. Its capacity is 1.5 ft³. When the tank was filled with oxygen to 2 000 lbf/in² at 25 °C, what percent of the total weight of the full tank was O_2? Assume the validity of the ideal gas law. (One atmosphere of pressure equals 14.7 lbf/in².)

Ans.   12%

6.33. Pure oxygen gas is not necessarily the most compact source of oxygen for confined fuel systems because of the weight of the cylinder necessary to confine the gas. Other compact sources are hydrogen peroxide and lithium peroxide. The oxygen-yielding reactions are

$$2H_2O_2 \rightarrow 2H_2O + O_2 \qquad 2Li_2O_2 \rightarrow 2Li_2O + O_2$$

Rate (a) 65% (by weight) H_2O_2 and (b) pure Li_2O_2 in terms of % of total weight which is "available" oxygen. Neglect the weights of the containers. Compare with Problem 6.32.

Ans.   (a) 31% (b) 35%

6.34. An iron meteorite was analyzed for its isotopic argon content. The amount of ^{36}Ar was 0.200 mm³ (S.T.P.) per kilogram of meteorite. If each ^{36}Ar atom had been formed by a single cosmic event, how many such events must there have been per kilogram of meteorite?

Ans. 5.4×10^{15}

6.35. Three volatile compounds of a certain element have gaseous densities calculated back to S.T.P. as follows: 6.75, 9.56, and 10.08 kg/m^3. The three compounds contain 96.0%, 33.9%, and 96.4% of the element in question, respectively. What is the most probable atomic weight of the element?

Ans.   72.6, although the data do not exclude 72.6/n, where n is a positive integer.

6.36. Chemical absorbers can be used to remove exhaled CO_2 of space travelers in short spaceflights. Li_2O is one of the most efficient in terms of absorbing capacity per unit weight. If the reaction is

$$Li_2O + CO_2 \rightarrow Li_2CO_3$$

what is the absorption efficiency of pure Li_2O in liters CO_2 (S.T.P.) per kilogram?

Ans.   752 L/kg

6.37. Argon gas liberated from crushed meteorites does not have the same isotopic composition as atmospheric argon. The gas density of a particular sample of meteoritic Ar was found to be 1.481 g/L at 27 °C and 740 torr. What is the average atomic weight of this sample of argon?

Ans.   37.5

6.38. Exactly 500 cm^2 of a gas at S.T.P. weighs 0.581 g. The composition of the gas is as follows: C = 92.24%, H = 7.76%. Derive its molcular formula.

Ans. C_2H_2

6.39. A hydrocarbon has the following composition: C = 82.66%, H = 17.34%. The density of the vapor is 0.2308 g/L at 30 °C and 75 torr. Determine its molecular weight and its molecular formula.

Ans.   58.2, C_4H_{10}

6.40. How many grams of oxygen are contained in 10.5 L of oxygen measured over water at 25 °C and 740 torr? Vapor pressure of water at 25 °C is 24 torr.

Ans.   12.9 g

6.41. The density of NO was very carefully determined to be 0.2579 kg/m^3 at a temperature and pressure at which oxygen's measured density was 0.2749 kg/m^3. On the basis of this information and the known atomic weight of oxygen, calculate the atomic weight of nitrogen.

Ans.   14.02

6.42. An empty flask open to the air weighed 24.173 g. The flask was filled with the vapor of an organic liquid and was sealed off at the barometric pressure and at 100 °C. At room temperature the flask then weighed 25.002 g. The flask was then opened and filled with water at room temperature, after which it weighed 176 g. The barometric reading was 725 mmHg. All weighings were done at the temperature of the room, 25 °C. What is the molecular weight of the organic vapor? Allow for the buoyancy of the air in the weighing of the sealed-off flask, using 1.18 g/L for the density of air at 25 °C and 1 atm.

Ans.   213 g/mol

6.43. A 50 cm^3 sample of a hydrogen-oxygen mixture was placed in a gas buret at 18 °C and confined at barometric pressure. A spark was passed through the sample so that the formation of water could go to completion. The resulting pure gas had a volume of 10 cm^3 at barometric pressure. What was the initial mole fraction of hydrogen in the mixture   (*a*) if the residual gas after sparking was hydrogen,   (*b*) if the residual gas was oxygen?

Ans.   (*a*) 0.73;   (*b*) 0.53

6.44. How much water vapor is contained in a cubic room 4.0 m along an edge if the relative humidity is 50 percent and the temperature is 27 °C? The vapor pressure of water at 27 °C is 26.7 torr. The relative humidity expresses the partial pressure of water as a percentage of the water vapor pressure.

Ans.   0.82 kg

6.45. A batch of wet clothes in a clothes dryer contains 0.983 kg of water. Assuming the air leaves the dryer saturated with water vapor at 48 °C and 738 torr total pressure, calculate the volume of dry air at 24 °C and 738 torr required to dry the clothes. The vapor pressure of water at 48 °C is 83.7 torr.

Ans. 1.070×10^4 L

6.46. Ethane gas, C_2H_6, burns in air as indicated by the equation $2C_2H_6 + 7O_2 \rightarrow 4CO_2 + 6H_2O$. Determine the number of   (a) moles of CO_2 and of H_2O formed when 1 mol of C_2H_6 is burned,   (b) liters of O_2 required to burn 1 L of C_2H_6,   (c) liters of CO_2 formed when 25 L of C_2H_6 is burned,   (d) liters (S.T.P.) of CO_2 formed when 1 mol of C_2H_6 is burned,   (e) moles of CO_2 formed when 25 L (S.T.P.) of C_2H_6 is burned, (f) grams of CO_2 formed when 25 L (S.T.P.) of C_2H_6 is burned.

Ans.   (a) 2 mol CO_2, 3 mol H_2O;   (b) 3.5 L;   (c) 50 L;   (d) 44.8 L;   (e) 2.23 mol;   (f) 98.2 g

6.47. In order to economize on the oxygen supply in spaceships, it has been suggested that the oxygen in exhaled CO_2 be converted to water by a reduction with hydrogen. The CO_2 output per astronaut has been estimated as 1.00 kg per 24-h day. An experimental catalytic converter reduces CO_2 at a rate of 500 cm³ (S.T.P.) per minute. What fraction of the time would such a converter have to operate in order to keep up with the CO_2 output of one astronaut?

Ans.   71 percent

6.48. How many grams of $KClO_3$ are needed to prepare 18 L of oxygen which is collected over water at 22 °C and 760 torr? Vapor pressure of water at 22 °C is 19.8 torr.

Ans.   59.2 g

6.49. A sample of pure calcium hydrogen carbonate, $Ca(HCO_3)_2$, was sealed in an evacuated tube and then heated very strongly until it decomposed completely to $CaO(s)$, $H_2O(g)$, and $CO_2(g)$. Calculate the mole fractions of the gases in the tube.

Ans. $x(H_2O) = 0.333$, $x(CO_2) = 0.667$

6.50. Fifty grams of aluminum is to be treated with a 10% excess of H_2SO_4. The chemical equation for the reaction is

$$2Al + 3H_2SO_4 \rightarrow Al_2(SO_4)_3 + 3H_2$$

(a) What volume of concentrated sulfuric acid, of density 1.80 g/cm³ and containing 96.5% H_2SO_4 by weight, must be taken?   (b) What volume of hydrogen would be collected over water at 20 °C and 785 torr? Vapor pressure of water at 20 °C is 17.5 torr.

Ans.   (a) 173 cm³;   (b) 66.2 L

6.51. A 0.750-g sample of solid benzoic acid, $C_7H_6O_2$, was placed in a 0.500-L pressurized reaction vessel filled with O_2 at 10.0 atm pressure and 25 °C. To the extent of the availability of oxygen, the benzoic acid was burned completely to water and CO_2. What were the final mole fractions of CO_2 and H_2O vapor in the resulting gas mixture brought to the initial temperature? The vapor pressure of water at 25 °C is 23.8 torr. Neglect both the volume occupied by nongaseous substances and the solubility of CO_2 in H_2O. (The water pressure in the gas phase cannot exceed the vapor pressure of water, so most of the water is condensed to the liquid.)

Ans. $x(CO_2) = 0.213$, $x(H_2O) = 0.003\,2$

6.52. Two gases in adjoining vessels were brought into contact by opening a stopcock between them. The one vessel measured 0.250 L and contained NO at 800 torr and 220 K; the other measured 0.100 L and contained O_2 at 600 torr and 220 K. The reaction to form N_2O_4 (*solid*) exhausts the limiting reactant completely. (*a*) Neglecting the vapor pressure of N_2O_4, what is the pressure and composition of the gas remaining at 220 K after completion of the reaction?   (*b*) What weight of N_2O_4 is formed?

Ans.   (*a*) 229 torr NO;   (*b*) 0.402 g N_2O_4

6.53. The industrial production of NH_3 by use of a natural gas feedstock can be represented by the following simplified set of reactions:

$$CH_4 + H_2O \rightarrow CO + 3H_2 \tag{1}$$
$$2CH_4 + O_2 \rightarrow 2CO + 2H_2 \tag{2}$$
$$CO + H_2O \rightarrow CO_2 + H_2 \tag{3}$$
$$N_2 + 3H_2 \rightarrow 2NH_3 \tag{4}$$

By assuming (i) that only the above reactions take place, plus the chemical absorption of CO_2, (ii) that natural gas consists just of CH_4, (iii) that air consists of 0.80 mole fraction N_2 and 0.20 mole fraction O_2, and (iv) that the ratio of conversion of CH_4 by processes (1) and (2) is controlled through admitting oxygen for reaction (2) by adding just enough air so that the mole ratio of N_2 to H_2 is exactly 1:3, consider the overall efficiency of a process in which 1 200 m^3 (S.T.P.) of natural gas yields 1.00 metric ton of NH_3.   (*a*) How many moles of NH_3 would be formed from each mole of natural gas if there were complete conversion of the natural gas subject to the stated assumptions?   (*b*) What percentage of the maximum yield calculated in (*a*) is the actual yield?

Ans.   (*a*) 2.29 mol NH_3/mol CH_4;   (*b*) 48 percent

KINETIC THEORY

6.54. Calculate the ratio of   (*a*) u_{rms} to u_{mp},   (*b*) u_{avg} to u_{mp}.

Ans.   (*a*) 1.22;   (*b*) 1.13

6.55. At what temperature would N_2 molecules have the same average speed as He atoms at 300 K?

Ans.   2 100 K

6.56. At what temperature would the most probable speed of CO molecules be twice that at 0 °C?

Ans.   819 °C

6.57. Show that the ideal gas law can be written $P = \frac{2}{3}\epsilon$, where ϵ is the kinetic energy per unit volume.

6.58. What is the ratio of the average molecular kinetic energy of UF_6 to that of He, both at 300 K?

Ans.   1.000

6.59. What is the kinetic energy of a mole of CO_2 at 400 K   (*a*) in kilojoules   (*b*) in kilocalories

Ans.   (*a*) 4.99 kJ;   (*b*) 1.192 kcal

6.60. Uranium isotopes have been separated by taking advantage of the different rates of diffusion of the two isotopic forms of UF_6. One form contains uranium of atomic weight 238, and the other of atomic weight 235. What are the relative rates of diffusion of these two molecules if Graham's law applies?

Ans. UF_6 with ^{235}U is faster by a factor of 1.004.

6.61. The pressure in a vessel that contained pure oxygen dropped from 2 000 torr to 1 500 torr in 47 min as the oxygen leaked through a small hole into a vacuum. When the same vessel was filled with another gas, the pressure dropped from 2 000 torr to 1 500 torr in 74 min. What is the molecular weight of the second gas?

Ans.   79

6.62. A large cylinder of helium filled at 2 000 lbf/in^2 has a small thin orifice through which helium escaped into an evacuated space at the rate of 3.4 millimoles per hour. How long would it take for 10 millimoles of CO to leak through a similar orifice if the CO were confined at the same pressure?

Ans.   7.8 h

Chapter 7

Thermochemistry

HEAT

Heat is a form of energy. Other forms of energy, such as mechanical, chemical, electric, etc., tend in natural processes to be transformed into heat energy. When any other kinds of energy are transformed into heat energy, or vice versa, the heat energy is exactly equivalent to the amount of transformed energy.

All substances require heat when their temperature is raised (other variables remaining constant), and yield the same amount of heat when they are cooled to the initial temperature. Heat is always absorbed when a solid melts or when a liquid evaporates.

The units most commonly used in expressing the quantity of heat are listed in Table 6-1. The use of the SI unit, the joule, based on quantities from mechanics, emphasizes the interconvertibility of the various forms of energy. Chemists have long used the calorie and the kilocalorie, but they now generally prefer the joule and kilojoule. Engineers have been the principal users of the Btu.

HEAT CAPACITY

The *heat capacity* of a substance is the amount of heat required to raise the temperature of that substance 1 K. The heat capacity itself may depend on temperature. Frequent use is made of such quantities as the *molar heat capacity* (heat capacity per mole) and the *specific heat capacity* (heat capacity per unit mass, sometimes called simply the *specific heat*).

The average specific heat capacity of water is

$$1.00 \text{ cal/g} \cdot \text{K} = 4.184 \text{ kJ/kg} \cdot \text{K}$$

and the deviations from this average are all less than 1 percent between the freezing and boiling points.

CALORIMETRY

An object undergoing a temperature change without a chemical reaction or change of state absorbs or discharges an amount of heat equal to its (average) heat capacity times the temperature change.

$$\text{Heat exchange} = (\text{heat capacity}) \times (\text{temperature change})$$

The heating or cooling of a body of matter of known heat capacity can be used in *calorimetry*, the measurement of quantities of heat.

ENERGY AND ENTHALPY

When a system absorbs heat, part of the absorbed energy may be used for doing work, such as lifting a weight, expanding against the atmosphere, or operating a battery, and part is stored within the system itself as the energy of the internal motions of the atoms and molecules themselves, as energy associated with rearrangements of the atoms that occur in chemical reactions, and as the energy of interaction among the atoms and molecules. This stored portion is known as the *internal energy*, E. The amount of heat absorbed by any system undergoing a modification such as an increase in temperature, a change in physical state, or a chemical reaction, depends somewhat on the conditions under which the process occurs. Specifically, the amount of heat absorbed is exactly equal to the increase in E if no work is done by the system. This would be the case in an ordinary chemical reaction, not linked to a battery, carried out in a closed reactor vessel so that no expansion against the outside atmosphere occurred. The increase in E

can be represented by ΔE. (The Δ symbol for a difference is borrowed from mathematics. ΔE is the difference in E accompanying the process, defined as the final value of E minus the initial value.)

Most chemical reactions, processes of heating or cooling, and changes of physical state are carried out in open, rather than closed, containers. Because of expansions or contractions in volume that might occur in the process, work might be done by the reacting system on the surroundings, or vice versa. In such a case, the principle of conservation of energy requires that the amount of heat absorbed must adjust itself to provide for the small but significant amount of this work. A new function, the *enthalpy*, H, can be defined which is related simply to the heat flow in an open, or constant-pressure, vessel. By definition, $H = E + PV$. The amount of heat absorbed in a constant-pressure process is exactly equal to ΔH, the increase in H.

In summary, if q is the amount of heat *absorbed* by the system of interest from its surroundings,

$$q \text{ (at constant volume)} = \Delta E$$

$$q \text{ (at constant pressure)} = \Delta H$$

These equations are exact in the absence of coupling to work-generating devices such as batteries. Any of the terms in these equations may have either sign. An *endothermic* process is one for which q is positive, i.e., heat is absorbed by the system of interest. An *exothermic* process is one for which q is negative, i.e., heat is evolved. Most of the thermochemical problems in this book will concern ΔH. Even though we may not know the absolute value of E or H, the above equations provide the experimental basis for measuring changes in these functions.

ENTHALPY CHANGES FOR VARIOUS PROCESSES

Change of temperature of a substance

If a substance of heat capacity C is heated or cooled through a temperature interval ΔT,

$$q = C \Delta T$$

provided, as is henceforth assumed, that C is independent of temperature. Subscripts are often used to designate a heat capacity measured at constant pressure, C_p, and at constant volume, C_v. Particularly,

$$\Delta H = C_p \Delta T$$

The quantities C and H are *extensive*, i.e., proportional to the amount of material undergoing the process. We shall use a lowercase symbol to refer to a unit mass. Thus c, c_p, and c_v will denote specific heat capacities; and, for a sample of material of mass w,

$$C = cw$$

Where the subscript is omitted in the problems of this book, C_p will be implied.

Change of phase of a substance

The heat that must be absorbed to melt a substance is sometimes called the *latent heat of fusion*, q(fusion). For the melting of ice at $0\,°C$, for example, the process may be written:

$$H_2O(s) \rightarrow H_2O(l) \qquad 0\,°C, 1 \text{ atm}$$

Here (s) and (l) are used to designate the solid and liquid states, respectively.

$$\Delta H = q\text{(fusion)} = (80 \text{ cal/g})(18.0 \text{ g/mol}) = 1.44 \text{ kcal/mol} = 6.02 \text{ kJ/mol}$$

Similarly, ΔH for a vaporization is equal to the *latent heat of vaporization* carried out at constant temperature and pressure. The latent heat of vaporization of water at $100\,°C$ and 1 atm is 540 cal/g or 9.72 kcal/mol $= 40.7$ kJ/mol. ΔH for a sublimation process is equal to the latent heat for conversion of a substance from the solid to the gaseous state.

Chemical reaction

A ΔH-value assigned to a chemical equation defines the enthalpy change occurring when the number of moles of each reactant consumed is equal to its coefficient in the balanced equation.

$$C(graphite) + O_2(g) \rightarrow CO_2(g) \qquad \Delta H = -393.51 \text{ kJ}$$

$$H_2(g) + I_2(s) \rightarrow 2HI(g) \qquad \Delta H = +52.72 \text{ kJ}$$

In the above reactions, 393.51 kJ is *liberated* when 1 mol of gaseous CO_2 is formed from graphite and oxygen, and 52.72 kJ is *absorbed* in the formation of 2 mol of gaseous HI from hydrogen and solid iodine.

The *standard enthalpy of formation* of a substance, ΔH_f°, is the enthalpy change accompanying the formation of *one mole* of the substance from its elements, each being in its normal stable state at the designated temperature and 1 bar. For example, at 25 °C the *standard state* of H_2, O_2, Cl_2, or N_2 is the pure gas at 1 bar pressure; the *standard state* of bromine or mercury is the liquid; the *standard state* of iron, sodium, or iodine is the solid; the *standard state* of carbon is graphite, which is more stable than diamond. With reference to the examples in the preceding paragraph, the standard enthalpy of formation of $CO_2(g)$ is -393.51 kJ/mol and of $HI(g)$ is $+26.36$ kJ/mol (half of the $+52.72$ kJ enthalpy increase on forming two moles of HI). The superscript $^\circ$ is used in ΔH_f°, and will be used elsewhere, to indicate the standard condition of pressure, 1 bar.

Table 7-1 lists some values of ΔH_f° at 25 °C. Note the absence of entries for H_2 (g), I_2 (s), C(*graphite*), and all other elements in their standard states. By definition their ΔH_f°-values are zero.

Table 7-1.   Standard Enthalpies of Formation at 25 °C

Substance	$\Delta H_f^\circ/\text{kcal} \cdot \text{mol}^{-1}$	$\Delta H_f^\circ/\text{kJ} \cdot \text{mol}^{-1}$	Substance	$\Delta H_f^\circ/\text{kcal} \cdot \text{mol}^{-1}$	$\Delta H_f^\circ/\text{kJ} \cdot \text{mol}^{-1}$
Al_2O_3 (s)	-400.50	-1675.7	HNO_3 (l)	-41.61	-174.10
B_2O_3 (s)	-304.20	-1272.8	H_2O (g)	-57.80	-241.81
Br (g)	$+26.73$	$+111.84$	H_2O (l)	-68.32	-285.83
C (*diamond*)	$+0.45$	$+1.88$	H_2O_2 (l)	-44.88	-187.8
CF_4 (g)	-220.9	-924.7	H_2S (g)	-4.93	-20.6
CH_3OH (g)	-47.96	-200.7	H_2S $(aq, undiss)$	-9.5	-39.7
C_9H_{20} (l)	-65.85	-275.5	I_2 (g)	$+14.92$	$+62.4$
$(CH_3)_2N_2H_2$ (l)	$+13.3$	$+55.6$	KCl (s)	-104.42	-436.9
$C(NO_2)_4$ (l)	$+8.8$	$+36.8$	$KClO_3$ (s)	-95.06	-397.7
CO (g)	-26.42	-110.53	$KClO_4$ (s)	-103.6	-433.5
CO_2 (g)	-94.05	-393.51	$LiAlH_4$ (s)	-24.21	-101.3
CaC_2 (s)	-14.2	-59.4	$LiBH_4$ (s)	-44.6	-186.6
CaO (s)	-151.6	-634.3	Li_2O (s)	-143.1	-598.7
$Ca(OH)_2$ (s)	-235.80	-986.6	N_2H_4 (l)	$+12.10$	$+50.63$
$CaCO_3$ (s)	-288.5	-1206.9	NO (g)	$+21.45$	$+89.75$
ClF_3 (l)	-45.3	-190	NO_2 (g)	$+8.60$	$+35.98$
Cl^- (aq)	-39.95	-167.16	N_2O_4 (g)	$+2.19$	$+9.16$
Cu^{2+} (aq)	$+15.49$	$+64.8$	N_2O_4 (l)	-4.66	-19.50
$CuSO_4$ (s)	-184.03	-769.98	O_3 (g)	$+34.1$	$+142.7$
Fe^{2+} (aq)	-21.3	-89.1	OH^- (aq)	-54.97	-229.99
Fe_2O_3 (s)	-197.0	-824.2	PCl_3 (l)	-76.4	-319.7
FeS (s)	-23.9	-100.0	PCl_3 (g)	-68.6	-287.0
H^+ (aq)	0.0	0.0	PCl_5 (g)	-89.6	-374.9
HBr (g)	-8.70	-36.38	$POCl_3$ (g)	-133.48	-558.48
HCl (g)	-22.06	-92.31	SO_2 (g)	-70.94	-296.81
HF (g)	-64.8	-271.1	SO_3 (g)	-94.58	-395.72
HI (g)	$+6.30$	$+26.36$	Zn^{2+} (aq)	-36.78	-153.89

The standard state had been 1 atm for many decades and was just recently changed to 1 bar. Consequently most available tabulations are based on 1 atm. Fortunately, since ΔH is not strongly dependent on pressure, and since 1 bar is very close to 1 atm, the numerical values show little or no differences and they can be overlooked so far as this chapter is concerned.

RULES OF THERMOCHEMISTRY

The internal energy and the enthalpy of a system depend only on the *state* of the system, as specified by the external parameters such as pressure and temperature. (For example, recall from Chapter 6 that the kinetic energy contribution to E for an ideal gas is uniquely determined by the temperature.) When the system goes from an initial to a final state, ΔE and ΔH depend only on those two states, and are independent of the path taken between them. This path-independence implies two important rules of thermochemistry.

1. ΔE and ΔH for reverse processes are exactly the negatives of the values for the corresponding forward processes.

EXAMPLE 1 ΔH for the fusion of a mole of ice is 1 440 cal/mol, since it is found experimentally that 1 440 cal is absorbed in the melting of 1 mol at constant temperature, 273 K, and at constant pressure, 1 atm. Then ΔH for freezing is $-$ 1 440 cal/mol; this amount of heat must be transferred from the water to the surroundings in order to freeze it.

2. If a process can be imagined to occur in successive stages, ΔH for the overall process is equal to the sum of the enthalpy changes for the individual steps. (This rule is called *Hess' Law of Constant Heat Summation.*)

EXAMPLE 2 It is impossible to measure accurately the heat liberated when C burns to CO, because the combustion cannot be stopped exactly at the CO stage. We can, however, measure accurately the heat liberated when C burns to CO_2 (393.5 kJ per mole), and also the heat liberated when CO burns to CO_2 (283.0 kJ per mole of CO). The enthalpy change for the burning of C to CO is determined by treating algebraically the two experimentally determined thermochemical equations. If two chemical equations are added or subtracted, their corresponding enthalpy changes are, by Hess' law, to be added or subtracted. Thus

$$2C(graphite) + 2O_2(g) \rightarrow 2CO_2(g) \qquad \Delta H° = (2 \text{ mole})(-393.5 \text{ kJ/mol}) = -787.0 \text{ kJ}$$

$$2CO(g) + O_2(g) \rightarrow 2CO_2(g) \qquad \Delta H° = (2 \text{ mol})(-283.0 \text{ kJ/mol}) = -566.0 \text{ kJ}$$

Subtracting the second equations (both chemical and enthalpic) from the first, and transposing $-2CO$ to the right, we have

$$2C(graphite) + O_2(g) \rightarrow 2CO(g) \qquad \Delta H° = -221.0 \text{ kJ}$$

Thus, $\Delta H_f°$, the standard enthalpy of formation of CO(g), is evaluated as

$$\frac{-221.0 \text{ kJ}}{2 \text{ mol CO}} = -110.5 \text{ kJ/mol}$$

EXAMPLE 3 The heat of sublimation of a substance is the sum of the heats of fusion and vaporization of that substance *at the same temperature.*

EXAMPLE 4 It follows from Hess' law that the enthalpy change of any reaction is equal to the sum of the enthalpies of formation of all products minus the sum of the enthalpies of formation of all reactants, each ΔH_f being multiplied by the number of moles of the substance in the balanced chemical equation. Thus, for the reaction

$$PCl_5(g) + H_2O(g) \rightarrow POCl_3(g) + 2HCl(g)$$

we have

$$\Delta H°$$

(1)	$P(white) + \frac{3}{2}Cl_2(g) + \frac{1}{2}O_2(g)$	$\rightarrow POCl_3(g)$	-558.5 kJ
(2)	$H_2(g) + Cl_2(g)$	$\rightarrow 2HCl(g)$	-184.6 kJ
(3)	$PCl_5(g)$	$\rightarrow P(white) + \frac{5}{2}Cl_2(g)$	$+374.9$ kJ
(4)	$H_2O(g)$	$\rightarrow H_2(g) + \frac{1}{2}O_2(g)$	$+241.8$ kJ

SUM $PCl_5(g) + H_2O(g) \rightarrow POCl_3(g) + 2HCl(g)$ -126.4 kJ

The $\Delta H°$ listing for (1) is $\Delta H°_f$ for $POCl_3$; for (2), twice $\Delta H°_f$ of HCl; for (3) and (4), the negative of $\Delta H°_f$ for PCl_5 and H_2O, respectively. The $\Delta H°_f$-values were all taken from Table 7-1. Note the use of fractional coefficients in the formation equations of some of the compounds; this is not uncommon in thermochemistry.

This example shows how a relatively small number of tabulated enthalpies of formation suffice for the calculation of the enthalpy change for many chemical reactions.

Solved Problems

HEAT CAPACITY AND CALORIMETRY

7.1. (a) How many calories are required to heat 100 g of copper ($c = 0.093$ cal/g $\cdot$ K) from 10 °C to 100 °C? (b) The same quantity of heat as in (a) is added to 100 g of aluminum ($c = 0.217$ cal/g $\cdot$ K) at 10 °C. Which gets hotter, the copper or aluminum?

(a) $\Delta H = C \, \Delta T = (0.093$ cal/g $\cdot$ K$)(100$ g$)[(100 - 10)$K$] = 840$ cal

(b) Since the specific heat capacity of copper is less than that of aluminum, less heat is required to raise the temperature of a mass of copper by 1 K than is required for an equal mass of aluminum. Hence the copper is hotter.

7.2. One kilogram of anthracite coal when burned evolves about 30 500 kJ. What amount of coal is required to heat 4.0 kg of water from room temperature (20 °C) to the boiling point (at 1 atm pressure), assuming that all the heat is available?

$$\Delta H(\text{heating the water}) = C \, \Delta T = (4.184 \text{ kJ/kg} \cdot \text{K})(4.0 \text{ kg})[(100 - 20)\text{K}] = 1\,339 \text{ kJ}$$

$$\text{Amount of coal required} = \frac{1\,339 \text{ kJ}}{30\,500 \text{ kJ/kg}} = 0.044 \text{ kg} = 44 \text{ g}$$

7.3. A steam boiler is made of steel and weighs 900 kg. The boiler contains 400 kg of water. Assuming that 70 percent of the heat is delivered to boiler and water, how much heat is required to raise the temperature of the whole from 10 °C to 100 °C? Specific heat capacity of steel is 0.11 kcal/kg $\cdot$ K.

$$\Delta H(\text{for heating}) = C(\text{total})\Delta T = [C(\text{boiler}) + C(\text{water})]\Delta T$$

$$= [(0.11)(900) \text{ kcal/K} + (1.00)(400)\text{kcal/K}](90 \text{ K}) = 44\,900 \text{ kcal}$$

$$\text{Input required} = \frac{44\,900 \text{ kcal}}{0.70} = 64\,000 \text{ kcal}$$

7.4. Exactly 3 g of carbon was burned to CO_2 in a copper calorimeter. The mass of the calorimeter is 1 500 g, and the mass of water in the calorimeter is 2 000 g. The initial temperature was 20.0 °C and the final temperature 31.3 °C. Calculate the heat of combustion of carbon in joules per gram. Specific heat capacity of copper is 0.389 J/g · K.

$$q(\text{calorimeter}) = C(\text{total})\Delta T = [C(Cu) + C(H_2O)]\Delta T$$

$$= [(0.389 \text{ J/g} \cdot \text{K})(1\,500 \text{ g}) + (4.184 \text{ J/g} \cdot \text{K})(2\,000 \text{ g})][(31.3 - 20.0)\text{K}]$$

$$= 1.012 \times 10^5 \text{ J}$$

$$\text{Heat value of carbon} = \frac{1.012 \times 10^5 \text{ J}}{3 \text{ g}} = 3.37 \times 10^4 \text{ J/g}$$

7.5. A 1.250-g sample of benzoic acid, $C_7H_6O_2$, was placed in a combustion bomb. The bomb was filled with an excess of oxygen at high pressure, sealed, and immersed in a pail of water which served as a calorimeter. The heat capacity of the entire apparatus was found to be 2 422 cal/K, including the bomb, the pail, a thermometer, and the water. The oxidation of the benzoic acid was triggered by passing an electric spark through the sample. After complete combustion of the sample, the thermometer immersed in the water registered a temperature 3.256 K greater than before the combustion. What is ΔE per mole of benzoic acid combusted in a bomb-type calorimeter? Assume that no correction must be made for the sparking process. (This is a constant-volume process.)

$$q(\text{acid}) = -q(\text{calorimeter}) = -(2\,422 \text{ cal/K})(3.256 \text{ K}) = -7.89 \text{ kcal}$$

$$\Delta E(\text{combustion}) = \frac{q(\text{acid})}{\text{number of moles acid}} = \frac{-7.89 \text{ kcal}}{(1.250 \text{ g})/(122.1 \text{ g/mol})} = -771 \text{ kcal/mol} = -3\,226 \text{ kJ/mol}$$

7.6. A 25.0-g sample of an alloy was heated to 100.0 °C and dropped into a beaker containing 90 g of water at 25.32 °C. The temperature of the water rose to a final value of 27.18 °C. Neglecting heat losses to the room and the heat capacity of the beaker itself, what is the specific heat of the alloy?

$$\text{Heat loss by alloy} = \text{heat absorbed by water}$$

$$(25.0 \text{ g})(c)[(100.0 - 27.2)\text{K}] = (90 \text{ g})(1 \text{ cal/g} \cdot \text{K})[(27.18 - 25.32)\text{K}]$$

from which $c = 0.092 \text{ cal/g} \cdot \text{K}$.

7.7 Determine the resulting temperature, t, when 150 g of ice at 0 °C is mixed with 300 g of water at 50 °C.

1. Consider the heat absorbed by the ice and by the water from it.

$$\Delta H(\text{fusion}) = (80 \text{ cal/g})(150 \text{ g}) = 1.20 \times 10^4 \text{ cal}$$

$$\Delta H(\text{heating 150 g water from 0 °C to final temperature}) = C\,\Delta T$$

$$= (1.00 \text{ cal/g} \cdot \text{K})(150 \text{ g})[(t - 0)\text{K}]$$

2. Now consider the change in enthalpy of the hot water.

$$\Delta H = C\,\Delta T = (1.00 \text{ cal/g} \cdot \text{K})(300 \text{ g})[(t - 50)\text{K}]$$

where, presumably, $t < 50$, consistent with a loss of heat from the hot water.

3. The sum of the ΔH's must equal zero since heat is assumed not to leak into or out of the total system treated in (1) and (2).

$$1.20 \times 10^4 + 150t + 300(t - 50) = 0$$

from which $t = 6.7 \text{ °C}$.

7.8. How much heat is given up when 20 g of steam at 100 °C is condensed and cooled to 20 °C?

The heat of vaporization of water at 100 °C is

$$(40.7 \text{ kJ/mol})\left(\frac{1 \text{ mol}}{18.02 \text{ g}}\right) = 2.26 \text{ kJ/g}$$

$$\Delta H(\text{condensation}) = -(\text{mass}) \times (\text{heat of vaporization}) = -(20 \text{ g})(2.26 \text{ kJ/g}) = -45.2 \text{ kJ}$$

$$\Delta H(\text{cooling}) = C\,\Delta T = 4.184 \text{ J/g} \cdot \text{K})(20 \text{ g})[(20 - 100)\text{K}] = -6.7 \text{ kJ}$$

$$\Delta H(\text{total}) = -45.2 - 6.7 = -51.9 \text{ kJ}$$

The amount of heat given up is 51.9 kJ.

7.9. How much heat is required to convert 40 g of ice ($c = 0.5 \text{ cal/g} \cdot \text{K}$) at -10 °C to steam ($c = 0.5 \text{ cal/g} \cdot \text{K}$) at 120 °C?

$$\Delta H(\text{heating ice from} -10\,°\text{C to ice at } 0\,°\text{C}) = C\,\Delta T = (0.5 \text{ cal/g} \cdot \text{K})(40 \text{ g})(10 \text{ K}) = 0.2 \text{ kcal}$$

$$\Delta H(\text{melting ice at } 0\,°\text{C}) = (\text{mass}) \times (\text{heat of fusion}) = (40 \text{ g})(80 \text{ cal/g}) = 3.2 \text{ kcal}$$

$$\Delta H(\text{heating water from } 0\,°\text{C to } 100\,°\text{C}) = C\,\Delta T = (1.00 \text{ cal/g} \cdot \text{K})(40 \text{ g})(100 \text{ K}) = 4.0 \text{ kcal}$$

$$\Delta H(\text{vaporization to steam at } 100\,°\text{C}) = (\text{mass}) \times (\text{heat of vaporization}) = (40 \text{ g})(540 \text{ cal/g}) = 21.6 \text{ kcal}$$

$$\Delta H(\text{heating steam from } 100\,°\text{C to } 120\,°\text{C}) = C\,\Delta T = (0.5 \text{ cal/g} \cdot \text{K})(40 \text{ g})(20 \text{ K}) = 0.4 \text{ kcal}$$

$$\Delta H(\text{total}) = (0.2 + 3.2 + 4.0 + 21.6 + 0.4)\text{kcal} = 29.4 \text{ kcal}$$

7.10. What is the heat of vaporization of water per gram at 25 °C and 1 atm?

We can write the thermochemical equation for the process:

$$\text{H}_2\text{O}(l) \rightarrow \text{H}_2\text{O}(g)$$

$\Delta H°$ can be evaluated by subtracting $\Delta H_f°$ of reactants from $\Delta H_f°$ of products, as tabulated in Table 7-1.

$$\Delta H° = \Delta H_f°(\text{products}) - \Delta H_f°(\text{reactants}) = -241.81 - (-285.83) = 44.02 \text{ kJ}$$

The enthalpy of vaporization per gram is

$$\frac{44.02 \text{ kJ/mol}}{18.02 \text{ g/mol}} = 2.44 \text{ kJ}$$

Note that the heat of vaporization at 25 °C is greater than the value (2.26 kJ/g) at 100 °C.

THERMOCHEMICAL EQUATIONS

7.11. The thermochemical equation for the combustion of ethylene gas, C_2H_4, is

$$\text{C}_2\text{H}_4(g) + 3\text{O}_2(g) \rightarrow 2\text{CO}_2(g) + 2\text{H}_2\text{O}(l) \qquad \Delta H° = -1\,410 \text{ kJ}$$

Assuming 70 percent efficiency, how many kilograms of water at 20 °C can be converted into steam at 100 °C by burning 1 m³ of C_2H_4 gas measured at S.T.P.?

$$n(\text{C}_2\text{H}_4) = \frac{(1 \text{ m}^3)(1\,000 \text{ L/m}^3)}{22.4 \text{ L/mol}} = 44.6 \text{ mol}$$

$$\Delta H(1 \text{ m}^3) = n(\text{C}_2\text{H}_4) \times \Delta H(1 \text{ mol}) = (44.6 \text{ mol})(-1\,410 \text{ kJ/mol}) = -6.29 \times 10^4 \text{ kJ}$$

The useful heat is then $(0.70)(6.29 \times 10^4 \text{ kJ}) = 4.40 \times 10^4 \text{ kJ}$.

For the overall process, consider two stages:

$$H_2O(l, 20\,°C) \rightarrow H_2O(l, 100\,°C) \qquad \Delta H = (4.184\ \text{kJ/kg} \cdot \text{K})(80\ \text{K})$$
$$= 335\ \text{kJ/kg}$$

$$H_2O(l, 100\,°C) \rightarrow H_2O(g, 100\,°C) \qquad \underline{\Delta H = 2\,260\ \text{kJ/kg}}$$

$$\Delta H(\text{total}) = 2\,595\ \text{kJ/kg}$$

The mass of water converted is then equal to the amount of heat available divided by the heat requirement per kilogram

$$m(H_2O) = \frac{4.40 \times 10^4\ \text{kJ}}{2\,595\ \text{kJ/kg}} = 17.0\ \text{kg}$$

7.12. Calculate $\Delta H°$ for reduction of ferric oxide by aluminum (thermite reaction) at 25 °C.

We must start with a balanced equation for the reaction. We may then write in parentheses under each formula the enthalpy of formation, taken from Table 7-1, and multiply each enthalpy by the corresponding number of moles in the balanced equation. Remember that $\Delta H_f°$ for any element in its standard state is zero, by definition.

$$2\text{Al} + \text{Fe}_2\text{O}_3 \rightarrow 2\text{Fe} + \text{Al}_2\text{O}_3$$
$$(n\ \Delta H_f°)/\text{kJ:}\quad 2(0)\quad 1(-824.2)\quad 2(0)\quad 1(-1\,675.7)$$

Then $\Delta H°$ of the reaction is given by

$$\Delta H° = (\text{sum of } n\ \Delta H_f° \text{ of products}) - (\text{sum of } n\ \Delta H_f° \text{ of reactants})$$
$$= -1\,675.7 - (-824.2) = -851.5\ \text{kJ}$$

This is $\Delta H°$ for the reduction of 1 mol Fe_2O_3.

7.13. $\Delta H_f°$ for N (g) (not the standard state of nitrogen) has been determined as 472.7 kJ/mol, and for O (g) as 249.2 kJ/mol. What is $\Delta H°$ for the hypothetical upper-atmosphere reaction

$$N\ (g) + O\ (g) \rightarrow NO\ (g)$$

both in kJ and in kcal?

$$N\ (g) + O\ (g) \rightarrow NO\ (g)$$
$$(n\ \Delta H_f°)/\text{kJ:}\quad 472.7\quad 249.2\quad 89.7$$

where $\Delta H_f°$ for NO (g) has been taken from Table 7-1. Then

$$\Delta H° = 89.7 - (472.7 + 249.2) = -632.2\ \text{kJ} = \frac{-632.2\ \text{kJ}}{4.184\ \text{kJ/kcal}} = -151.1\ \text{kcal}$$

7.14. Calculate the enthalpy of decomposition of $CaCO_3$ into CaO and CO_2.

$$CaCO_3\ (s) \rightarrow CaO\ (s) + CO_2\ (g)$$
$$(n\ \Delta H_f°)/\text{kcal:}\quad -288.4\quad -151.6\quad -94.0$$
$$\Delta H° = (-151.6 - 94.0) - (-288.4) = 42.8\ \text{kcal} = 179.1\ \text{kJ}$$

This is the enthalpy change for the decomposition of 1 mol $CaCO_3$. A positive value signifies an endothermic reaction.

7.15. (a) Calculate the enthalpy of neutralization of a strong acid by a strong base in water (b) The heat liberated on neutralization of HCN (weak acid) by NaOH is 12.1 kJ/mol. How many kilojoules are absorbed in ionizing 1 mol of HCN in water?

(a) The basic equation for neutralization is as follows:

$$H^+ (aq) + OH^- (aq) \rightarrow H_2O (l)$$

$(n \, \Delta H_f^\circ)/kJ:$ 0 -230.0 -285.8

Thus $\Delta H^\circ = -285.8 - (-230.8) = -55.8$ kJ.

(b) The neutralization of HCN (aq) by NaOH (aq) may be thought of as the sum of two processes, ionization of HCN (aq) and neutralization of H^+ (aq) with OH^- (aq). [Since NaOH is a strong base, NaOH (aq) implies complete ionization, and a separate thermochemical equation for the ionization need not be written.] We may therefore construct the following thermochemical cycle:

$$HCN (aq) \rightarrow H^+ (aq) + CN^- (aq) \qquad \Delta H^\circ(\text{ionization}) \quad = x$$
$$H^+ (aq) + OH^- (aq) \rightarrow H_2O (l) \qquad \Delta H^\circ \qquad\qquad = -55.8 \text{ kJ}$$

SUM $HCN (aq) + OH^- (aq) \rightarrow H_2O (l) + CN^- (aq)$ $\Delta H^\circ(\text{experimental}) = -12.1$ kJ

ΔH°(experimental) is negative because heat is liberated on neutralization.

From the principle of additivity,

$$x + (-55.8) = -12.1 \quad \text{or} \quad x = 43.7 \text{ kJ}$$

The ionization process is endothermic to the extent of 43.7 kJ/mol.

7.16. The heat evolved on combustion of acetylene gas, C_2H_2, at 25 °C is 1 299.1 kJ/mol. Determine the enthalpy of formation of acetylene gas.

The complete combustion of an organic compound involves the formation of CO_2 and H_2O.

$$C_2H_2 (g) + \tfrac{5}{2}O_2 (g) \rightarrow 2CO_2 (g) + H_2O (l) \qquad \Delta H^\circ = -1\,299.1 \text{ kJ}$$

$(n \, \Delta H_f^\circ)/kJ:$ x 0 $2(-393.5)$ -285.8

Thus $-1\,299.1 = [2(-393.5) + (-285.8)] - x$

Solving, $x = \Delta H_f^\circ$ for C_2H_2 (g) = 226.3 kJ/mol.

7.17. How much heat will be required to make 1 kg of CaC_2 according to the reaction given below?

$$CaO (s) + 3C (s) \rightarrow CaC_2 (s) + CO (g)$$

$(n \, \Delta H_f^\circ)/kJ$ -634.3 0 -59.4 -110.5

$$\Delta H^\circ = -(59.4 + 110.5) - (-634.3) = +464.4 \text{ kJ}$$

This is the heat required to make 1 mol CaC_2; 1 kg CaC_2 will require

$$\left(\frac{1\,000 \text{ g } CaC_2}{64.10 \text{ g } CaC_2/\text{mol}} \right)(464.4 \text{ kJ/mol}) = 7\,245 \text{ kJ}$$

7.18 How many kilojoules of heat will be evolved in making 22.4 L at S.T.P. (1 mol) of H_2S from FeS and dilute hydrochloric acid?

$$FeS (s) + 2H^+ (aq) \rightarrow Fe^{2+} (aq) + H_2S (g)$$

$(n \, \Delta H_f^\circ)/kJ:$ -95.4 0 -87.9 -20.6

Since HCl and $FeCl_2$ are strong electrolytes, chloride can be omitted from the balanced equation.

$$\Delta H^\circ = -(87.9 + 20.6) + 95.4 = -13.1 \text{ kJ/mol } H_2S$$

Supplementary Problems

HEAT CAPACITY AND CALORIMETRY

7.19. How many calories are required to heat each of the following from 15 °C to 65 °C: (*a*) 1.0 g water, (*b*) 5.0 g Pyrex glass, (*c*) 20 g platinum? Specific heat capacity of Pyrex glass, 0.20 cal/g·K; of platinum, 0.032 cal/g·K.

Ans.   (*a*) 50 cal;   (*b*) 50 cal;   (*c*) 32 cal

7.20. The combustion of 5.00 g of coke raised the temperature of 1 kg of water from 10 °C to 47 °C. Calculate the heat value of coke in kcal/g.

Ans.   7.4 kcal/g

7.21. Assuming that 50 percent of the heat is useful, how many kilograms of water at 15 °C can be heated to 95 °C by burning 200 L of methane, CH_4, measured at S.T.P.? The heat of combustion of methane is 891 kJ/mol.

Ans.   11.9 kg

7.22. The heat of combustion of ethane gas, C_2H_6 is 1 561 kJ/mol. Assuming that 60 percent of the heat is useful, how many liters of ethane measured at S.T.P. must be burned to supply enough heat to convert 50 kg of water at 10 °C to steam at 100 °C?

Ans   3 150 L

7.23. A 45.0-g sample of an alloy was heated to 90.0 °C and then dropped into a beaker containing 82.0 g of water at 23.50 °C. The temperature of the water rose to a final 26.25 °C. What is the specific heat capacity of the alloy?

Ans.   0.329 J/g·K = 0.079 cal/g·K

7.24. If the specific heat capacity of a substance is *h* cal/g·K, what is its specific heat in Btu/lb·°F?

Ans.   *h* Btu/lb·°F

7.25. Determine the resulting temperature when 1 kg of ice at 0 °C is mixed with 9 kg of water at 50 °C. Heat of fusion of ice is 80 cal/g.

Ans.   37 °C

7.26. How much heat is required to change 10 g of ice at 0 °C to steam at 100 °C? Heat of vaporization of water at 100 °C is 540 cal/g.

Ans.   7.2 kcal

7.27. What is the heat of sublimation of solid iodine at 25 °C?

Ans.   14.92 kcal/mol I_2 = 62.4 kJ/mol I_2

7.28. Is the process of dissolving H_2S gas in water endothermic or exothermic? To what extent?

Ans.   exothermic, 4.6 kcal/mol = 19.1 kJ/mol

7.29. How much heat is released on dissolving 1 mol of HCl (*g*) in a large amount of water? (*Hint*: HCl is completely ionized in dilute solution.)

Ans.   17.9 kcal = 74.8 kJ

7.30. In an ice calorimeter, a chemical reaction is allowed to occur in thermal contact with an ice-water mixture at 0 °C. Any heat liberated by the reaction is used to melt some ice; the volume change of the ice-water mixture indicates the amount of melting. When solutions containing 1.00 millimole each of $AgNO_3$ and NaCl were mixed in such a calorimeter, both solutions having been precooled to 0 °C, 0.20 g of ice melted. Assuming complete reaction in this experiment, what is ΔH for the reaction $Ag^+ + Cl^- \rightarrow AgCl$?

Ans. -67 kJ $= -16$ kcal

THERMOCHEMICAL EQUATIONS

7.31. The standard enthalpy of formation of H (*g*) has been determined to be 218.0 kJ/mol. Calculate $\Delta H°$ in kilojoules for the following two reactions:

$$(a)\quad H\ (g) + Br\ (g) \rightarrow HBr\ (g) \qquad (b)\quad H\ (g) + Br_2\ (l) \rightarrow HBr\ (g) + Br\ (g)$$

Ans.   (*a*) -366.2 kJ;   (*b*) -142.6 kJ

7.32. Given

$$N_2\ (g) + 3H_2\ (g) \rightarrow 2NH_3\ (g) \qquad \Delta H° = -22.0 \text{ kcal}$$

What is the standard enthalpy of formation of NH_3 gas?

Ans. -11.0 kcal/mol

7.33. Determine $\Delta H°$ of decomposition of 1 mol of solid $KClO_3$ into solid KCl and gaseous oxygen.

Ans. -10.9 kcal $= -45.5$ kJ

7.34. The heat released on neutralization of CsOH with all strong acids is 13.4 kcal/mol. The heat released on neutralization of CsOH with HF (weak acid) is 16.4 kcal/mol. Calculate $\Delta H°$ of ionization of HF in water.

Ans. -3.0 kcal/mol

7.35. Calculate $\Delta H°$ for the reaction $CuSO_4\ (aq) + Zn\ (s) \rightarrow ZnSO_4\ (aq) + Cu\ (s)$.

Ans. -52.2 kcal $= -218.3$ kJ

7.36. Find the heat evolved in slaking 1 kg of quicklime (CaO) according to the reaction

$$CaO\ (s) + H_2O\ (l) \rightarrow Ca\ (OH)_2\ (s)$$

Ans.   282 kcal $= 1\,180$ kJ

7.37. The heat liberated on complete combustion of 1 mol of CH_4 gas to CO_2 (*g*) and H_2O (*l*) is 890 kJ. Determine the enthalpy of formation of 1 mol of CH_4 gas.

Ans. -75 kJ/mol

7.38. The heat evolved on combustion of 1 g of starch, $(C_6H_{10}O_5)_x$, into CO_2 (*g*) and H_2O (*l*) is 17.48 kJ. Compute the standard enthalpy of formation of 1 g of starch.

Ans. -5.88 kJ

7.39. The amount of heat evolved in dissolving $CuSO_4$ is 74.9 kJ/mol. What is $\Delta H_f°$ for SO_4^{2-} (*aq*)?

Ans. -909.7 kJ/mol

7.40. The heat of solution of $CuSO_4 \cdot 5H_2O$ in a large amount of water is 5.4 kJ/mol (endothermic). Calculate the heat of reaction for

$$CuSO_4 \, (s) + 5H_2O \, (l) \rightarrow CuSO_4 \cdot 5H_2O \, (s)$$

Use data from Problem 7.39.

Ans.   80.3 kJ (exothermic) = 19.2 kcal

7.41. The heat evolved on combustion into $CO_2 \, (g)$ and $H_2O \, (l)$ of 1 mol C_2H_6 is 1 559.8 kJ, and of 1 mol C_2H_4 is 1 410.8 kJ. Calculate ΔH of the following reaction: $C_2H_4 + H_2 \, (g) \rightarrow C_2H_6$.

Ans. -136.8 kJ

7.42. The solution of $CaCl_2 \cdot 6H_2O$ in a large volume of water is endothermic to the extent of 14.6 kJ/mol. For the reaction

$$CaCl_2 \, (s) + 6H_2O \, (l) \rightarrow CaCl_2 \cdot 6H_2O \, (s)$$

$\Delta H = -97.0$ kJ. What is the heat of solution of $CaCl_2$ (anhydrous) in a large volume of water?

Ans.   82.4 kJ/mol (exothermic)

7.43. The thermochemical equation for the dissociation of hydrogen gas into atoms may be written:

$$H_2 \rightarrow 2H \qquad \Delta H = 436 \text{ kJ}$$

What is the ratio of the energy yield on combustion of hydrogen atoms to steam to the yield on combustion of an equal mass of hydrogen molecules to steam?

Ans.   2.80

7.44. The commerical production of water gas utilizes the reaction $C + H_2O \, (g) \rightarrow H_2 + CO$. The required heat for this endothermic reaction may be supplied by adding a limited amount of air and burning some carbon to carbon dioxide. How many grams of carbon must be burned to CO_2 to provide enough heat for the water gas conversion of 100 g of carbon? Neglect all heat losses to the environment.

Ans.   33.4 g

7.45. The reversible reaction

$$Na_2SO_4 \cdot 10H_2O \rightarrow Na_2SO_4 + 10H_2O \qquad \Delta H = +18.8 \text{ kcal}$$

goes completely to the right at temperatures above 32.4 °C, and remains completely on the left below this temperature. This system has been used in some solar houses for heating at night with the energy absorbed from the sun's radiation during the day. How many cubic feet of fuel gas could be saved per night by the reversal of the dehydration of a fixed charge of 100 lb $Na_2SO_4 \cdot 10H_2O$? Assume that the fuel value of the gas is 2 000 Btu/ft^3.

Ans.   5.3 ft^3

7.46. An important criterion for the desirability of fuel reactions for rockets is the fuel value in kilojoules per gram of reactant or per cubic centimeter of reactant. Compute both of these quantities for each of the following reactions:

     (a) $N_2H_4 \, (l) + 2H_2O_2 \, (l) \rightarrow N_2 \, (g) + 4H_2O \, (g)$
     (b) $2LiBH_4 \, (s) + KClO_4 \, (s) \rightarrow Li_2O \, (s) + B_2O_3 \, (s) + KCl \, (s) + 4H_2 \, (g)$
     (c) $6LiAlH_4 \, (s) + 2C(NO_2)_4 \, (l) \rightarrow 3Al_2O_3 \, (s) + 3Li_2O \, (s) + 2CO_2 \, (g) + 4N_2 \, (g) + 12H_2 \, (g)$
     (d) $4HNO_3 \, (l) + 5N_2H_4 \, (l) \rightarrow 7N_2 \, (g) + 12H_2O \, (g)$
     (e) $7N_2O_4 \, (l) + C_9H_{20} \, (l) \rightarrow 9CO_2 \, (g) + 10H_2O \, (g) + 7N_2 \, (g)$
     (f) $4ClF_3 \, (l) + (CH_3)_2N_2H_2 \, (l) \rightarrow 2CF_4 \, (g) + N_2 \, (g) + 4HCl \, (g) + 4HF \, (g)$

Use the following density values: N_2H_4 (l), 1.01 g/cm^3; H_2O_2 (l), 1.46 g/cm^3; $LiBH_4$ (s), 0.66 g/cm^3; $KClO_4$ (s), 2.52 g/cm^3; $LiAlH_4$ (s), 0.92 g/cm^3; $C(NO_2)_4$ (l), 1.65 g/cm^3; HNO_3 (l), 1.50 g/cm^3; N_2O_4 (l), 1.45 g/cm^3; C_9H_{20} (l), 0.72 g/cm^3; ClF_3 (l), 1.77 g/cm^3; $(CH_3)_2N_2H_2$ (l), 0.78 g/cm^3. In computing the volume of each reaction mixture, assume that the reactants are present in stoichiometric proportions.

Ans.   (a) 6.4 kJ/g, 8.2 kJ/cm^3;   (b) 8.3 kJ/g, 12.4 kJ/cm^3;   (c) 11.4 kJ/g, 14.6 kJ/cm^3;   (d) 6.0 kJ/g, 7.5 kJ/cm^3;   (e) 7.2 kJ/g, 8.9 kJ/cm^3;   (f) 6.0 kJ/g, 9.1 kJ/cm^3

7.47.   An early-model Concorde supersonic airplane consumed 4 700 gallons of aviation fuel per hour at cruising speed. The density of the fuel was 6.65 pounds per gallon and ΔH of combustion was $-10\,500$ kcal/kg. Express the power consumption in megawatts (1 MW = 10^6 W = 10^6 J/s) during supersonic cruising.

Ans.   173 MW

7.48.   Two solutions, initially at 25.08 °C, were mixed in an insulated bottle. One consisted of 400 mL of a weak monoprotic acid solution of concentration 0.200 mol/L. The other consisted of 100 mL of a solution having 0.800 mol NaOH/L. After mixing, the temperature rose to 26.25 °C. How much heat is evolved in the neutralization of one mole of the acid? Assume that the densities of all solutions are 1.00 g/cm^3 and that their specific heat capacities are all 4.2 J/g·K. (These assumptions are in error by several percent, but the subsequent errrors in the final result partly cancel each other.)

Ans.   31 kJ/mol

Chapter 8

Atomic Structure and the Periodic Law

ABSORPTION AND EMISSION OF LIGHT

Historically important in the development of modern atomic theory was the recognition that although polyatomic molecules show more or less broad *bands* of absorption and emission in the visible and ultraviolet regions of the spectrum, the characteristic light absorption or emission by individual atoms occurs at fairly narrow *lines* of the spectrum, which correspond to sharply defined wavelengths. The line spectrum of each element is so uniquely characteristic of that element that atomic spectroscopy can be used for precise elementary analysis of many types of chemically complex materials.

The exploitation of spectral information requires a recognition of two aspects of light, its wave character and its particle character.

Wave character of light

A light beam is associated with an electromagnetic disturbance, which is propagated periodically in the direction of the beam. The distance between two successive maxima in the intensity of the disturbance is called *wavelength* and is usually given the symbol lambda, λ. The *frequency* is the number of such maxima passing a given point per second and is usually given the symbol nu, v. The common unit for frequency is the *hertz*, having the symbol Hz, where $1 \text{ Hz} = 1 \text{ s}^{-1}$. The product of the wavelength and the frequency is equal to the velocity of propagation, usually designated by c.

$$c = \lambda v$$

The velocity of light in vacuum is the same for all wavelengths, 2.998×10^8 m/s, so that a simple inverse proportionality exists between wavelength and frequency. The velocity of light in the atmosphere is reduced by less than one part in a thousand below the value for vacuum, so that the above value is acceptable for most laboratory applications. Another term is often used, the *wave number*, $\tilde{v}$, defined as $1/\lambda$, or as v/c. The most common unit for wave number is cm^{-1}.

Particle character of light

The energy of light is absorbed, emitted, or converted to other forms of energy in individual units, or *quanta* (singular, *quantum*). The unit itself, often referred to as the particle of light, is called a *photon*. The energy of a photon is proportional to the frequency:

$$\varepsilon = hv = (6.626 \times 10^{-34} \text{ J} \cdot \text{s})v$$

The universal proportionality constant, h, is called *Planck's constant*.

Although visible light, to which the human eye is sensitive, is confined to the wavelength range 400 to 700 nm, the above discussion refers to radiation of all wavelengths. Chemists in their studies of matter routinely use radiation whose wavelength ranges from 0.1 nm (x-rays) to several centimeters (microwaves).

INTERACTION OF LIGHT WITH MATTER

A major step toward the understanding of atomic structure was Bohr's explanation of the hydrogen spectrum. His postulates were the following:

1. The electron of the hydrogen atom revolves around the nucleus in stable circular orbits.
2. In each stable orbit the electrostatic attraction between the negatively charged electron and the positively charged nucleus provides the centripetal force needed for the circular motion of the

electron. The energy of the atom is the sum of the potential energy of electrostatic interaction between the nucleus and the electron and the kinetic energy of motion of the electron.

3. Only certain stable orbits are allowed, those for which the angular momentum of the electron is an integer, n, times the constant $h/2\pi$. (This is the same h that occurs in the equation for the energy of a photon of light.)

4. An electron can move from one stable orbit to another only by absorbing or releasing an amount of energy exactly equal to the difference between the energies of the two orbits. If this energy is absorbed or released as light, a single photon of absorbed or emitted light must account for the required energy difference, so that

$$h\nu = |\Delta E|$$

where ΔE is the difference between the energies of the final and initial orbits.

This theory of Bohr satisfactorily accounted for the observed series of spectral lines for hydrogen. The predicted orbit energies are given by

$$E(n) = -\frac{me^4 Z^2}{8\epsilon_0^2 h^2 n^2}$$

where m and e are the mass and charge of the electron, ϵ_0 is the permittivity of free space, and Z is the atomic number of the nucleus (1 in the case of hydrogen). Note that the energy is negative with respect to the state in which the electron and nucleus are very widely separated, which is the zero energy state. In SI units,

$$m = 9.109\,5 \times 10^{-31}\,\text{kg} \qquad e = 1.602 \times 10^{-19}\,\text{C} \qquad \epsilon_0 = 8.854 \times 10^{-12}\,\text{C}^2/\text{N}\cdot\text{m}^2$$

The predicted allowed wave numbers observed in the spectrum are:

$$\tilde{\nu} = \frac{|E(n_2) - E(n_1)|}{hc} = \frac{me^4 Z^2}{8\epsilon_0^2 h^3 c}\left|\frac{1}{n_1^2} - \frac{1}{n_2^2}\right|$$

$$= RZ^2\left|\frac{1}{n_1^2} - \frac{1}{n_2^2}\right|$$

where $n_1 < n_2$ in absorption and $n_1 > n_2$ in emission. The quantity R, defined purely in terms of universal constants,

$$R = \frac{me^4}{8\epsilon_0^2 h^3 c}$$

is called the *Rydberg constant*. It has the value $109\,737\,\text{cm}^{-1}$. The predicted radii of the orbits are:

$$r(n) = \frac{n^2}{Z}a_0 \qquad \text{where} \qquad a_0 = \frac{h^2\epsilon_0}{\pi me^2}$$

For hydrogen ($Z = 1$), the predicted value of the first Bohr orbit ($n = 1$) is $a_0 = 5.29 \times 10^{-11}\,\text{m} = 0.529\,\text{Å}$.

The Bohr theory, even when elaborated and improved, applies only to hydrogen and hydrogen-like atoms; it explains neither the spectra of atoms containing even as few as two electrons nor the existence and stability of chemical compounds. The next advance required an understanding of the wave nature of matter.

PARTICLES AND WAVES

De Broglie proposed that not only does light have the dual properties of waves and particles but also particles of matter have properties of waves. The wavelength of those particle waves is given by

$$\lambda = \frac{h}{mv}$$

where m and v are the mass and velocity of the particle. Planck's constant is so small that the wavelengths are in observable range only for particles of atomic or subatomic dimensions.

An experimental confirmation of the de Broglie relation for a beam of uniformly energetic electrons was followed by a theoretical development of *quantum mechanics*, or *wave mechanics*, which reproduced Bohr's successful prediction of the stable energy levels of the hydrogen atom by means of a powerful new formalism which is applicable in principle to many-electron atoms and to many-atom molecules. The Bohr postulates are replaced by the *Schrödinger equation*, which must be solved by the methods of partial differential equations. The equation has mathematical similarities to descriptions of physical waves, and the arbitrary introduction of integers in the Bohr theory receives justification in quantum mechanics in the requirement that wavelike solutions of the equation must be continuous, finite, and single-valued. The steady-state solutions of the wave equation correspond to states of fixed energy, as in the Bohr theory, and in many cases, including the hydrogen atom, of fixed angular momentum. Many other descriptive properties, like the location of an electron, are not definitely fixed but are represented by probabilities of distribution over a range of numerical values. In the hydrogen atom the electron is not confined to a two-dimensional orbit but is represented by a wave that extends over three-dimensional space. The wave represents the varying probabilities of finding the electron at different locations with respect to the nucleus.

Orbitals

A solution to the Schrödinger equation for an electron must satisfy three quantum conditions corresponding to the three dimensions of space. Each quantum condition introduces an integer, called a *quantum number*, into the solution. A separate solution, describing a probability distribution of finding the electron at various locations, exists for each allowed set of the three quantum numbers. Such a solution is called an *orbital*; it is similar to a hypothetical time-exposure photograph of an electron taken over a time interval large enough so that each region of space is represented by the weighted probability of finding the electron in that region. These three quantum numbers are usually designated as follows:

1. n, *principal quantum number*: This number almost exclusively determines the energy of the orbital in one-electron systems, and is still the principal determinant of energy in many-electron systems.
2. l, *angular momentum quantum number*: This number defines the shape of the orbital and together with n defines the average distance of the electron from the nucleus.
3. m_l, *magnetic quantum number*: This number determines the angular orientation of the orbital in space.

In addition to the three dimensions in space describing the relative positions of the electron and the nucleus, there is a fourth dimension internal to the electron itself, characterized by the spin of the electron around an internal axis and by the magnetic moment associated with this spin. The quantum number associated with electron spin is usually designated as m_s.

Each of the four quantum numbers may have only certain values.

(a) n may be any positive integer. Electrons having a given n-value are said to be in the same *shell*. Shells are designated by capital letters as follows:

n	1	2	3	4	5	6	7
Designation	K	L	M	N	O	P	Q

(b) l may have any integral value from 0 to $n - 1$. For historical reasons, orbitals with a given l-value are designated by a characteristic letter.

l	0	1	2	3	4
Designation	s	p	d	f	g

An orbital is usually referred to by its principal quantum number, followed by a letter corresponding to its l value; e.g., $1s$, $2p$, $3s$.

(c) m_l may have any integral value from $-l$ to $+l$. This rule gives the correct *number* of orbitals, $2l + 1$, associated with a given (n, l) combination. Thus there are three "p" orbitals, corresponding to $m_l = 1, 0,$ and -1. However, it is usually more convenient in chemistry to use a new set of three orbitals (oriented with respect to the x, y, and z axes, respectively) which are obtained from the basic set by a mathematical mixing process. Similar procedures apply to a set of five "d" orbitals, and so forth.

(d) m_s may be either $+\frac{1}{2}$ or $-\frac{1}{2}$.

The probabilities of finding an electron at various distances r from the hydrogen nucleus are given in Fig. 8-1 for several (n, l) combinations. The shapes of s, p, and d orbitals are shown in Fig. 8-2. An inspection of Fig. 8-1 shows a detail which emerges as an exact mathematical consequence of the theory: when $l = n - 1$, the most probable distance of the electron from the nucleus is exactly equal to the radius of the Bohr orbit, $n^2 a_0$. In general, the average distance of the electron from the nucleus increases with increasing n. This same figure shows another interesting feature for all allowed l-values except the maximum for a given n; namely, the existence of minima of zero probability, corresponding to spherical *nodal surfaces* around the nucleus, at which the electron will not be found. Figure 8-2 shows that s orbitals

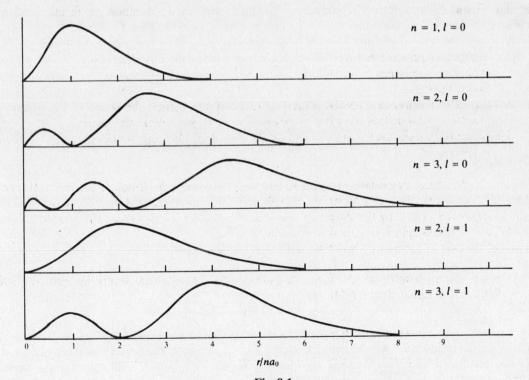

r/na_0

Fig. 8-1

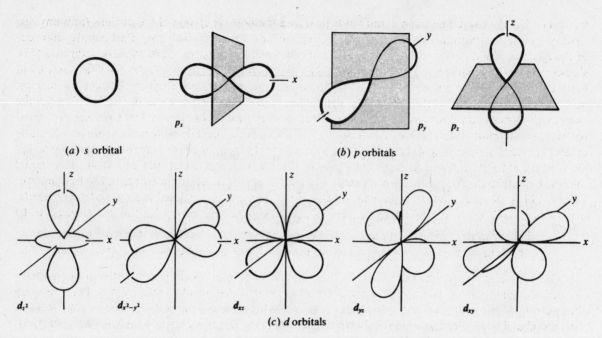

Fig. 8-2   Angular Dependence of Orbitals

are spherically symmetrical around the nucleus. Each p orbital is concentrated along the + and − portions of one of the Cartesian axes, with a *nodal plane* of zero probability perpendicular to that axis at the nucleus. Of the five allowed d orbitals for a given n, only four have the same shape. Each of these four looks like a three-dimensional four-leaf clover, with the highest probability along the x and y axes in the case of the $d_{x^2-y^2}$ and in between the axes in the case of the d_{xy}, d_{xz}, and d_{yz}. Each of these orbitals has two perpendicular nodal planes through the nucleus. The fifth d orbital, d_{z^2}, has maxima in the $+z$ and $-z$ directions and a secondary concentration in the (x, y) plane. This orbital has two conical nodal surfaces through the nucleus, one projecting above the (x, y) plane and one below, separating the $+z$ and $-z$ lobes from the doughnut-shaped ring of probability concentrated along the (x, y) plane.

THE PAULI PRINCIPLE AND THE PERIODIC LAW

The *Pauli exclusion principle* states that no two electrons in an atom may have the same set of all four quantum numbers. This principle places the following limits on the number of electrons for various (n, l) combinations:

Orbital Type	Maximum Number of Electrons, $2(2l + 1)$
s	2
p	6
d	10
f	14
g	18

The electrons for a given atom will preferentially occupy those orbitals having the lowest energy. The *buildup (Aufbau)* principle describes in detail the way in which each additional electron for each succeeding element of increasing atomic number is orbitally distributed. Although n is the principal determinant of orbital energy, the importance of l cannot be overlooked in polyelectron atoms. The reason for this is that the overall electron cloud tends to screen the positive charge of the nucleus from any test electron in question, and this screening works differently for different orbital types. Particularly, that part of the electron cloud closer to the nucleus than the test electron has the more effective screening. The shapes of the orbitals shown in Fig. 8-2 give a clue to the differences in energy of orbitals of the same n but of different l-values. The higher the l-value, the greater the number of nodal surfaces through the nucleus; an s orbital has none, a p orbital has one, a d orbital has two, and so on. Since a nodal surface through the nucleus imposes a compulsory avoidance by the electron of those close-in regions where the electron could receive its maximum stabilization by attraction to the nucleus, orbitals with high l-values will allow greater amounts of screening of the nuclear charge from the electron. Increased screening means decreased electrostatic stabilization and increased energy (i.e., the energy is less negative). Conversely, an s orbital allows an electron to come close to the nucleus, to avoid screening by other electrons, and to have a lower energy. This type of qualitative argument, details of the quantum mechanical theory, and experimental observation have led to the following sequence of increasing orbital energy for the last electron to be added to form neutral atoms:

$$1s < 2s < 2p < 3s < 3p < 4s < 3d < 4p < 5s < 4d < 5p < 6s < 5d = 4f < 6p < 7s < 6d = 5f$$

These principles are at the heart of the *periodic law*, according to which the properties of the elements are periodic functions of their atomic numbers. The chemical behavior of an atom is based primarily on the electrons of highest n, only secondarily on the $(n-1)$-electrons, and tertiarily on the $(n-2)$-electrons. Greatest chemical similarities are thus shown among elements having a common assignment of electrons in their outermost shells.

ATOMIC RADII

The electron cloud around an atom makes the concept of atomic size somewhat imprecise. Nevertheless it is useful to refer to an atomic size or an atomic radius. Operationally, one can arbitrarily divide the experimentally determined distance between centers of two chemically bonded atoms to arrive at two radii, on the crude picture that the two bonded atoms are spheres in contact. If the bonding is covalent (see Chapter 9), the radius is called a *covalent radius*, if it is ionic, the radius is an *ionic radius*. The radius for a nonbonded situation may also be defined in terms of the distance of closest nonbonding approach; such a measure is called the *van der Waals radius*. The following generalizations are observed:

1. Within a given group of the periodic table, the radius increases with increasing atomic number. This fact is related to the increased n of the outermost shell.
2. Within a given period of the periodic table, the covalent radii generally decrease with increasing atomic number. This is related to the facts that (i) the size of an atom depends on the average distance of its outermost electron(s), (ii) there is no change in n of the outermost electron(s) within a given period, and (iii) there is increasing nuclear charge with increasing atomic number.
3. Ionic radii of cations are fairly small compared to the covalent radii for the corresponding atoms since all the outermost electrons (highest n) are usually removed. Anionic radii are only slightly larger than the van der Waals radii for the corresponding atoms since the extra electron(s) have the same n. However the covalent radii of these atoms are appreciably less since they are bonded to their neighbors by electron sharing.
4. Within a given group, ionic radii increase with increasing atomic number. Within a given period, cationic radii decrease with increasing atomic number as do anionic radii, but there is a discontinuous increase in going from the last cation in a period to the first anion.

IONIZATION ENERGIES

The Bohr formula for the hydrogen atom energy levels predicts that higher energy levels get closer and closer together and asymptotically approach a limit of zero as $n \to \infty$. At this limit the atom has been ionized. The minimum energy required to ionize an isolated gaseous atom may be determined spectroscopically, thermochemically, or electrically. In the electrical method, a measurement is made of the accelerating potential that will impart to a projectile electron (not the electron within the atom in question) an amount of kinetic energy just sufficient to dislodge the bound electron from its atom. Thus the ionization energy can be measured directly in electrical terms. The *electron volt* is the energy imparted to an electron accelerated by a potential difference of 1 V.

$$1 \text{ eV} = (\text{charge on electron}) \times (\text{potential difference}) = (1.602\,2 \times 10^{-19} \text{ C})(1 \text{ V}) = 1.602\,2 \times 10^{-19} \text{ J}$$

Note that $1 \text{ V} = 1 \text{ J/C}$.

Ionization energies (I.E.) have been measured for all the atoms. They are all positive, corresponding to an ionization process that is endothermic. Also, ionization energies have been measured for removing successive electrons from polyelectron atoms. Some interesting trends have been observed.

1. Within a given group of the periodic table, the first ionization energy decreases with increasing atomic number. This is related to the increase in atomic size and the decreasing hold which the nucleus has for the increasingly distant outermost electron. (This rule is not uniformly obeyed for the transition metals.)
2. Within a given period, the general trend is an increase in ionization energy with increasing atomic number. Atoms just beginning a new subshell or the second half of a subshell usually have a smaller ionization energy than the previous atom.
3. The ionization energy for each succeeding stage of ionization is always greater than for the preceding stage. For instance, the second ionization energy of magnesium is considerably greater than the first (about twice as great), because the residual Mg^{2+} has twice the electrostatic attraction for the removed electron as does Mg^+. However, the second ionization energy of sodium is many times the first, because the second electron must be removed from the $n = 2$ shell rather than the $n = 3$ shell.

ELECTRON AFFINITY

Some free atoms can capture an extra electron to form a stable gaseous anion, particularly with elements having almost-completed p subshells. For example,

$$Cl(g) + e^-(g) \rightarrow Cl^-(g) \qquad \Delta H = -3.61 \text{ eV}$$

Chlorine is said to have an electron affinity (E.A.) of 3.61 eV. In this book a positive electron affinity implies an exothermic electron attachment as in the above example.

MAGNETIC PROPERTIES

The magnetic properties of bulk matter depend on the properties of the individual atoms. We have noted that electron spin has a magnetic moment associated with it. Two electrons occupying the same orbital (same n, l, and m_l) have their magnetic moments canceled out because the two values of m_s correspond to equal but oppositely directed spin angular momenta and thus to equal but opposed magnetic moments. Atoms, ions, or compounds in which at least one orbital is singly occupied can thus have a net magnetic moment. Such a substance is said to be *paramagnetic* and is attracted into a magnetic field. The magnitude of the magnetic moment, and thus the number of unpaired electron spins, can be determined experimentally by measuring the force of attraction of an externally imposed magnetic field for the substance. Substances without unpaired electron spins do not have magnetic moments and are repelled by a magnetic field; they are called *diamagnetic*. The repulsion of diamagnetic substances is much smaller in magnitude than the attraction of paramagnetic substances.

Magnetic measurements are an important tool for determining the electron assignment into orbitals for atoms, ions, and compounds. Several rules have been developed for the detailed assignment of electrons within a subshell.

1. Electrons within a subshell for which $l > 0$ will tend to avoid pairing within the same orbital. This rule reflects the relatively greater electrostatic repulsion between two electrons in the same orbital within a subshell as compared with occupancy of two orbitals having differing m_l-values.
2. Electrons in singly occupied orbitals tend to have their spins in the same direction so as to maximize the net magnetic moment.

Solved Problems

ENERGY RELATIONSHIPS

8.1. Determine the frequencies of electromagnetic radiation of the following wavelengths:
(a) 1.0 Å,   (b) 5 000 Å,   (c) 4.4 μm,   (d) 89 m,   (e) 562 nm.

The basic equation for all these problems is $v = (2.998 \times 10^8 \text{ m/s})/\lambda$.

(a) $$v = \frac{3.0 \times 10^8 \text{ m/s}}{(1.0 \text{ Å})(10^{-10} \text{ m/Å})} = 3.0 \times 10^{18} \text{ s}^{-1} = 3.0 \times 10^{18} \text{ Hz}$$

(b) $$v = \frac{2.998 \times 10^8 \text{ m/s}}{(5\,000 \text{ Å})(10^{-10} \text{ m/Å})} = 5.996 \times 10^{14} \text{ Hz} = 599.6 \text{ THz}$$

(c) $$v = \frac{3.00 \times 10^8 \text{ m/s}}{4.4 \times 10^{-6} \text{ m}} = 6.8 \times 10^{13} \text{ Hz} = 68 \text{ THz}$$

(d) $$v = \frac{3.00 \times 10^8 \text{ m/s}}{89 \text{ m}} = 3.4 \times 10^6 \text{ Hz} = 3.4 \text{ MHz}$$

(e) $$v = \frac{2.998 \times 10^8 \text{ m/s}}{562 \times 10^{-9} \text{ m}} = 5.33 \times 10^{14} \text{ Hz} = 533 \text{ THz}$$

8.2. (a) What change in molar energy would be associated with an atomic transition giving rise to radiation at 1 Hz? (b) What is the relationship between the electron volt and the wavelength in nanometers of the energetically equivalent photon?

(a) If each of N_A atoms gives off one 1-Hz photon,

$$\Delta E = N_A(hv) = (6.022 \times 10^{23} \text{ mol}^{-1})(6.626 \times 10^{-34} \text{ J} \cdot \text{s})(1 \text{ s}^{-1}) = 3.990 \times 10^{-10} \text{ J} \cdot \text{mol}^{-1}$$

Since ΔE and v are proportional, we may treat the ratio

$$\frac{3.990 \times 10^{-10} \text{ J} \cdot \text{mol}^{-1}}{1 \text{ Hz}}$$

as a "conversion factor" between Hz and J · mol^{-1}, provided we understand just what this "conversion" means. Thus, for 1-MHz radiation,

$$\Delta E = (10^6 \text{ Hz})(3.990 \times 10^{-10} \text{ J} \cdot \text{mol}^{-1} \cdot \text{Hz}^{-1}) = 3.990 \times 10^{-4} \text{ J} \cdot \text{mol}^{-1}$$

(b) First let us find the frequency equivalent of 1 eV from the Planck equation, and then find the wavelength from the frequency.

$$v = \frac{\epsilon}{h} = \frac{1.602\,2 \times 10^{-19}\,\text{J}}{6.626 \times 10^{-34}\,\text{J} \cdot \text{s}} = 2.418\,0 \times 10^{14}\,\text{s}^{-1}$$

$$\lambda = \frac{c}{v} = \frac{2.998 \times 10^{8}\,\text{m} \cdot \text{s}^{-1}}{2.418\,0 \times 10^{14}\,\text{s}^{-1}} = (1.239\,8 \times 10^{-6}\,\text{m})(10^{9}\,\text{nm/m}) = 1\,239.8\,\text{nm}$$

Because of the inverse proportionality between wavelength and energy, the relationship may be written

$$\lambda\epsilon = hc = 1\,239.8\,\text{nm} \cdot \text{eV}$$

8.3. In the photoelectric effect, an absorbed quantum of light results in the ejection of an electron from the absorber. The kinetic energy of the ejected electron is equal to the energy of the absorbed photon minus the energy of the longest-wavelength photon that causes the effect. Calculate the kinetic energy of a photoelectron produced in cesium by 400-nm light. The critical (maximum) wavelength for the photoelectric effect in cesium is 660 nm.

Using the result of Problem 8.2(b),

$$\text{Kinetic energy of electron} = hv - hv_{\text{crit}} = \frac{hc}{\lambda} - \frac{hc}{\lambda_{\text{crit}}}$$

$$= \frac{1\,240\,\text{nm} \cdot \text{eV}}{400\,\text{nm}} - \frac{1\,240\,\text{nm} \cdot \text{eV}}{660\,\text{nm}} = 1.22\,\text{eV}$$

8.4. It has been found that gaseous iodine molecules dissociate into separate atoms after absorption of light at wavelengths less than 499.5 nm. If each quantum is absorbed by one molecule of I_2, what is the minimum input, in kJ/mol, needed to dissociate I_2 by this photochemical process?

$$E(\text{per mole}) = N_A(hv) = \frac{N_A hc}{\lambda} = \frac{(6.022 \times 10^{23}\,\text{mol}^{-1})(6.626 \times 10^{-34}\,\text{J} \cdot \text{s})(2.998 \times 10^{8}\,\text{m} \cdot \text{s}^{-1})}{499.5 \times 10^{-9}\,\text{m}}$$

$$= 239.5\,\text{kJ/mol}$$

8.5. A beam of electrons accelerated through 4.64 V in a tube containing mercury vapor was partly absorbed by the vapor. As a result of absorption, electronic changes occurred within a mercury atom and light was emitted. If the full energy of a single incident electron was converted into light, what was the wave number of the emitted light?

Using the result of Problem 8.2(b)

$$\tilde{v} = \frac{1}{\lambda} = \frac{v}{c} = \frac{hv}{hc}$$

$$= \frac{4.64\,\text{eV}}{1\,240\,\text{nm} \cdot \text{eV}} = 0.003\,74\,\text{nm}^{-1} = 37\,400\,\text{cm}^{-1}$$

8.6. An electron diffraction experiment was performed with a beam of electrons accelerated by a potential difference of 10 kV. What was the wavelength of the electron beam?

We can use the de Broglie equation, taking the mass of an electron as 0.911×10^{-30} kg. The velocity of the electron is found by equating its kinetic energy, $\frac{1}{2}mv^2$, to its loss of electric potential energy, 10 keV.

$$\tfrac{1}{2}mv^2 = (10^4\,\text{eV})(1.602 \times 10^{-19}\,\text{J/eV})$$

$$= 1.602 \times 10^{-15}\,\text{J} = 1.602 \times 10^{-15}\,\text{kg} \cdot \text{m}^2 \cdot \text{s}^{-2}$$

$$v = \left(\frac{2 \times 1.602 \times 10^{-15}\,\text{kg} \cdot \text{m}^2 \cdot \text{s}^{-2}}{0.911 \times 10^{-30}\,\text{kg}}\right)^{1/2} = (35.17 \times 10^{14})^{1/2}\,\text{m/s} = 5.93 \times 10^{7}\,\text{m/s}$$

Now the de Broglie equation gives

$$\lambda = \frac{h}{mv} = \frac{6.63 \times 10^{-34} \text{ J} \cdot \text{s}}{(0.911 \times 10^{-30} \text{ kg})(5.93 \times 10^{7} \text{ m/s})}$$

$$= \frac{1.23 \times 10^{-11} \text{ kg} \cdot \text{m}^2 \cdot \text{s}^{-1}}{\text{kg} \cdot \text{m} \cdot \text{s}^{-1}} = (1.23 \times 10^{-11} \text{ m})(10^9 \text{ nm/m}) = 0.012\,3 \text{ nm}$$

The results calculated above are somewhat in error because of the law of relativity, which becomes more and more relevant as the velocity approaches the speed of light. For instance for $E = 300 \text{ kV}$, the velocity calculated as above would exceed c, a meaningless result since no particle can have a velocity greater than the speed of light.

ATOMIC PROPERTIES AND THE PERIODIC LAW

8.7. The Rydberg constant for deuterium (^{2}H) is 109 707 cm^{-1}. (This value reflects a refinement of the simple Bohr theory, wherein the Rydberg constant and the orbital radii depend on the so-called *reduced mass* rather than the electron mass. The reduced mass, in turn, varies slightly with the mass of the nucleus.) Calculate   (a) the shortest wavelength in the absorption spectrum of deuterium,   (b) the ionization energy of deuterium, and   (c) the radii of the first three Bohr orbits.

(a) The shortest-wavelength transition would correspond to the highest frequency and to the highest energy. The transition would thus be from the lowest energy state (the *ground state*), for which $n = 1$, to the highest, for which $n = \infty$.

$$\tilde{v} = R\left(\frac{1}{1^2} - \frac{1}{\infty^2}\right) = R = 109\,707 \text{ cm}^{-1}$$

$$\lambda = \frac{1}{109\,707 \text{ cm}^{-1}} = (0.911\,52 \times 10^{-5} \text{ cm})(10^7 \text{ nm/cm}) = 91.152 \text{ nm}$$

(b) The transition computed in (a) is indeed the ionization of the atom in its ground state. From the result of Problem 8.2(b),

$$\text{I.E.} = \frac{1\,239.8 \text{ nm} \cdot \text{eV}}{91.152 \text{ nm}} = 13.601 \text{ eV}$$

This value is slightly greater than the value for ^{1}H.
(c) From the equation ($Z = 1$)

$$r = n^2 a_0 = n^2(5.29 \times 10^{-11} \text{ m})$$

the radii are 1, 4, and 9 times a_0, or 0.529, 2.116, and 4.76 Å. The reduced-mass correction, involving an adjustment of 3 parts in 10^4, is not significant, and the a_0 for the first Bohr orbit of ^{1}H is a perfectly satisfactory substitution.

8.8. (a) Neglecting reduced-mass effects, what optical transition in the He$^+$ spectrum would have the same wavelength as the first Lyman transition of hydrogen ($n = 2$ to $n = 1$)?
(b) What is the second ionization energy of He?   (c) What is the radius of the first Bohr orbit for He$^+$?

(a) He$^+$ has only one electron. It is thus classified as a hydrogen-like species with $Z = 2$, and the Bohr equations may be applied. From the equation

$$\tilde{v} = RZ^2\left(\frac{1}{n_1^2} - \frac{1}{n_2^2}\right)$$

the first Lyman transition for hydrogen would be given by

$$\tilde{v} = R\left(\frac{1}{1^2} - \frac{1}{2^2}\right)$$

The assumption regarding mass effects is equivalent to considering R for He$^+$ the same as for ^{1}H. The Z^2-term can just be compensated by increasing n_1 and n_2 by a factor of 2 each.

$$\tilde{v} = R(2^2)\left(\frac{1}{2^2} - \frac{1}{4^2}\right)$$

The transition in question is thus the transition from $n = 4$ to $n = 2$.

(b) The second ionization energy for He is the same as the first ionization energy for He$^+$, and the Bohr equations may be applied to the ground state of He$^+$, for which $Z = 2$ and $n = 1$.

$$\tilde{v} = RZ^2\left(\frac{1}{n_1^2} - \frac{1}{n_2^2}\right) = R(2)^2\left(\frac{1}{1^2} - \frac{1}{\infty^2}\right) = 4R$$

This result is 4 times the result for deuterium in Problem 8.7. Since $\tilde{v}$ is proportional to energy, the ionization energy will likewise be 4 times that for deuterium.

$$\text{I.E. (He}^+\text{)} = 4 \times \text{I.E. (}^2\text{H)} = 4 \times 13.6 \text{ eV} = 54.4 \text{ eV}$$

(c) $$r = \frac{n^2 a_0}{Z} = \frac{0.529 \text{ Å}}{2} = 0.264 \text{ Å}$$

8.9.   (a) Write the electron configurations for the ground states of N, Ar, Fe, Fe^{2+}, Pr^{3+}.   (b) How many unpaired electrons would there be in each of these isolated particles?

(a) The atomic number of N is 7. The first (or K) shell will contain its maximum of 2 electrons. The next 5 electrons will be distributed in the L shell, 2 filling the lower-energy s subshell and the remaining 3 the p subshell. A common notation is

$$1s^2 2s^2 2p^3$$

Another notational form, $(\text{He})2s^2 2p^3$, shows only those electrons beyond the complement in the previous rare gas.

The atomic number of Ar is 18. The order of filling is that of the successive K, L, and M shells.

$$1s^2 2s^2 2p^6 3s^2 3p^6$$

The atomic number of Fe is 26. Beyond the Ar structure the order of filling is $4s$, then $3d$, until 26 electrons are assigned in all.

$$1s^2 2s^2 2p^6 3s^2 3p^6 3d^6 4s^2 \quad \text{or} \quad (\text{Ar})3d^6 4s^2$$

Fe^{2+} has 24 electrons, two fewer than the neutral Fe atom. Although $4s$ is of lower energy than $3d$ for atomic numbers 19 and 20 (K and Ca), this order reverses for higher nuclear charges. In general, the electrons most easily lost from any atom are those with the largest principal quantum number.

$$(\text{Ar})3d^6$$

Pr^{3+} has 56 electrons, 3 fewer than the neutral Pr atom. Beyond Xe, the previous rare gas, the order of filling for the next period of elements is $6s^2$, then one $5d$ electron, then the whole $4f$ subshell, then the rest of the $5d$ subshell, then the $6p$ subshell. There are frequent replacements of the first $5d$ assignment with an additional $4f$, or of one of the $6s$ with an additional $5d$, but these irregularities are of no consequence in the assignment of electrons in Pr^{3+}. The 3 electrons removed from the neutral atom to form the ion follow the general rule of removal first from the outermost shell and then from the next-to-outermost; in this case the $6s$ electrons are removed first and no $5d$ electrons would remain even if there were one in the neutral atom.

$$1s^2 2s^2 2p^6 3s^2 3p^6 3d^{10} 4s^2 4p^6 4d^{10} 4f^2 5s^2 5p^6 \quad \text{or} \quad (\text{Xe})4f^2$$

(b) Completed subshells have no unpaired electrons; thus, it is necessary to examine only the electrons past a rare gas core.

For N, the $2p$ is the only incomplete subshell. The three electrons in this subshell will singly occupy the three available orbitals; there will be 3 unpaired electrons.

Ar has no incomplete subshells and thus no unpaired electrons.

Fe has six $3d$ electrons in the only unfilled subshell. The maximum unpairing occurs with double occupancy of one of the available d orbitals and single occupancy of the remaining four. There will thus be 4 unpaired electrons.

Fe^{2+} also has 4 unpaired electrons, for the same reason as Fe.

Pr^{3+} has 2 unpaired electrons, the two $4f$ electrons that singly occupy 2 of the 7 available $4f$ orbitals.

8.10. Nickel has the electron configuration $(Ar)3d^84s^2$. How do you account for the fact that the configuration of the next element, Cu, is $(Ar)3d^{10}4s$?

In the hypothetical procedure of making up the electronic complement of Cu by adding 1 electron to the configuration of the preceding element, Ni, one might have expected only a ninth $3d$ electron. For atomic number 19, the $3d$ subshell is decidedly of higher energy than $4s$; thus potassium has a single $4s$ electron and no $3d$ electron. After the $3d$ subshell beings to fill, however, beginning with element 21 (after $4s$ has already been filled), the addition of each succeeding $3d$ electron is accompanied by a lowering of the average energy of the $3d$ level. This is because each succeeding element has an increased nuclear charge which is only partly screened from a test $3d$ electron by the additional electron in the same subshell. The energy of the $3d$ subshell decreases gradually as the subshell undergoes filling and drops below the level of the $4s$ toward the end of the transition series.

Another factor is that the configuration $3d^{10}4s^1$ has a spherically symmetrical distribution of electron density, a stabilizing arrangement characteristic of all filled or half-filled subshells. On the other hand, the configuration $3d^94s^2$ has a "hole" (a missing electron) in the $3d$ shell, destroying the symmetry and any extra stabilization.

8.11. The ionization energies of Li and K are 5.4 and 4.3 eV, respectively. What do you predict for the ionization energy of Na?

The first I.E. of Na should be intermediate between that of Li and K. The I.E. of Na should be close to the arithmetic average of the two, or 4.9 eV. (The observed value is 5.1 eV.)

8.12. The (first) ionization energies of Li, Be, and C are 5.4, 9.3, and 11.3 eV. What do you predict for the ionization energies of B and N?

There is a general increasing trend of I.E. with increasing atomic number in a given period. Note, however, a larger increase in going from Li ($Z = 3$) to Be ($Z = 4$) than in going from Be to C ($Z = 6$). The filling of the $2s$ subshell gives Be a greater stability than would be suggested by a smooth progression across the periodic table. The next element, B ($Z = 5$), would have an I.E. that represents a balancing of the two oppositely directed factors, an increase with respect to Be because of the increased Z and a decrease because a new subshell is beginning to be filled in the case of B. One might guess that the I.E. for B might actually be less than for Be, and this turns out to be the case. The observed I.E. for B is 8.3 eV.

The increase in going from $Z = 5$ to $Z = 6$ is 3.0 eV. One expects the increase in going to $N(Z = 7)$ to be similar, bringing the I.E. of N to about 14.3 eV. In fact because of the extra stability of the half-filled p subshell the I.E. is even greater; 14.5 eV is observed.

8.13. In the ionic compound KF, the K^+ and F^- ions are found to have practically identical radii, about 0.134 nm each. What do you predict about the relative covalent radii of K and F?

The covalent radius of K should be much greater than 0.134 nm and that of F much smaller, since atomic cations are smaller than their parent atoms, while atomic anions are bigger than their parents. The observed covalent radii of K and F are 0.20 and 0.06 nm respectively.

8.14. The single covalent radius of P is 0.11 nm. What do you predict for the single covalent radius of Cl?

P and Cl are members of the same period. Cl should have a smaller radius in keeping with the usual trend across a period. The experimental value is 0.10 nm.

8.15. The first ionization energy for Li is 5.4 eV and the electron affinity of Cl is 3.61 eV. Compute ΔH for the reaction

$$Li(g) + Cl(g) \rightarrow Li^+(g) + Cl^-(g)$$

carried out at such low pressures that the resulting ions do not combine with each other.

The overall reaction may be decomposed into two partial reactions

$$(1) \quad Li(g) \rightarrow Li^+(g) + e^-(g) \qquad \Delta E = N_A(\text{I.P.})$$
$$(2) \quad Cl(g) + e^-(g) \rightarrow Cl^-(g) \qquad \Delta E = N_A(-\text{E.A.})$$

where e^- stands for an electron. Although ΔH for each of the above partial reactions differs slightly from ΔE (by the term $p\,\Delta V$), the ΔH of the overall reaction is the sum of the two ΔE's (the overall volume change is zero). Since the values of I.E. and E.A. are given on a per-atom basis, the factor 6.02×10^{23} atoms/mol is required to obtain ΔH on the conventional per-mole basis.

$$\Delta H(\text{total reaction}) = \Delta E(1) + \Delta E(2) = N_A(\text{I.P.} - \text{E.A.})$$
$$= (6.02 \times 10^{23})(1.8 \text{ eV})(1.60 \times 10^{-19} \text{ J/eV}) = 170 \text{ kJ}$$

Supplementary Problems

ENERGY RELATIONSHIPS

8.16. Find the wavelength λ in the indicated units for radiation of the following frequencies: (a) 55 MHz (λ in meters), (b) 1 000 Hz (λ in centimeters), (c) 7.5×10^{15} Hz (λ in nanometers).

Ans. (a) 5.5 m; (b) 2.998×10^7 cm; (c) 40 nm

8.17. The critical wavelength for producing the photoelectric effect in tungsten is 260 nm (a) What is the energy of a quantum at this wavelength in joules and in electron volts? (b) What wavelength would be necessary to produce photoelectrons from tungsten having twice the kinetic energy of those produced at 220 nm?

Ans. (a) 7.65×10^{-19} J $= 4.77$ eV; (b) 191 nm

8.18. In a measurement of the quantum efficiency of photosynthesis in green plants, it was found that 8 quanta of red light at 685 nm were needed to evolve one molecule of O_2. The average energy storage in the photosynthetic process is 469 kJ per mole of O_2 evolved. What is the energy conversion efficiency in this experiment?

Ans. 33.5 percent

8.19. O_2 undergoes photochemical dissociation into one normal oxygen atom and one oxygen atom 1.967 eV more energetic than normal. The dissociation of O_2 into two normal oxygen atoms is known to require 498 kJ/mol O_2. What is the maximum wavelength effective for the photochemical dissociation of O_2?

Ans. 174 nm

8.20. The dye acriflavine, when dissolved in water, has its maximum light absorption at 453 nm, and its maximum fluorescence emission at 508 nm. The number of fluorescence quanta is, on the average, 53 percent of the number of quanta absorbed. Using the wavelengths of maximum absorption and emission, what percentage of absorbed energy is emitted as fluorescence?

Ans. 47 percent

8.21. The prominent yellow line in the spectrum of a sodium vapor lamp has a wavelength of 590 nm. What minimum accelerating potential is needed to excite this line in an electron tube containing sodium vapor?

Ans.   2.10 V

8.22. Show by substitution in the formula given in the text that a_0, the radius of the first Bohr orbit for hydrogen, is 5.29×10^{-11} m.

8.23. What accelerating potential is needed to produce an electron beam with an effective wavelength of 0.025 6 nm?

Ans.   2.30 kV

8.24. A 1.0-g projectile is shot from a gun with a velocity of 100 m/s. What is the de Broglie wavelength?

Ans. 6.6×10^{-33} m

8.25. What accelerating potential must be imparted to a proton beam to give it an effective wavelength of 0.005 0 nm?

Ans.   33 V

ATOMIC PROPERTIES AND THE PERIODIC LAW

8.26. The Rydberg constant for Li^{2+} is 109 729 cm^{-1}. (*a*) What is the long-wavelength limit in the absorption spectrum of Li^{2+} at ordinary temperatures (where all the ions are in their ground state)? (*b*) What would be the shortest wavelength in the emission line spectrum of Li^{2+} within the visible (400 to 750 nm) region? (*c*) What would be the orbital radius of the ground state of Li^{2+}? (*d*) What is the ionization energy of Li^{2+}?

Ans.   (*a*) 13.5 nm;   (*b*) 415.4 nm ($n = 8 \rightarrow n = 5$);   (*c*) 0.176 Å;   (*d*) 122.4 eV

8.27. The first ionization energy of Li is found experimentally to be 5.363 eV. If the L electron ($n = 2$) is assumed to move in a central field of an effective nuclear change, Z_{eff}, consisting of the nucleus and the other electrons, by how many units of charge is the nucleus shielded by the other electrons? Assume that the ionization energy can be calculated from Bohr theory.

Ans. $Z_{eff} = 1.26$; since the nuclear charge is $3+$, the effective shielding by the two 1s electrons is 1.74 charge units.

8.28. What are the electron configurations of Ni^{2+}, Re^{3+}, and Ho^{3+}? How many unpaired electron spins are in each of these ions?

Ans. Ni^{2+}: (Ar)$3d^8$, 2 unpaired spins. Re^{3+}: (Xe)$4f^{14}5d^4$, 4 unpaired spins. Ho^{3+}: (Xe)$4f^{10}$, 4 unpaired spins.

8.29. Which shell would be the first to have a g subshell?

Ans.   O, $n = 5$

8.30. What would you predict for the atomic number of the rare gas beyond Rn, if such an element had sufficient stability to be prepared or observed? Assume that g orbitals are still not occupied in the ground states of the preceding elements.

Ans.   118

8.31. All the lanthanides form stable compounds containing the $3+$ cation. Of the few other ionic forms known, Ce forms the stablest $4+$ series of ionic compounds and Eu the stablest $2+$ series. Account for these unusual ionic forms in terms of their electronic configurations.

Ans. Ce^{4+} has the stable electronic configuration of the rare gas Xe. Eu^{2+}, with 61 electrons, could have the configuration $(Xe)4f^7$, with the added stability of a half-filled $4f$ subshell.

8.32. For the gaseous reaction $K + F \rightarrow K^+ + F^-$, ΔH was calculated to be 91 kJ under conditions where the cations and anions were prevented by electrostatic separation from combining with each other. The ionization energy of K is 4.34 eV. What is the electron affinity of F?

Ans.   3.40 eV

8.33. The ionic radii of S^{2-} and Te^{2-} are 1.84 and 2.21 Å, respectively. What would you predict for the ionic radius of Se^{2-} and for P^{3-}?

Ans.   Since Se falls in between S and Te, one expects an intermediate value; observed value is 1.98 Å. Since P is just to the left of S a slightly larger value is expected; observed value is 2.12 Å

8.34. Van der Waals radii for S and Cl are 1.85 and 1.80 Å respectively. What would you predict for the van der Waals radius of Ar?

Ans.   The observed value is 1.54 Å.

8.35. The first ionization energy of C is 11.2 eV. Would you expect the first ionization energy of Si to be greater or less than this amount?

Ans.   Less because Si is larger; the observed value is 8.1 eV.

8.36. The first ionization energies of Al, Si, and S are 6.0, 8.1, and 10.3 eV, respectively. What do you predict for the first ionization energy of P?

Ans.   A value slightly greater than that of S is predicted because of the stability of the half-filled shell in P. The observed value is 10.9 eV.

8.37. Several experimenters have attempted to synthesize superheavy elements by bombarding atoms of the actinide series with heavy ions. Pending confirmation and acceptance of the results, some investigators in the early 1970's referred to elements 104 and 105 as *eka*-hafnium and *eka*-tantalum, respectively. Why were these names chosen?

Ans.   Mendeleev had used the prefix *eka*- (Sanskrit word for first) to name elements whose existence he predicted, applying the prefix to a known element in the same periodic group as the predicted element. His *eka*-boron, *eka*-aluminum, and *eka*-silicon were later discovered, confirmed, and named scandium, gallium, and germanium, respectively. Elements 104 and 105 are predicted to have electronic structures analogous to Hf and Ta, respectively.

Chapter 9

Chemical Bonding and Molecular Structure

The formulas of the chemical compounds are no accident. There is an NaCl, but no $NaCl_2$; there is a CaF_2, but no CaF. On the other hand, certain pairs of elements form two, or even more, different compounds, e.g. Cu_2O, CuO; N_2O, NO, NO_2. The power of an atom to combine with others is called *valence*, and we may distinguish between two main types, *ionic valence* and *covalence*. A few simple rules of valence given here help to systematize the formulas of many simple compounds.

IONIC VALENCE

Ionic bonding describes the combination between oppositely charged particles, or *ions*. The principal forces are the classical electrical forces operating between any two charged particles. Some of the more common ions are listed in Table 9-1. The charges of the elementary ions can be understood in terms of the electronic structure of atoms. For example, elements whose atomic numbers are within 2 or 3 of a rare gas tend to form ionic compounds containing a *cation* (positive ion) or *anion* (negative ion) *isoelectronic with* (having the same number of electrons as) the neighboring rare gas atom. For the transition metals, the lanthanides, and the actinides, such a simple rule is not very helpful, and a full rationalization of ionic charges requires a sophisticated understanding of chemical bonding which has not yet been reached for all

Table 9-1

Ion	Ionic Radius	Ion	Name
H^+		Cu^+	copper(I), or cuprous
Li^+	60 pm	Cu^{2+}	copper(II), or cupric
Na^+	95 pm	Fe^{2+}	iron(II), or ferrous
K^+	133 pm	Fe^{3+}	iron(III), or ferric
Cs^+	169 pm	Cr^{2+}	chromium(II), or chromous
Ag^+	126 pm	Cr^{3+}	chromium(III), or chromic
Mg^{2+}	65 pm	Hg_2^{2+}	mercury(I), or mercurous
Ca^{2+}	99 pm	Hg^{2+}	mercury(II), or mercuric
Sr^{2+}	113 pm	NH_4^+	ammonium
Ba^{2+}	135 pm	OH^-	hydroxide
Zn^{2+}	74 pm	HCO_3^-	bicarbonate, or hydrogen carbonate
Cd^{2+}	97 pm	CO_3^{2-}	carbonate
Ni^{2+}	69 pm	NO_3^-	nitrate
Al^{3+}	50 pm	NO_2^-	nitrite
H^-	208 pm	PO_4^{3-}	(ortho)phosphate
F^-	136 pm	SO_4^{2-}	sulfate
Cl^-	181 pm	SO_3^{2-}	sulfite
Br^-	195 pm	ClO_4^-	perchlorate
I^-	216 pm	ClO_3^-	chlorate
O^{2-}	140 pm	ClO_2^-	chlorite
S^{2-}	184 pm	ClO^-	hypochlorite
		$Cr_2O_7^{2-}$	dichromate
		CrO_4^{2-}	chromate
		MnO_4^-	permanganate

cases. If we accept the ionic charges as chemical facts, we can easily write the empirical formulas for ionic compounds in conformity to the requirement that the compound as a whole be neutral. The charge of an ion is often called the *ionic valence*, since it determines the number of opposite charges the ion can combine with to form a neutral compound.

The names of the elementary cations listed in the left-hand column of Table 9-1 are simply the names of the elements, e.g. *calcium ion*. The names of the anions listed in the left-hand column are derived from the names of the elements by replacing the last (or last two) syllable(s) of the element names by the suffix *-ide*, e.g. *chloride, oxide*. The significance of the ionic radius for the elementary ions will be discussed in the next chapter. When two cations exist for the same element, the accepted convention is to write the charge *per atom* in Roman numerals within parentheses after the name of the metal. There is still some remnant of the older mode of differentiating the two states by use of the suffix *-ous* for the lower of the two charge states and *-ic* for the higher. Distinctive suffixes are used for the complex oxy-anions in a partly systematic way, *-ate* for the most common or most stable and *-ite* for the ion containing less oxygen. The prefix *per-* is added to an *-ate* name to indicate even more oxygen, and the prefix *hypo-* is added to an *-ite* name to indicate a lower oxygen content than in the *-ite* ion.

Only a small fraction of the known cases of chemical bonding involve ionic bonding, the attraction between separate and oppositely charged chemical entities. The next section deals with the more common type of chemical bonding, *covalence*, the attraction within a molecule or complex ion of constituent atoms which do not have completely separate existences because of the overlap of their electron clouds.

COVALENCE

The covalent force between atoms *sharing* two or more "bonding" electrons is related to the *delocalization*, or smearing out, of the electrons over a wider region of space in the bond than they occupy in the separated atoms. The sharing of electrons in a covalent bond brings the bonded atoms closer together than in the absence of bonding. Thus, the distance between the two H nuclei in H_2, 74 pm, is much less than the sum of the van der Waals (nonbonding) radii of two H atoms, 240 pm.

This binding of the two atoms at close approach is reflected also in energetic considerations. The energy of a bonded pair of atoms is less than the sum of the energies of the separated atoms. The so-called *bond energy* is the magnitude of this energy lowering; from another viewpoint, the bond energy is the amount of energy, ΔE, required to break a chemical bond into two nonbonded fragments. Bond formation is exothermic and bond rupture is endothermic. Covalent bond energies, almost equal to ΔH of gaseous bond-rupturing chemical reactions, range from about 145 to 565 kJ/mol for single electron-pair bonds formed among elements in the first three periods at normal temperatures. (Under these conditions, the difference between ΔE and ΔH is less than 4 kJ/mol.)

Another normal feature of the single covalent bond is that the spins of the bond-forming electrons, unpaired in the separate atoms, become paired in bond formation.

VALENCE-BOND REPRESENTATION

The formulas of many covalent compounds, especially those involving only the elements of the first few periods of the periodic table, were brought within the framework of the *octet rule*, according to which a total of eight *valence*, or outer-shell, electrons, either shared or unshared, should be in the region of each atom beyond the first period. For hydrogen the desired number is two. Electrons that are shared between two atoms are the bonding electrons and are to be counted toward the octet (or duplet in the case of hydrogen) of each of the bonding atoms. A single covalent bond consists of a pair of shared electrons, a double bond has two shared pairs, and a triple bond has three shared pairs. Bond distances are shorter and bond energies are greater for multiple bonds than for single bonds.

Structural formulas, such as shown in Fig. 9-1, represent the electron distributions in covalent molecules and ions. These structures are not meant to indicate actual bond angles in three-dimensional varieties such as CH_3Cl, NH_3, and NH_4^+; they merely show the number of bonds connecting the various atoms. In these so-called *Lewis formulas*, a *line* between two atoms represents a *pair* of shared electrons

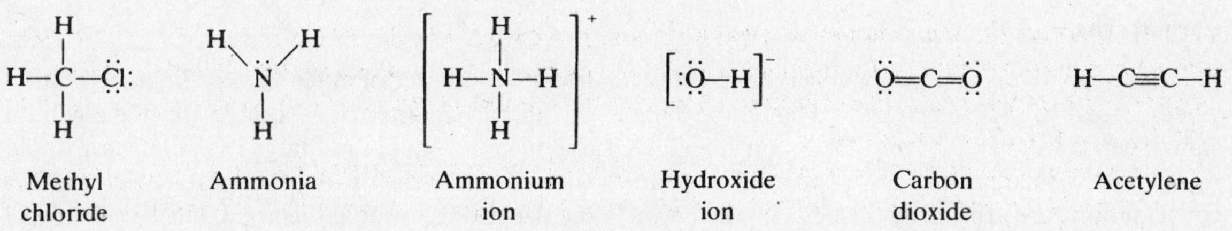

Methyl chloride   Ammonia   Ammonium ion   Hydroxide ion   Carbon dioxide   Acetylene

Fig. 9-1

and a *dot* represents an unshared electron. *Two lines* constitute a *double bond*, and *three lines* constitute a *triple bond*. The total number of electrons shown in such a molecular structure is equal to the sum of the numbers of valence (outer-shell) electrons in the free atoms: 1 for H, 4 for C, 5 for N, 6 for O, and 7 for Cl. For an ionic structure, one additional electron must be added to this sum for each unit of negative charge on the whole ion, as in OH^-, and one electron must be subtracted from the sum for each unit of positive charge on the ion, as in NH_4^+. The number of pairs of electrons shared by an atom is called its *covalence*.

The covalence of hydrogen is always one. The covalence of oxygen is practically always one or two. The covalence of carbon is four in almost all its stable compounds. Thus each carbon is expected to form either four single bonds, a double bond and two single bonds, two double bonds, or a single and a triple bond. Although the octet rule is not a rigid rule of chemical bonding, it is obeyed for C, N, O, and F in almost all their compounds. The octet is exceeded commonly for elements in the third and higher periods of the periodic table.

Resonance

Sometimes more than one satisfactory structure can be written and there is no reason to select one over another. In such cases a single structural formula is inadequate to represent a substance correctly, and several such diagrams must be written. The true structure is then said to be a *resonance* hybrid of the several diagrams.

EXAMPLE 1 Experiment has shown that the two terminal oxygens in ozone are equivalent; that is, they are equidistant from the central oxygen. If only one of the resonance diagrams in Fig. 9-2(*a*) were written, it would appear that one of the terminal oxygens is bonded more strongly to the central oxygen (by a double bond) than is the other (by a single bond) and that the more strongly bonded atom should be closer to the central atom. The hybrid of the two ozone structures gives equal weight to the extra bonding of the two terminal oxygen atoms. Similarly, the three resonance structures of carbonate in Fig. 9-2(*b*) are needed to account for the experimental fact that all three oxygens are equidistant from the central carbon.

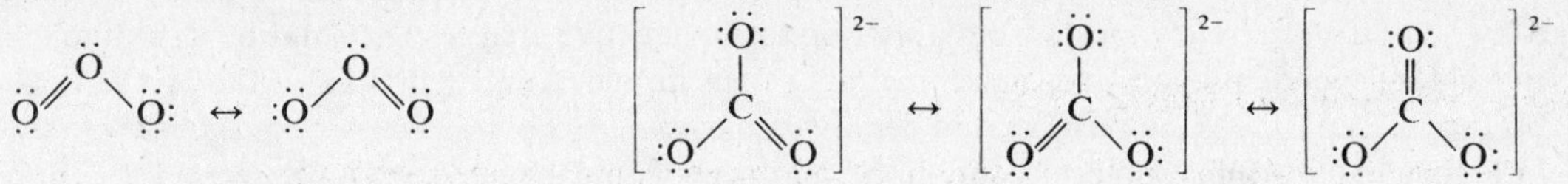

(*a*) Ozone   (*b*) Carbonate ion

Fig. 9-2

The total bond energy of a substance for which resonance structures are written is greater than would be expected if there were only one formal Lewis structure. This *additional* stabilization is called *resonance energy*. It arises from the sample principle that is responsible for covalent bond energy, the delocalization of electrons about the atoms forming the bond. As a result of resonance in ozone, for example, the electrons constituting the second pair of the double bond are delocalized around the 3 oxygen atoms. The writing of two or more resonance structures is a way of overcoming the inability of a single valence-bond structure to show this delocalization.

Formal charge

Although a molecule as a whole is electrically neutral, it is a matter of much current interest to know whether there are local charges which can be identified with particular parts of a molecule, the algebraic sum of which would equal zero. In an ion, the algebraic sum would equal the charge of the ion as a whole. In one approximate method of apportioning charges within a molecule or ion, the shared electrons in a covalent bond are arbitrarily divided equally between the two atoms forming the bond. Unshared valence electrons on an atom are assigned exclusively to that atom. Each atom is then assigned a *formal charge* which is equal to the number of valence electrons possessed by that atom in the neutral free state minus the number of valence electrons assigned to it in the structure. These charges may be written near the atoms on the structural diagrams.

EXAMPLE 2   Figure 9-3 shows a single resonance structure for ozone. The central oxygen is assigned just five electrons (two in the unshared pair plus half of the three pairs in the bonds); this atom, being one electron short of the complement of six in a free oxygen atom, is thus assigned a formal charge of $+1$. The terminal oxygen connected by a single bond is assigned 7 (6 in the unshared pairs plus half of one pair in the bond); having one electron more than a neutral oxygen atom, this atom is assigned a formal charge of -1. The other terminal oxygen has no formal charge because six electrons are assigned to it (four in the unshared pairs plus half of the two shared pairs).

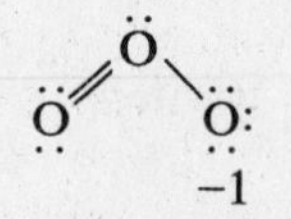

Fig. 9-3

A rough rule useful in choosing one Lewis structure over another is that structures which minimize formal charge separation are favored. Especially to be avoided are formal charges of magnitude greater than 1 and structures in which appreciable formal charges of the same sign are located on adjacent atoms.

Dipole moments and electronegativity

There are some experimental procedures that give information about the *actual* distribution of charges within a molecule (as distinct from the arbitrary assignment of formal charges). One such is the measurement of *dipole moment*. An electric dipole is a neutral object that has a positive charge of magnitude q and a separately located, equal but opposite negative charge. The rotation of a dipole by an electric field is dependent upon, and thus a measure of, the dipole moment, defined as the product of q and the distance, d, separating the positive and negative charges.

In a covalent diatomic molecule, the dipole moment would be expected to be zero if the bonding electrons were shared truly equally by the two atoms. This is indeed the case in molecules of the type X_2, where two identical atoms are bonded. In the more general type, XY, two different kinds of atoms are bonded, and a dipole moment is usually observed. This is explained by hypothesizing that one of the atoms, say Y, has a greater attraction for the shared electrons in the bond than does X. Y is said to have a greater *electronegativity* than X. Electronegativity correlates with other atomic properties; in general, atoms with high ionization energies and/or high electron affinities tend to have high electronegativities. The most electronegative elements in order of decreasing electronegativity are

$$F > O > N = Cl$$

C is more electronegative than H. Metals are less electronegative than nonmetals.

Hybrid orbitals

The assignment of electrons to the various atomic orbitals, as discussed in Chapter 8, concerned the electron distribution in the *ground state* (the state having the lowest energy) of an isolated *free* atom. From the following ground-state configurations of elements in the second period,

$$\underline{B} \quad (He)2s^22p$$
$$\underline{C} \quad (He)2s^22p^2$$
$$\underline{N} \quad (He)2s^22p^3$$
$$\underline{O} \quad (He)2s^22p^4$$

one might predict that the maximum covalence of an element would equal the number of unpaired electrons, since an unpaired electron from each of the bonding atoms participates in the covalent bond formation that we have described. The maximum covalences of B, C, N, and O should then be 1, 2, 3, and 2, respectively. These values account for most of the known compounds and complex ions of N (e.g., NH_3, NO_3^-) and of O (e.g., H_2O, CH_3OH, $HOOH$), but not for the commonly observed trivalence of B (e.g., BF_3), the quadrivalence of C (e.g., CH_4, CCl_4, CH_3OH), or the occasional quadrivalence of N (e.g., NH_4^+).

We can explain the higher-than-predicted covalences of B and C by rearranging the electron configuration to an energy level higher than the ground state before considering compound formation. For example, B might go into the configuration $(He)2s2p^2$, where there are three unpaired electrons available for the formation of three covalent bonds. Although a boron atom could achieve this preparatory state only if it gained the excitation energy needed to transform the atom from its ground state, the principle of overall energy minimization could still be obeyed if the bond energies of the two additional bonds that could be formed exceeded in magnitude this required excitation energy. Similarly, a carbon atom could be excited to the configuration $(He)2s2p^3$, with four unpaired electrons, if the energy released in the formation of the two additional bonds exceeded in magnitude the energy required for excitation to the new electron configuration.

The above explanation accounts for the number of covalent bonds formed by boron and carbon but not for the equivalence of these bonds. The difference in spatial character of s and p orbitals (Fig. 8-2), and in their energies, might suggest differences in the bonds they formed, as measured by bond energy, bond distance, or angles between bonds. Experiment shows, on the contrary, that all three bonds in BF_3 are equivalent, that the angles between any two bonds are the same, and that in fact the three fluorine atoms lie at the corners of an equilateral triangle with the boron atom in the center. Similarly, in CH_4 all four bonds are equivalent, the angles between any two bonds are the same, and the four hydrogen atoms lie at the corners of a regular tetrahedron (Fig. 9-4) with the carbon atom at the center.

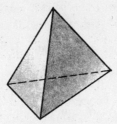

Fig. 9-4

The equivalence of the bonds in BF_3, CH_4, and similar compounds was first explained by Linus Pauling. To accommodate the surrounding atoms, s and p orbitals of a given atom may mix with each other, or *hybridize*. The mathematical formulations of the *hybrid orbitals* are simple linear combinations of the mathematical formulations of the separate s and p orbitals, and the geometrical description of each hybrid orbital is a kind of superposition of the mappings of the separate s and p orbitals. Pauling showed that when one s and three p orbitals are hybridized so as to give a maximum concentration of the probability distribution along some particular direction (so as best to form a bond along this direction), the four resulting so-called sp^3 hybrid orbitals indeed point to the corners of a regular tetrahedron. The angle made by any two lines (i.e., bonds) connecting the center of a regular tetrahedron with a corner is 109°28′ (Problem 9.18). This is the observed situation in CH_4, CCl_4, SiF_4, and many other compounds of the Group IV elements.

A similar hybridization of one s with two p orbitals leads to a set of sp^2 hybrid orbitals which have their maximum concentrations along a set of lines in one plane, forming angles of 120° with each other. This is the observed situation in BF_3; in C_2H_4, each of the two carbon atoms forms a set of sp^2 hybrid orbitals (Fig. 9-5). Two other important types of hybrid orbitals are the sp type, in which the two orbitals point in directions leading to a 180° bond angle (Fig. 9-6), the situation in C_2H_2 and in the high-temperature form of $BeCl_2$, and the d^2sp^3 type, in which the six orbitals point to the corners of a regular octahedron (Fig. 9-7), the situation in SF_6 and in many coordination compounds.

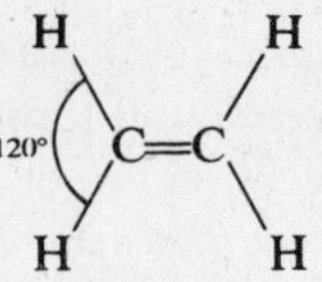

Fig. 9-5 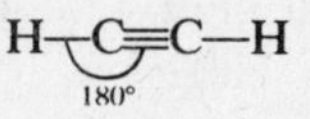   Fig. 9-6 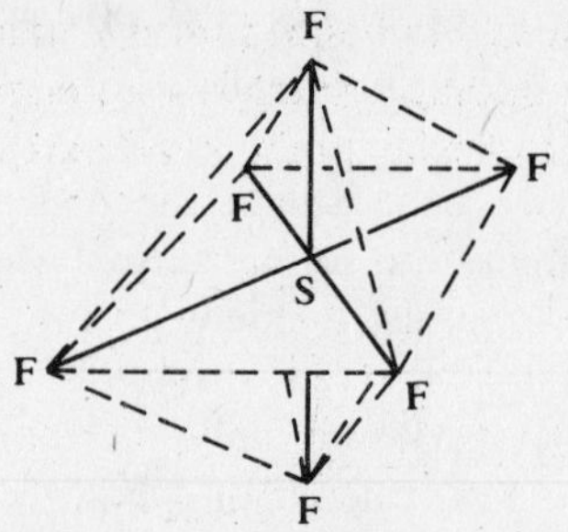   Fig. 9-7

MOLECULAR-ORBITAL REPRESENTATION

In Lewis structures, valence electrons are described as being either unshared electrons localized on particular atoms or shared electrons assigned to the bonds linking particular pairs of atoms. In an alternative representation, all the valence electrons are assigned to *molecular orbitals* (m.o.s) appropriate to the molecule as a whole. Just as an atomic orbital is the mathematical solution of a Schrödinger equation describing the probability distribution of the various positions that an electron having a given set of quantum numbers may occupy around an atomic nucleus, so a molecular orbital describes the distribution of positions in a molecule available to an electron having a given set of quantum numbers. Molecular orbitals may be approximated by writing mathematical combinations of the atomic orbitals of the constituent atoms or they may be pictured, at least in qualitative terms, as geometrical combinations of the contributing atomic orbitals. Several simple rules may be stated for molecular orbitals.

1. The total number of molecular orbitals (m.o.s) equals the sum of the numbers of atomic orbitals (a.o.s) of the constituent atoms.
2. Each orbital can hold 0, 1, or 2 electrons, corresponding to the possibility of two different directions of electron spin and the application of the Pauli principle.
3. When several molecular orbitals of equal or almost equal energy exist, electrons tend to fill these orbitals so as to maximize the number of unpaired electron spins. The more nearly equal the energy levels, the greater is this tendency. (*Hund's rule*)
4. The directions in space describing the orientation of the orbitals, although arbitrary in the case of a free atom (Fig. 8-2), are related to the positions of neighboring atoms in the case of molecules or complex ions.
5. A molecular orbital is most likely to be composed of atomic orbitals of similar energy levels.
6. In diatomic molecules, or more generally in localized two-centered bonds, orbitals of the two atoms that may be combined to form a molecular orbital are those that have the same symmetry about the internuclear axis. (In constructing molecular orbitals that extend over

three or more atoms, more complicated rules of symmetry must be considered, involving the symmetry of the extended nuclear framework.)

7. A two-centered orbital with high electron probability in the region between the nuclei is a *bonding* orbital. One with zero or low electron probability between the nuclei is an *antibonding* orbital. A stable chemical bond exists when the number of electrons in bonding orbitals exceeds the number in antibonding orbitals. Similar but more complex rules apply to orbitals that extend over more than two atoms.

The directional properties of molecular orbitals are governed by quantum numbers analogous to the atomic quantum numbers l and m_l. Greek letters are used in molecular orbital notation to designate increasing values of the l-type quantum number, σ, π, δ, etc., analogous to the Latin letters s, p, d, etc., for atomic orbitals. Only σ and π orbitals will be considered in this book. There are two kinds of σ orbitals, bonding and antibonding (Fig. 9-8). (An antibonding orbital is designated with a superscript asterisk applied to the Greek letter.) A bonding orbital has a region of electron overlap, or high probability, between the bonded atoms, and has a lower energy than either of the component atomic orbitals. An antibonding orbital has a nodal plane, or region of zero probability, between the bonding atoms and perpendicular to the bond axis; its energy greater than that of either constituent atomic orbital.

σ_s^* *antibonding* σ_s *bonding*

Fig. 9-8

If the bond axis is designated by x, σ bonds can be formed by combining any two of the following atomic orbitals, each of which has a region of high electron probability lying along the x axis and cylindrically symmetrical about it: s, p_x, $d_{x^2-y^2}$, or hybrid orbitals pointing along the x axis. If the bond axis is z, σ orbitals may be formed from s, p_z, d_{z^2}, or appropriate hybrids. If the bond axis is y, the component atomic orbitals for σ orbitals are s, p_y, $d_{x^2-y^2}$, or appropriate hybrids.

A π orbital related to a bond in the x direction is characterized by having a zero value at all points along the x axis. π_y orbitals, having their maximum probabilities in the $+y$ and $-y$ directions (above and below the xz plane), may be formed from p_y atomic orbitals. Similarly, π_z orbitals, having their maximum probabilities above and below the xy plane, may be formed from p_z atomic orbitals. π orbitals may also be bonding or antibonding (Fig. 9-9). A d_{xy} orbital may combine with a p_y orbital, or a d_{xz} with a p_z, to form π orbitals.

π_p^* *antibonding* π_p *bonding*

Fig. 9-9

Nonbonding atomic orbitals are those which do not interact with orbitals of other atoms for one of the following reasons:

1. The two atoms are too far apart for good orbital overlap (as in the case of nonadjacent atoms).
2. The energy of the nonbonding orbital is not close to that of any orbital on the adjacent atom (e.g., the $3s$ orbital of Cl is of much lower energy than the $1s$ orbital of H in HCl).

3. The nonbonding orbital is in an inner shell and would not overlap with an orbital even of the neighboring atom (e.g., the K electrons in F_2).
4. The nonbonding orbital does not match its symmetry with any available orbital of the adjacent atom (e.g., the $3p_y$ of Cl does not have a symmetry match with $1s$ of H in HCl, where x is the bonding direction, and the $2p_y$ orbital of H is too high in energy to enter the picture).

A nonbonding orbital has the same energy as in the free atom. Electrons occupying nonbonding orbitals correspond to the unshared electrons in Lewis structures.

A buildup principle analogous to that for atoms exists for molecules. The order of filling molecular orbitals from the valence shells in the case of *homonuclear* diatomic molecules, where x is the bond axis, is as follows:

$$\sigma_s < \sigma_s^* < \pi_y = \pi_z < \sigma_{p_x} < \pi_y^* = \pi_z^* < \sigma_{p_x}^*$$

The ordering may change for *heteronuclear* diatomic molecules, and for homonuclear molecules at the point where this complete set of orbitals is about half-filled.

The *bond order* in a diatomic molecule is defined as one-half the difference between the number of electrons in bonding orbitals and the number in antibonding orbitals. The factor one-half preserves the concept of the electron pair and makes the bond order correspond to the multiplicity in the valence-bond formulation: one for a single bond, two for a double bond, and three for a triple bond. Fractional bond orders are allowed.

π Bonding and multicenter π bonds

An explanation for the bonding in ethylene, C_2H_4, is as follows. The basic framework is established by combining five, two-centered, bonding, σ orbitals, four of which are made from a $1s$ orbital on a hydrogen and an sp^2 hybrid orbital on a carbon, and one of which is made from one sp^2 orbital on each of the carbons. (The combinations of these two-centered σ orbitals are bonding molecular orbitals that extend over the whole molecular framework. These extended molecular orbitals, represented by the lightly shaded regions of Fig. 9-10, may be referred to as σ-*type* because their electron density is concentrated mainly along the axes connecting pairs of adjacent atoms.) If the plane of each H_2C—C group is designated as the xy plane, the p_x and p_y atomic orbitals are used to form the sp^2 hybrid orbitals. The p_z orbitals on the two carbons (the darkly shaded regions of Fig. 9-10) are then available for forming π_z orbitals (as indicated by the dashed lines in Fig. 9-10, which represent the overlap of the darkly shaded orbitals). After filling of the five σ-type bonding orbitals, the remaining electron pair (of the total of 12 valence electrons on the two carbons and four hydrogens) goes into the π_z orbital. The two carbons are thus bonded partly by the electrons in the σ-type framework, constituting the equivalent of a single bond between the two carbons (plus the equivalent of a single bond connecting each of the hydrogens to its adjacent carbon), and partly by the pair of π_z electrons forming the second part of the double bond indicated in Fig. 9-5. The π bond prevents rotation about the C—C axis and constrains all six atoms in C_2H_4 to the same plane. In acetylene, C_2H_2, the σ-type skeleton of three bonding orbitals (represented by the lightly shaded regions of Fig. 9-11) is formed from the $1s$ orbitals on the hydrogen atoms and the hybrid sp orbitals on the carbon atoms. The p_x atomic orbitals are used to form the hybrid orbitals which point along the bonding x direction, and the remaining p orbitals are free to form the π_y and π_z orbitals (represented by the dashed lines joining the darkly shaded regions of Fig. 9-11). The ten valence electrons (one from each hydrogen and four from each carbon) fill the three σ-type orbitals and the π_y and π_z orbitals. The carbons are held together by the equivalent of a triple bond (one σ bond and two π bonds), as indicated in Fig. 9-6.

In ozone, O_3, the atoms are held together in the first instance by a σ-type framework (represented by the lightly shaded regions of Fig. 9-12) in the xy plane of the molecule, with its electron density concentrated mainly along the axes connecting pairs of nearest-neighbor oxygen atoms. The p_z orbitals of all three oxygen atoms have the same symmetry with respect to the plane of the molecule and can thus combine to form π orbitals. The p_z orbital of the central oxygen atom overlaps with the p_z orbitals of both

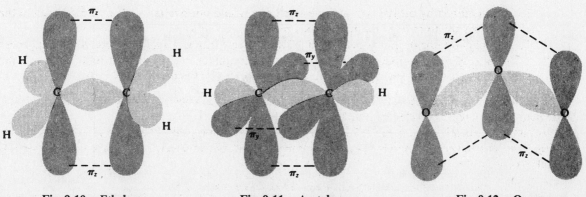

Fig. 9-10. Ethylene   Fig. 9-11. Acetylene   Fig. 9-12. Ozone

the terminal oxygens. As a result, a π bonding orbital and a π^* antibonding orbital extend over all three atoms in the molecule, and a π nonbonding orbital involves the two terminal atoms. Two electrons in ozone occupy the π bonding orbital, spending some of their time near one side of the molecule and some near the other; this representation is an alternative to resonance, which was conceptualized to preserve the supposed sanctity of the octet rule. The π nonbonding orbital is also occupied in ozone. The occupied π bonding orbital is represented by the dashed lines joining the darkly shaded regions of Fig. 9-12.

Multicentered π orbitals are involved in the molecular-orbital representation of most structures for which resonance must be invoked in the valence-bond representation. In a long chain of atoms bonded within a planar configuration, such as the plant pigment carotene, $C_{40}H_{56}$, or in planar ring compounds like naphthalene, $C_{10}H_8$, each π orbital extends over many carbon atoms, all those in the basic molecular plane, since the overlap of the p_z orbital of any nonterminal carbon with those of its two or three neighbors allows for the buildup of long chains or rings of extensively overlapping electron probability distributions.

SHAPES OF MOLECULES

Bond lengths

Bond lengths between a given pair of atoms are approximately constant from compound to compound if the order of the bond (single, double, or triple) is the same. If it is assumed that the length of a single covalent bond is the sum of the *covalent radii* of the two bonding atoms, then quick and fairly reliable estimates of bond lengths can be obtained from a compilation such as Table 9-2.

Table 9-2. Single-Bond Covalent Radii

C	77 pm	O	66 pm
Si	117 pm	S	104 pm
N	70 pm	F	64 pm
P	110 pm	Cl	99 pm
Sb	141 pm	I	133 pm

Double bonds are shorter. From the precisely measured lengths 154 pm, 133 pm, and 120 pm, for $H_3C—CH_3$, $H_2C=CH_2$, and $HC\equiv CH$, respectively, a rule of thumb has been developed to allow a 21 pm shortening for any double bond and 34 pm shortening for any triple bond. In the case of resonance or multicentered π orbitals, a bond length is intermediate between the values it would have in the separate resonance structures or between the values it would have in the absence of π bonding and in the presence of a localized two-centered π bond. Some applications of this rule are shown in Table 9-3.

Table 9-3

Substance	Bond	Predicted Length	Observed Length
CH_3Cl	C—Cl	$r_C + r_{Cl} =$ $77 + 99 = 176$ pm	177 pm
$(CH_3)_2O$	C—O	$r_C + r_O =$ $77 + 66 = 143$ pm	143 pm
H_2CO	C=O	$r_C + r_O - 21 =$ $143 - 21 = 122$ pm	122 pm
ICN	C≡N	$r_C + r_N - 34 =$ $77 + 70 - 34 = 113$ pm	116 pm

Bond angles

Starting with the Lewis structure it is possible to predict fairly accurately the bond angles in a molecule. The VSEPR method (acronym for Valence Shell Electron Pair Repulsion) focuses on some central atom and counts the number of atoms bonded to it plus the number of unshared pairs. (Multiple bonds count the same as single bonds.) This characteristic VSEPR number is the number of orbitals (each occupied by an electron pair) that must emanate from the central atom. The angles between them are determined by the principle that the electron pairs will repel each other as far as possible. The VSEPR method is a simple unsophisticated technique that does not require explicit identification of the orbitals. Invoking the hybrid orbital technique results in the same angles, but in a formal manner linked to the mathematical treatments of bonding. VSEPR numbers, corresponding angles, and hybrid orbital sets are listed in Table 9-4.

Table 9-4

VSEPR Number	Nominal Bond Angles	Hybrid Orbital Set
2	180°	sp
3	120°	sp^2
4	109°28′	sp^3
5	90°, 120°, 180°	dsp^3
6	90°, 180°	d^2sp^3

Deviations from the above angles arise mostly from unshared pairs, which repel more strongly than shared pairs. For example, the VSEPR number is 4 for CH_4, $:NH_3$, and $H\ddot{O}H$. The HCH angle in CH_4 is a perfect tetrahedral angle (109°28′), but the HNH bond in $:NH_3$ is compressed to about 107° by the unshared pair, and in $H\ddot{O}H$ the two unshared pairs compress the HOH angle further to 104°. Double bonds also contribute extra repulsion which compresses adjacent bond angles slightly. Deviations arise also if the various atoms around the central atom are very different in size.

COORDINATION COMPOUNDS

The electrons of an electron-pair bond need not be contributed by both of the bonding atoms, as is demonstrated in the formation of NH_4^+ by the addition of a proton to NH_3. The Lewis structures for NH_3 and NH_4^+ are shown in Fig. 9-13; H^+ has no electrons. Such a bond is often called a *coordinate* covalent bond, but is essentially no different from any other covalent bond. In this particular case, once the

$$H\overset{\displaystyle ..}{—N}—H \qquad \left[H—\overset{\displaystyle \overset{H}{|}}{\underset{\displaystyle \underset{H}{|}}{N}}—H \right]^{+}$$

$$\overset{|}{H}$$

Fig. 9-13

coordinate covalent bond is formed, it becomes indistinguishable from the other three N—H bonds; in fact, the structure is a regular tetrahedron.

Coordinate covalence is the common bonding type in coordination compounds, in which a central metal atom or ion is bonded to one or more neutral or ionic *ligands*. A typical ligand, such as NH_3, Cl^-, or CO, has an unshared electron pair that forms the coordinate covalent bond by interacting with the unfilled orbitals of the central metal. The overall charge of a complex ion is the algebraic sum of the charge of the central metal and the charges of the ligands.

A number of rules have been adopted internationally for the naming of coordination compounds.

1. If the compound itself is ionic, the cation is named first.
2. A complex ion or nonionic molecule carries the name of the central metal last, with its oxidation state (charge per atom) in parenthesized Roman numerals (or zero). (A detailed discussion of oxidation state is given in Chapter 11.)
3. Anionic ligands are named with the suffix *-o*, as in *chloro, oxalato, cyano*.
4. The number of ligands of a given type is given by a Greek prefix, like *mono-* (often omitted), *di-*, *tri-, tetra-, penta-, hexa-*.
5. If the name of the ligand itself contains a Greek prefix, the number of ligands is indicated by such prefixes as *bis-, tris-, tetrakis-*, for 2, 3, or 4, respectively, and the name of the ligand is enclosed in parentheses.
6. When the complex ion is anionic, the name of the metal is given in its Latin form with the suffix *-ate*.
7. Some neutral ligands are given special names, like *ammine* for NH_3, *aqua* for H_2O, *carbonyl* for CO.
8. When several ligands occur in the same complex, they are named in alphabetical order (ignoring numerical prefixes).

EXAMPLE 3

$[Co(NH_3)_6](NO_3)_3$	Hexaamminecobalt(III) nitrate
$Ni(CO)_4$	Tetracarbonylnickel(0)
$K[Ag(CN)_2]$	Potassium dicyanoargentate(I)
$[Cr(NH_3)_4Cl_2]Cl$	Tetraamminedichlorochromium(III) chloride
$[Co(NH_2C_2H_4NH_2)_3]Br_3$	Tris(ethylenediamine)cobalt(III)bromide

The last formula may be abbreviated as $[Co(en)_3]Br_3$.

Structure, properties, and bonding

In many coordination compounds, the ligands are arranged around the central metal in regular geometrical forms, such as the octahedron, the tetrahedron, or the square. In the formulas listed above, the brackets define the complex made up of the central metal and its ligands. These brackets are often omitted when there is no ambiguity as to the nature of the complex. Many of the compounds are colored. Some are paramagnetic because of the presence of unpaired electrons, while other complexes of the same metal are not paramagnetic. The molecular-orbital theory has been applied to these complex compounds and ions in an attempt to explain these properties, as well as the stability of the complexes. Here we shall consider only the six-coordinated octahedral complexes.

The hybridization on a metal atom leading to octahedral bonding is d^2sp^3. If the bonding axes are x, y, and z, the two d orbitals that are used for the hybridization are those that point along one or more of these axes, d_{z^2} and $d_{x^2-y^2}$ (Fig. 8-2). Each of the resulting six hybrid orbitals mixes with an orbital directed along a bonding axis on the ligand, such as p or a tetrahedral hybrid, to form σ and σ^* molecular orbitals. Each bonding σ orbital is occupied by a pair of electrons.

The full set of molecular orbitals for the complex may be constructed from the following atomic orbitals: all five d orbitals of the next-to-outermost shell of the metal, the s and the three p orbitals of the metal's outer shell, and one orbital from each of the six ligands pointed along a bond axis, such as a p or a hybrid orbital. The total number of participating atomic orbitals is 15, and the total number of orbitals remaining after molecular-orbital formation is still 15. The electrons to be accommodated in these orbitals include the six previously unshared pairs provided by the six ligands for bond formation to the metal, plus the d electrons of the metal or its ionic form appropriate to the particular oxidation state. Note that the ligand electrons that remain nonbonding within the complex (the electrons used for bonding within the ligand themselves), as well as the inner-shell electrons of both metal and ligands, need not be taken into account, at least to a first approximation, since they are unaffected by complex formation.

A schematic molecular-orbital diagram is shown in Fig. 9-14, with the contributing metal atomic orbitals shown on the left and the ligand orbitals on the right. (A more complicated diagram would be needed if the ligands also had π orbitals that participate in the bonding to the metal.) The actual energy spacings will depend on the particular case, but the relative ordering of the $(n-1)d$, ns, and np metal

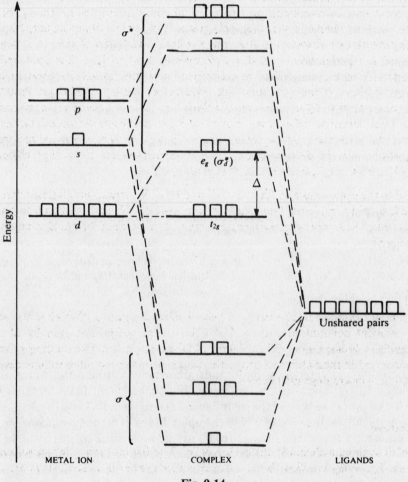

Fig. 9-14

orbitals (where n is the principal quantum number of the outermost shell) is general. Also it is common for the ligand orbitals to be at lower energy than the metal orbitals. Note that of the nine metal orbitals, only six contribute to primary bonding in the complex. These six are the same ones that can be hybridized to form the octahedral hybrids pointing along the $\pm x$, $\pm y$, and $\pm z$ directions; namely, the s, the three p, and the d_{z^2} and $d_{x^2-y^2}$. The six metal orbitals contributing to bonding mix with the six ligand orbitals to form six σ-type bonding and six σ^*-type antibonding molecular orbitals delocalized over the whole complex. The bonding molecular orbitals have lower energies than the ligand orbitals, and the antibonding have higher energies, as is usual. Also as usual, the lower the energy of a bonding orbital, the higher that of its corresponding antibonding orbital. The three remaining metal d orbitals, not having σ character with respect to the bond directions, remain nonbonding and their energy is unchanged to a first approximation. Several consequences can easily be seen from this diagram.

1. The 12 electrons supplied by the ligands can be thought to occupy and fill the six bonding σ-type orbitals.
2. The next-lowest-lying orbitals, available for the valence electrons supplied by the metal, are the three nonbonding orbitals (the d_{xy}, d_{xz}, and d_{yz} orbitals, which do not have maxima in their probability distributions along any of the bonding axes). These orbitals are referred to as t_{2g}.
3. Of the σ^*-type antibonding orbitals, the lowest-lying are the two whose metal components are $d_{x^2-y^2}$ and d_{z^2}, designated either as σ_d^* or as e_g. The energy difference between this level and the nonbonding level is denoted as Δ. (The symbol for this energy difference is not standard; various textbooks use X, $10\,Dq$ or Δ_o.)
4. The first three valence electrons of the metal will occupy the nonbonding t_{2g} orbitals.
5. The next two valence electrons of the metal could occupy either the t_{2g} or the e_g orbitals, depending on whether Δ is respectively greater than or less than the energy increase associated with pairing two electrons in the same orbital, in violation of rule 1 on page 105. If the t_{2g} orbitals are preferentially filled, the complexes tend to have low spin and the d^3 and d^6 configurations are stabilizing, in accordance with the features of filled or half-filled equal-energy orbitals. If the e_g orbitals are filled before the t_{2g} orbitals are doubly occupied, the complexes tend to have high spin and the d^5 and d^{10} are stabilizing configurations.
6. The weak electron transitions responsible for the color of coordination complexes are correlated with the t_{2g} to e_g transition, of energy Δ. (The origin of the strong transitions responsible for the deep colors of some complexes of metals in high oxidation states, like MnO_4^- and CrO_4^{2-}, has a different explanation.)

With regard to the options in rule 5, CN^-, CO, and NO_2^- are *strong-field* ligands that tend to increase Δ and promote low-spin complexes; OH^- and Cl^- are *weak-field* ligands that lead to smaller values of Δ and high-spin complexes. A more complete listing of ligands in order of their field strength is known as the *spectrochemical series*.

ISOMERISM

There are many examples in chemistry of two or more substances that have the same numbers of atoms of each element per molecule or ion but differ in the spatial arrangement of the atoms. Such substances, known as *isomers*, may differ in their physical properties (like melting point, boiling point, density, and color) and in their chemical properties (reactivities toward other substances). Three different categories of isomerism are described below.

Structural isomerism

One way of describing a chemical compound or ion is to list for each atom the numbers of each kind of atom connected to it by covalent bonds. Isomers which differ in these listings are called *structural isomers*.

EXAMPLE 4   As shown in Fig. 9-15, two different structures can be drawn for C_4H_{10}, both conforming to the octet rule for carbon. In *n*-butane, two of the carbons are bonded to one carbon each and two are bonded to two each. In isobutane, three of the carbons are bonded to one carbon each and the fourth to three carbons. *n*-Butane melts at $-135\,°C$ and boils at $0\,°C$, while isobutane melts at $-145\,°C$ and boils at $-10\,°C$. Within any one molecule there is a continuous, practically unrestricted, rotation of the atoms about any C—C bond, so that all structures which on paper differ only by the angular position of an atom or group of atoms linked by a single bond are really the same, as shown in Fig. 9-16. These diagrams are all representations of the same substance, *n*-butane. In each, we can number the carbon atoms 1 to 4 as we move from one end of the molecule to the other through successive C—C bonds. We see that the kinds of neighbors of a given numbered carbon atom are the same in all three diagrams.

n-Butane

Isobutane

Fig. 9-15

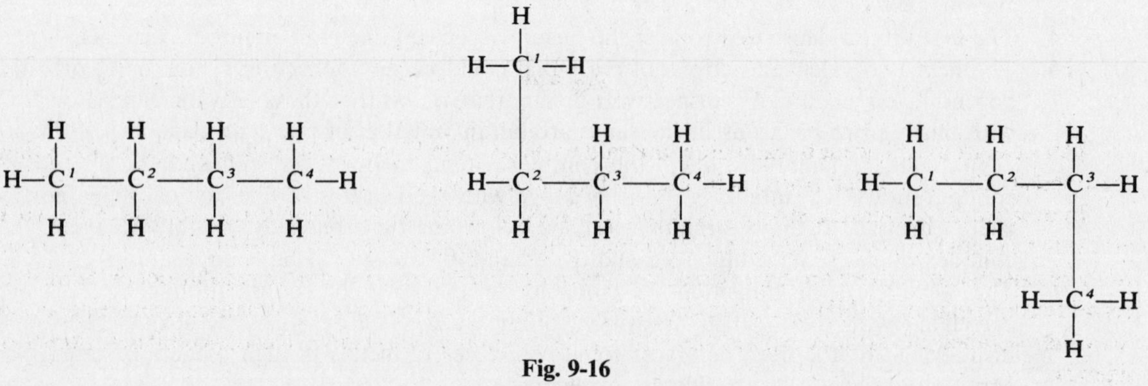

Fig. 9-16

Structural isomers also exist for some coordination compounds. One example is a set of compounds in which a ligand in one isomer may occupy a position outside the coordination sphere in another isomer. Thus

$$[Co(NH_3)_5Br]SO_4 \quad \text{and} \quad [Co(NH_3)_5SO_4]Br$$

are structural isomers, each recognizable by its own color and complete set of distinctive properties.

Geometrical isomerism

In some isomers the listing of bonded atoms around each atom is the same, but the compounds differ because at least two atoms, bonded to the same or to adjacent atoms but not to each other, are at different distances in the several forms. Such isomers are called *geometrical isomers*.

An important set of geometrical isomers occurs in compounds containing a carbon-carbon double bond. The extra strength of the double bond, together with the coplanarity of the atoms bonded to the doubly bonded carbons, which is forced by the π electron overlap of the doubly bonded carbons, restricts the motion of groups with respect to the double bond. Consequently, diagrams showing different angular positions represent different substances.

EXAMPLE 5   In Fig. 9-17 structures (*a*) and (*b*) are *structural* isomers of each other, as are (*a*) and (*c*), since in (*a*) both chlorines are attached to the same carbon and in (*b*) or (*c*) one chlorine is attached to each carbon. Structures (*b*)

Cl\C=C/H   Cl\C=C/Cl   Cl\C=C/H
Cl/   \H   H/   \H   H/   \Cl

(a)   (b)   (c)

Fig. 9-17

and (c) are geometrical isomers of each other because they differ in the spatial arrangement of their atoms but not in the listing of the mere numbers of atoms of each kind to which each atom is bonded. All three structures are therefore different substances.

EXAMPLE 6 Geometrical isomerism in compounds with square symmetry is illustrated in Fig. 9-18. Because of the rigidity of the square planar arrangement of the Pt, N and Cl atoms, the two forms are distinctive. For example, the distance between the two chlorine atoms in (a) is greater than in (b). If this compound had been tetrahedral with Pt in the center, isomerism would not have occurred, because any corner of a regular tetrahedron is equidistant to the other three corners.

        NH$_3$ NH$_3$
         | |
Cl—Pt—Cl   Cl—Pt—NH$_3$
         | |
        NH$_3$ Cl

      (a)   (b)

Fig. 9-18

Identical atoms adjacent to each other are said to be *cis*, as in Figs. 9-17(b) and 9-18(b), and when they are opposite they are called *trans*, as in Figs. 9-17(c) and 9-18(a).

In octahedral complexes, structures of the type MX_4Y_2 may exist as geometrical isomers. The two Y atoms (or groups) can occupy either adjacent sites [Fig. 9-19(a)] or opposite sites [Fig. 9-19(b)]; only two isomers exist because there are only two different distances between corners of a regular octahedron. The *cis* and *trans* forms of $Pt(NH_3)_2Cl_4$ are isomers of this type. Possibilities for geometrical isomerism occur for other octahedral formula types, like MX_3Y_3, or complexes with more than two different kinds of ligands.

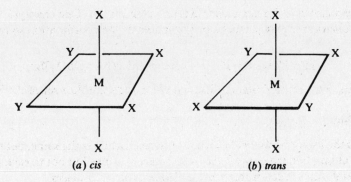

(a) cis   (b) trans

Fig. 9-19

Optical isomerism

Optical isomers, occurring in pairs, are distinct structures with groups bonded to a central atom in arrangements which are mirror images of each other. Among the important categories of optical isomers are compounds containing four different groups singly bonded to a given carbon atom (Fig. 9-20) and octahedral complexes with three different kinds of ligands or with several multivalent ligands (Fig. 9-21).

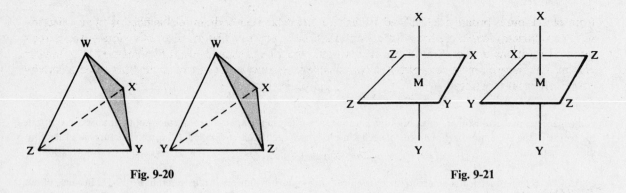

Fig. 9-20   Fig. 9-21

BONDING IN METALS

Molecular orbital theory has been invoked to explain the bonding in metallic crystals, such as pure sodium or pure aluminum. Each m.o., however, instead of subtending just a few atoms as in a typical molecule, must in a metal cover the entire crystal (of perhaps 10^{20} atoms!). Following the rule that the number of m.o.s must equal the number of a.o.s combined, this many m.o.s must be so close on an energy level diagram that they form a continuous band of energies. Thus the theory is known as the "band" theory.

Consider the case of sodium metal. Of the 11 electrons, the 10 forming the neon core are localized around each Na nucleus, leaving one per atom to fill the m.o.s that pervade the crystal. If there were N atoms in the crystal N m.o.s could be formed by use of one $3s$ orbital from each. $N/2$ of these would be bonding, and these would just be filled (at two electrons per orbital) by the N available electrons.

In metals beyond Group I the picture becomes more complicated as both s and p orbitals are used to form the band of m.o.s, which then contains many more orbitals than the number of electron pairs available. The criterion for electrical conductivity, a property of all metals, is that the energy band be only partially filled.

A simple alternative model, consistent with band theory, is the "electron sea" picture, illustrated in Fig. 9-22 for sodium. The circles represent the sodium ions, which occupy regular lattice positions (the second and fourth lines of atoms shown are in a plane below the first and third). The eleventh electron

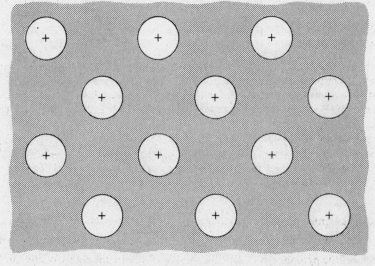

Fig. 9-22

from each atom is broadly delocalized so that the space between sodium ions is filled with an "electron sea" of sufficient density to keep the crystal electrically neutral. The massive ions vibrate about their nominal positions in the electron sea, which holds them in place somewhat like cherries in a bowl of gelatin. This simple model successfully accounts for the unusual properties of metals such as conductivity and mechanical toughness.

Solved Problems

FORMULAS

9.1. Write formulas for the following ionic compounds: (a) barium oxide, (b) aluminum chloride, (c) magnesium orthophosphate.

    (a) The formula is BaO since the $+2$ charge on one barium ion just balances the -2 charge on the oxide ion.

    (b) Three chloride ions, -1 charge on each, are needed to balance the $+3$ charge of one aluminum ion. The formula is $AlCl_3$.

    (c) Since neither 2 (the positive charge on magnesium ion) nor 3 (the negative charge on orthophosphate) is an integral multiple of the other, the smallest number must be found that is a multiple of both. The number is 6. The formula, $Mg_3(PO_4)_2$, shows 6 units of positive charge (3×2) and 6 units of negative charge (2×3), per empirical formula unit.

9.2. Name the following compounds: (a) Mg_3P_2, (b) $Hg_2(NO_3)_2$, (c) NH_4TcO_4.

    (a) Although the ion of phosphorus is not included in Table 9-1, it must be an anion, since magnesium forms only a cation. Application of the usual rule for naming binary compounds gives the name *magnesium phosphide*.

    (b) Since the charge on two NO_3^- (nitrate ions) is -2, the total cationic charge (Hg_2) must be $+2$. Since the average charge per Hg is $+1$, the name is *mercury(I) nitrate* (or *mercurous nitrate*).

    (c) The charge on the anion must be -1 in order to balance the $+1$ charge on the ammonium ion. Since Tc belongs to the same group of the periodic table as Mn, TcO_4^- is analogous to MnO_4^-, permanganate. The name is thus *ammonium pertechnetate*.

9.3. Determine the charges of the complex ions in parentheses [or brackets] in the following formulas:

    (a) $Na_2(MnO_4)$ (c) $NaCd_2(P_3O_{10})$ (e) $Ca_3(CoF_6)_2$ (g) $(UO_2)Cl_2$

    (b) $H_4[Fe(CN_6)]$ (d) $Na_2(B_4O_7)$ (f) $Mg_3(BO_3)_2$ (h) $(SbO)_2SO_4$

    (a) The charge of (MnO_4) must balance that of two Na^+, i.e., $(MnO_4)^{2-}$. (This ion is called *manganate* and is different from permanganate.)

    (b) The ion in brackets must balance the charge of four H^+, i.e., $[Fe(CN)_6]^{4-}$.

    (c) The charge of (P_3O_{10}) must balance the charge of one Na^+ and two Cd^{2+}, i.e., $(P_3O_{10})^{5-}$.

    (d) The charge of (B_4O_7) must balance the charge of two Na^+, i.e., $(B_4O_7)^{2-}$.

    (e) The charge of two (CoF_6) ions must balance the charge of three Ca^{2+}, i.e., $(CoF_6)^{3-}$.

    (f) The charge of two (BO_3) ions must balance the charge of three Mg^{2+}, i.e., $(BO_3)^{3-}$.

    (g) The charge of the (UO_2) ion must balance the charge of two Cl^-, i.e., $(UO_2)^{2+}$.

    (h) The charge of two (SbO) ions must balance the charge of SO_4^{2-}, i.e., $(SbO)^+$.

9.4. The formula of calcium pyrophosphate is $Ca_2P_2O_7$. Determine the formulas of sodium pyrophosphate and iron(III) pyrophosphate (ferric pyrophosphate).

    The charge of the pyrophosphate ion must be -4 to balance the charge of two Ca^{2+}. We can then write $Na_4P_2O_7$ and $Fe_4(P_2O_7)_3$.

9.5.   Write all possible octet structural formulas for   (a) CH_4O,   (b) C_2H_3F,   (c) N_3^-.

(a)   Since the valence level of hydrogen is saturated with two electrons, each hydrogen can form only one covalent bond. Thus hydrogen cannot serve as a bridge between C and O. The only possibility of providing four bonds to the C is to have 3 H's and the O bonded directly to it. Only one structure is possible, as shown in Fig. 9-23. Note that the total number (14) of valence electrons in the structure is the sum of the numbers of valence electrons in the free component atoms: 4 (in C) + 6 (in O) + 4 (in 4 H).

(b)   The only way of providing four bonds for each C within the limitation of 18 valence electrons is to have a C=C bond, as in Fig. 9-24. The reader should determine by trial and error that no other structure is possible.

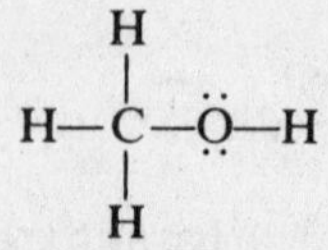

Fig. 9-23

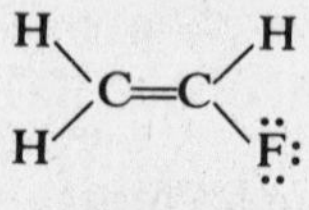

Fig. 9-24

(c)   The total number of available valence electrons is 16 (5 in each of the 3 free N atoms plus 1 arising from the net ionic charge). From this it can easily be seen that a linear structure without multiple bonds would not satisfy the octet rule. For example, in Fig 9-25, the available electrons provide octets for the terminal nitrogen atoms, but four less than an octet for the central one. The only way to have the same number of electrons provide octets for more atoms is to reduce the number of unshared electrons by four and increase the number of shared electrons accordingly, by forming either two double bonds or one triple bond. The possible linear structures are shown in Fig. 9-26.

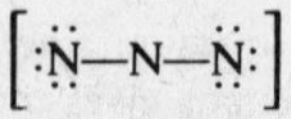

Incorrect structure   (a)   (b)   (c)

Fig. 9-25   **Fig. 9-26**

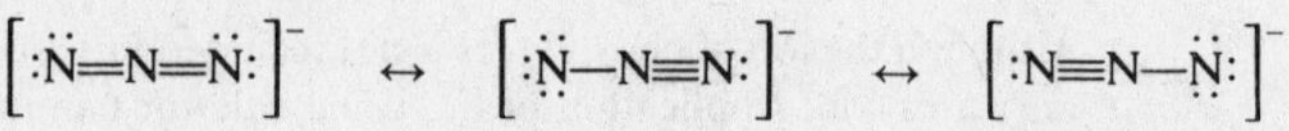

Two resonance structures are written with the triple bond because there is no reason why one of the terminal nitrogens in an isolated azide ion should be different from the other. A different type of structure, involving a ring, also satisfies the octet rule. This structure (Fig. 9-27) is ruled out because the bond angles of 60° or less demanded by the structure require considerable strain from the normal angles of bonding, especially with a double bond in the ring.

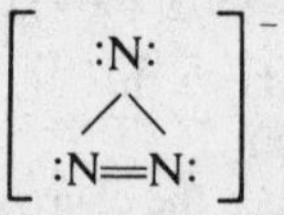

Fig. 9-27

9.6.   Experimentally the azide ion, N_3^-, is found to be linear, with each adjacent nitrogen-nitrogen distance equal to 116 pm.   (a) Evaluate the formal charge at each nitrogen in each of the three linear octet structures depicted in Fig. 9-26.   (b) Predict the relative importance of these three resonance structures in N_3^-.

(a)   In structure (a) of Fig. 9-26, the central N is assigned $\frac{1}{2}$ of the four shared pairs, or four electrons. This number is one less than the number of valence electrons in a free N atom, and thus this atom has a formal charge of +1. Each terminal N is assigned four unshared electrons plus $\frac{1}{2}$ of the two shared pairs, or six altogether. The formal charge is thus −1. The net charge of the ion, −1, is the sum 2(−1) + 1.

In structures (b) and (c) of Fig. 9-26, the central N again has four assigned electrons with a resulting formal charge of $+1$. The terminal triply bonded N has 2 plus $\frac{1}{2}$ of 3 pairs, or a total of 5, with a formal charge of 0. The singly bonded terminal N has 6 plus $\frac{1}{2}$ of 1 pair, or a total of 7, with a formal charge of -2. The net charge of the ion, -1, is the sum of $+1$ and -2.

(b) Structure (a) is the most important because it has no formal charge of magnitude greater than 1. For structure (a) the nitrogen-nitrogen bond distance is predicted to be $(70 + 70)$ minus the double bond shortening of 21 pm, or 119 pm. The observed bond length is 116 pm.

9.7. The sulfate ion is tetrahedral, with four equal S—O distances of 149 pm. Draw a reasonable structural formula consistent with these facts.

It is possible to place the 32 valence electrons (6 for each of the 5 Group VI atoms plus 2 for the net negative charge) in an octet structure having only single bonds. There are two objections to this formula (Fig. 9-28). (1) The predicted bond distance, $r_S + r_O = 104 + 66 = 170$ pm, is much too high. (2) The calculated formal charge on the sulfur, $+2$, is rather high. Nonetheless this structure appears widely in texts, and it is rationalized that the short bond length is the result of the strong attraction between the formally $+2$ sulfur and -1 oxygens. An alternative is to write resonance structures containing double bonds. A structure like that in Fig. 9-29 places zero formal charge on the sulfur and -1 on each of the singly bonded oxygens. The shrinkage in bond length due to double bond formation helps to account for the low observed bond distance. Other resonance structures (making a total of six) with alternate locations of the double bonds are, of course, implied. Structures like this, with an expanded valence level beyond the octet, are generally considered to involve d orbitals of the central atom. This is the reason that second-period elements (C, N, O, F) do not form compounds requiring more than eight valence electrons per atom; there is no such thing as a $2d$ subshell.

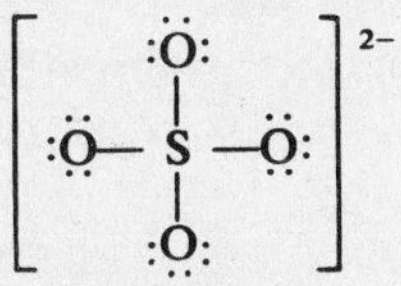

Fig. 9-28 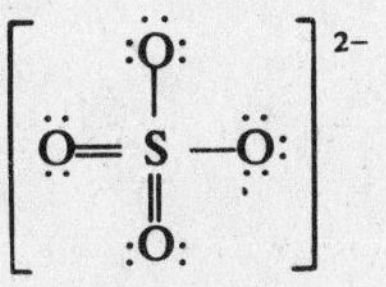   Fig. 9-29

9.8. Draw all the octet resonance structures of (a) benzene, C_6H_6, and (b) naphthalene, $C_{10}H_8$. Benzene is known to have hexagonal symmetry, and the carbon framework of naphthalene consists of two coplanar fused hexagons. Invoke bonding only between adjacent atoms.

(a) The hydrogens are distributed uniformly, one on each carbon, to conform with the hexagonal symmetry (Fig. 9-30). This leaves three carbon-carbon bonds to be formed at each carbon to bring its total covalence to four. Alternating single and double bonds form the only scheme for writing formulas within all the above restrictions. Hydrocarbons containing planar rings in which every ring carbon forms one double bond and two single bonds are called *aromatic* hydrocarbons. A shorthand notation has been developed for writing aromatic structures by the use of polygons. A carbon atom is assumed to lie at each corner of the polygons. Carbon-carbon bonds are written in the polygons but carbon-hydrogen bonds are not explicitly written. Figure 9-31 shows the same two structures as Fig. 9-30.

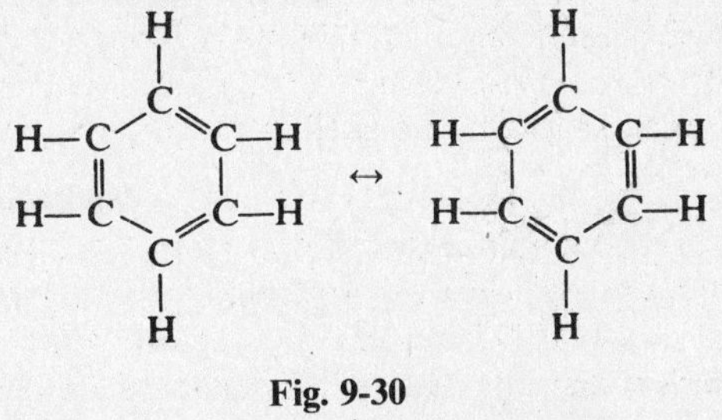

Fig. 9-30

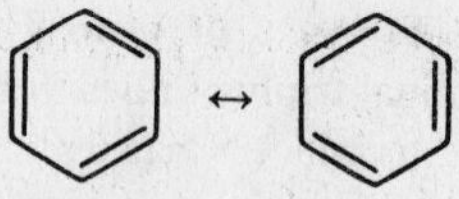

Fig. 9-31

(b) The coplanarity of naphthalene indicates its aromatic character. Note in Fig. 9-32 that the two carbons at the fusion of the two rings reach their covalence of four without bonding to hydrogen. The shorthand notation of these structures is shown in Fig. 9-33. A C—H bond is assumed at each of those carbons possessing only three bonds within the carbon skeleton.

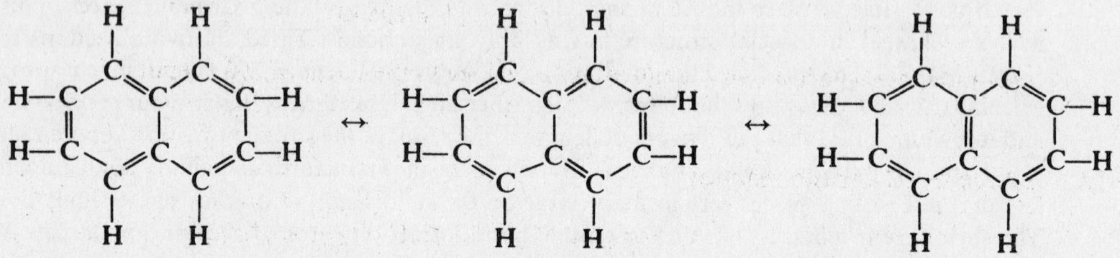

Fig. 9-32

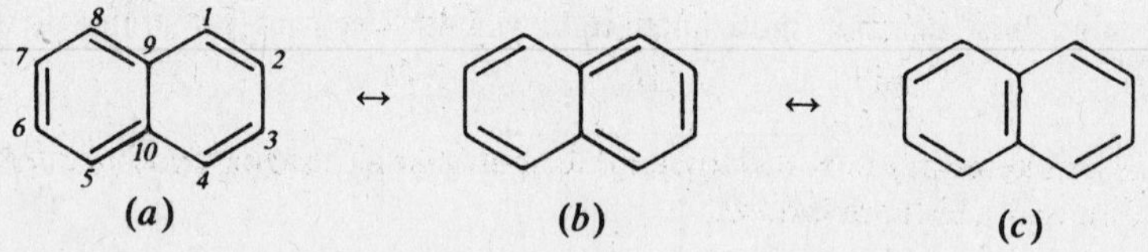

Fig. 9-33

Fig. 9-34

It is common practice to use the shorthand notation shown in Fig. 9-34 (a) and (b) to represent benzene and naphthalene, respectively. The circle emphasizes the delocalization of the electrons without the necessity of showing the resonance forms.

BONDING PROPERTIES

9.9. From the data of Problem 7.31, express the H—H bond energy in kJ/mol.

The bond energy is the energy needed to dissociate gaseous H_2 into separated atoms.

$$H_2 \rightarrow 2H$$

ΔH for this reaction is twice ΔH_f for 1 mol of H.

$$\Delta H = 2(218 \text{ kJ/mol}) = 436 \text{ kJ/mol}$$

9.10. Does Table 7-1 give sufficient data for the evaluation of the Br—Br bond energy?

No; the energy of Br is given with respect to the standard state for Br_2, which is the liquid and not the gas.

9.11. Enthalpies of hydrogenation of ethylene (C_2H_4) and benzene (C_6H_6) have been measured, where all reactants and products are gases.

$$C_2H_4 + H_2 \rightarrow C_2H_6 \qquad \Delta H = -137 \text{ kJ}$$
$$C_6H_6 + 3H_2 \rightarrow C_6H_{12} \qquad \Delta H = -206 \text{ kJ}$$

Estimate the resonance energy of benzene.

If C_6H_6 had three isolated carbon-carbon double bonds, ΔH of hydrogenation would be close to three times ΔH of hydrogenation of C_2H_4, with one double bond, or -411 kJ. The fact that benzene's hydrogenation is less exothermic by

$$411 - 206 = 205 \text{ kJ}$$

means that benzene has been stabilized by resonance to the extent of 205 kJ/mol.

9.12. Estimate ΔH for the reaction:

$$C_2H_6(g) + Cl_2(g) \rightarrow C_2H_5Cl(g) + HCl(g)$$

given the following bond energies, in kJ/mol:

C—C	C—H	Cl—Cl	C—Cl	H—Cl
348	414	242	327	431

It is first necessary to write structural formulas for all reactants and products to determine which bonds are broken and which are formed.

$$
\begin{array}{c}
\quad \text{H} \;\; \text{H} \qquad\qquad\qquad \text{H} \;\; \text{H} \\
\quad | \;\;\; | \qquad\qquad\qquad\; | \;\;\; | \\
\text{H—C—C—H} + \text{Cl—Cl} \rightarrow \text{H—C—C—Cl} + \text{H—Cl} \\
\quad | \;\;\; | \qquad\qquad\qquad\; | \;\;\; | \\
\quad \text{H} \;\; \text{H} \qquad\qquad\qquad \text{H} \;\; \text{H}
\end{array}
$$

Bonds broken		Bonds formed	
C—H	414	C—Cl	327
Cl—Cl	242	H—Cl	431

Since bond breaking is endothermic (positive ΔH) and bond formation is exothermic (negative ΔH), ΔH for the reaction is

$$414 + 242 - 327 - 431 = -102 \text{ kJ/mol}$$

(The accurate answer, calculated from ΔH_f° values, is -113 kJ/mol.)

9.13. The dipole moment of LiH is 1.964×10^{-29} C·m, and the interatomic distance between Li and H in this molecule is 159.6 pm. What is the *percent ionic character* in LiH?

Let us calculate the dipole moment of a hypothetical, 100 percent ionized Li^+H^- ion pair with a separation of 159.6 pm.

$$\mu(\text{hypothetical}) = (1 \text{ electronic charge}) \times (\text{separation})$$
$$= (1.602 \times 10^{-19} \text{ C})(1.596 \times 10^{-10} \text{ m}) = 2.557 \times 10^{-29} \text{ C·m}$$

The percent ionic character equals 100 percent times the fraction: actual dipole moment over the hypothetical.

$$\text{Percent ionic character} = 100\% \times \frac{1.964 \times 10^{-29} \text{ C·m}}{2.557 \times 10^{-29} \text{ C·m}} = 76.8\% \text{ ionic}$$

9.14. The dipole moments of SO_2 and CO_2 are 5.37×10^{-30} C·m and zero, respectively. What can be said of the shapes of the two molecules?

Oxygen is considerably more electronegative than either sulfur or carbon. Thus each sulfur-oxygen and carbon-oxygen bond should be polar, with a net negative charge residing on the oxygen.

If CO_2 has no net dipole moment, the two C—O bond moments must exactly cancel. This can occur only if the two bonds are in a straight line [Fig. 9-35(a)]. (The net moment of the molecule is the vector sum of the bond moments.) The existence of a dipole moment in SO_2 must mean that the molecule is not linear, but bent [Fig. 9-35(b)].

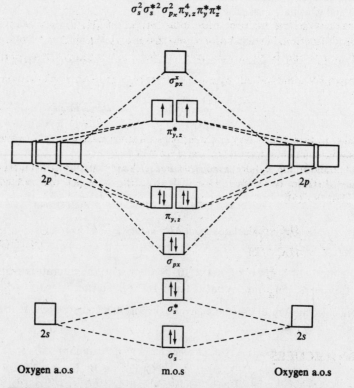

Fig. 9-35

9.15. By the use of molecular-orbital considerations, account for the fact that oxygen gas is paramagnetic. What is the bond order in O_2?

Each oxygen atom has the $1s^2 2s^2 2p^4$ configuration in the ground state. Aside from the K electrons of the two atoms in O_2, which are so deeply imbedded in their respective atoms as not to overlap with other electrons, the remaining 12 electrons (6 from each of the atoms) will fill the lowest of the available molecular orbitals as shown in Fig. 9-36.

$$\sigma_s^2 \sigma_s^{*2} \sigma_{p_x}^2 \pi_{y,z}^4 \pi_y^* \pi_z^*$$

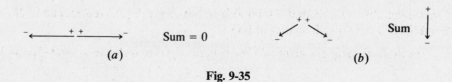

Fig. 9-36. Electrons are indicated by arrows, orbitals by squares. K electrons are omitted. Note: The above energy level diagram is not relevant to the molecules B_2 and C_2, in which the order of the σ_{p_x} and $\pi_{y,z}$ levels is reversed.

The last two electrons went into the antibonding equi-energetic π^* orbitals, one into π_y^* and one into π_z^*, so as to maximize electron spin in accordance with Hund's rule. These two unpaired electrons confer paramagnetism upon the molecule.

Bond order

$$= \frac{\text{(number of electrons in bonding orbitals)} - \text{(number of electrons in antibonding orbitals)}}{2}$$

$$= \frac{8-4}{2} = 2$$

9.16. Explain the observations that the bond length in N_2^+ is 2 pm greater than in N_2, while the bond length in NO^+ is 9 pm less than in NO.

The electron configurations should be written for the four molecules according to the buildup principle

$$
\begin{array}{ll}
N_2 & (K \text{ electrons)}\ \sigma_s^2 \sigma_s^{*2} \sigma_{p_x}^2 \pi_{y,z}^4 \\
N_2^+ & (K \text{ electrons)}\ \sigma_s^2 \sigma_s^{*2} \sigma_{p_x}^2 \pi_{y,z}^3 \\
NO & (K \text{ electrons)}\ \sigma_s^2 \sigma_s^{*2} \sigma_{p_x}^2 \pi_{y,z}^4 \pi_{y,z}^* \\
NO^+ & (K \text{ electrons)}\ \sigma_s^2 \sigma_s^{*2} \sigma_{p_x}^2 \pi_{y,z}^4
\end{array}
$$

The computed bond orders for N_2 and N_2^+ are 3 and $2\frac{1}{2}$, respectively. N_2 therefore has the stronger bond and should have the shorter bond length. The computed bond orders for NO and NO^+ are $2\frac{1}{2}$ and 3, respectively. NO^+ has the stronger bond and should have the shorter bond length. As opposed to ionization of N_2, which involves the loss of an electron in a *bonding* orbital, ionization of NO involves the loss of an electron in an antibonding orbital.

9.17. Two substances having the same molecular formula, C_4H_8O, were examined in the gaseous state by electron diffraction. The carbon-oxygen distance was found to be 143 pm in compound A, and 124 pm in compound B. What can you conclude about the structures of these two compounds?

In compound A, the carbon-oxygen distance is the sum of the single-bond covalent radii of carbon and oxygen.

$$77 \text{ pm} + 66 \text{ pm} = 143 \text{ pm}$$

The oxygen must therefore be nonterminal. One such structure, the actual one selected for this experiment, is the heterocyclic compound tetrahydrofuran [Fig. 9-37 (a)].

In compound B, the carbon-oxygen distance is close to that predicted for a double bond, 122 pm. The oxygen must therefore be terminal. One such structure, the actual one selected for this experiment, is 2-butanone [Fig. 9-37(b)].

Fig. 9-37

SHAPES OF MOLECULES

9.18. Verify the value $\theta = 109°28'$ for the central angles in a regular tetrahedron.

A simple way of constructing a regular tetrahedron is to select alternating corners of a cube and to connect each of the selected corners with each of the other three, as in Fig. 9-38(a). Figure 9-38(b) shows

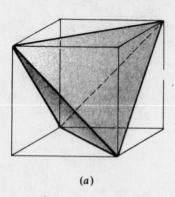

(a)   (b)

Fig. 9-38

triangle OAB, determined by the center of the cube, which is also the center of the tetrahedron, and two corners of the tetrahedron. If P is the midpoint of AB, we see from right triangle OPA that

$$\tan \frac{\theta}{2} = \frac{a\sqrt{2}/2}{a/2} = \sqrt{2}$$

$$\frac{\theta}{2} = 54°44'$$

$$\theta = 109°28'$$

9.19. The C—C single-bond distance is 154 pm. What is the distance between the terminal carbons in propane, C_3H_8? Assume that the four bonds of any carbon atom are pointed toward the corners of a regular tetrahedron.

With reference to Fig. 9-38(b), two terminal carbons can be thought of as lying at A and B, while the central carbon is at O. Then

$$\overline{AB} = 2\overline{AP} = 2\left(\overline{AO} \sin \frac{\theta}{2}\right) = 2(154 \text{ pm})(\sin 54°44') = 251 \text{ pm}$$

9.20. Compute the chlorine-to-chlorine distances in each of the isomeric $C_2H_2Cl_2$ compounds shown in Fig. 9-17. Use Table 9-2 and the average double-bond shortening for estimating bond distances.

The starting point of the solution is a set of diagrams, Fig. 9-39, based on the 120° angle in double-bond compounds.

(a) From Table 9-2, we estimate the C—Cl bond length to be 176 pm. This is the hypotenuse AB of a 60° right triangle, ABD in Fig. 9-40, whose side AD is half the nonbonded chlorine-chlorine distance AC. Then $\overline{AC} = 2\overline{AD} = 2(176 \sin 60°) = 305 \text{ pm}$.

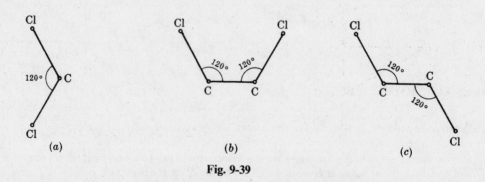

(a)   (b)   (c)

Fig. 9-39

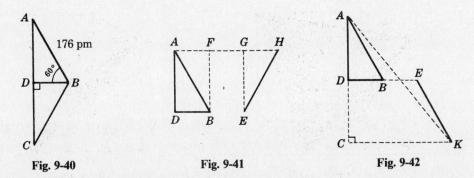

Fig. 9-40   Fig. 9-41   Fig. 9-42

(b) This is related to the preceding case. The chlorine-chlorine distance (Fig. 9-41) is

$$\overline{AH} = \overline{AF} + \overline{FG} + \overline{GH}$$

Now, $\overline{AF} = \overline{GH} = \overline{BD} = 176 \cos 60° = 88$ pm; $\overline{FG}$ is the C=C double-bond distance, 133 pm. Thus $\overline{AH} = 88 + 133 + 88 = 309$ pm.

(c) This can be solved by reference to parts (a) and (b). The chlorine-chlorine distance is AK, the hypotenuse of right triangle ACK. The legs of this triangle ACK (Fig. 9-42) are AC and CK. $\overline{AC}$ was evaluated in (a) to be 305 pm. $\overline{CK}$ equals $\overline{AH}$, evaluated in (b) to be 309 pm. Then

$$\overline{AK} = \sqrt{(305)^2 + (309)^2} = 434 \text{ pm}$$

9.21. What are the O—N—O bond angles in the nitrate and nitrite ions?

In both cases examination of the Lewis structure reveals a VSEPR number of 3 for the central N. (In NO_3^- there are three σ-bonded neighbors and no unshared pairs, and in NO_2^- two neighbors and one unshared pair.) Thus the nominal bond angle is 120°, exactly what is found in nitrate, NO_3^-. In NO_2^-, however, the unshared pair repels the bonding pairs more than they repel each other, forcing them closer to each other. We would expect a slightly smaller angle than 120°; 115° is found.

9.22. The $POCl_3$ molecule has the shape of an irregular tetrahedron with the P atom located centrally. The Cl—P—Cl angle is found to be 103.5°. Give a qualitative explanation for the deviation of this structure from a regular tetrahedron.

The VSEPR number for the P atom is 4 so that the nominal bond angles are 109°28'. However the Lewis structure for $POCl_3$ shows a double bond between P and O. (P is allowed to exceed the octet because of the availability of $3d$ orbitals.) The increased electron density in the P=O bond would make the intrinsic repulsion between the P=O bond and a P—Cl bond greater than between two P—Cl bonds. Thus the Cl—P—Cl angle is lowered and the Cl—P=O angle raised.

9.23. PCl_5 has the shape of a trigonal bipyramid [Fig. 9-43(a)] whereas IF_5 has the shape of a square pyramid [Fig. 9-43(b)]. Account for this difference.

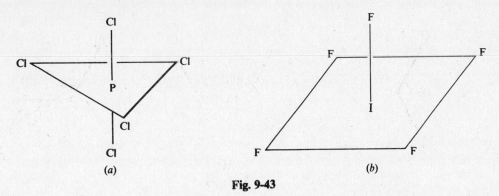

(a)   (b)

Fig. 9-43

$$\begin{array}{cc}
:\ddot{C}l: \quad :\ddot{C}l: & :\ddot{F}: \quad :\ddot{F}: \\
P & F{-}I{-}F: \\
:\ddot{C}l \quad | \quad \ddot{C}l: & :\ddot{F}:
\end{array}$$

Fig. 9-44

The Lewis structures of the singly bonded compounds, Fig. 9-44, reveal a VSEPR number of 5 in PCl_5^-, for which the trigonal bipyramid structure with 90°, 120°, and 180° angles is predicted (Table 9-4). However, the unshared pair on the iodine raises its VSEPR number to 6 so that the nominal bond angles are 90°. The square pyramid structure of IF_5 may be thought of as an octahedron with the unshared pair pointing to one of the corners. Because of the repulsion of the unshared pair one expects the I atom to be slightly below the base plane of the pyramid.

9.24. Which of the four carbon-carbon bond types in naphthalene would you predict to be the shortest? Refer to Fig. 9-33.

The four different kinds of carbon-carbon bonds are represented by 1–2, 1–9, 2–3, and 9–10. (Every other carbon-carbon bond is equivalent to one of these four. For example, 6–7 is equivalent to 2–3, 7–8 is equivalent to 1–2, and so on.) The bond with the greatest double-bond character should be the shortest. Among the three resonance structures shown in Fig. 9-33, the frequency of double bonds for the various bond types is as follows: 2 in *1–2* [in (*a*) and (*c*)], 1 in *1–9* [in (*b*)], 1 in *2–3* [in (*b*)], and 1 in *9–10* [in (*c*)]. Bond *1–2* is expected to have the greatest double-bond character and thus the shortest length. This prediction is found to be true experimentally. The above four bonds are found to have lengths of 136.5, 142.5, 140.4, and 139.3 pm, respectively.

Note that the method of counting the number of resonance structures containing a double bond between a given pair of carbon atoms is very crude and cannot distinguish among the last three of the listed bond types, each of which shows a double bond in only one resonance structure. Even within the framework of the limited resonance theory, it would be necessary to know the relative weighting of each of the two equivalent structures (*a*) and (*b*) with the nonequivalent structure (*c*).

9.25. What is the shape of the triple iodide ion, I_3^-?

The Lewis structure reveals a VSEPR number of 5 for the central I atom, two bonded neighbors and three unshared pairs. To determine which corners of the trigonal bipyramid are occupied by the terminal I atoms, find the arrangement which maximizes the angles between the unshared pairs. The preferred arrangement [Fig. 9-45(*a*)] must be the one in which the unshared pairs are all at 120° because any other alternative [Fig. 9-45 (*b*) and (*c*)] would have two sets of pairs at 90°. Thus the two terminal I atoms must occupy the axial positions, at 180° to each other, and the molecule is perfectly linear.

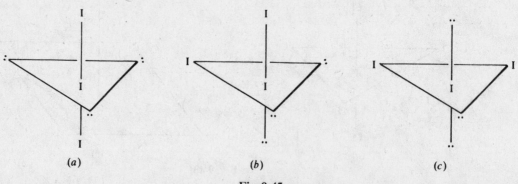

(*a*)   (*b*)   (*c*)

Fig. 9-45

COORDINATION. COMPOUNDS

9.26. Soluble compounds of the complex ion $Co(NH_3)_6^{3+}$ have a maximum in absorption of visible light at 437 nm.   (*a*) What is the value of Δ for this complex ion, expressed in cm^{-1}, and what is the ion's color?   (*b*) How many unpaired electrons would you expect this ion to have if it is considered low-spin, and how many if it is considered high-spin?

(*a*)
$$\Delta = \left(\frac{1}{437 \text{ nm}}\right)\left(\frac{10^9 \text{ nm}}{10^2 \text{ cm}}\right) = 22\,900 \text{ cm}^{-1}$$

The color of the ion is complementary to the light absorbed. Although the prediction of the color from the data is not absolute, since the color depends not only on the wavelength of the absorption maximum but also on the shape of the whole absorption band and on the color sensitivity of the human eye, fairly reliable conclusions can be drawn. The absorption, which peaks in the blue-violet region of the spectrum, would be expected to cover most of the blue region and part of the green. The color would be expected to be yellow.

(*b*) The outer-electron configuration of Co^{3+} is $3s^2 3p^6 3d^6$. The spin is due to unpaired d electrons. For low spin, the six d electrons would all be paired in the three t_{2g} orbitals, and the spin would be zero. For high spin, the two e_g molecular orbitals would be available as well (see Fig. 9-14). Four of the available orbitals would be singly occupied and one doubly occupied, in order to maintain the maximum number of unpaired electrons, four in this case. The Δ-value in this case is large enough to rule out high spin, and the ion is diamagnetic.

9.27. Predict the magnetic properties of   (*a*) $Rh(NH_3)_6^{3+}$, and   (*b*) CoF_6^{3-}.

(*a*) This problem can be approached by comparison with Problem 9.26. For analogous complexes of two different members of the same group in the periodic table, Δ increases with increasing atomic number. Since Δ for $Co(NH_3)_6^{3+}$ is already so high that the ion is low-spin, $Rh(NH_3)_6^{3+}$ would certainly be low-spin and diamagnetic. [The observed Δ for $Rh(NH_3)_6^{3+}$ is 34 000 cm^{-1} and the ion is diamagnetic.]

(*b*) F^- is a weak-field ligand, tending to form complexes with a low Δ-value, so that the ion would be expected to be high-spin, with four unpaired and parallel electron spins [compare with Problem 9.26(*b*)]. The measured Δ is 13 000 cm^{-1}, a low figure, and the ion is indeed paramagnetic.

ISOMERISM

9.28. Write all the structural isomeric formulas for C_4H_9Cl.

The molecular composition resembles butane, C_4H_{10}, with the exception that one H is replaced by Cl. The formulas can be found by picking all the differently located hydrogens in the two isomeric formulas of C_4H_{10} (Fig. 9-15).

Note that the two end carbons of *n*-butane are identical and the two interior carbons are identical. Thus a chlorine substituted on the left-hand carbon would have given the same compound as (*a*) in Fig. 9-46, only viewed from a different end. Similarly, substitution on the carbon next to the left would have given a

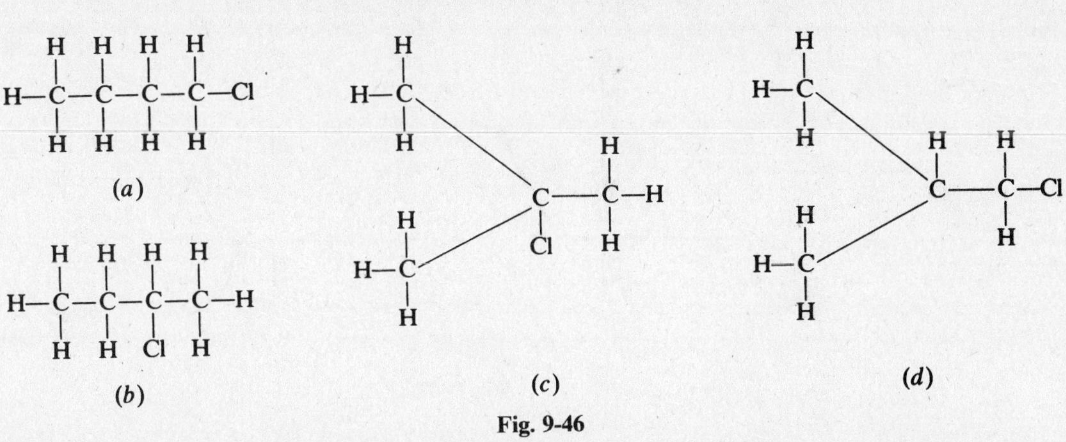

Fig. 9-46

compound identical with (*b*). In isobutane, the three terminal carbons are identical. Because of the free rotation around the C—C single bond axis, it is immaterial which of the several hydrogens attached to a given C is substituted for; all positions become averaged in time anyway, because of the free rotation.

9.29. Write formulas for all the structural and geometrical isomers of C_4H_8.

If the four carbons are in a row, there must be one double bond in order to satisfy the tetracovalence of carbon. The double bond occurs either in the center of the molecule or toward an end. In the former case, two geometrical isomers occur with different positions of the terminal carbons relative to the double bond; in the latter case, two structural isomers occur, differing in the extent of branching within the skeleton. Additional possibilities are ring structures without double bonds. Unlike N (Problem 9.5*c*), C can form three-atom rings.

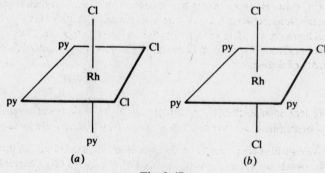

Note the shorthand notation of grouping two or three hydrogens with the carbon to which they are bonded. It is understood, of course, that each hydrogen is bonded to the carbon of its group and that the carbon of the group is bonded to the next carbon or to the carbon of the adjoining group or groups.

9.30. Which of the isomeric C_4H_9Cl compounds would you expect to be optically active? Refer to Fig. 9-46.

Compound (*b*) is the only one which would exist in optically active isomeric forms, because it is the only one which has a carbon atom (the one bonded to Cl) which is bonded to four different groups. Every other carbon atom in this or the other structures is bonded to at least two hydrogen atoms or to two CH_3 groups.

9.31. How many geometrical isomers could $[Rh(py)_3Cl_3]$ have? (py is an abbreviation for the ligand pyridine.)

There are only two possibilities, one [Fig. 9-47(*a*)] with the three chlorines occupying bonding positions *cis* to each other on one face of the octahedron and the three pyridines on the opposite face, and the other [Fig. 9-47(*b*)] in which two of the chlorines are *trans* to each other and two of the pyridines in turn are *trans* to each other.

Fig. 9-47

9.32. Some ligands are multifunctional; that is, they have two or more atoms that can bind to the central metal atom or ion. Each binding site occupies a different corner on the coordination surface. Ethylenediamine (abbreviated en) is such a ligand; the two binding atoms are nitrogens and the two binding sites must be *cis* to each other. How many geometrical isomers of $[Cr(en_2)Cl_2]^+$ should exist, and which isomer(s) might display optical activity?

Two geometrical isomers exist, *cis* and *trans* (Fig. 9-48). Each en can be represented by an arc terminating at the two binding sites. By drawing other arrangements of the arcs while preserving the positions of the chlorines, one can see that (*b*) is a distinct mirror image of (*a*), whereas there is no mirror image of (*c*) except (*c*) itself. In other words, only the *cis* isomer can be optically active; the *trans* isomer is its own mirror image.

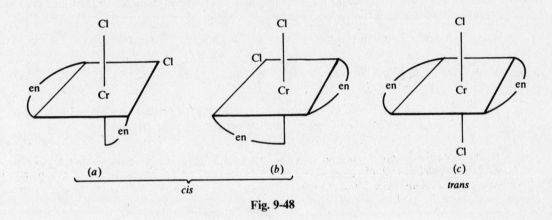

Fig. 9-48

BONDING IN METALS

9.33. Explain why metals are lustrous (mirror-like).

In the band model there is a continuum of empty energy levels rather than discrete energy levels. Thus light quanta of all energies within a wide range of wavelengths will be absorbed equally, and then the energized electrons will re-emit the light when they fall back into their ground-state orbitals.

9.34. Predict how the Group II metals differ from Group I in density, melting point, and mechanical strength.

In any given period the Group II ions are smaller and may thus approach each other more closely. At the same time twice as many electrons are present in the electron sea. The closer approach and the much greater electrostatic interactions between the 2 + ions and the sea of high negative charge density lead to greater density and much greater bonding energy, which in turn leads to much higher melting point and hardness. Actually going from Group I to Group II increases the density by a factor of about 2 and raises the melting point by hundreds of degrees.

9.35. Metals usually feel cool to the touch compared with other materials because they are such good conductors of heat. How can we explain this unusual thermal conductivity?

In most materials heat is conducted by atom-to-atom transfer of vibrational motion from the hot end to the cold end. In metals the thermal energy is transferred primarily by the motion of the electron sea's free electrons, which are very mobile.

Supplementary Problems

FORMULAS

9.36. Determine the ionic valences of the groups in parentheses in the following formulas: (a) $Ca(C_2O_4)$, (b) $Ca(C_2H_3O_2)_2$, (c) $Mg_3(AsO_3)_2$, (d) $(MoO)Cl_3$, (e) $(CrO_2)F_2$, (f) $(PuO_2)Br$, (g) $(PaO)_2S_3$.

Ans.   (a) -2;   (b) -1;   (c) -3;   (d) $+3$;   (e) $+2$;   (f) $+1$;   (g) $+3$

9.37. Write formulas for the following ionic compounds: (a) lithium hydride, (b) calcium bromate, (c) chromium(II) oxide, (d) thorium(IV) perchlorate, (e) nickel orthophosphate, (f) zinc sulfate.

Ans.   (a) LiH; (b) $Ca(BrO_3)_2$; (c) CrO; (d) $Th(ClO_4)_4$; (e) $Ni_3(PO_4)_2$; (f) $ZnSO_4$

9.38. Name the following compounds: (a) $Mg(IO)_2$,   (b) $Fe_2(SO_4)_3$,   (c) $CaMnO_4$,   (d) $KReO_4$, (e) $CaWO_4$,   (f) $CoCO_3$.

Ans.   (a) magnesium hypoiodite; (b) iron(III) sulfate or ferric sulfate; (c) calcium manganate; (d) potassium perrhenate; (e) calcium tungstate; (f) cobalt(II) carbonate

9.39. The formula of potassium arsenate is K_3AsO_4. The formula of potassium ferrocyanide is $K_4Fe(CN)_6$. Write the formulas of (a) calcium arsenate, (b) ferric arsenate, (c) barium ferrocyanide, (d) aluminum ferrocyanide.

Ans.   (a) $Ca_3(AsO_4)_2$; (b) $FeAsO_4$; (c) $Ba_2Fe(CN)_6$; (d) $Al_4[Fe(CN)_6]_3$

9.40. Draw Lewis structures for each of the following: (a) C_2HCl,   (b) C_2H_6O,   (c) C_2H_4O, (d) NH_3O, (e) NO_2^- (both oxygens terminal),   (f) N_2O_4 (all oxygens terminal),   (g) OF_2.

Ans.   (a) $H-C\equiv C-Cl$ (b) $H_3C-\overset{\displaystyle H}{\underset{\displaystyle H}{\overset{|}{\underset{|}{C}}}}-\ddot{O}-H$ (c) $H_3C-\overset{\displaystyle H}{\overset{|}{C}}=\ddot{O}$

or $H_3C-\ddot{O}-CH_3$ *or* $H_2C=\overset{\displaystyle H}{\overset{|}{C}}-\ddot{O}-H$

(d) $H-\overset{\displaystyle H}{\overset{|}{N}}-\ddot{O}-H$

(e) $\left[\ddot{O}\diagup^{\ddot{N}}\diagdown\ddot{O}:\right]^- \leftrightarrow \left[:\ddot{O}\diagdown_{\ddot{N}}\diagup\ddot{O}\right]^-$

(f) (structures as drawn)

(g) $:\ddot{F}-\ddot{O}-\ddot{F}:$

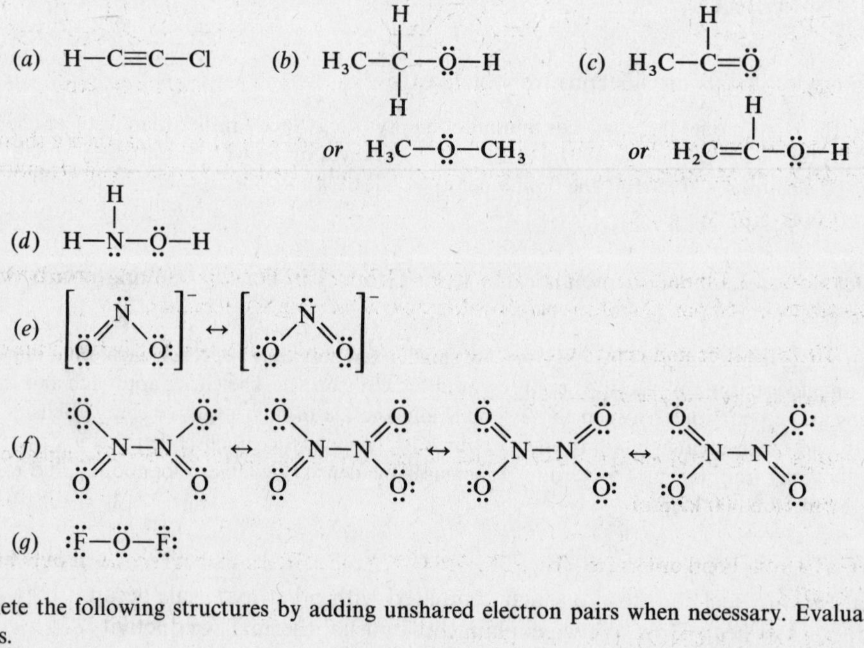

9.41. Complete the following structures by adding unshared electron pairs when necessary. Evaluate the formal charges.

(a) $N\equiv C-C\equiv N$ (b) $N=C=C=N$ (c) $Cl-C\equiv N$

(d) $Cl=C=N$ (e) $N=N=O$ (f) $N\equiv N-O$

(g) $\underset{\text{Cl}}{\overset{\text{Cl}}{\diagdown}} C = O$
(h) $\underset{\text{Cl}}{\overset{\text{Cl}}{\diagdown}} C - O$
(i) $Cl - N \overset{O}{\underset{O}{<}}$

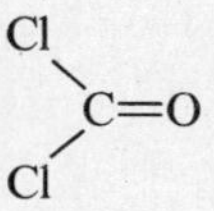

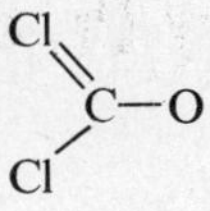

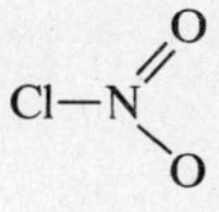

(j) $\left[\underset{\overset{\|}{O}}{\overset{\overset{O}{\|}}{O - Cl - O}} \right]^{-}$

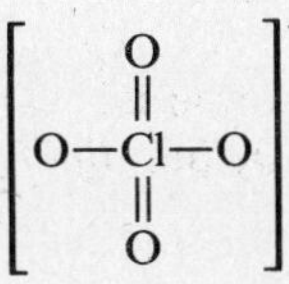

(k)

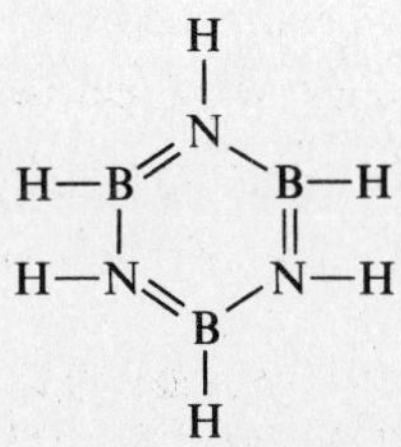

(l)

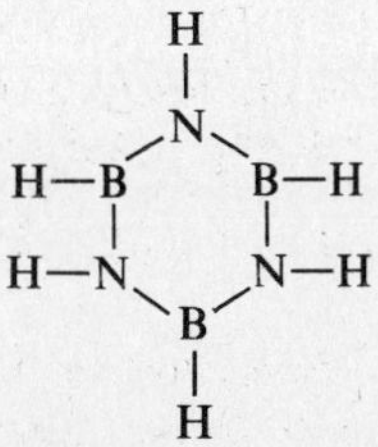

Ans. (a) all zero; (b) +1 on one N (which does not have an octet), −1 on the other; (c) all zero; (d) +1 on Cl, −1 on N; (e) −1 on terminal N, +1 on central N; (f) +1 on central N, −1 on O; (g) all zero; (h) +1 on doubly bonded Cl, −1 on O; (i) +1 on N, −1 on singly bonded O; (j) +1 on Cl, −1 on each singly bonded O; (k) +1 on each N, −1 on each B; (l) all zero

BONDING PROPERTIES

9.42. The chlorine-to-oxygen bond distance in ClO_4^- is 144 pm. What do you conclude about the valence-bond structures for this ion?

Ans. There must be considerable double-bond character in the bonds.

9.43. How many resonance structures can be written for each of the following aromatic hydrocarbons?

(a) Anthracene   (b) Phenanthrene   (c) Naphthacene

Consider double bonds only between adjacent carbons. (The circle inside the hexagon is a shorthand notation used to designate an aromatic ring without having to write out all the valence-bond structures.)

Ans. (a) 4; (b) 5; (c) 5

9.44. The structure of 1,3-butadiene is often written as $H_2C{=}CH{-}CH{=}CH_2$. The distance between the central carbon atoms is 146 pm. Comment on the adequacy of the assigned structure.

Ans. There must be non-octet resonance structures involving double bonding between the central carbons, such as $\overset{+}{C}H_2{-}CH{=}CH{-}\overset{..}{C}H_2^-$.

9.45. The average C—C bond energy is 347 kJ/mol. What do you predict for the Si—Si single-bond energy?

Ans. less than 300 kJ/mol

9.46. (a) What are the bond orders for CN^-, CN, and CN^+? (b) Which of these species should have the shortest bond length?

Ans. (a) CN^-, 3; CN, $2\frac{1}{2}$; CN^+, 2; (b) CN^-

9.47. Which homonuclear diatomic molecule(s) of second-period elements, besides O_2, should be paramagnetic?

Ans. B_2

9.48. Which supposed homonuclear diatomic molecules of the second-period elements should have zero bond order?

> *Ans.* Be_2, Ne_2

9.49. Dipole moments are sometimes expressed in *debyes*, where

$$1 \text{ debye} = 10^{-18} \text{ (esu of charge)} \cdot \text{cm}$$

The electrostatic unit (esu) of charge is defined by $1 \text{ C} = 2.998 \times 10^9$ esu. What is the value of 1 debye in SI units?

> *Ans.* 3.336×10^{-30} C·m

9.50 Helium can be excited to the $1s2p$ configuration by light of 58.44 nm. The lowest excited singlet state, with the configuration $1s2s$, lies $4\,857$ cm^{-1} below the $1s2p$ state. What would the average He—H bond energy have to be in order that HeH_2 could form nonendothermically from the He and H_2? Assume that the compound would form from the lowest excited singlet state of helium, neglect any differences between ΔE and ΔH, and use information given in Problem 7.31.

> *Ans.* $1\,212$ kJ/mol

9.51. The dipole moment of HBr is 2.60×10^{-30} C·m, and the interatomic spacing is 141 pm. What is the percent ionic character of HBr?

> *Ans.* 11.5%

9.52. The dipole moments of NH_3, AsH_3, and BF_3 are (4.97, 0.60, and 0.00) $\times 10^{-30}$ C·m, respectively. What can be concluded about the shapes of these molecules?

> *Ans.* NH_3 and AsH_3 are both pyramidal and BF_3 is planar. From the dipole moments alone, nothing can be concluded about the relative flatness of the NH_3 and AsH_3 pyramids, because the electronegativities of N and As are different.

9.53. The As—Cl bond distance in $AsCl_3$ is 217 pm. Estimate the single-bond covalent radius of As.

> *Ans.*   118 pm

9.54. The carbon-carbon double-bond energy in C_2H_4 is 615 kJ/mol and the carbon-carbon single-bond energy in C_2H_6 is 347 kJ/mol. Why is the double-bond energy appreciably less than twice the single-bond energy?

> *Ans.*   A σ orbital has greater electron overlap between the atoms because its component atomic orbitals are directed toward each other, whereas the component p orbitals making up the π orbital are directed perpendicularly to the internuclear axis, and have only side-to-side overlap.

9.55. Estimate ΔH for the reaction:

$$C_2H_5Cl(g) \rightarrow HCl(g) + C_2H_4(g)$$

given the following bond energies in kJ/mol:

C—C	C—H	Cl—Cl	C—Cl	H—Cl	C=C
348	414	242	327	431	615

> *Ans.* $+43$ kJ

9.56. Estimate the bond energy of the F—F bond given that ΔH_f^0 of HF(g) is -271 kJ/mol, and given the following bond energies in kJ/mol:

H—F	H—H
565	435

> *Ans.*   153 kJ/mol

SHAPES OF MOLECULES

9.57. The platinum-chlorine distance has been found to be 232 pm in several crystalline compounds. If this value applies to both of the compounds shown in Fig. 9-18, what is the chlorine-chlorine distance in   (a) structure (a),   (b) structure (b)?

Ans.   (a) 464 pm,   (b) 328 pm

9.58. What is the length of a polymer molecule containing 1 001 carbon atoms singly bonded in a line, if the molecule could be stretched to its maximum length consistent with maintenance of the normal tetrahedral angle within any C—C—C group?

Ans.   126 nm

9.59. A plant virus was examined by the electron microscope and was found to consist of uniform cylindrical particles 15.0 nm in diameter and 300 nm long. The virus has a specific volume of $0.73 \text{ cm}^3/\text{g}$. If the virus particle is considered to be one molecule, what is its molecular weight?

Ans. 4.4×10^7

9.60. Assuming the additivity of covalent radii in the C—Cl bond, what would be the chlorine-chlorine distance in each of the three dichlorobenzenes (Fig. 9-49)? Assume that the ring is a regular hexagon and that each C—Cl bond lies on a line through the center of the hexagon. The distance between adjacent carbons is 140 pm.

Ans.   (a) 316 pm;   (b) 547 pm;   (c) 632 pm

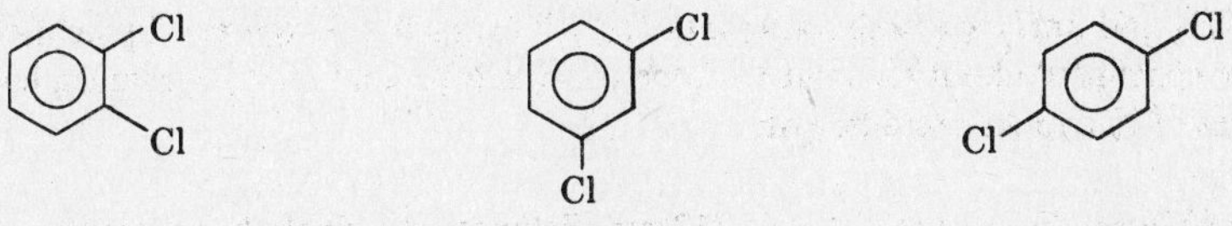

Fig. 9-49

9.61. Estimate the length and width of the carbon skeleton in anthracene [see Problem 9-43(a) for a diagram]. Assume hexagonal rings with equal carbon-carbon distances of 140 pm.

Ans.   730 pm long, 280 pm wide

9.62. There are two structural isomers of $C_2H_4Cl_2$.   (a) In the isomer in which both chlorines are attached to the same carbon, find the chlorine-chlorine distance.   (b) In the other isomer find the minimum and maximum distances between the two chlorines as one CH_2Cl group rotates about the C—C bond as an axis. Assume tetrahedral angles and additivity of covalent bond radii. (*Hint*: Refer to Problems 9-19 and 9-20.)

Ans.   (a) 287 pm;   (b) minimum 271 pm, maximum 428 pm

9.63. BBr_3 is a symmetrical planar molecule, all B—Br bonds lying at 120° to each other. The distance between Br atoms is found to be 324 pm. From this fact and from information in Table 9-2, estimate the covalent radius of boron, assuming that the bonds are all single bonds.

Ans.   73 pm

9.64. What is the VSEPR number for the central atom in each of the following species? (*Hint*: Draw the Lewis structures first.)   (a) SO_2,   (b) SO_3,   (c) SO_3^{2-},   (d) SO_4^{2-},   (e) SF_4,   (f) SCl_2.

Ans.   (a) 3,   (b) 3,   (c) 4,   (d) 4,   (e) 5,   (f) 4

9.65.   What are the bond angles in each of the species in the preceding problem?

> *Ans.*   (*a*) slightly less than 120°;   (*b*) 120°;   (*c*) slightly less than 109°28′;   (*d*) 109°28′;   (*e*) 120°, 90°, 180° (all slightly less because the unshared pair occupies a trigonal position);   (*f*) slightly less than 109°28′

9.66.   For each of the following species give the VSEPR number, the general shape, and the bond angles. (*a*) XeF_4,   (*b*) XeO_3,   (*c*) XeF_2.

> *Ans.*   (*a*) VSEPR number 6, square planar, 90°;   (*b*) VSEPR number 4, trigonal pyramid with unshared pair at apex, slightly less than 109°28′;   (*c*) VSEPR number 5, linear, 180°

COORDINATION COMPOUNDS

9.67.   Name the following (en = ethylenediamine, py = pyridine):
(*a*) $[Co(NH_3)_5Br]SO_4$ (*c*) $[Pt(py)_4][PtCl_4]$ (*e*) $K_3[Fe(CN)_5CO]$
(*b*) $[Cr(en)_2Cl_2]Cl$ (*d*) $K_2[NiF_6]$ (*f*) $CsTeF_5$

> *Ans.*   (*a*) pentaamminebromocobalt(III) sulfate
> (*b*) dichlorobis(ethylenediamine)chromium(III) chloride
> (*c*) tetrapyridineplatinum(II) tetrachloroplatinate(II)
> (*d*) potassium hexafluoronickelate(IV)
> (*e*) potassium carbonylpentacyanoferrate(II)
> (*f*) cesium pentafluorotellurate(IV)

9.68.   Write formulas for the following compounds, using brackets to enclose the complex ion portion: (*a*) triamminebromoplatinum(II) nitrate,   (*b*) dichlorobis(ethylenediamine)cobalt(II) monohydrate,   (*c*) pentaamminesulfatocobalt(III) bromide,   (*d*) potassium hexafluoroplatinate(IV),   (*e*) tetraaquadibromochromium(III) chloride,   (*f*) ammonium heptafluorozirconate(IV).

> *Ans.*   (*a*) $[Pt(NH_3)_3Br]NO_2$ (*c*) $[Co(NH_3)_5SO_4]Br$ (*e*) $[Cr(H_2O)_4Br_2]Cl$
> (*b*) $[Co(en)_2Cl_2] \cdot H_2O$ (*d*) K_2PtF_6 (*f*) $(NH_4)_3[ZrF_7]$

9.69. Δ for $IrCl_6^{3-}$ is 27 600 cm^{-1} (*a*) What is the wavelength of maximum absorption?   (*b*) What would you predict for the magnetic behavior of this ion?

> *Ans.*   (*a*) 362 nm;   (*b*) diamagnetic

9.70.   (*a*) What is the maximum number of unpaired electrons which a high-spin complex of the first transition series could possess in the ground state?   (*b*) What elements could show this maximum, and in what oxidation states?

> *Ans.*   (*a*) 5;   (*b*) Mn(II), Fe(III)

9.71.   If the complexing metal of the first transition series has a d^i configuration, for what values of *i* could magnetic properties alone distinguish between strong-field and weak-field ligands?

> *Ans.*   4, 5, 6, 7

9.72.   Both $Fe(CN)_6^{4-}$ and $Fe(H_2O)_6^{2+}$ appear colorless in dilute solutions. The former ion is low-spin and the latter is high-spin.   (*a*) How many unpaired electrons are in each of these ions?   (*b*) Why should both ions be colorless, in view of the apparent significant difference in their Δ-values?

> *Ans.*   (*a*) 0 in $Fe(CN)_6^{4-}$, 4 in $Fe(H_2O)_6^{2+}$.   (*b*) Δ for $Fe(CN)_6^{4-}$ is so large that the absorption peak is in the ultraviolet; Δ for $Fe(H_2O)_6^{2+}$ is so small that its absorption peak is in the infrared. Practically no visible light is absorbed by either.

9.73.   Hexaaquairon(III) ion is practically colorless. Its solutions become red when NCS^- is added. Explain. (Compare with Problem 9.72)

Ans. H_2O is not a strong-field ligand, as noted in Problem 9.72. NCS^- has vacant π^* orbitals that overlap the t_{2g} orbitals of the metal and so can accept electron density from the metal. This "back-bonding" increases the strength of the metal-ligand bond and lowers the t_{2g} energy level, thus making NCS^- a strong-field ligand. A strong-field ligand causes an increase in Δ and thus a lowering of the wavelength of maximum d-d absorption from the near-infrared well into the visible region (actually into the blue-green).

ISOMERISM

9.74. How many structural isomers can be drawn for each of the following: (*a*) C_5H_{12}, (*b*) C_3H_7Cl,   (*c*) $C_3H_6Cl_2$,   (*d*) $C_4H_8Cl_2$,   (*e*) $C_5H_{11}Cl$,   (*f*) C_6H_{14},   (*g*) C_7H_{16}?

Ans.   (*a*) 3;   (*b*) 2;   (*c*) 4;   (*d*) 9;   (*e*) 8;   (*f*) 5;   (*g*) 9

9.75. Among the paraffin hydrocarbons (C_nH_{2n+2}, where n is an integer), what is the empirical formula of the compound of lowest molecular weight which could demonstrate optical activity in at least one of its structural isomers?

Ans. C_7H_{16}

9.76. How many structural and geometrical isomers can be written for the following without counting ring compounds: (*a*) C_3H_5Cl,   (*b*) $C_3H_4Cl_2$,   (*c*) C_4H_7Cl,   (*d*) C_5H_{10}?

Ans.   (*a*) 4;   (*b*) 7;   (*c*) 11;   (*d*) 6

9.77. For the square coplanar complex $[Pt(NH_3)(NH_2OH)py(NO_2)]^+$, how many geometrical isomers are possible? Draw them.

Ans.   There are three isomers. Any one ligand can be *trans* to any of the other three; the two ligands not *trans* to the first have their positions automatically fixed as *trans* to each other.

9.78. Predict whether $[Ir(en)_3]^{3+}$ should exhibit optical isomerism. If so, prove by diagrams that the two optical isomers are not simple rotational aspects of the same compound.

Ans.   There are two optical isomers, which may be represented as in Fig. 9-50.

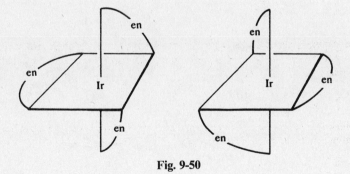

Fig. 9-50

BONDING IN METALS

9.79. In silicon there is a gap in energy between the band of bonding m.o.s and a band of an equal number of antibonding m.o.s, all derived from the $3s$ and $3p$ atomic orbitals. Is silicon a metallic electrical conductor? Explain.

Ans.   No. In a crystal of N atoms there will be $4N$ m.o.s, $2N$ of them bonding. The $4N$ valence electrons will just fill these $2N$ bonding m.o.s; but metallic conductivity requires that the band be only partially filled.

9.80. How does a rise in temperature affect the electrical conductivity of a metal? Explain.

   Ans.   A rise in temperature lowers the conductivity because vigorous atomic motion disrupts the long-range order of the lattice, thus obstructing the m.o.s that pervade the whole crystal.

9.81. Metals are ductile and malleable, in contrast to the brittleness of most other solids. Account for these properties in terms of the electron-sea model.

   Ans.   Since the atoms are rather far apart and the electron sea offers little resistance to deformation, it does not require much energy to cause one layer of atoms to slide past another. Thus the crystal is deformed rather than shattered by an external stress.

Chapter 10

Solids and Liquids

Solids cannot be understood as well as the much simpler units of matter, the atoms and molecules. There are some simplifying features of solids, however, which allow considerable insight into their nature. One of these features is the regularity in the structure of all crystalline solids, as a consequence of which an entire macrocrystal consists of a three-dimensional replication of a basic unit of atoms, ions, or molecules arranged in a fixed way.

CRYSTALS

The arrangement of the simpler particles in a crystalline array is called a *lattice*. Every lattice is a three-dimensional stacking of identical building blocks called *unit cells*. The properties of the entire crystal, including its overall symmetry, can be understood in terms of the unit cell.

The description of crystals has been systematized by the mathematical finding that there is a rather small number of arrangements of identical points, 14 to be specific, fulfilling the requirement that the arrangement, extended by pure translation, can fill out all of three-dimensional space. These 14 arrangements are called *Bravais lattices*.

In this book only the three Bravais lattices having cubic symmetry will be considered in detail: the *simple cubic*, the *face-centered cubic*, and the *body-centered cubic*. The corresponding unit cells are represented in Fig. 10-1. Actually, the lattice points (corners of the cubes and centers of the edges or faces) represent the *centers* of the atoms or ions occupying the lattice. The atoms or ions by themselves are not points but three-dimensional objects which usually are in contact with each other. The representations of these objects in Fig. 10-1 are purposely shrunk for clarity of visualization. The length of the cube edge is designated by a. A crystal having any of these three lattices can be thought of as a three-dimensional stack of unit cell cubes, packed face to face so as to fill all of the space occupied by the crystal.

Crystal classes less symmetrical than the cubic have unit cells which may be thought of as more or less distorted cubes, the opposite faces of which are parallel to each other. (The name of such a general geometric shape is a *parallelepiped*.) Thus, crystals with hexagonal symmetry, like snow (ice), have unit cells that are prisms with a vertical axis perpendicular to a rhombus-shaped base, the equal edges of which are at 60° and 120° with respect to each other. A typical hexagonal unit cell is shown in Fig. 10-2. The length of the rhombus edge is designated by a, and the height of the unit cell by c. Note that a hexagonal prism can be formed by joining three unit cells.

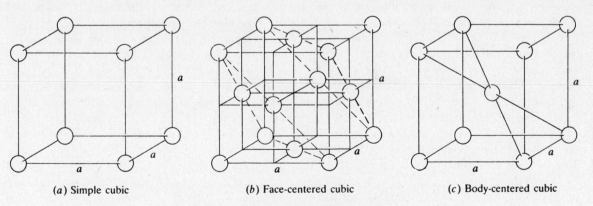

(*a*) Simple cubic   (*b*) Face-centered cubic   (*c*) Body-centered cubic

Fig. 10-1. Unit Cells of Cubic Symmetry

149

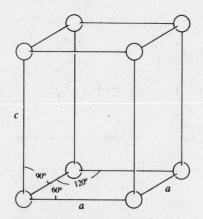

Fig. 10-2.   A Hexagonal Unit Cell

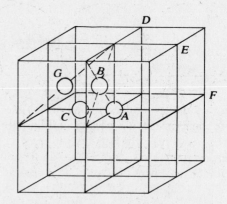

Fig. 10-3.   Stack of Eight Cubic Unit Cells

The bulk density of a perfect crystal can be computed from the properties of the unit cell. It is necessary to apportion the mass of the crystal among the various unit cells, and then to divide the mass apportioned to one unit cell by the volume of the unit cell. In computing the mass of a unit cell, it is necessary to assign to the cell only that fraction of an atom which lies in that cell alone. If the unit cell is cubic, we see from Fig. 10-3 that

Mass per unit cell $= \frac{1}{8}$ (mass of atoms, like A, on corners of the unit cell)

$+ \frac{1}{4}$ (mass of noncorner atoms, like C, on unit cell edges)

$+ \frac{1}{2}$ (mass of nonedge atoms, like B, on the faces of the unit cell)

$+$ (mass of interior atoms, like G, within the unit cell)

since a corner atom is shared by eight unit cells, etc. Note that the formula holds whether or not all the atoms belong to the same species. It also holds for noncubic cells.

Coordination number

The *coordination number* of an atom in a crystal is the number of nearest-neighbor atoms. The coordination number is constant for a given lattice [see Problems 10.1(b) and 10.12(d)].

Close packing

There are two simple lattice structures that allow a high degree of packing of the atoms in a crystal. A structure in which identical spheres (the atoms) occupy the greatest fraction of the total space is called a *close-packed* structure. It can be achieved by packing on top of each other two-dimensional close-packed layers, in each of which every sphere is surrounded by a regular hexagonal arrangement of six other spheres, as in Fig. 10-4. In the figure, the large circles represent the spheres in one such layer, and the squares surround the *centers* of spheres in the adjoining superimposed layer. If the third and every odd-numbered layer are made up of spheres directly over the spheres in the first layer and if every even-numbered layer is made up of spheres directly over the spheres in the second layer, the structure is called *hexagonal close-packed*; the corresponding unit cell for the centers of the spheres is given in Problem 10.12.

If instead the structure is a regular alternation of three kinds of layers, the third being made up of spheres whose centers are enclosed by the small circles in Fig. 10-4, the structure is *cubic close-packed*; its unit cell is just the face-centered cubic. The close-packed layers are perpendicular to a body diagonal of the unit cell [see the dashed triangles in Fig. 10-1(b)].

Both the hexagonal close-packed and the cubic close-packed lattices have the same 74 percent filling of space by spheres in contact [see Problem 10.1(d)], and the two lattices also have the same coordination number, 12.

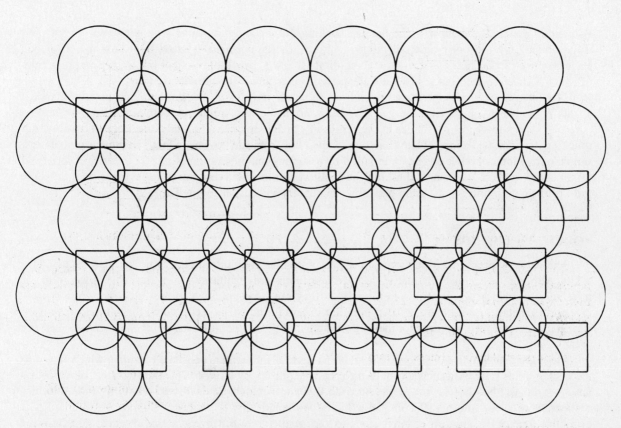

Fig. 10-4

The void spaces, or *holes*, in a crystal can sometimes admit foreign particles of smaller size. Thus an understanding of the geometry of these holes becomes important. In the cubic close-packed structure, the two major types of holes are the *tetrahedral* and the *octahedral* holes. In Fig. 10-1(*b*), tetrahedral holes are in the centers of the indicated minicubes of side $a/2$. Each tetrahedral hole has four nearest-neighbor occupied sites. The octahedral holes are in the body center and on the centers of the edges of the indicated unit cell. Each octahedral hole has six nearest-neighbor occupied sites.

CRYSTAL FORCES

The strength of the forces holding crystalline substances together varies considerably. In *molecular crystals*, like CO_2 and benzene, each molecule is almost independent of all the others and retains practically the same internal geometry (bond lengths and angles) that it has in the gaseous or liquid state. The crystal is held together by the relatively weak van der Waals forces, and the melting point is never very high. For substances that can undergo intermolecular hydrogen bonding, the intermolecular forces in the crystal can be great enough to impose a noticeable change in the molecular geometry. *Hydrogen bonding* is the attraction, largely electrostatic in nature, between the positive charge-carrying hydrogen of the polar bond of one molecule (or part of a molecule) and the negative charge-carrying atom of the polar bond of another molecule (or part of the same molecule). The most important polar bonds that allow hydrogen bonding are those between hydrogen and the most electronegative elements: fluorine, oxygen, nitrogen, and chlorine. Water is an example; the angle between the two O—H bonds in the vapor is about 105 ° but equals the tetrahedral 109 °28′ in the crystal, in conformity to the crystalline, rather than molecular, spatial requirements.

In *metals*, a special type of crystalline force comes into play, characterized by a nondirectional nature. Fixed bond angles do not play a major role in metals, and the stablest crystal structures for most elementary metals are those with the densest packing. Important exceptions are the Group I metals and iron, which have the body-centered cubic structure. In the *covalent crystals*, like diamond or silicon carbide, the crystal is held together by a three-dimensional network of covalent bonds, the mutual angles of which are determined largely by the valence requirements of the individual atoms.

The attractive forces operating in *ionic crystals* are mostly electrostatic in nature, the classical attraction between oppositely charged particles. To avoid the repulsion between similarly charged particles, ionic substances crystallize in structures in which a positive and a negative ion can come within touching distance, while ions of like charge are kept away from each other. In fact, the dimensions of most simple purely ionic crystals can be understood by assuming an ionic radius (Chapter 9) for each ion, valid for all compounds of that ion, and taking the smallest cation-anion separation to be the sum of the ionic radii of the cation and anion. Radii for some elementary ions are listed in Table 9-1.

FORCES IN LIQUIDS

The kinds of forces that bind atoms, ions, and molecules in liquids are the same as those discussed for solids. Although just above the melting point these forces are insufficient to restrain the atoms, ions, or molecules to their lattice positions, in most cases they are strong enough to prevent vaporization.

Liquid metals and salts are not very familiar except in industrial technology. Since the forces are so strong the temperatures required for melting are high and the temperature range for the liquid state is very great.

Most familiar liquids consist of molecular substances, e.g., water, benzene, bromine, which freeze to molecular solids. The molecules are bound to their neighbors by weak forces, the strongest of which, by far, are hydrogen bonds. Others are known collectively as van der Waals forces. *All* atoms and molecules are attracted to one another. If a molecule has a permanent dipole moment (Chapter 9), dipole-dipole attraction makes a significant contribution to the weak force. But even in the absence of permanent dipoles, weak forces known as London forces exist, due to transient dipoles of very short duration, as explained by Fritz London. Generally London forces between molecules are greater, the greater the atomic number of the atoms in contact and the greater the area of contact.

The strength of the weak forces is revealed by the volatility of the substance, the greater the force, the higher the boiling point. Thus among the noble gases there is a steady increase in boiling point as atomic number increases and the London force increases. Comparing $SiCl_4$(b.p. 57.6 °C) to its neighbor PCl_3(b.p. 75.5 °C) we see the contribution of the dipole-dipole attractions in the latter case. The contributions of dipole-dipole attractions and hydrogen bonding are very important in comparing organic compounds, as illustrated by C_2H_6(b.p. −89 °C, London forces only), CH_3F(b.p. −78 °C, dipole), and CH_3OH(b.p. 65 °C, H bonding). Among the compounds of formula C_nH_{2n+2} the straight-chain compound always has a boiling point higher than any of its branched-chain isomers, because of the increased area of contact of such a molecule with its surroundings.

Solved Problems

CRYSTAL DIMENSIONS

10.1. Metallic gold crystallizes in the face-centered cubic lattice. The length of the cubic unit cell [Fig. 10-1(b)] is $a = 407.0$ pm. (a) What is the closest distance between gold atoms? (b) How many "nearest neighbors" does each gold atom have at the distance calculated in (a)? (c) What is the density of gold? (d) Prove that the *packing factor* for gold, the fraction of the total volume occupied by the atoms themselves, is 0.74.

(a) Consider a corner gold atom in Fig. 10-1(b). The closest distance to another corner atom is a. The distance to an atom at the center of a face is one-half the diagonal of that face, i.e.,

$$\frac{1}{2}(a\sqrt{2}) = \frac{a}{\sqrt{2}}, \text{ which is less than } a$$

Thus the closest distance between atoms is

$$\frac{407.0}{\sqrt{2}} = 287.8 \text{ pm}$$

(b) The problem is to find how many face-centers are equidistant from a corner atom. Point A in Fig. 10-3 may be taken as the reference corner atom. In that same figure, B is one of the face-center points at the nearest distance to A. In plane ABD in the figure there are three other points equally close to A: the centers of the squares in the upper right, lower left, and lower right quadrants of the plane, measured around A. Plane ACE, parallel to the plane of the paper, also has points in the centers of each of the squares in the four quadrants around A. Also, plane ACF, perpendicular to the plane of the paper, has points in the centers of each of the squares in the four quadrants around A. Thus there are 12 nearest neighbors in all, the number expected for a close-packed structure.

The same result would have been obtained by counting the nearest neighbors around B, a face-centered point.

(c) For the face-centered cubic structure, with eight corners and six face-centers,

$$\text{Mass per unit cell} = \tfrac{1}{8}(8m) + \tfrac{1}{2}(6m) = 4m$$

where m is the mass of a single gold atom, 197.0 u. Converting to grams,

$$m = (197.0 \text{ u})\left(\frac{1 \text{ g}}{6.023 \times 10^{23} \text{ u}}\right) = 3.27 \times 10^{-22} \text{ g}$$

and

$$\text{Density} = \frac{4m}{a^3} = \frac{4(3.27 \times 10^{-22} \text{ g})}{(4.070 \times 10^{-8} \text{ cm})^3} = 19.4 \text{ g/cm}^3$$

The reverse of this type of calculation can be used for a precise determination of Avogadro's number, provided the lattice dimension, the density, and the atomic weight are known precisely.

(d) Since the atoms at closest distance are in contact in a close-packed structure, the closest distance between centers calculated in (a), $a/\sqrt{2}$, must equal the sum of the radii of the two spherical atoms, $2r$. Thus, $r = a/2^{3/2}$. From (c), there are four gold atoms per unit cell. Then,

$$\text{Volume of 4 gold atoms} = 4\left(\frac{4}{3}\pi r^3\right) = 4\left(\frac{4}{3}\pi\right)\left(\frac{a}{2^{3/2}}\right)^3 = \frac{\pi a^3}{3\sqrt{2}}$$

and

$$\text{Packing fraction} = \frac{\text{volume of 4 gold atoms}}{\text{volume of unit cell}} = \frac{\pi a^3/3\sqrt{2}}{a^3} = \frac{\pi}{3\sqrt{2}} = 0.740\,5$$

Note that the parameter a for the gold unit cell canceled, so that the result holds for any cubic close-packed structure.

10.2. Show that the tetrahedral and octahedral holes in gold are appropriately named. Find the closest distance between an impurity atom and a gold atom if the impurity atom occupies a tetrahedral hole, an octahedral hole. How many holes of each type are there per gold atom?

Examine Fig. 10-1(b), and imagine a hole in the center of the upper left front minicube. This hole is equidistant from the four occupied corners of the minicube, the common distance being half a body diagonal of the minicube, or

$$\frac{1}{2}\sqrt{\left(\frac{a}{2}\right)^2 + \left(\frac{a}{2}\right)^2 + \left(\frac{a}{2}\right)^2} = \frac{a\sqrt{3}}{4}$$

These four occupied corners define a regular tetrahedron (see Problem 9.18), and the center of the tetrahedron is the point equidistant from the corners, which we showed is the location of the hole. This justifies the name "tetrahedral hole." As the unit cell contains 8 tetrahedral holes (one in each minicube) and 4 gold atoms (Problem 10.1), there are $8 \div 4 = 2$ tetrahedral holes per gold atom.

Now consider the hole at the center of the unit cell of Fig. 10-1(b). This hole is equidistant from the centers of all six faces of the unit cell, all of which are the nearest occupied sites to the hole. These six points are the vertices of an eight-faced figure; the faces are congruent equilateral triangles (whose edges are face diagonals of the minicubes). Such a figure is a regular octahedron, and the hole is at its center; hence the name "octahedral hole." The distance between the hole and a nearest-neighbor atom is $a/2$. A similar proof can be made for an octahedral hole on the center of an edge of the unit cell in Fig. 10-1(b) if we note that the actual crystal lattice consists of a three-dimensional stack of unit cells, as in Fig. 10-3. Each such edge-center hole is shared by four unit cells, and there are 12 edges in a cube, so that the number of octahedral holes per unit cell is

$$1 + \tfrac{1}{4}(12) = 4$$

The ratio of octahedral holes to gold atoms is thus 4 to 4, or 1.

Note the competing advantages of tetrahedral and octahedral holes for housing impurities or second components of an alloy. If the crystal forces, whatever their nature, depend mostly on interactions between nearest neighbors, the octahedral hole has an advantage in having more nearest neighbors to interact with (6, as opposed to 4). However, the tetrahedral hole has a shorter nearest-neighbor distance ($a\sqrt{3}/4 = 0.433a$, as opposed to $0.500a$), giving it the advantage of a greater potential interaction with any one host atom.

10.3. CsCl crystallizes in a cubic structure that has a Cl^- at each corner and a Cs^+ at the center of the unit cell. Use the ionic radii listed in Table 9-1 to predict the lattice constant a and compare with the value of a calculated from the observed density of CsCl, 3.97 g/cm^3.

Figure 10-5(a) shows a schematic view of the unit cell, where the open circles represent Cl^- and the filled circle Cs^+. The circles are made small with respect to the unit cell length a in order to show more clearly the locations of the various ions. Figure 10-5(b) is a more realistic representation of the right triangle ABC, showing anion-cation-anion contact along the body diagonal AC.

Let us assume that the closest Cs^+-to-Cl^- distance is the sum of the ionic radii of Cs^+ and Cl^-, or

$$169 + 181 = 350 \text{ pm}$$

This distance is one-half the cube diagonal, or $a\sqrt{3}/2$. Then

$$\frac{a\sqrt{3}}{2} = 350 \text{ pm} \qquad \text{or} \qquad a = \frac{2(350 \text{ pm})}{\sqrt{3}} = 404 \text{ pm}$$

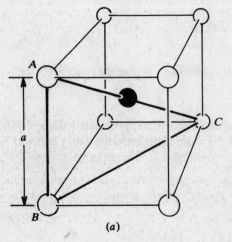

(a)

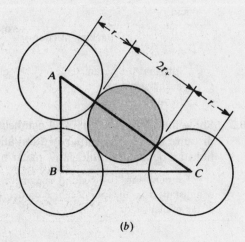

(b)

Fig. 10-5

The density can be used to compute a if we count the number of ions of each type per unit cell. The number of assigned Cl^- ions per unit cell is one-eighth of the number of corner Cl ions, or $\frac{1}{8}(8) = 1$. The only Cs^+ in the unit cell is the center Cs^+, so that the assigned number of cesium ions is also 1. (This type of assignment of ions or atoms in a compound must always agree with the empirical formula of the compound, as the 1:1 ratio in this problem does.) The assigned mass per unit cell is thus that of 1 formula unit of CsCl,

$$\frac{132.9 + 34.5}{6.02 \times 10^{23}}\, g = 2.797 \times 10^{-22}\, g$$

Volume of unit cell $= a^3 = \dfrac{mass}{density} = \dfrac{2.797 \times 10^{-22}\, g}{3.97\, g/cm^3} = 70.4 \times 10^{-24}\, cm^3$

whence $a = \sqrt[3]{70.4 \times 10^{-24}\, cm^3} = 4.13 \times 10^{-8}\, cm = 413\, pm$

This value, based on the experimental density, is to be considered the more reliable since it is based on a measured property of CsCl, while the ionic radii are based on averages over many different compounds. Unit cell dimensions can be measured accurately by X-ray diffraction, and from them the theoretical density can be calculated. The measured density is usually lower because most samples which are large enough to measure are not perfect single crystals and contain empty spaces in the form of grain boundaries and various crystalline imperfections.

The CsCl structure is *not* described as body-centered, since the particle occupying the center is different from the particles occupying the corners of the unit cell. There are two ways of describing the structure. One way is to say that Cs^+ occupies the central holes of the Cl^- simple cubic lattice. Another way is to say that the structure is made up of two interpenetrating simple cubic lattices, one made up of Cl^- and one of Cs^+. The Cs^+ lattice is displaced from the Cl^- lattice along the direction of the unit cell diagonal by one-half the length of the unit cell diagonal.

10.4.   The CsCl structure (Fig. 10-5) is observed in alkali halides only when the radius of the cation is sufficiently large to keep its eight nearest-neighbor anions from touching. What minimum value for the ratio of cation to anion radii, r_+/r_-, is needed to prevent this contact?

In the CsCl structure, the nearest cation-anion distance occurs along the diagonal of the unit cell cube, while the nearest anion-anion distance occurs along a unit cell edge. This relationship is shown in Fig. 10-5(b). In the figure,

$$\overline{AB} = a \qquad \overline{BC} = a\sqrt{2} \qquad \overline{AC} = a\sqrt{3}$$

If we assume anion-cation contact along AC, then $\overline{AC} = 2(r_+ + r_-) = a\sqrt{3}$. In the limiting case, where anions touch along the edge of the unit cell, $2r_- = a$. Dividing the former equation by the latter,

$$\frac{r_+}{r_-} + 1 = \sqrt{3} \qquad or \qquad \frac{r_+}{r_-} = \sqrt{3} - 1 = 0.732$$

If the ratio were less than this critical value, anions would touch (increasing the repulsive forces) and cation and anion would be separated (decreasing the attractive forces). Both effects would tend to make the structure unstable.

10.5.   Ice crystallizes in a hexagonal lattice. At the low temperature at which the structure was determined, the lattice constants were $a = 453$ pm and $c = 741$ pm (Fig. 10-2). How many H_2O molecules are contained in a unit cell?

The volume V of the unit cell in Fig. 10-2 is

$$V = (area\ of\ rhombus\ base) \times (height\ c)$$
$$= (a^2 \sin 60\,°)c = (453\ pm)^2(0.866)(741\ pm) = 132 \times 10^6\ pm^3 = 132 \times 10^{-24}\ cm^3$$

Although the density of ice at the experimental temperature is not stated, it could not be very different from the value at $0\,°C$, $0.92\ g/cm^3$.

Mass of unit cell $= V \times$ density $= (132 \times 10^{-24}\ cm^3)(0.92\ g/cm^3)(6.02 \times 10^{23}\ u/g) = 73\ u$

This is close to 4 times the molecular weight of H_2O; we conclude that there are four molecules of H_2O per unit cell. The discrepancy between 73 u and the actual mass of four molecules, 72 u, is undoubtedly due to the uncertainty in the density at the experimental temperature.

10.6. $BaTiO_3$ crystallizes in the perovskite structure. This structure may be described as a barium-oxygen face-centered cubic lattice, with barium ions occupying the corners of the unit cell, oxide ions occupying the face centers, and titanium ions occupying the centers of the unit cells. (*a*) If titanium is described as occupying holes in the Ba-O lattice, what type of hole does it occupy? (*b*) What fraction of the holes of this type does it occupy? (*c*) Can you suggest a reason why it occupies those holes of this type which it does but not the other holes of the same type?

(*a*) These are octahedral holes.

(*b*) The octahedral holes at the centers of unit cells constitute just one-fourth of all the octahedral holes in a face-centered cubic lattice. (See Problem 10.2.)

(*c*) An octahedral hole at the center of a unit cell, occupied by a titanium ion, has six nearest-neighbor oxide ions. The other octahedral holes, located at the centers of the edges of the unit cell, have six nearest neighbors each, as is the case with any octahedral hole, but two of the six neighbors are barium ions (at the unit-cell corners terminating the given edge) and four are oxide ions. The proximity of two cations, Ba^{2+} and Ti^{4+}, would be electrostatically unfavorable.

CRYSTAL FORCES

10.7. The melting point of quartz, one of the crystalline forms of SiO_2, is $1\,610\,°C$, and the sublimation point of CO_2 is $-79\,°C$. How similar do you expect the crystal structures of these two substances to be?

The big difference in melting points suggests a difference in type of crystal binding. The intermolecular forces in solid CO_2 must be very low to be overcome by a low-temperature sublimation. CO_2 is actually a molecular lattice held together only by the weak van der Waals forces between discrete CO_2 molecules. SiO_2, on the other hand, is a covalent lattice, with a three-dimensional network of bonds; each silicon atom is bonded tetrahedrally to four oxygens and each oxygen is bonded to two silicons.

10.8. In the hexagonal ice structure (Fig. 10-6), each oxygen is coordinated tetrahedrally with four other oxygens, with an intervening hydrogen between adjoining oxygens. ΔH of sublimation of ice at $0\,°C$ is $51.0\,kJ/mol\ H_2O$. It has been estimated by comparison with non-hydrogen-bonded solids having intermolecular van der Waals forces similar to those in ice that the ΔH of sublimation would be only $15.5\,kJ/mol\ H_2O$ if ice were not hydrogen-bonded. From these data estimate the strength of the hydrogen bond in ice.

The excess of ΔH of sublimation above that of a non-hydrogen-bonded solid can be attributed to hydrogen bonds.

$$\Delta H(\text{excess}) = 51.0 - 15.5 = 35.5\ kJ/mol\ H_2O$$

Each H_2O is hydrogen-bonded to four other H_2O molecules, through O—H---O linkages (indicated in Fig. 10-6 only for the two interior molecules). Each such hydrogen-bonded linkage is shared by two H_2O molecules (to which the two oxygen atoms belong). Thus, on the average, each H_2O can be assigned 4 halves, or 2 hydrogen bonds.

$$\Delta H(\text{hydrogen bond}) = \frac{35.5\ kJ/mol\ H_2O}{2\ mol\ H\text{-bonds/mol}\ H_2O}$$

$$= 17.8\ kJ/mol\ \text{hydrogen bond}$$

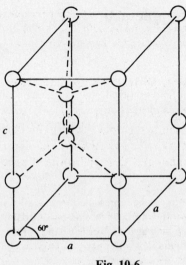

Fig. 10-6

Observe from Fig. 10-6 that

$$8(\tfrac{1}{8}) + 4(\tfrac{1}{4}) + 2 = 4$$

H_2O molecules are uniquely assigned to the unit cell. This is in agreement with the result of Problem 10.5.

10.9. Which would have the higher melting point (a) V or Ca, (b) MgO or KCl? Explain.

(a) V; vanadium would have a much greater charge density in the electron sea since it has three $3d$ electrons in addition to the two $4s$ electrons to contribute. Also the ionic cores, being smaller, will be closer together. Thus the crystal forces will be greater. The actual melting points are V 1 890 °C, Ca 845 °C. The relative closeness of the cores is indicated by the densities:

$$\text{V } 6.11 \text{ g/cm}^3, \text{ Ca } 1.55 \text{ g/cm}^3.$$

(b) MgO; the crystal forces will be greater on two accounts: (1) the smallness of both cation and anion in MgO allows closer approach, (2) both cation and anion carry twice as much charge. (Recall that electrostatic force is proportional to the product of the charges and inversely proportional to the square of the distance between them.) Actual melting points are MgO 2 800 °C, KCl 776 °C.

FORCES IN LIQUIDS

10.10. In each case indicate which liquid will have the higher boiling point and explain why.

(a) CO_2 or SO_2　　　(b) $(CH_3)_2CHCH(CH_3)_2$ or $CH_3CH_2CH_2CH_2CH_2CH_3$
(c) Cl_2 or Br_2　　　(d) C_2H_5SH or C_2H_5OH

(a) SO_2. The molecule is bent and has a permanent dipole moment, in contrast to CO_2 which is straight. (*Note*: CO_2 is never liquid at 1 atm pressure but sublimes directly from the solid to the gas.)
(b) $CH_3CH_2CH_2CH_2CH_2CH_3$. The straight-chain isomer has a greater area of contact with its neighbors.
(c) Br_2. The higher atomic number leads to a greater force of attraction between molecules.
(d) C_2H_5OH. Hydrogen bonding is present in the oxygen compound but is negligible in the sulfur compound.

Supplementary Problems

CRYSTAL DIMENSIONS

10.11. Potassium crystallizes in a body-centered cubic lattice, with a unit cell length $a = 520$ pm. (a) What is the distance between nearest neighbors? (b) What is the distance between next-nearest neighbors? (c) How many nearest neighbors does each K atom have? (d) How many next-nearest neighbors does each K have? (e) What is the calculated density of crystalline K?

    Ans.   (a) 450 pm;   (b) 520 pm;   (c) 8;   (d) 6;   (e) 0.924 g/cm³

10.12. The hexagonal close-packed lattice can be represented by Fig. 10-2, if $c = a\sqrt{\frac{8}{3}} = 1.633a$. There is an atom at each corner of the unit cell and another atom which can be located by moving one-third the distance along the diagonal of the rhombus base, starting at the lower left-hand corner, and moving perpendicularly upward by $c/2$. Mg crystallizes in this lattice and has a density of 1.74 g/cm³. (a) What is the volume of the unit cell? (b) What is a? (c) What is the distance between nearest neighbors? (d) How many nearest neighbors does each atom have?

    Ans.   (a) 46.4×10^6 pm³;   (b) 320 pm;   (c) 320 pm;   (d) 12

10.13. The NaCl lattice has the cubic unit cell shown in Fig. 10-7. KBr crystallizes in this lattice. (a) How many K^+ ions and how many Br^- ions are in each unit cell? (b) Assuming the additivity of ionic radii, what is a? (c) Calculate the density of a perfect KBr crystal. (d) What minimum value of r_+/r_- is needed to prevent anion-anion contact in this structure?

    Ans.   (a) 4 each;   (b) 656 pm;   (c) 2.80 g/cm³;   (d) 0.414

10.14. MgS and CaS both crystallize in the NaCl-type lattice (Fig. 10-7). From ionic radii listed in Table 9-1, what conclusion can you draw about anion-cation contact in these crystals?

    Ans. Ca^{2+} and S^{2-} can be in contact, but Mg^{2+} and S^{2-} cannot. In MgS, if Mg^{2+} and S^{2-} were in contact there would not be enough room for the sulfide ions along the diagonal of a square constituting one-quarter of a unit cell face.

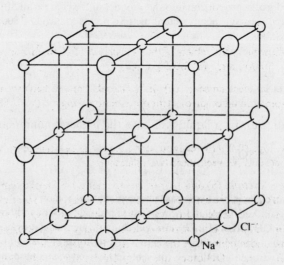

Fig. 10-7.   NaCl Unit Cell

10.15. Each rubidium halide crystallizing in the NaCl-type lattice has a unit cell length 30 pm greater than that for the corresponding potassium salt of the same halogen. What is the ionic radius of Rb^+ computed from these data?

Ans.   148 pm

10.16. Iron crystallizes in several modifications. At about 910 °C, the body-centered cubic α-form undergoes a transition to the face-centered cubic γ-form. Assuming that the distance between nearest neighbors is the same in the two forms at the transition temperature, calculate the ratio of the density of γ-iron to that of α-iron at the transition temperature.

Ans.   1.09

10.17. The ZnS zinc blende structure is cubic. The unit cell may be described as a face-centered sulfide ion sublattice with zinc ions in the centers of alternating minicubes made by partitioning the main cube into eight equal parts. (a) How many nearest neighbors does each Zn^{2+} have? (b) How many nearest neighbors does each S^{2-} have? (c) What angle is made by the lines connecting any Zn^{2+} to any two of its nearest neighbors? (d) What minimum r_+/r_- ratio is needed to avoid anion-anion contact, if closest cation-anion pairs are assumed to touch?

Ans.   (a) 4; (b) 4; (c) 109°28′; (d) 0.225

10.18. Why does ZnS not crystallize in the NaCl structure? (*Hint*: Refer to Problem 10.13.)

Ans.   The r_+/r_- ratio is 0.402, too low to avoid anion-anion contact in the NaCl structure.

10.19. Compute the packing factor for spheres occupying (a) a body-centered cubic, and (b) a simple cubic structure, where closest neighbors in both cases are in contact.

Ans.   (a) 0.680; (b) 0.524

10.20. Many oxide minerals can be visualized as a face-centered oxide ion lattice, with cations distributed within the tetrahedral and octahedral holes. Calculate the lattice constant, a, for a face-centered O^{2-} lattice. If cations occupy all the octahedral holes in MgO and CaO, calculate a for these minerals. Use data in Table 9-1.

Ans.   For an oxide lattice $a = 396$ pm. With Mg^{2+} and Ca^{2+} in the octahedral holes anion-anion contact is broken and a expands to 410 pm and 478 pm, respectively. (Actual values are a little greater.)

CRYSTAL FORCES

10.21. In each of the following cases, which of the two crystals has the higher melting point, and why? (a) Cs, Ba; (b) Si, P_4; (c) Xe, Kr; (d) MgF_2, $CaCl_2$.

Ans.   (a) Ba, denser electron sea; (b) Si, covalently bonded network versus molecular crystal; (c) Xe, higher atomic number gives stronger London forces; (d) MgF_2, cations and anions are both smaller.

10.22. With reference to Fig. 10-8, how do you account for the differences in melting point between (a) and (b), between (c) and (d), and between these two differences?

Ans.   The crystal forces in (b) and (d) are largely van der Waals. Compounds (a) and (c), containing the polar —OH group, are capable of hydrogen bonding. In the case of (c), the hydrogen bonding is from the —OH of one molecule to the doubly bonded oxygen of the neighboring molecule; and the resulting *inter*molecular attraction leads to an increase in melting point as compared with (d), the non-hydrogen-bonded control. In the case of (a), the molecular structure allows *intra*molecular hydrogen bonding from the —OH group of each molecule to the doubly bonded oxygen of the same molecule; in the absence of strong *inter*molecular hydrogen bonding the difference in melting point as compared

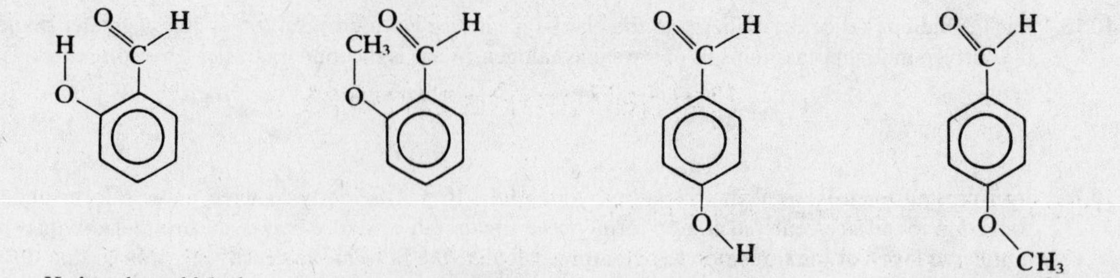

o-Hydroxybenzaldehyde	o-Methoxybenzaldehyde	p-Hydroxybenzaldehyde	p-Methoxybenzaldehyde
m.p. = 266 °C	m.p. = 309 °C	m.p. = 388 °C	m.p. = 273 °C
(a)	(b)	(c)	(d)

Fig. 10-8

with the reference substance (b) should be small, related perhaps to differences in the crystal structure or to the van der Waals forces, which should be slightly larger for (b) than for (a) on account of the extra CH_3 group.

FORCES IN LIQUIDS

10.23. Which one of each of the following pairs of liquids has the higher boiling point? (a) $CH_3CH_2CH_2OH$, $HOCH_2CH_2OH$; (b) $CH_3CH_2CH_3$, CH_3CH_2F; (c) Xe, Kr; (d) H_2O, H_2S; (e) $CH_3CH_2CH_2CH_2CH_2CH_3$, $CH_3CH_2CH_2CH_3$.

Ans. (a) $HOCH_2CH_2OH$, (b) CH_3CH_2F, (c) Xe, (d) H_2O, (e) $CH_3CH_2CH_2CH_2CH_2CH_3$

10.24. Explain why water dissolves in acetone (CH_3COCH_3), but not in hexane (C_6H_{14}).

Ans. When water dissolves much energy is required to break its hydrogen bonds. In acetone this is compensated for by hydrogen-bonding between water and the oxygen atom of CH_3COCH_3, but there is no compensating strong interaction in hexane.

Chapter 11

Oxidation-Reduction

OXIDATION-REDUCTION REACTIONS

Oxidation is a chemical change in which electrons are lost by an atom or group of atoms, and *reduction* is a chemical change in which electrons are gained by an atom or group of atoms. These definitions can be applied most simply in the case of elementary substances and their ions. A transformation that converts a neutral atom to a positive ion must be accompanied by the loss of electrons and must, therefore, be an oxidation. Consider the following example:

$$Fe \rightarrow Fe^{2+} + 2e^-$$

Electrons (symbol e^-) are written explicitly on the right side and provide equality of total charge on the two sides of the equation. Similarly, the transformation of a neutral element to an anion must be accompanied by electron gain and is classified as a reduction, as in the following case:

$$Cl_2 + 2e^- \rightarrow 2Cl^-$$

Oxidation and reduction always occur simultaneously, and the total number of electrons lost in the oxidation must equal the number of electrons gained in the reduction.

OXIDATION STATE

It is not immediately apparent from the ionic charge alone whether a compound substance is undergoing oxidation or reduction. For example, MnO_2 reacts with hydrochloric acid to produce, among other things, chlorine gas and the Mn^{2+} ion. It is obvious that neutral chlorine is produced by oxidation of Cl^-. We infer, therefore, that MnO_2 is undergoing reduction, in spite of the fact that a neutral substance, MnO_2 is converted partly to a cationic substance, Mn^{2+}. In another example, arsenious acid, H_3AsO_3, reacts with I_2 to form, among other things, the arsenate ion, $HAsO_4^{2-}$, and the iodide ion, I^-. Since the iodine is reduced (neutral halogen to anion), the arsenious acid must be oxidized, in spite of the conversion of neutral H_3AsO_3 to anionic $HAsO_4^{2-}$.

Oxidation state is a concept that is helpful in diagnosing quickly the state of oxidation or reduction of particular atoms in such compound species as MnO_2, H_3AsO_3, and $HAsO_4^{2-}$. Oxidation state of an atom in some chemical combination is the electrical charge assigned to that atom according to a prescribed set of rules. Other expressions often used to refer to oxidation state are *oxidation number* or *valence state*. In what follows, Roman numerals (or 0) will be used to designate oxidation state.

It should be made clear that oxidation state is not the same as formal charge (Chapter 9). Formal charge is based on an attempted mapping out of the real charge distribution of a molecule or ion among the constituent atoms in accordance with detailed knowledge of structure and electronic binding. Oxidation state, however, is a simpler assignment that does not require information about such electronic variables as single or multiple bonding and octet versus nonoctet structures. Oxidation state is computed directly from the compositional formula itself, such as MnO_2 or H_3AsO_3. The two basic rules for oxidation state assignment follow.

1. In ionic binary compounds the oxidation state is the charge per atom.

EXAMPLE 1 $CdCl_2$ is an ionic compound and may be designated $Cd^{2+}(Cl^-)_2$ to show its ionic character. The cadmium possesses a true charge of $+2$ and an oxidation state of $+II$. Each chloride ion possesses a true charge of -1 and an oxidation state of $-I$. In Hg_2Cl_2, each mercury in Hg_2^{2+} has an average charge of $+1$ and an oxidation state of $+I$. The chlorine in Cl^- is again $-I$.

2. In covalent or nonionic compounds, the electrons involved in bond formation are not completely transferred from one element to the other, but are shared more or less equally by the bonding atoms. For purposes of oxidation state computation, however, it is conventional to assign each bonding electron to some particular atom. If the atoms are of the same kind, half of the bonding electrons are assigned to each of the two atoms. If the atoms are different, all electrons in the bond are assigned to that atom which has the greater electronegativity (Chapter 9).

Rules 1 and 2 have several important corollaries.

(a) The oxidation state of a free and uncombined element is zero.

EXAMPLE 2 Hg in Hg, H in H_2, O in O_2, S in S_8, etc. In H_2, of the two electrons in the molecule, one is assigned to each hydrogen atom. A hydrogen with one electron is the same as a neutral free hydrogen atom. Thus the oxidation state is zero.

(b) The oxidation state of hydrogen in compounds is usually $+I$, except in the case of the metallic hydrides, where it is $-I$.

EXAMPLE 3 In NH_3, the nitrogen atom is bonded directly to each of the three hydrogen atoms. Since nitrogen is more electronegative than hydrogen, all the bonding electrons are assigned to nitrogen. Each hydrogen is thus left with zero assigned electrons, one less than a free hydrogen atom. Thus the hydrogen has an apparent charge, or an oxidation state of $+I$. The arbitrary nature of this charge designation is apparent in NH_3, in which the true charge separation between the nitrogen and the hydrogens is very slight. NH_3 never ionizes in water, for example, to produce hydrogen ions.

In CaH_2, on the contrary, each hydrogen, being more electronegative than calcium, is assigned two electrons, one more than a free hydrogen atom. The oxidation state of hydrogen is therefore $-I$.

(c) The oxidation state of oxygen in compounds is usually $-II$, except in peroxides, where it is $-I$, or in fluorine compounds, where it may be positive.
(d) The algebraic sum of the positive and negative oxidation states of all atoms in a neutral molecule is equal to zero.
(e) The algebraic sum of the positive and negative oxidation states of all atoms in an ion is equal to the charge of the ion.

EXAMPLE 4 In H_2SiO_3, the oxidation state of Si is $+IV$, in accordance with corollary (d).
In PO_4^{3-}, the oxidation state of P is $+V$, in accordance with corollary (e):

$$+5 + 4(-2) = -3$$

Only the phosphate ion as a whole has a true charge, or ionic valence. Each atom within this complex ion has only an arbitrarily assigned charge, its oxidation state.

The definitions of oxidation and reduction may now be generalized as follows: *The increase in oxidation state is oxidation and the decrease in oxidation state is reduction.* In the reduction of MnO_2 to Mn^{2+}, for example, the oxidation state of manganese is changed from $+IV$ to $+II$; in the oxidation of H_3AsO_3 to $HAsO_4^{2-}$, the oxidation state of arsenic is changed from $+III$ to $+V$.

IONIC NOTATION FOR EQUATIONS

In writing oxidation-reduction equations, care should be taken to write formulas only for compounds or ions that have true chemical existence, like MnO_2, H_3AsO_3, $HAsO_4^{2-}$. Mn(IV), As(III), and As(V) are not true species and should not be written as such in equations.

When the ionic convention is to be used for writing formulas, the following rules are customarily observed:

1. Ionic substances are written in the ionic form only if the ions are separated from each other in the reaction medium. Thus, NaCl would be the conventional notation for reactions involving solid salt, because the ions in the solid are bound together in the crystal. Salt reactions in solution, however, would be indicated by Na^+ and Cl^-, or by either of these ions alone if only the sodium *or* the chlorine undergoes a change in oxidation state. Fairly insoluble salts, like CuS, are always written in the neutral form.

2. Partially ionized substances are written in the ionic form only if the extent of ionization is appreciable (about 20 percent or more). Thus water, which is ionized to the extent of less than one part in a hundred million, is written as H_2O. Strong acids, like hydrochloric acid and nitric acid, may be written in the ionized form; but weak acids, like nitrous, acetic, and sulfurous acids, are always written in the molecular form. Ammonia, a weak base, is written NH_3. Sodium hydroxide, a strong base, is written in the ionized form Na^+ and OH^-, for water solutions.

3. Some complex ions are so stable that one or more of the groups from which they are formed do not exist in appreciable amounts outside the complex. In such a case the formula for the entire complex is written. Thus, the ferricyanide ion is always written as $[Fe(CN)_6]^{3-}$, never as separate ferric and cyanide ions. $[Cu(NH_3)_4]^{2+}$ is the notation for the common blue complex ion formed by cupric salts in ammonia solutions.

4. A mixed convention will be employed in this chapter as an aid to indicate whether a given compound may be written in the ionized form. Thus, $Ba^{2+}(NO_3^-)_2$ should indicate to the student that barium nitrate has the overall composition of $Ba(NO_3)_2$, that it is soluble and ionized, and that either the barium ion, Ba^{2+}, or the nitrate ion, NO_3^-, may be written separately if desired. The absence of an ionic notation, as in As_2S_5, means that only the neutral formula of the whole compound should be used. This mixed convention will not be uniformly used in subsequent chapters. Experienced chemists may write $Ba(NO_3)_2$ in a reaction in aqueous solution, recognizing the ionic character of this salt even without an explicit ionic notation.

BALANCING OXIDATION-REDUCTION EQUATIONS

The principles of oxidation-reduction are the basis of two simple systematic methods for balancing these equations. If all the products of reaction are known, the balancing may be done either by the *ion-electron method* or by the *oxidation-state method*. (The two methods are compared following Problem 11.8.) After students have acquired more experience, they will be able to predict some or all of the principal products if they keep in mind such facts as the following:

(a) If a free halogen is reduced, the product of reduction must be the halogenide ion (charge $= -1$).

(b) If a metal that has only one positive valence is oxidized, the oxidation state of the product is obvious.

(c) Reductions of concentrated nitric acid lead to NO_2, whereas the reduction of dilute nitric acid may lead to NO, N_2, NH_4^+, or other products, depending on the nature of the reducing agent and the extent of the dilution.

(d) Permanganate ion, MnO_4^-, is reduced to Mn^{2+} in distinctly acid solution, as is MnO_2. The reduction product of permanganate in neutral or alkaline solution may be MnO(OH), MnO_2, or MnO_4^{2-}.

(e) If a peroxide is reduced, the product of reduction must contain oxygen in the $-II$ oxidation state, as in H_2O or OH^-. If a peroxide is oxidized, molecular oxygen is formed.

(f) Dichromate, $Cr_2O_7^{2-}$, is reduced in acid solution to Cr^{3+}.

Ion-electron method

1. Write a *skeleton equation* that includes those reactants and products that contain the elements undergoing a change in oxidation state.
2. Write a partial skeleton equation for the oxidizing agent, with the element undergoing a reduction in oxidation state on each side of the equation. The element should not be written as a free atom or ion unless it really exists as such. It should be written as part of a real molecular or ionic species.
3. Write another partial skeleton equation for the reducing agent, with the element undergoing an increase in oxidation state on each side.
4. Balance each partial equation as to number of atoms of each element. In neutral or acid solution, H_2O and H^+ may be added for balancing oxygen and hydrogen atoms. The oxygen atoms are balanced first. For each excess oxygen atom on one side of an equation, balance is secured by adding one H_2O to the *other* side. Then H^+ is used to balance hydrogens. Note that O_2 and H_2 are not used to balance the oxygen and hydrogen atoms unless they are known to be principal participants in the reaction.

    If the solution is alkaline OH^- may be used. For each excess oxygen on one side of an equation, balance is secured by adding one H_2O to the *same* side and $2OH^-$ to the *other* side. If hydrogen is still unbalanced after this is done, balance is secured by adding one OH^- for each excess hydrogen on the *same* side as the excess and one H_2O on the *other* side. If both oxygen and hydrogen are in excess on the same side of the skeleton equation, an OH^- can be written on the *other* side for each paired excess of H and O.
5. If an element undergoing a change in oxidation state is complexed in one of its states with some other element, balance the complexing groups with a species of that element in the same oxidation state occurring in the complex.
6. Balance each partial equation as to the number of charges by adding electrons to either the left or the right side of the equation. If the preceding steps have been followed carefully, it will be found that electrons must be added to the left in the partial equation for the oxidizing agent and to the right in the partial equation for the reducing agent.
7. Multiply each partial equation by a number chosen so that the total number of electrons lost by the reducing agent equals the number of electrons gained by the oxidizing agent.
8. Add the two partial equations resulting from the multiplications. In the sum equation, cancel any terms common to both sides. All electrons should cancel.
9. For an understanding of the nature of the reaction, step 8 can be considered the last step. For calculations involving the masses of reactants or products, transform the ionic equation of step 8 into a molecular equation. This is done by adding to each side of the equation equal numbers of the ions which do not undergo electron transfer but which are present together with the reactive components in neutral chemical substances. Proper pairs of ions may be combined to give a molecular formula.
10. Check the final equation by counting the number of atoms of each element on each side of the equation. If step 9 was omitted, also check the net charge on each side.

Oxidation-state method

1. Write a skeleton equation that includes, as principal formulas, those reactants and products that contain the elements undergoing a change in oxidation state.
2. Determine the *change* in oxidation state which some element in the oxidizing agent undergoes. The number of electrons gained is equal to this change times the number of atoms undergoing the change.
3. Determine the same for some element in the reducing agent.
4. In the skeleton equation, multiply the principal formulas by such numbers as to make the total number of electrons lost by the reducing agent equal to the number of electrons gained by the oxidizing agent.

5.   By inspection supply the proper coefficients for the rest of the equation.
6.   Check the final equation by counting the number of atoms of each element on both sides of the equation.

Solved Problems

FORMULAS AND OXIDATION STATE

11.1.   Given that the oxidation state of hydrogen is $+I$, of oxygen $-II$, and of fluorine $-I$, determine the oxidation states of the other elements in   (a) PH_3,   (b) H_2S,   (c) CrF_3,   (d) H_2SO_4, (e) H_2SO_3,   (f) Al_2O_3.

(a) H_3 represents an oxidation-state sum of $+3(+1$ for each of the three hydrogens). Then the oxidation state of P must be $-III$, since the algebraic sum of the oxidation states of all atoms in a compound must equal zero.

(b)   Oxidation-state sum of H_2 is $+2$; then oxidation state of S is $-II$.

(c)   Oxidation-state sum of F_3 is -3; then oxidation state of Cr is $+III$.

(d)   Oxidation-state sum of H_2 is $+2$ and of O_4 is -8, or a total of -6; then oxidation state of S is $+VI$. The sulfate radical, SO_4, has an ionic valence of -2, since the positive ionic valence of H_2 (i.e., $2H^+$) is $+2$.

(e)   Oxidation-state sum of H_2 is $+2$ and of O_3 is -6, or a total of -4; then oxidation state of S is $+IV$. The sulfite radical, SO_3, has an ionic valence of -2, since the ionic charge of H_2 is $+2$.

(f)   Oxidation-state sum of O_3 is -6; then oxidation-state sum of Al_2 is $+6$, and oxidation state of Al is $\frac{1}{2}(+6) = +III$.

BALANCING OXIDATION-REDUCTION EQUATIONS

11.2.   Balance the oxidation-reduction equation

$$H^+NO_3^- + H_2S \rightarrow NO + S + H_2O$$

Ion-Electron Method.

1.   The skeleton equation is given above. (H_2O is not necessary to the skeleton equation.)
2.   The oxidizing agent is nitrate ion, NO_3^-, since it contains the element, N, which undergoes a decrease in oxidation state. The reducing agent is H_2S, since it contains the element, S, which undergoes an increase in oxidation state. (S^{2-} might have been selected as the reducing agent, but H_2S is preferable because of its very slight degree of ionization in acid solution.)
3.   The partial skeleton equation for the oxidizing agent is

$$NO_3^- \rightarrow NO$$

4.   The partial skeleton equation for the reducing agent is

$$H_2S \rightarrow S$$

5.   (a)   In the first partial equation, $2H_2O$ must be added to the right side in order to balance the oxygen atoms. Then $4H^+$ must be added to the left side to balance the H atoms.

$$4H^+ + NO_3^- \rightarrow NO + 2H_2O \tag{a}$$

(b)   The second partial equation can be balanced by adding $2H^+$ to the right side.

$$H_2S \rightarrow S + 2H^+ \tag{b}$$

6. (a) In equation 5(a) above, the net charge on the left is $+4 - 1 = +3$, and on the right it is 0. Therefore 3 electrons must be added to the left side.

$$4H^+ + NO_3^- + 3e^- \rightarrow NO + 2H_2O \qquad (a)$$

(b) In equation 5(b), the net charge on the left is 0, and on the right is $+2$. Hence 2 electrons must be added to the right side.

$$H_2S \rightarrow S + 2H^+ + 2e^- \qquad (b)$$

7. Equation 6(a) must be multiplied by 2, and equation 6(b) by 3.

$$8H^+ + 2NO_3^- + 6e^- \rightarrow 2NO + 4H_2O \qquad (a)$$
$$3H_2S \rightarrow 3S + 6H^+ + 6e^- \qquad (b)$$

8. Addition of equations 7(a) and 7(b) results in

$$8H^+ + 2NO_3^- + 3H_2S + 6e^- \rightarrow 2NO + 4H_2O + 3S + 6H^+ + 6e^-$$

Since $6H^+$ and $6e^-$ are common to both sides, they may be canceled.

$$2H^+ + 2NO_3^- + 3H_2S \rightarrow 2NO + 4H_2O + 3S$$

This form of the equation shows all the reactant ions and compounds in the proper form.

9. If we want to know how much HNO_3 is required, we merely combine H^+ with NO_3^-.

$$2H^+NO_3^- + 3H_2S \rightarrow 2NO + 4H_2O + 3S$$

Oxidation-State Method.

1. Note that the oxidation state of N changes from $+V$ in NO_3^- to $+II$ in NO.
2. The oxidation state of S changes from $-II$ in H_2S to 0 in S.
3. Electron balance *diagrams* can be written as follows. (These are not *equations*.)

$$N(+V) + 3e^- \rightarrow N(+II) \qquad (1)$$
$$S(-II) \rightarrow S(0) + 2e^- \qquad (2)$$

4. In order that the number of electrons lost shall equal the number gained, we must multiply (1) by 2, and (2) by 3.

$$2N(+V) + 6e^- \rightarrow 2N(+II)$$
$$3S(-II) \rightarrow 3S(0) + 6e^-$$

Hence the coefficient of $H^+NO_3^-$ and of NO is 2, and of H_2S and of S is 3. Part of the skeleton equation can now be filled in.

$$2H^+NO_3^- + 3H_2S \rightarrow 2NO + 3S$$

5. The 8 atoms of H on the left (2 from $H^+NO_3^-$ plus 6 from H_2S) must form $4H_2O$ on the right. The final and complete equation is

$$2H^+NO_3^- + 3H_2S \rightarrow 2NO + 3S + 4H_2O$$

Note that the oxygen atoms were balanced automatically, without special attention.

The first partly filled skeleton equation could have been written in terms of NO_3^- rather than $H^+NO_3^-$. For variety in the subsequent solved problems, the neutral-compound notation will be used in the oxidation-state method, and the ionic notation in the ion-electron method.

11.3. Balance the following oxidation-reduction equation:

$$K^+MnO_4^- + K^+Cl^- + (H^+)_2SO_4^{2-} \rightarrow Mn^{2+}SO_4^{2-} + (K^+)_2SO_4^{2-} + H_2O + Cl_2$$

Ion-Electron Method.

The skeleton partial equations may be written

$$MnO_4^- \rightarrow Mn^{2+} \tag{a}$$

$$Cl^- \rightarrow Cl_2 \tag{b}$$

Partial (a) requires $4H_2O$ on the right to balance the oxygen atoms; then $8H^+$ on the left to balance the hydrogen. Partial (b) balances in routine fashion.

$$8H^+ + MnO_4^- \rightarrow Mn^{2+} + 4H_2O \tag{a}$$

$$2Cl^- \rightarrow Cl_2 \tag{b}$$

The net charge on the left of partial (a) is $+8 - 1 = +7$, and on the right it is $+2$; therefore 5 electrons must be added to the left. In partial (b) the net charge on the left is -2, and on the right it is 0; therefore 2 electrons must be added to the right.

$$8H^+ + MnO_4^- + 5e^- \rightarrow Mn^{2+} + 4H_2O \tag{a}$$

$$2Cl^- \rightarrow Cl_2 + 2e^- \tag{b}$$

The multiplying factors are seen to be 2 and 5, respectively.

$$16H^+ + 2MnO_4^- + 10e^- \rightarrow 2Mn^{2+} + 8H_2O$$
$$10Cl^- \rightarrow 5Cl_2 + 10e^-$$

$$\overline{16H^+ + 2MnO_4^- + 10Cl^- \rightarrow 2Mn^{2+} + 8H_2O + 5Cl_2}$$

Finally, check charges on each side: $+4$ on the left $(+16 - 2 - 10)$, $+4$ on the right $(+2 \times 2)$.

Since MnO_4^- was added as $KMnO_4$, $2MnO_4^-$ introduces $2K^+$ to the left side of the equation; and since K^+ does not react, the same number will appear on the right side. Since Cl^- was added as KCl, the $10Cl^-$ introduces $10K^+$ to each side of the equation. Since H^+ was added as H_2SO_4, $16H^+$ introduces $8SO_4^{2-}$ to each side of the equation. Then

$$16H^+ + 8SO_4^{2-} + 2K^+ + 2MnO_4^- + 10K^+ + 10Cl^- \rightarrow 2Mn^{2+} + 8H_2O + 5Cl_2 + 12K^+ + 8SO_4^{2-}$$

Pairs of ions may be grouped on the left side to show that the chemical reagents were H_2SO_4, $KMnO_4$, and KCl. Pairs of ions may be grouped on the right to show that evaporation of the solution, after reaction has taken place, would cause the crystallization of $MnSO_4$ and K_2SO_4.

$$8(H^+)_2SO_4^{2-} + 2K^+MnO_4^- + 10K^+Cl^- \rightarrow 2Mn^{2+}SO_4^{2-} + 6(K^+)_2SO_4^{2-} + 5Cl_2 + 8H_2O$$

Oxidation-State Method.

Mn undergoes a change in oxidation state from $+VII$ in MnO_4^- to $+II$ in Mn^{2+}. Cl undergoes a change in oxidation state from $-I$ in Cl^- to 0 in Cl_2. The electron balance diagrams are

$$Mn(+VII) + 5e^- \rightarrow Mn(+II) \tag{1}$$

$$2Cl(-I) \rightarrow 2Cl(0) + 2e^- \tag{2}$$

Diagram (2) was written in terms of two Cl atoms because these atoms occur in pairs in the product Cl_2. The multiplying factors are 2 and 5, just as in the previous method.

$$2Mn(+VII) + 10e^- \rightarrow 2Mn(+II)$$
$$10Cl(-I) \rightarrow 10Cl(0) + 10e^-$$

Hence the coefficient of $K^+MnO_4^-$ and of $Mn^{2+}SO_4^{2-}$ is 2, of K^+Cl^- is 10, of Cl_2 is 5 $(\frac{1}{2} \times 10)$.

$$2K^+MnO_4^- + 10K^+Cl^- \rightarrow 2Mn^{2+}SO_4^{2-} + 5Cl_2 \quad \text{(incomplete)}$$

So far no provision has been made for the H_2O, $(H^+)_2SO_4^{2-}$, and $(K^+)_2SO_4^{2-}$. The 8 atoms of oxygen from $2K^+MnO_4^-$ form $8H_2O$. For $8H_2O$ we need 16 atoms of hydrogen, which can be furnished by

$8(H^+)_2SO_4^{2-}$. The 12 atoms of $K(10K^+Cl^- + 2K^+MnO_4^-)$ yields $6(K^+)_2SO_4^{2-}$. Note that all the oxygen in the oxidizing agent is converted to water. The sulfate radical retains its identity throughout the reaction.

$$2K^+MnO_4^- + 10K^+Cl^- + 8(H^+)_2SO_4^{2-} \rightarrow 2Mn^{2+}SO_4^{2-} + 5Cl_2 + 6(K^+)_2SO_4^{2-} + 8H_2O$$

11.4. Balance the following oxidation-reduction equation:

$$(K^+)_2Cr_2O_7^{2-} + H^+Cl^- \rightarrow K^+Cl^- + Cr^{3+}(Cl^-)_3 + H_2O + Cl_2$$

Ion-Electron Method.

The balancing of the partial equation for the oxidizing agent proceeds as follows:

$$Cr_2O_7^{2-} \rightarrow Cr^{3+}$$
$$Cr_2O_7^{2-} \rightarrow 2Cr^{3+}$$
$$Cr_2O_7^{2-} \rightarrow 2Cr^{3+} + 7H_2O$$
$$14H^+ + Cr_2O_7^{2-} \rightarrow 2Cr^{3+} + 7H_2O$$
$$14H^+ + Cr_2O_7^{2-} + 6e^- \rightarrow 2Cr^{3+} + 7H_2O \quad \text{(balanced)}$$

The balancing of the partial equation for the reducing agent is as follows.

$$Cl^- \rightarrow Cl_2$$
$$2Cl^- \rightarrow Cl_2$$
$$2Cl^- \rightarrow Cl_2 + 2e^- \quad \text{(balanced)}$$

The overall equation is

$$1 \times [14H^+ + Cr_2O_7^{2-} + 6e^- \rightarrow 2Cr^{3+} + 7H_2O]$$
$$\underline{3 \times [\qquad\qquad 2Cl^- \rightarrow Cl_2 + 2e^- \qquad]}$$
$$14H^+ + Cr_2O_7^{2-} + 6Cl^- \rightarrow 2Cr^{3+} + 7H_2O + 3Cl_2$$

The $14H^+$ was added as $14H^+Cl^-$, and 6 of the 14 chloride ions were oxidized. To each side of the equation 8 more Cl^- can be added to represent those Cl^- which were not oxidized. Similarly, $2K^+$ may be added to each side to show that $Cr_2O_7^{2-}$ came from $(K^+)_2CrO_7^{2-}$.

$$14H^+ + 6Cl^- + 8Cl^- + 2K^+ + Cr_2O_7^{2-} \rightarrow 2Cr^{3+} + 2K^+ + 8Cl^- + 3Cl_2 + 7H_2O$$
$$14H^+Cl^- + (K^+)_2Cr_2O_7^{2-} \rightarrow 2Cr^{3+}(Cl^-)_3 + 2K^+Cl^- + 3Cl_2 + 7H_2O$$

Oxidation-State Method.

The electron balance diagrams are written in terms of 2 atoms of Cr and 2 atoms of Cl because of the appearance of pairs of atoms of these kinds in $(K^+)_2Cr_2O_7^{2-}$ and Cl_2.

For the oxidizing agent:

$$2Cr(+VI) \rightarrow 2Cr(+III)$$
$$2Cr(+VI) + 6e^- \rightarrow 2Cr(+III)$$

For the reducing agent:

$$2Cl(-I) \rightarrow 2Cl(0)$$
$$2Cl(-I) \rightarrow 2Cl(0) + 2e^-$$

Multiplying and adding:

$$1 \times [2Cr(+VI) + 6e^- \rightarrow 2Cr(+III) \quad]$$
$$\underline{3 \times [\qquad 2Cl(-I) \rightarrow 2Cl(0) + 2e^-]}$$
$$2Cr(+VI) + 6Cl(-I) \rightarrow 2Cr(+III) + 6Cl(0)$$

Hence $(K^+)_2Cr_2O_7^{2-} + 6H^+Cl^- \rightarrow 2Cr^{3+}(Cl^-)_3 + 3Cl_2 \quad \text{(incomplete)}$

The equation is still unbalanced, as no provision has been made for K^+Cl^-, H_2O, or the H^+Cl^- that acts as an acid (as opposed to the H^+Cl^- that acts as reducing agent).

By inspection, the 7 atoms of oxygen in $(K^+)_2Cr_2O_7^{2-}$ form $7H_2O$. For $7H_2O$ we need 14 atoms of H, which can be furnished by $14H^+Cl^-$. Since 6 of the chloride ions were oxidized to Cl_2, the remaining $8(14-6)$ should appear on the right as K^+Cl^- or $Cr^{3+}(Cl^-)_3$. Moreover, one $(K^+)_2Cr_2O_7^{2-}$ yields $2K^+Cl^-$. Thus the coefficient of H^+Cl^- is 14, of H_2O is 7, and of K^+Cl^- is 2.

$$(K^+)_2Cr_2O_7^{2-} + 14H^+Cl^- \rightarrow 2Cr^{3+}(Cl^-)_3 + 3Cl_2 + 7H_2O + 2K^+Cl^- \qquad \text{(balanced)}$$

Note that here again all the oxygen in the oxidizing agent is converted to water.

11.5. Balance the following oxidation-reduction equation by the oxidation-state method.

$$FeS_2 + O_2 \rightarrow Fe_2O_3 + SO_2$$

The two special features of this problem are that both the iron and sulfur in FeS_2 undergo a change in oxidation state and that the reduction product of oxygen gas occurs in combination with both iron and sulfur. The electron balance diagram for oxidation must contain Fe and S atoms in the ratio of 1 to 2, since this is the ratio in which they are oxidized. The two diagrams, with their multiplying factors, are

$$4 \times [Fe(+II) + 2S(-I) \rightarrow Fe(+III) + 2S(+IV) + 11e^-]$$
$$11 \times [\qquad 2O(0) + 4e^- \rightarrow 2O(-II) \qquad\qquad]$$

$$\overline{\quad 4Fe(+II) + 8S(-I) + 22O(0) \rightarrow 4Fe(+III) + 8S(+IV) + 22O(-II)}$$

Hence
$$4FeS_2 + 11O_2 \rightarrow 2Fe_2O_3 + 8SO_2 \qquad \text{(balanced)}$$

In problems like the above, involving unusual oxidation states, the student may ask, "What will happen if I choose the wrong oxidation numbers?" The answer is that the same result will be obtained so long as the method is followed consistently. If one chooses to assign $+IV$ to the Fe and $-II$ to the S in the above example the total oxidation state increase is still 11. An advantage of the ion-electron method is that such decisions are avoided, since the oxidation state is not used to determine the electron change.

11.6. Balance the following oxidation-reduction equation by the the ion-electron method:

$$Zn + Na^+NO_3^- + Na^+OH^- \rightarrow (Na^+)_2ZnO_2^{2-} + NH_3 + H_2O$$

The skeleton partial for the oxidizing agent is

$$NO_3^- \rightarrow NH_3$$

In alkaline solutions, each excess oxygen atom is balanced by adding $1H_2O$ to the same side of the equation and $2OH^-$ to the opposite side. Each excess hydrogen atom is balanced by adding $1OH^-$ to the same side and $1H_2O$ to the opposite side. In this example we must add $3H_2O$ to the left and $6OH^-$ to the right to balance the excess oxygen of the NO_3^-. Also, we must add $3OH^-$ to the right and $3H_2O$ to the left to balance the excess hydrogen of the NH_3. In all, $9OH^-$ must be added to the right and $6H_2O$ to the left.

$$NO_3^- + 6H_2O \rightarrow NH_3 + 9OH^-$$
$$NO_3^- + 6H_2O + 8e^- \rightarrow NH_3 + 9OH^- \qquad \text{(balanced)}$$

The balancing of the partial equation for the reducing agent is as follows:

$$Zn \rightarrow ZnO_2^{2-}$$
$$4OH^- + Zn \rightarrow ZnO_2^{2-} + 2H_2O$$
$$4OH^- + Zn \rightarrow ZnO_2^{2-} + 2H_2O + 2e^- \qquad \text{(balanced)}$$

The overall equation is

$$1 \times [NO_3^- + 6H_2O + 8e^- \rightarrow NH_3 + 9OH^- \qquad]$$
$$4 \times [\qquad 4OH^- + Zn \rightarrow ZnO_2^{2-} + 2H_2O + 2e^-]$$

$$\overline{\quad NO_3^- + 6H_2O + 4Zn + 16OH^- \rightarrow NH_3 + 9OH^- + 4ZnO_2^{2-} + 8H_2O}$$

Canceling,
$$NO_3^- + 4Zn + 7OH^- \rightarrow NH_3 + 4ZnO_2^{2-} + 2H_2O$$

Adding ions and combining,

$$Na^+NO_3^- + 4Zn + 7Na^+OH^- \rightarrow NH_3 + 4(Na^+)_2ZnO_2^{2-} + 2H_2O$$

11.7. Balance the following equation by the ion-electron method:

$$HgS + H^+Cl^- + H^+NO_3^- \rightarrow (H^+)_2HgCl_4^{2-} + NO + S + H_2O$$

The partial equation for the oxidizing agent is balanced as in Problem 11.2.

$$4H^+ + NO_3^- + 3e^- \rightarrow NO + 2H_2O$$

The skeleton for the reducing agent is

$$HgS \rightarrow S$$

The above skeleton contains both reduced and oxidized forms of sulfur, the only element undergoing a change in oxidation state. The unbalance is not one of hydrogen or oxygen atoms but of mercury. According to the overall equation, the form in which mercury exists among the products is the ion $HgCl_4^{2-}$. If such an ion is added to the right to balance the mercury, then chloride ions must be added to the left to balance the chlorine. In general, it is always allowable to add ions necessary for complex formation when such an addition does not require the introduction of new oxidation states (rule 5, page 164).

$$HgS + 4Cl^- \rightarrow S + HgCl_4^{2-} + 2e^- \qquad \text{(balanced)}$$

The overall equation is

$$2 \times [4H^+ + NO_3^- + 3e^- \rightarrow NO + 2H_2O \qquad\quad]$$
$$3 \times [\qquad\quad HgS + 4Cl^- \rightarrow S + HgCl_4^{2-} + 2e^-]$$
$$8H^+ + 2NO_3^- + 3HgS + 12Cl^- \rightarrow 2NO + 4H_2O + 3S + 3HgCl_4^{2-}$$

Adding ions and combining,

$$3HgS + 2H^+NO_3^- + 12H^+Cl^- \rightarrow 3S + 3(H^+)_2HgCl_4^{2-} + 2NO + 4H_2O$$

11.8. Complete and balance the following skeleton equation for a reaction in acid solution:

$$H_2O_2 + MnO_4^- + \qquad\qquad \rightarrow$$

The products must be deduced from chemical experience. MnO_4^- contains manganese in its highest oxidation state. Hence, if reaction occurs at all, MnO_4^- is reduced. Since the solution is acid the reaction product will be Mn^{2+}. The H_2O_2 must therefore act as a reducing agent in this reaction, and its only possible oxidation product is O_2. The usual procedure may be followed from this point, leading to the following solution:

$$2 \times [MnO_4^- + 8H^+ + 5e^- \rightarrow Mn^{2+} + 4H_2O \quad]$$
$$5 \times [\qquad\qquad H_2O_2 \rightarrow O_2 + 2H^+ + 2e^-]$$
$$\overline{2MnO_4^- + 16H^+ + 5H_2O_2 \rightarrow 2Mn^{2+} + 8H_2O + 5O_2 + 10H^+}$$

After canceling,

$$2MnO_4^- + 6H^+ + 5H_2O_2 \rightarrow 2Mn^{2+} + 8H_2O + 5O_2$$

The above equation is as definite a statement as can be made within the given overall skeleton. If we are to write an equation in terms of neutral substances, we are free to decide which salt of MnO_4^- and which acid to use. If we use $K^+MnO_4^-$ and $(H_2^+)SO_4^{2-}$, we obtain

$$2K^+MnO_4^- + 3(H^+)_2SO_4^{2-} + 5H_2O_2 \rightarrow 2Mn^{2+}SO_4^{2-} + 5O_2 + (K^+)_2SO_4^{2-} + 8H_2O$$

COMPARISON OF THE TWO METHODS OF BALANCING

Both methods lead to the correct form of the balanced equation. The ion-electron method has two advantages:

1. It differentiates those components of a system which react from those which do not. In Problem 11.8, any permanganate and any strong acid could have been used, not necessarily $K^+MnO_4^-$ and $(H^+)_2SO_4^{2-}$. The use of complete formulas for neutral substances is helpful only for the calculation of mass relations. Since permanganate ion cannot be weighed on a balance, it is necessary to choose one particular permanganate, like $K^+MnO_4^-$, and weigh an amount of $K^+MnO_4^-$ that will give the correct weight of MnO_4^-.

2. The half-reactions given by the partial equations of the ion-electron method can actually be made to occur independently. Many oxidation-reduction reactions can be carried out as galvanic cell processes for producing an electrical potential. This can be done by placing the reducing agent and oxidizing agent in separate vessels and making electrical connections between the two. It has been found that the reaction taking place in each beaker corresponds exactly to a partial equation written according to the rules for the ion-electron method.

Some chemists prefer to use the ion-electron method for oxidation-reduction reactions carried out in dilute aqueous solutions, where free ions have more or less independent existence, and to use the oxidation-state method for oxidation-reduction reactions between solid chemicals or for reactions in concentrated acid media.

STOICHIOMETRY IN OXIDATION-REDUCTION

11.9. In Problem 11.3, how much Cl_2 could be produced by the reaction of 100 g $KMnO_4$?

This mass problem is solved like all other mass problems if the balanced molecular equation is used; it shows that 2 mol $KMnO_4$ can produce 5 mol Cl_2. The problem can also be solved directly from the balanced ionic equation showing that 2 MnO_4^- ions yield 5 molecules of Cl_2. We need only the chemical equivalence of $2MnO_4^-$ with $2KMnO_4$.

In either case, the solution is

$$x \text{ g } Cl_2 = (100 \text{ g } KMnO_4)\left(\frac{1 \text{ mol } KMnO_4}{158 \text{ g } KMnO_4}\right)\left(\frac{5 \text{ mol } Cl_2}{2 \text{ mol } KMnO_4}\right)\left(\frac{70.9 \text{ g } Cl_2}{1 \text{ mol } Cl_2}\right) = 112 \text{ g } Cl_2$$

Supplementary Problems

11.10. Determine the oxidation state of the italicized element in　(a) $K_4P_2O_7$,　(b) $NaAuCl_4$, (c) $Rb_4Na[HV_{10}O_{28}]$,　(d) ICl,　(e) Ba_2XeO_6,　(f) OF_2,　(g) $Ca(ClO_2)_2$.

Ans.　(a) +V;　(b) +III;　(c) +V;　(d) +I;　(e) +VIII;　(f) +II;　(g) +III

Write balanced ionic and molecular equations for the following:

11.11. $CuS + H^+NO_3^-$ *(dilute)* $\rightarrow Cu^{2+}(NO_3^-)_2 + S + H_2O + NO$

11.12. $K^+MnO_4^- + H^+Cl^- \rightarrow K^+Cl^- + Mn^{2+}(Cl^-)_2 + H_2O + Cl_2$

11.13. $Fe^{2+}(Cl^-)_2 + H_2O_2 + H^+Cl^- \rightarrow Fe^{3+}(Cl^-)_3 + H_2O$

11.14. $As_2S_5 + H^+NO_3^-$ *(conc.)* $\rightarrow H_3AsO_4 + H^+HSO_4^- + H_2O + NO_2$

11.15. $Cu + H^+NO_3^-$ $(conc.) \rightarrow Cu^{2+}(NO_3^-)_2 + H_2O + NO_2$

11.16. $Cu + H^+NO_3^-$ $(dilute) \rightarrow Cu^{2+}(NO_3^-)_2 + H_2O + NO$

11.17. $Zn + H^+NO_3^-$ $(dilute) \rightarrow Zn^{2+}(NO_3^-)_2 + H_2O + NH_4^+NO_3^-$

11.18. $(Na^+)_2C_2O_4^{2-} + K^+MnO_4^- + (H^+)_2SO_4^{2-} \rightarrow (K^+)_2SO_4^{2-} + (Na^+)_2SO_4^{2-} + H_2O + Mn^{2+}SO_4^{2-} + CO_2$

11.19. $CdS + I_2 + H^+Cl^- \rightarrow Cd^{2+}(Cl^-)_2 + H^+I^- + S$

11.20. $MnO + PbO_2 + H^+NO_3^- \rightarrow H^+MnO_4^- + Pb^{2+}(NO_3^-)_2 + H_2O$

11.21. $Cr^{3+}(I^-)_3 + K^+OH^- + Cl_2 \rightarrow (K^+)_2CrO_4^{2-} + K^+IO_4^- + K^+Cl^- + H_2O$
(Note that both the chromium and the iodide are oxidized in this reaction.)

11.22. $(Na^+)_2HAsO_3^{2-} + K^+BrO_3^- + H^+Cl^- \rightarrow Na^+Cl^- + K^+Br^- + H_3AsO_4$

11.23. $(Na^+)_2TeO_3^{2-} + Na^+I^- + H^+Cl^- \rightarrow Na^+Cl^- + Te + H_2O + I_2$

11.24. $U^{4+}(SO_4^{2-})_2 + K^+MnO_4^- + H_2O \rightarrow (H^+)_2SO_4^{2-} + (K^+)_2SO_4^{2-} + Mn^{2+}SO_4^{2-} + UO_2^{2+}SO_4^{2-}$

11.25. $I_2 + (Na^+)_2S_2O_3^{2-} \rightarrow (Na^+)_2S_4O_6^{2-} + Na^+I^-$

11.26. $Ca^{2+}(OCl^-)_2 + K^+I^- + H^+Cl^- \rightarrow I_2 + Ca^{2+}(Cl^-)_2 + H_2O + K^+Cl^-$

11.27. $Bi_2O_3 + Na^+OH^- + Na^+OCl^- \rightarrow Na^+BiO_3^- + Na^+Cl^- + H_2O$

11.28. $(K^+)_3Fe(CN)_6^{3-} + Cr_2O_3 + K^+OH^- \rightarrow (K^+)_4Fe(CN)_6^{4-} + (K^+)_2CrO_4^{2-} + H_2O$

11.29. $H^+NO_3^- + H^+I^- \rightarrow NO + I_2 + H_2O$

11.30. $Mn^{2+}SO_4^{2-} + (NH_4^+)_2S_2O_8^{2-} + H_2O \rightarrow MnO_2 + (H^+)_2SO_4^{2-} + (NH_4^+)_2SO_4^{2-}$

11.31. $(K^+)_2Cr_2O_7^{2-} + Sn^{2+}(Cl^-)_2 + H^+Cl^- \rightarrow Cr^{3+}(Cl^-)_3 + SnCl_4 + K^+Cl^- + H_2O$

11.32. $Co^{2+}(Cl^-)_2 + (Na^+)_2O_2^{2-} + Na^+OH^- + H_2O \rightarrow Co(OH)_3 + Na^+Cl^-$

11.33. $Cu(NH_3)_4^{2+}(Cl^-)_2 + K^+CN^- + H_2O \rightarrow NH_3 + NH_4^+Cl^- + (K^+)_2Cu(CN)_3^{2-} + K^+CNO^- + K^+Cl^-$

11.34. $Sb_2O_3 + K^+IO_3^- + H^+Cl^- + H_2O \rightarrow HSb(OH)_6 + K^+Cl^- + ICl$

11.35. $Ag + K^+CN^- + O_2 + H_2O \rightarrow K^+Ag(CN)_2^- + K^+OH^-$

11.36. $WO_3 + Sn^{2+}(Cl^-)_2 + H^+Cl^- \rightarrow W_3O_8 + (H^+)_2SnCl_6^{2-} + H_2O$

11.37. $Co^{2+}(Cl^-)_2 + K^+NO_2^- + H^+C_2H_3O_2^- \rightarrow (K^+)_3Co(NO_2)_6^{3-} + NO + K^+C_2H_3O_2^- + K^+Cl^- + H_2O$

11.38. $V(OH)_4^+Cl^- + Fe^{2+}(Cl^-)_2 + H^+Cl^- \rightarrow VO^{2+}(Cl^-)_2 + Fe^{3+}(Cl^-)_3 + H_2O$

Balance the following equations by the oxidation-state method:

11.39. $NH_3 + O_2 \rightarrow NO + H_2O$

11.40. $CuO + NH_3 \rightarrow N_2 + H_2O + Cu$

11.41. $KClO_3 + H_2SO_4 \rightarrow KHSO_4 + O_2 + ClO_2 + H_2O$

11.42. $Sn + HNO_3 \rightarrow SnO_2 + NO_2 + H_2O$

11.43. $I_2 + HNO_3 \rightarrow HIO_3 + NO_2 + H_2O$

11.44. $KI + H_2SO_4 \rightarrow K_2SO_4 + I_2 + H_2S + H_2O$

11.45. $KBr + H_2SO_4 \rightarrow K_2SO_4 + Br_2 + SO_2 + H_2O$

11.46. $Cr_2O_3 + Na_2CO_3 + KNO_3 \rightarrow Na_2CrO_4 + CO_2 + KNO_2$

11.47. $P_2H_4 \rightarrow PH_3 + P_4H_2$

11.48. $Ca_3(PO_4)_2 + SiO_2 + C \rightarrow CaSiO_3 + P_4 + CO$

Complete and balance the following skeleton equations by the ion-electron method:

11.49. $I^- + NO_2^- \rightarrow I_2 + NO$ (acid solution)

11.50. $Au + CN^- + O_2 \rightarrow Au(CN)_4^-$ (aqueous solution)

11.51. $MnO_4^- \rightarrow MnO_4^{2-} + O_2$ (alkaline solution)

11.52. $P \rightarrow PH_3 + H_2PO_2^-$ (alkaline solution)

11.53. $Zn + As_2O_3 \rightarrow AsH_3$ (acid solution)

11.54. $Zn + ReO_4^- \rightarrow Re^-$ (acid solution)

11.55. $ClO_2 + O_2^{2-} \rightarrow ClO_2^-$ (alkaline solution)

11.56. $Cl_2 + IO_3^- \rightarrow IO_4^-$ (alkaline solution)

11.57. $V \rightarrow HV_6O_{17}^{3-} + H_2$ (alkaline solution)

11.58. In Problem 11.8, what volume of O_2 at S.T.P. is produced for each gram of H_2O_2 consumed?

    Ans.   0.66 L

11.59. How much $KMnO_4$ is needed to oxidize 100 g $Na_2C_2O_4$? See the equation for the reaction in Problem 11.18.

    Ans.   52 g

Chapter 12

Concentrations of Solutions

SOLUTE AND SOLVENT

In a solution of one substance in another substance, the dissolved substance is called the *solute*. The substance in which the solute is dissolved is called the *solvent*. When the relative amount of one substance in a solution is much greater than that of the other, the substance present in greater amount is generally regarded as the solvent. When the relative amounts of the two substances are of the same order of magnitude, it becomes difficult, in fact arbitrary, to specify which substance is the solvent.

CONCENTRATIONS EXPRESSED IN PHYSICAL UNITS

When physical units are employed, the concentrations of solutions are generally expressed in one of the following ways:

1. By the mass of solute per unit volume of solution (e.g., 20 g of KCl per liter of solution)
2. By the percentage composition, or the number of mass units of solute per 100 mass units of solution

EXAMPLE 1   A 10% aqueous NaCl solution contains 10 g of NaCl per 100 g of solution. Ten grams of NaCl is mixed with 90 g of water to form 100 g of solution.

3. By the mass of solute per mass of solvent (e.g., 5.2 g of NaCl in 100 g of water)

CONCENTRATIONS EXPRESSED IN CHEMICAL UNITS

Molar concentration

The *molar concentration* (M) is the number of moles of the solute contained in one liter of solution.

EXAMPLE 2   A 0.5 molar (0.5 M) solution of H_2SO_4 contains 49.04 g of H_2SO_4 per liter of solution, since 49.04 is half of the molecular weight of H_2SO_4, 98.08. A one molar (1 M) solution contains 98.08 g H_2SO_4 per liter of solution.

Note that M is the symbol for a quantity, the molar concentration, and M is the symbol for a unit, mol/L. (The term *molarity* is often used to refer to molar concentration; it will not be used in this book, in order to avoid confusion with *molality*, define below.)

Normality

The *normality* of a solution (N) is the number of *gram-equivalents* of the solute contained in one liter of solution. The *equivalent weight* is that fraction of the molecular weight which corresponds to one defined unit of chemical reaction, and a *gram-equivalent* is this same fraction of a mole. (The gram-equivalent is referred to in some books as the *equivalent* or the *gram-equivalent weight*.) Equivalent weights are determined as follows:

1. The defined unit of reaction for acids and bases is the neutralization reaction

$$H^+ + OH^- \rightarrow H_2O$$

174

The equivalent weight of an acid is that fraction of the molecular weight which contains or can supply for reaction one acid H^+. (In other words, the equivalent weight is the molecular weight divided by the number of H^+ furnished per molecule.) A gram-equivalent (g-eq) is then that amount which contains or can supply for reaction 1 mol of H^+.

EXAMPLE 3 The equivalent weights of HCl and $HC_2H_3O_2$ are the same as their molecular weights, since each contains one acidic hydrogen per molecule. A g-eq of each of these molecules is the same as 1 mol. The equivalent weight of H_2SO_4 is usually half the molecular weight and a g-eq is $\frac{1}{2}$ mol, since both hydrogens are replaceable in most reactions of dilute sulfuric acid. A g-eq of H_3PO_4 may be 1 mol, $\frac{1}{2}$ mol, or $\frac{1}{3}$ mol, depending on whether one, two, or three hydrogen atoms per molecule are replaced in a particular reaction. A g-eq of H_3BO_3 is always 1 mol, since only one hydrogen is replaceable in neutralization reactions. The equivalent weight of SO_3 is one-half the molecular weight, since SO_3 can react with water to give two H^+.

$$SO_3 + H_2O \rightarrow 2H^+ + SO_4^{2-}$$

There are no simple rules for predicting how many hydrogens of an acid will be replaced in a given neutralization.

2. The equivalent weight of a base is that fraction of the formula weight which contains or can supply one OH^-, or can react with one H^+.

EXAMPLE 4 The equivalent weights of NaOH, NH_3 (which can react with water to give $NH_4^+ + OH^-$), $Mg(OH)_2$, and $Al(OH)_3$, are equal to $\frac{1}{1}$, $\frac{1}{1}$, $\frac{1}{2}$, and $\frac{1}{3}$ of their formula weights, respectively.

3. The equivalent weight of an oxidizing or reducing agent for a particular reaction is equal to its formula weight divided by the total number of electrons gained or lost when the reaction of that formula unit occurs. A given oxidizing or reducing agent may have more than one equivalent weight, depending on the reaction for which it is used.

Equivalent weights are so defined that equal numbers of gram-equivalents of two substances react exactly with each other. This is true for neutralization because one H^+ neutralizes one OH^-, and for oxidation-reduction because the number of electrons lost by the reducing agent equals the number of electrons gained by the oxidation agent.

EXAMPLE 5 1 mol HCl, $\frac{1}{2}$ mol H_2SO_4, and $\frac{1}{6}$ mol $K_2Cr_2O_7$ (as oxidizing agent), each in 1 L of solution, give normal (1 N) solutions of these substances. A normal (1 N) solution of HCl is also a molar (1 M) solution. A normal (1 N) solution of H_2SO_4 is also a one-half molar (0.5 M) solution.
Note that N is the symbol for a quantity, the normality, and N is the symbol for a unit, g-eq/L.

Molality

The *molality* of a solution is the number of moles of the solute per kilogram of solvent contained in a solution. The molality (m) cannot be computed from the molar concentration (M) unless the density of the solution is known (see Problem 12.58).

EXAMPLE 6 A solution made up of 98.08 g of pure H_2SO_4 and 1 000 g of water would be a 1 molal solution of H_2SO_4. (Some books use the symbol m to designate both the quantity, the molality, and the unit, mol/kg. In this book, m is used to refer to the quantity only and not to the unit.)

Mole fraction

The *mole fraction* (x) of any component in a solution is defined as the number of moles (n) of that component, divided by the total number of moles of all components in the solution. The sum of the mole fractions of all components of a solution is 1. In a two-component solution,

$$x(\text{solute}) = \frac{n(\text{solute})}{n(\text{solute}) + n(\text{solvent})} \qquad x(\text{solvent}) = \frac{n(\text{solvent})}{n(\text{solute}) + n(\text{solvent})}$$

Expressed as a percentage, mole fraction is called *mole percent*.

COMPARISON OF THE CONCENTRATION SCALES

The *molar concentration* and *normality* scales are useful for volumetric experiments, in which the amount of solute in a given portion of solution is related to the measured volume of solution. As will be seen in subsequent chapters, the normality scale is very convenient for comparing the relative volumes required for two solutions to react chemically with each other. A limitation of the normality scale is that a given solution may have more than one normality, depending on the reaction for which it is used. The molar concentration of a solution, on the other hand, is a fixed number because the molecular weight of a substance does not depend on the reaction for which the substance is used, as may the equivalent weight.

The *molality* scale is useful for experiments in which physical measurements (as of freezing point, boiling point, vapor pressure, etc.) are made over a wide range of temperatures. The molality of a given solution, which is determined solely by the masses of the solution components, is independent of temperature. The molar concentration or the normality of a solution, on the other hand, being defined in terms of the volume, may vary appreciably as the temperature is changed, because of the temperature dependence of the volume.

The *mole fraction* scale finds use in theoretical work because many physical properties of solutions (Chapter 14) are expressed most clearly in terms of the relative numbers of solvent and solute molecules. (The number of moles of a substance is proportional to the number of molecules.)

SUMMARY OF CONCENTRATION UNITS

$$\text{Molar concentration of a solution} = M = \frac{\text{number of moles of solute}}{\text{number of liters of solution}}$$

$$\text{Normality of a solution} = N = \frac{\text{number of gram-equivalents of solute}}{\text{number of liters of solution}}$$

$$= \frac{\text{number of milliequivalents of solute}}{\text{number of cubic centimeters of solution}}$$

$$\text{Molality of a solution} = m = \frac{\text{number of moles of solute}}{\text{number of kilograms of solvent}}$$

$$\text{Mole fraction of any component} = x = \frac{\text{number of moles of that component}}{\text{total number of moles of all components}}$$

The second expression for the normality is obtained from the first expression by multiplying numerator and denominator by 1 000.

DILUTION PROBLEMS

The volumetric scales of concentration are those, like molar concentration and normality, in which the concentration is expressed as the amount of solute per fixed volume of solution. When the concentration is expressed on a volumetric scale, the amount of solute contained in a given volume of solution is equal to the product of the volume and the concentration:

$$\text{Amount of solute} = \text{volume} \times \text{concentration}$$

If a solution is diluted, the volume is increased and the concentration is decreased, but the total amount of solute is constant. Hence, two solutions of different concentrations but containing the same amounts of solute will be related to each other as follows:

$$\text{Volume}_1 \times \text{concentration}_1 = \text{volume}_2 \times \text{concentration}_2$$

If any three terms in the above equation are known, the fourth can be calculated. The quantities on both sides of the equation must be expressed in the same units.

Solved Problems

CONCENTRATIONS EXPRESSED IN PHYSICAL UNITS

12.1. Explain how you would prepare 60 cm³ of an aqueous solution of $AgNO_3$ of strength 0.030 g $AgNO_3$ per cubic centimeter.

Since each cubic centimeter of solution is to contain 0.030 g $AgNO_3$, 60 cm³ of solution should contain

$$(0.030 \text{ g/cm}^3)(60 \text{ cm}^3) = 1.8 \text{ g } AgNO_3$$

Thus, dissolve 1.8 g $AgNO_3$ in about 50 cm³ of water. Then add sufficient water to make the volume exactly 60 cm³. Stir thoroughly to insure uniformity. (Note that 60 cm³ is to be the volume of the final solution, not of the water used to make up the solution.)

12.2. How many grams of a 5.0% by weight NaCl solution are necessary to yield 3.2 g NaCl?

A 5.0% NaCl solution contains 5.0 g NaCl in 100 g of solution. Then:

$$1 \text{ g NaCl} \quad \text{is contained in} \quad \frac{100}{5.0} \text{ g solution}$$

and

$$3.2 \text{ g NaCl} \quad \text{is contained in} \quad (3.2)\left(\frac{100}{5.0} \text{ g solution}\right) = 64 \text{ g solution}$$

Or, by proportion, letting x = required number of grams of solution,

$$\frac{5.0 \text{ g NaCl}}{100 \text{ g solution}} = \frac{3.2 \text{ g NaCl}}{x} \qquad \text{whence} \qquad x = 64 \text{ g solution}$$

Or, by the use of a quantitative factor,

$$x \text{ g solution} = (3.2 \text{ g NaCl})\left(\frac{100 \text{ g solution}}{5.0 \text{ g NaCl}}\right) = 64 \text{ g solution}$$

12.3. How much $NaNO_3$ must be weighed out to make 50 cm³ of an aqueous solution containing 70 mg Na^+ per cm³?

Mass of Na^+ in 50 cm³ of solution $= (50 \text{ cm}^3)(70 \text{ mg/cm}^3) = 3\,500 \text{ mg} = 3.5 \text{ g } Na^+$

Formula weight of $NaNO_3$ is 85; atomic weight of Na is 23. Then:

$$23 \text{ g Na}^+ \quad \text{is contained in} \quad 85 \text{ g } NaNO_3$$

$$1 \text{ g Na}^+ \quad \text{is contained in} \quad \tfrac{85}{23} \text{ g } NaNO_3$$

and $\qquad\qquad 3.5 \text{ g Na}^+ \quad \text{is contained in} \quad (3.5)(\tfrac{85}{23} \text{ g}) = 12.9 \text{ g } NaNO_3$

Or, by direct use of quantitative factors,

$$x \text{ g } NaNO_3 = (50 \text{ cm}^3 \text{ solution})\left(\frac{70 \text{ mg Na}^+}{1 \text{ cm}^3 \text{ solution}}\right)\left(\frac{85 \text{ g } NaNO_3}{23 \text{ g Na}^+}\right)\left(\frac{1 \text{ g}}{1\,000 \text{ mg}}\right) = 12.9 \text{ g } NaNO_3$$

12.4. Calculate the mass of $Al_2(SO_4)_3 \cdot 18H_2O$ needed to make up 50 cm^3 of an aqueous solution of strength 40 mg Al^{3+} per cm^3.

Mass of Al^{3+} in 50 cm^3 of solution = $(50 \text{ cm}^3)(40 \text{ mg/cm}^3) = 2\,000 \text{ mg} = 2.00 \text{ g } Al^{3+}$

Atomic weight of Al is 27; formula weight of $Al_2(SO_4)_3 \cdot 18H_2O$ is 666.

$$x \text{ g } Al_2(SO_4)_3 \cdot 18H_2O = (2.00 \text{ g } Al^{3+})\left(\frac{666 \text{ g } Al_2(SO_4)_3 \cdot 18H_2O}{54 \text{ g } Al^{3+}}\right) = 24.7 \text{ g } Al_2(SO_4)_3 \cdot 18H_2O$$

A solution of identical composition could be prepared from the appropriate amount of anhydrous $Al_2(SO_4)_3$. To compute the appropriate amount in this case, the formula weight of $Al_2(SO_4)_3$, 342, would be used instead of 666 above. The experimenter would find that more water is needed to make up the 50 cm^3 of solution from the anhydrous salt than from the hydrate. In general, hydrated salts differ from the anhydrous salts only in the crystalline state. In solution, the water of hydration and the water of the solvent become indistinguishable from each other.

12.5. Describe how you would prepare 50 g of a 12.0% $BaCl_2$ solution, starting with $BaCl_2 \cdot 2H_2O$ and pure water.

A 12.0% $BaCl_2$ solution contains 12.0 g $BaCl_2$ per 100 g of solution, or 6.0 g $BaCl_2$ in 50 g of solution. Formula weight of $BaCl_2$ is 208; of $BaCl_2 \cdot 2H_2O$, 244. Therefore,

$$208 \text{ g } BaCl_2 \quad \text{is contained in} \quad 244 \text{ g } BaCl_2 \cdot 2H_2O$$

$$1 \text{ g } BaCl_2 \quad \text{is contained in} \quad \tfrac{244}{208} \text{ g } BaCl_2 \cdot 2H_2O$$

and $\qquad\qquad 6.0 \text{ g } BaCl_2 \quad \text{is contained in} \quad (6.0)(\tfrac{244}{208} \text{ g}) = 7.0 \text{ g } BaCl_2 \cdot 2H_2O$

The desired solution is prepared by dissolving 7.0 g $BaCl_2 \cdot 2H_2O$ in 43 g (43 cm^3) of water.

12.6. Calculate the mass of anhydrous HCl in 5.00 cm^3 of concentrated hydrochloric acid (density 1.19 g/cm^3) containing 37.23% HCl by weight.

The mass of 5.00 cm^3 of solution is $(5.00 \text{ cm}^3)(1.19 \text{ g/cm}^3) = 5.95 \text{ g}$. The solution contains 37.23% HCl by weight. Hence the mass of HCl in 5.95 g solution is

$$(0.372\,3)(5.95 \text{ g}) = 2.22 \text{ g anhydrous HCl}$$

12.7. Calculate the volume of concentrated sulfuric acid (density 1.84 g/cm^3), containing 98% H_2SO_4 by weight, that would contain 40.0 g of pure H_2SO_4.

One cm^3 of solution has a mass of 1.84 g and contains $(0.98)(1.84 \text{ g}) = 1.80 \text{ g}$ of pure H_2SO_4. Then 40.0 g H_2SO_4 is contained in

$$\left(\frac{40.0}{1.80}\right)(1 \text{ cm}^3 \text{ solution}) = 22.2 \text{ cm}^3 \text{ solution})$$

Or, by direct use of conversion factors,

$$(40.0 \text{ g H}_2\text{SO}_4)\left(\frac{100 \text{ g solution}}{98 \text{ g H}_2\text{SO}_4}\right)\left(\frac{1 \text{ cm}^3 \text{ solution}}{1.84 \text{ g solution}}\right) = 22.2 \text{ cm}^3 \text{ solution}$$

12.8. Exactly 4.00 g of a solution of sulfuric acid was diluted with water, and an excess of $BaCl_2$ was added. The washed and dried precipitate of $BaSO_4$ weighed 4.08 g. Find the percent H_2SO_4 in the original acid solution.

First determine the mass of H_2SO_4 required to precipitate 4.08 g $BaSO_4$ by the reaction

$$H_2SO_4 + BaCl_2 \rightarrow 2HCl + BaSO_4$$

The equation shows that 1 mol $BaSO_4$ (233.4 g) requires 1 mol H_2SO_4 (98.08 g). Therefore 4.08 g $BaSO_4$ requires

$$\left(\frac{4.08 \text{ g BaSO}_4}{233.4 \text{ g BaSO}_4}\right)(98.08 \text{ g H}_2\text{SO}_4) = 1.72 \text{ g H}_2\text{SO}_4$$

and

$$\text{Fraction H}_2\text{SO}_4 \text{ by weight} = \frac{\text{mass of H}_2\text{SO}_4}{\text{mass of solution}} = \frac{1.72 \text{ g}}{4.00 \text{ g}} = 0.430 = 43.0\%$$

CONCENTRATIONS EXPRESSED IN CHEMICAL UNITS

12.9. How many grams of solute are required to prepare 1 L of 1 M $Pb(NO_3)_2$? What is the molar concentration of the solution with respect to each of the ions?

A molar solution contains 1 mol of solute in 1 L of solution. The formula weight of $Pb(NO_3)_2$ is 331.2; hence 331.2 g of $Pb(NO_3)_2$ is needed.

A 1 M solution of $Pb(NO_3)_2$ is 1 M with respect to Pb^{2+} and 2 M with respect to NO_3^-.

12.10. What is the molar concentration of a solution containing 16.0 g CH_3OH in 200 cm^3 of solution?

Molecular weight of CH_3OH is 32.0.

$$M = \text{molar concentration} = \frac{n(\text{solute})}{\text{volume of solution in L}} = \frac{\dfrac{16.0 \text{ g}}{32.0 \text{ g/mol}}}{0.200 \text{ L}} = 2.50 \text{ mol/L} = 2.50 \text{ M}$$

12.11. Determine the molar concentration of each of the following solutions: (a) 18.0 g $AgNO_3$ per liter of solution, (b) 12.00 g $AlCl_3 \cdot 6H_2O$ per liter of solution. Formula weight of $AgNO_3$ is 169.9; of $AlCl_3 \cdot 6H_2O$, 241.4.

(a)
$$\frac{18.0 \text{ g/L}}{169.9 \text{ g/mol}} = 0.106 \text{ mol/L} = 0.106 \text{ M}$$

(b)
$$\frac{12.00 \text{ g/L}}{241.4 \text{ g/mol}} = 0.049\,7 \text{ mol/L} = 0.049\,7 \text{ M}$$

12.12. How much $(NH_4)_2SO_4$ is required to prepare 400 cm^3 of M/4 solution? (The notation M/4 is sometimes used in place of $\frac{1}{4}$M.)

Formula weight of $(NH_4)_2SO_4$ is 132.1. One liter of M/4 solution contains $\frac{1}{4}(132.1 \text{ g}) = 33.02$ g $(NH_4)_2SO_4$. Then 400 cm^3 of M/4 solution requires

$$(0.400 \text{ L})(33.02 \text{ g/L}) = 13.21 \text{ g } (NH_4)_2SO_4$$

Another Method.

$$\text{Mass} = (\text{molar concentration}) \times (\text{formula weight}) \times (\text{volume of solution})$$
$$= (\tfrac{1}{4}\,\text{mol/L})(132.1\,\text{g/mol})(0.400\,\text{L}) = 13.21\,\text{g (NH}_4)_2\text{SO}_4$$

12.13. What is the molality of a solution which contains 20.0 g of cane sugar, $C_{12}H_{22}O_{11}$, dissolved in 125 g of water?

Molecular weight of $C_{12}H_{22}O_{11}$ is 342.

$$m = \text{molality} = \frac{n(\text{solute})}{\text{mass of solvent in kg}} = \frac{(20.0\,\text{g})/(342\,\text{g/mol})}{0.125\,\text{kg}} = 0.468\,\text{mol/kg}$$

12.14. The molality of a solution of ethyl alcohol, C_2H_5OH, in water is 1.54 mol/kg. How many grams of alcohol are dissolved in 2.5 kg of water?

Molecular weight of C_2H_5OH is 46.1. Since the molality is 1.54, 1 kg water dissolves 1.54 mol alcohol. Then 2.5 kg water dissolves $(2.5)(1.54) = 3.85$ mol alcohol, and

$$\text{Mass of alcohol} = (3.85\,\text{mol})(46.1\,\text{g/mol}) = 177\,\text{g alcohol}$$

12.15. Calculate the (a) molar concentration and (b) molality of a sulfuric acid solution of density 1.198 g/cm^3, containing 27.0% H_2SO_4 by weight.

(a) Each cubic centimeter of acid solution has a mass of 1.198 g and contains $(0.270)(1.198) = 0.324$ g H_2SO_4. Since the molecular weight of H_2SO_4 is 98.1,

$$M = \frac{n(H_2SO_4)}{\text{volume of solution in L}} = \frac{(0.324\,\text{g})/(98.1\,\text{g/mol})}{(1\,\text{cm}^3)(10^{-3}\,\text{L/cm}^3)} = 3.30\,\text{mol/L} = 3.30\,\text{M}$$

(b) From (a), there is 324 g, or 3.30 mol, of solute per liter of solution. The amount of water in 1 L of solution is $1\,198\,\text{g} - 324\,\text{g} = 874\,\text{g}\,H_2O$. Hence

$$m = \frac{n(\text{solute})}{\text{mass of solvent in kg}} = \frac{3.30\,\text{mol}\,H_2SO_4}{0.874\,\text{kg}\,H_2O} = 3.78\,\text{mol/kg}$$

12.16. Determine the mole fractions of both substances in a solution containing 36.0 g of water and 46 g of glycerin, $C_3H_5(OH)_3$.

Molecular weight of $C_3H_5(OH)_3$ is 92; of H_2O, 18.0.

$$n(\text{glycerin}) = \frac{46\,\text{g}}{92\,\text{g/mol}} = 0.50\,\text{mol} \qquad n(\text{water}) = \frac{36.0\,\text{g}}{18.0\,\text{g/mol}} = 2.00\,\text{mol}$$

$$\text{Total number of moles} = 0.50 + 2.00 = 2.50\,\text{mol}$$

$$x(\text{glycerin}) = \text{mole fraction of glycerin} = \frac{n(\text{glycerin})}{\text{total number of moles}} = \frac{0.50}{2.50} = 0.20$$

$$x(\text{water}) = \text{mole fraction of water} = \frac{n(\text{water})}{\text{total number of moles}} = \frac{2.00}{2.50} = 0.80$$

Check: Sum of mole fractions $= 0.20 + 0.80 = 1.00$.

12.17. How many gram-equivalents of solute are contained in   (*a*) 1 L of 2 N solution,   (*b*) 1 L of 0.5 N solution,   (*c*) 0.5 L of 0.2 N solution?

A normal solution contains 1 g-eq of solute in 1 L of solution. Amount of solute = (volume) × (concentration).

(*a*)   1 L of 2 N contains 2 g-eq of solute.
(*b*)   1 L of 0.5 N contains 0.5 g-eq of solute.
(*c*)   0.5 L of 0.2 N contains (0.5 L)(0.2 g-eq/L) = 0.1 g-eq of solute.

12.18. How many   (*a*) gram-equivalents and   (*b*) milliequivalents of solute are present in 60 cm^3 of 4.0 N solution?

(*a*) Number of gram-equivalents = (number of liters) × (normality) = (0.060 L)(4.0 g-eq/L) = 0.24 g-eq

(*b*)   (0.24 g-eq)(1 000 meq/g-eq) = 240 meq

Another Method.

Number of meq = (number of cm^3) × (normality) = (60 cm^3)(4.0 meq/cm^3) = 240 meq

12.19. How many grams of solute are required to prepare 1 L of 1 N solution of   (*a*) LiOH,   (*b*) Br_2 (as oxidizing agent),   (*c*) H_3PO_4 (for a reaction in which three H are replaceable)?

A normal solution contains 1 g-eq of solute in 1 L of solution. Formula weight of LiOH is 23.95; of Br_2, 159.82; of H_3PO_4, 97.99.

(*a*)   One liter of 1 N LiOH requires (23.95/1) g = 23.95 g LiOH.
(*b*)   Note from the partial equation $Br_2 + 2e^- \rightarrow 2Br^-$ that two electrons react per Br_2. Thus, the equivalent weight of Br_2 is *half* its molecular weight, and 1 L of 1 N Br_2 requires (159.82/2) g = 79.91 g Br_2.
(*c*)   One liter of 1 N H_3PO_4 requires (97.99/3) g = 32.66 g H_3PO_4.

12.20. Calculate the normality of each of the following solutions: (*a*) 7.88 g of HNO_3 per liter of solution,   (*b*) 26.5 g of Na_2CO_3 per liter of solution (if neutralized to form CO_2).

(*a*)   Equivalent weight of HNO_3 (when used as an acid, not as an oxidizing agent) = formula weight = 63.02

$$N = \text{normality} = \frac{(7.88 \text{ g})/(63.02 \text{ g/g-eq})}{1 \text{ L}} = 0.125 \text{ 1 g-eq/L} = 0.125 \text{ 1 N}$$

(*b*)   The reaction is $CO_3^{2-} + 2H^+ \rightarrow CO_2 + H_2O$.
Equivalent weight of Na_2CO_3 = $(\frac{1}{2})$ × (formula weight) = $(\frac{1}{2})$ (106.0) = 53.0

$$N = \frac{(26.5 \text{ g})/(53.0 \text{ g/g-eq})}{1 \text{ L}} = 0.500 \text{ N}$$

12.21. How many cubic centimeters of 2.00 M $Pb(NO_3)_2$ contain 600 mg Pb^{2+}?

One liter of 1 M $Pb(NO_3)_2$ contains 1 mol Pb^{2+}, or 207 g. Then a 2 M solution contains 2 mol Pb^{2+} or 414 g Pb^{2+} per liter, or 414 mg Pb^{2+} per cubic centimeter. Hence 600 mg Pb^{2+} is contained in

$$\frac{600 \text{ mg}}{414 \text{ mg/}cm^3} = 1.45 \text{ } cm^3$$

of 2.00 M $Pb(NO_3)_2$.

12.22. Given the unbalanced equation

$$K^+MnO_4^- + K^+I^- + (H^+)_2SO_4^{2-} \rightarrow (K^+)_2SO_4^{2-} + Mn^{2+}SO_4^{2-} + I_2 + H_2O$$

(a) How many grams of $KMnO_4$ are needed to make $500\ cm^3$ of 0.250 N solution? (b) How many grams of KI are needed to make $25\ cm^3$ of 0.36 N solution?

(a) In this oxidation-reduction reaction, the oxidation state of Mn changes from $+VII$ in MnO_4^- to $+II$ in Mn^{2+}. Hence

$$\text{Equivalent weight of } KMnO_4 = \frac{\text{formula weight}}{\text{oxidation state change}} = \frac{158}{5} = 31.6$$

Then 0.500 L of 0.250 N requires

$$(0.500\ L)(0.250\ g\text{-eq/L})(31.6\ g/g\text{-eq}) = 3.95\ g\ KMnO_4$$

If the $KMnO_4$ made above were to be used in the reaction:

$$2MnO_4^- + 3H_2O_2 + 2H^+ \rightarrow 2MnO_2 + 3O_2 + 4H_2O$$

the label 0.250 N would no longer be appropriate because the oxidation number change in this case is 3, not 5. The appropriate normality would be

$$\left(\frac{0.250\ N}{5}\right)(3) = 0.150\ N$$

(b) The oxidation state of I changes from $-I$ in I^- to 0 in I_2. Hence

$$\text{Equivalent weight of KI = formula weight} = 166$$

Then 0.025 L of 0.36 N requires

$$(0.025\ L)(0.36\ g\text{-eq/L})(166\ g/g\text{-eq}) = 1.49\ g\ KI$$

12.23. Calculate the molar concentration of an NaBr solution which has a density of $1.167\ g/cm^3$ and a molality of 2.28 mol/kg.

The formula weight of NaBr is 102.9. For each kilogram of water there is $(2.28)(102.9) = 235\ g$ of NaBr, or a total mass of $1\,000 + 235 = 1\,235\ g$. The mass of solution occupies a volume of $1\,235\ g/(1.167\ g/cm^3) = 1\,058\ cm^3$ or 1.058 L. The molar concentration then is

$$2.28\ mol/1.058\ L = 2.16\ mol/L\ \text{or}\ 2.16\ M$$

12.24. A solution of a certain organic halide in benzene has a mole fraction of halide of 0.082 1. Express its concentration in terms of molality.

The formula for benzene is C_6H_6 and its molecular weight is 78.1. The 0.082 1 mol of halide is mixed with $1 - 0.082\,1 = 0.917\,9$ mol of benzene, which has a mass of $(0.917\,9\ mol)(78.1\ g/mol) = 71.7\ g = 0.071\,7\ kg$. Hence

$$\text{Molality} = \frac{0.082\,1\ mol}{0.071\,7\ kg} = 1.145\ mol/kg$$

DILUTION PROBLEMS

12.25. To what extent must a solution of concentration 40 mg $AgNO_3$ per cubic centimeter be diluted to yield one of concentration 16 mg $AgNO_3$ per cubic centimeter?

Let V be the volume to which $1\ cm^3$ of the original solution must be diluted to yield a solution of concentration 16 mg $AgNO_3$ per cubic centimeter. Because the amount of solute does not change with dilution,

$$\text{Volume}_1 \times \text{concentration}_1 = \text{volume}_2 \times \text{concentration}_2$$
$$1\ cm^3 \times 40\ mg/cm^3 \quad = \quad V \times 16\ mg/cm^3$$

Solving, $V = 2.5$ cm^3. Each cubic centimeter of the original solution must be diluted to a volume of 2.5 cm^3.

Another Method.

The amount of solute per cubic centimeter of diluted solution will be $\frac{16}{40}$ as much as in the original solution. Hence $\frac{40}{16}$ cm^3 = 2.5 cm^3 of the diluted solution will contain as much solute as 1 cm^3 of the original solution.

Note that 2.5 is not the number of cubic centimeters of water to be added, but the final volume of the solution after water has been added to 1 cm^3 of the original solution. The dilution formula always gives answers in terms of the *total* volume of solution. If we can assume that there is no volume shrinkage or expansion on dilution, the amount of water to be added in this problem is 1.5 cm^3 per cubic centimeter of original solution. Unless this assumption is made, the answers in the subsequent problems will be left in terms of the *total* volumes of the solutions.

12.26. To what extent must a 0.50 M BaCl$_2$ solution be diluted to yield one of concentration 20 mg Ba^{2+} per cubic centimeter?

The original solution contains 0.50 mol of BaCl$_2$ or of Ba^{2+} per liter. The mass of Ba^{2+} in 0.50 mol is

$$(0.50 \text{ mol})(137.3 \text{ g/mol}) = 68.6 \text{ g Ba}^{2+}$$

Thus 0.50 M BaCl$_2$ contains 68.6 g Ba^{2+} per liter, or 68.6 mg Ba^{2+} per cubic centimeter.

The problem now is to find the extent to which a solution of strength 68.6 mg Ba^{2+} per cubic centimeter must be diluted to yield one of concentration 20 mg Ba^{2+} per cubic centimeter.

$$\text{Volume}_1 \times \text{concentration}_1 = \text{volume}_2 \times \text{concentration}_2$$
$$1 \text{ cm}^3 \times 68.6 \text{ mg/cm}^3 \quad = \quad V \times 20 \text{ mg/cm}^3$$

Solving, $V = 3.43$ cm^3. Each cubic centimeter of 0.50 M BaCl$_2$ must be diluted with water to a volume of 3.43 cm^3.

12.27. A procedure calls for 100 cm^3 of 20% H$_2$SO$_4$, density 1.14 g/cm^3. How much of the concentrated acid, of density 1.84 g/cm^3 and containing 98% H$_2$SO$_4$ by weight, must be diluted with water to prepare 100 cm^3 of acid of the required strength?

The concentrations must first be changed from a mass basis to a volumetric basis, so that the dilution equation will apply.

$$\text{Mass of H}_2\text{SO}_4 \text{ per cm}^3 \text{ of } 20\% \text{ acid} = (0.20)(1.14 \text{ g/cm}^3) = 0.228 \text{ g/cm}^3$$
$$\text{Mass of H}_2\text{SO}_4 \text{ per cm}^3 \text{ of } 98\% \text{ acid} = (0.98)(1.84 \text{ g/cm}^3) = 1.80 \text{ g/cm}^3$$

Let V be the volume of 98% acid required for 100 cm^3 of 20% acid.

$$\text{Volume}_1 \times \text{concentration}_1 = \text{volume}_2 \times \text{concentration}_2$$
$$100 \text{ cm}^3 \times 0.228 \text{ g/cm}^3 \quad = \quad V \times 1.80 \text{ g/cm}^3$$

Solving, $V = 12.7$ cm^3 of the concentrated acid.

12.28. What volumes of N/2 and of N/10 HCl must be mixed to give 2 L of N/5 HCl?

Let x = volume of N/2 required; $2L - x$ = volume of N/10 required.

$$\text{Number of g-eq of N/5} = (\text{number of g-eq of N/2}) + (\text{number of g-eq of N/10})$$
$$(2 \text{ L})(\tfrac{1}{5}\text{N}) = x(\tfrac{1}{2}\text{N}) + (2 \text{ L} - x)(\tfrac{1}{10}\text{N})$$

Solving, $x = 0.5$ L. Thus 0.5 L of N/2 and 1.5 L of N/10 are required.

12.29. How many cubic centimeters of concentrated sulfric acid, of density 1.84 g/cm^3 and containing 98% H$_2$SO$_4$ by weight, should be taken to make   (a) 1 L of normal solution,   (b) 1 L of 3.00 N solution,   (c) 200 cm^3 of 0.500 N solution?

$$\text{Equivalent weight of H}_2\text{SO}_4 = \tfrac{1}{2}(\text{formula weight}) = \tfrac{1}{2}(98.1) = 49.0$$

The H_2SO_4 content of 1 L of the concentrated acid is $(0.98)(1\,000\text{ cm}^3)(1.84\text{ g/cm}^3) = 1\,800\text{ g }H_2SO_4$. The normality of the concentrated acid is

$$\frac{1\,800\text{ g }H_2SO_4/L}{49.0\text{ g }H_2SO_4/\text{g-eq}} = 36.7\text{ g-eq/L}$$

The dilution formula, $V_{\text{conc.}} \times N_{\text{conc.}} = V_{\text{dil.}} \times N_{\text{dil.}}$, can now be applied to each case.

(a) $$V_{\text{conc.}} = \frac{(1\text{ L})(1.00\text{ N})}{36.7\text{ N}} = 0.027\,2\text{ L} = 27.2\text{ cm}^3\text{ conc. acid}$$

(b) $$V_{\text{conc.}} = \frac{(1\text{ L})(3.00\text{ N})}{36.7\text{ N}} = 0.081\,7\text{ L} = 81.7\text{ cm}^3\text{ conc. acid}$$

(c) $$V_{\text{conc.}} = \frac{(200\text{ cm}^3)(0.500\text{ N})}{36.7\text{ N}} = 2.72\text{ cm}^3\text{ conc. acid}$$

Supplementary Problems

CONCENTRATIONS EXPRESSED IN PHYSICAL UNITS

12.30. How much NH_4Cl is required to prepare 100 cm^3 of a solution of strength 70 mg NH_4Cl per cubic centimeter?

Ans.   7.0 g

12.31. How many grams of concentrated hydrochloric acid, containing 37.9% HCl by weight, will give 5.0 g HCl?

Ans.   13.2 g

12.32. It is required to prepare 100 g of a 19.7% by weight solution of NaOH. How many grams each of NaOH and H_2O are required?

Ans.   19.7 g NaOH, 80.3 g H_2O

12.33. How much $CrCl_3 \cdot 6H_2O$ is needed to prepare 1 L of solution containing 20 mg Cr^{3+} per cubic centimeter?

Ans.   102 g

12.34. How many grams of Na_2CO_3 are needed to prepare 500 cm^3 of a solution containing 10.0 mg CO_3^{2-} per cubic centimeter?

Ans.   8.83 g

12.35. Calculate the volume occupied by 100 g of sodium hydroxide solution of density 1.20 g/cm^3.

Ans.   83.3 cm^3

12.36. What volume of dilute nitric acid, of density 1.11 g/cm^3 and 19% HNO_3 by weight, contains 10 g HNO_3?

Ans.   47 cm^3

12.37. How many cubic centimeters of a solution containing 40 g $CaCl_2$ per liter are needed to react with 0.642 g of pure Na_2CO_3? $CaCO_3$ is formed in the reaction.

Ans.   16.8 cm^3

12.38. Ammonia gas is passed into water, yielding a solution of density of 0.93 g/cm^3 and containing 18.6% NH_3 by weight. What is the mass of NH_3 per cubic centimeter of solution?

Ans.   173 mg/cm^3

12.39. Hydrogen chloride gas is passed into water, yielding a solution of density 1.12 g/cm^3 and containing 30.5% HCl by weight. What is the mass of HCl per cubic centimeter of solution?

Ans.   342 mg/cm^3

12.40. Given 100 cm^3 of pure water at 4 °C, what volume of a solution of hydrochloric acid, density 1.175 g/cm^3 and containing 34.4% HCl by weight, could be prepared?

Ans.   130 cm^3

12.41. A volume of 105 cm^3 of pure water at 4 °C is saturated with NH_3 gas, yielding a solution of density 0.90 g/cm^3 and containing 30% NH_3 by weight. Find the volume of the ammonia solution resulting, and the volume of the ammonia gas at 5 °C and 775 torr which was used to saturate the water.

Ans.   167 cm^3, 59 L

12.42. An excellent solution for cleaning grease stains from cloth or leather consists of the following: carbon tetrachloride 80% (by volume), ligroin 16%, amyl alcohol 4%. Calculate how many cubic centimeters of each should be taken to make up 75 cm^3 of solution, assuming no volume change on mixing. *Caution: Carbon tetrachloride is a health hazard. Avoid contact with skin or breathing of fumes.*

Ans.   60 cm^3, 12 cm^3, 3 cm^3

12.43. A liter of milk weighs 1 032 g. The butterfat which it contains to the extent of 4.0% by volume has a density of 0.865 g/cm^3. What is the density of the fat-free "skimmed" milk?

Ans.   1.039 g/cm^3

12.44. To make a benzene-soluble cement, melt 49 g of rosin in an iron pan and add 28 g each of shellac and beeswax. How much of each component should be taken to make 75 kg of cement?

Ans.   35 kg rosin, 20 kg shellac, 20 kg beeswax

12.45. How much $CaCl_2 \cdot 6H_2O$ and water must be weighed out to make 100 g of a solution that is 5.0% $CaCl_2$?

Ans.   9.9 g $CaCl_2 \cdot 6H_2O$, 90.1 g water

12.46. How much $BaCl_2$ would be needed to make 250 cm^3 of a solution having the same concentration of Cl^- as one containing 3.78 g NaCl per 100 cm^3?

Ans.   16.8 g $BaCl_2$

CONCENTRATIONS EXPRESSED IN CHEMICAL UNITS

12.47. What is the molar concentration of a solution containing 37.5 g $Ba(MnO_4)_2$ per liter, and what is the molar concentration with respect to each type of ion?

Ans.   0.100 M $Ba(MnO_4)_2$, 0.100 M Ba^{2+}, 0.200 M MnO_4^-

12.48. How many grams of solute are required to prepare 1 L of 1 M $CaCl_2 \cdot 6H_2O$?

Ans.   219.1 g

12.49. Exactly 100 g of NaCl is dissolved in sufficient water to give 1 500 cm^3 of solution. What is the molar concentration?

Ans.   1.14 M

12.50. Calculate the molality of a solution containing   (*a*) 0.65 mol glucose, $C_6H_{12}O_6$, in 250 g water,   (*b*) 45 g glucose in 1 kg water,   (*c*) 18 g glucose in 200 g water.

Ans.   (*a*) 2.6 molal;   (*b*) 0.25 molal;   (*c*) 0.50 molal

12.51. How many grams of $CaCl_2$ should be added to 300 mL of water to make up a 2.46 molal solution.

Ans.   82 g

12.52. A solution contains 57.5 cm^3 of ethyl alcohol (C_2H_5OH) and 600 cm^3 of benzene (C_6H_6). How many grams of alcohol are in 1 000 g benzene? What is the molality of the solution? Density of C_2H_5OH is 0.80 g/cm^3; of C_6H_6, 0.90 g/cm^3.

Ans.   85 g, 1.85 mol/kg

12.53. A solution contains 10.0 g of acetic acid, CH_3COOH, in 125 g of water. What is the concentration of the solution expressed as   (*a*) mole fractions of CH_3COOH and H_2O.   (*b*) molality?

Ans.   (*a*) x(acid) = 0.024, x(water) = 0.976;   (*b*) 1.33 mol/kg

12.54. A solution contains 116 g acetone (CH_3COCH_3), 138 g ethyl alcohol (C_2H_5OH), and 126 g water. Determine the mole fraction of each.

Ans. x(acetone) = 0.167, x(alcohol) = 0.250, x(water) = 0.583

12.55. What is the mole fraction of the solute in a 1.00 molal aqueous solution?

Ans.   0.0177

12.56. An aqueous solution labeled 35.0% $HClO_4$ had a density 1.251 g/cm^3. What are the molar concentration and molality of the solution?

Ans.   4.36 M, 5.36 mol/kg

12.57. A sucrose solution was prepared by dissolving 13.5 g $C_{12}H_{22}O_{11}$ in enough water to make exactly 100 cm^3 of solution, which was then found to have a density of 1.050 g/cm^3. Compute the molar concentration and molality of the solution.

Ans.   0.395 M, 0.431 mol/kg

12.58. For a solute of molecular weight $\mathcal{M}$, show that the molar concentration M and molality m of the solution are related by

$$M\left(\frac{\mathcal{M}}{1\,000} + \frac{1}{m}\right) = d$$

where d is the solution density in g/cm^3. (*Hint*: Show that each cubic centimeter of solution contains $M\mathcal{M}/1\,000$ grams of solute and M/m grams of solvent.) Use this relation to check the answers to Problems 12.56 and 12.57.

12.59. What volume of a 0.232 N solution contains   (*a*) 3.17 meq of solute,   (*b*) 6.5 g-eq of solute?

Ans.   (*a*) 13.7 cm^3;   (*b*) 28.0 L

12.60. Determine the molar concentration of each of the following solutions: (a) 166 g KI per liter of solution, (b) 33.0 g $(NH_4)_2SO_4$ in 200 cm^3 of solution, (c) 12.5 g $CuSO_4 \cdot 5H_2O$ in 100 cm^3 of solution, (d) 10.0 mg Al^{3+} per cubic centimeter of solution.

Ans.   (a) 1.00 M;   (b) 1.25 M;   (c) 0.500 M;   (d) 0.371 M

12.61. What volume of 0.200 M $Ni(NO_3)_2 \cdot 6H_2O$ contains 500 mg Ni^{2+}?

Ans.   42.6 cm^3

12.62. Compute the volume of concentrated H_2SO_4 (density 1.835 g/cm^3, 93.2% H_2SO_4 by weight) required to make up 500 cm^3 of 3.00 N acid.

Ans.   43.0 cm^3

12.63. Compute the volume of concentrated HCl (density 1.19 g/cm^3, 38% HCl by weight) required to make up 18 L of N/50 acid.

Ans.   29 cm^3

12.64. Determine the mass of $KMnO_4$ required to make 80 cm^3 of N/8 $KMnO_4$ when the latter acts as an oxidizing agent in acid solution and Mn^{2+} is a product of the reaction.

Ans.   0.316 g

12.65. Given the unbalanced equation

$$Cr_2O_7^{2-} + Fe^{2+} + H^+ \rightarrow Cr^{3+} + Fe^{3+} + H_2O$$

(a) What is the normality of a $K_2Cr_2O_7$ solution 35.0 cm^3 of which contains 3.87 g of $K_2Cr_2O_7$? (b) What is the normality of a $FeSO_4$ solution 750 cm^3 of which contains 96.3 g of $FeSO_4$?

Ans.   (a) 2.25 N;   (b) 0.845 N

12.66. What mass of $Na_2S_2O_3 \cdot 5H_2O$ is needed to make up 500 cm^3 of 0.200 N solution for the following reaction?

$$2S_2O_3^{2-} + I_2 \rightarrow S_4O_6^{2-} + 2I^-$$

Ans.   24.8 g

DILUTION PROBLEMS

12.67. A solution contains 75 mg NaCl per cubic centimeter. To what extent must it be diluted to give a solution of concentration 15 mg NaCl per cubic centimeter of solution?

Ans.   Each cubic centimeter of original solution is diluted with water to a volume of 5 cm^3.

12.68. How many cubic centimeters of a solution of concentration 100 mg Co^{2+} per cubic centimeter are needed to prepare 1.5 L of solution of concentration 20 mg Co^{2+} per cubic centimeter?

Ans.   300 cm^3

12.69. Calculate the approximate volume of water that must be added to 250 cm^3 of 1.25 N solution to make it 0.500 N (neglecting volume changes).

Ans.   375 cm^3

12.70. Determine the volume of dilute nitric acid (density 1.11 g/cm^3, 19.0% HNO_3 by weight) that can be prepared by diluting with water 50 cm^3 of the concentrated acid (density 1.42 g/cm^3, 69.8% HNO_3 by weight). Calculate the molar concentrations and molalities of the concentrated and dilute acids.

    Ans.   235 cm^3; molar concentrations, 15.7 and 3.35; molalities, 36.7 and 3.72

12.71. What volume of 95.0% alcohol by weight (density 0.809 g/cm^3) must be used to prepare 150 cm^3 of 30.0% alcohol by weight (density 0.957 g/cm^3)?

    Ans.   56.0 cm^3

12.72. What volumes of 12 N and 3 N HCl must be mixed to give 1 L of 6 N HCl?

    Ans. $\frac{1}{3}$ liter 12 N, $\frac{2}{3}$ liter 3 N

Chapter 13

Reactions Involving Standard Solutions

ADVANTAGES OF VOLUMETRIC STANDARD SOLUTIONS

Solutions of specified molar concentrations, as defined in Chapter 12, are called *standard* solutions and can be used conveniently for reactions involved in quantitative procedures. In a procedure called *titration* a standard solution is added slowly from a calibrated container to a reaction vessel until a first indication appears that the reactant in the vessel is completely consumed. Measured volumes of these solutions contain precisely determined amounts of solutes, according to the basic relation

$$\text{Number of moles} = (\text{number of liters}) \times (\text{molar concentration})$$

or $$\text{Number of millimoles} = (\text{number of milliliters}) \times (\text{molar concentration})$$

Stoichiometric calculations involving solutions of specified normalities are even simpler. By the definition of equivalent weight (Chapter 12), two solutions will react exactly with each other if they contain the same number of gram-equivalents; that is, if

$$\text{Volume}_1 \times \text{normality}_1 = \text{volume}_2 \times \text{normality}_2$$

In this relation the two normalities must be expressed in the same unit, as must the two volumes; the units may be chosen arbitrarily.

Solutions of given normalities are useful even when only one of the reactants is dissolved. In this case, the number of gram-equivalent (or milliequivalents) of the nondissolved reactant is computed in the usual way, by dividing the mass of the sample in grams (or milligrams) by the equivalent weight. The number of g-eq (or meq) of one reactant must still equal the number of g-eq (or meq) of the other.

Solved Problems

13.1. What volume of 1.40 M H_2SO_4 solution is needed to react exactly with 100 g Al?

The balanced molecular equation for the reaction is

$$2\,Al + 3\,H_2SO_4 \rightarrow Al_2(SO_4)_3 + 3\,H_2$$

Mole Method.

$$\text{Number of moles of Al in 100 g Al} = \frac{100\ \text{g}}{27.0\ \text{g/mol}} = 3.70\ \text{mol}$$

$$\text{Number of moles of } H_2SO_4 \text{ required for 3.70 mol Al} = \tfrac{3}{2}(3.70) = 5.55\ \text{mol}$$

$$\text{Volume of 1.40 M } H_2SO_4 \text{ containing 5.55 mol} = \frac{5.55\ \text{mol}}{1.40\ \text{mol/L}} = 3.96\ \text{L}$$

Factor-Label Method.

$$x \text{ liters solution} = \left(\frac{100\ \text{g Al}}{27.0\ \text{g Al/mol Al}}\right)\left(\frac{3\ \text{mol } H_2SO_4}{2\ \text{mol Al}}\right)\left(\frac{1\ \text{L solution}}{1.40\ \text{mol } H_2SO_4}\right) = 3.96\ \text{L solution}$$

13.2. In standardizing a solution of $AgNO_3$ it was found that 40.0 cm^3 was required to precipitate all the chloride ions contained in 36.0 cm^3 of 0.520 M NaCl. How many grams of Ag could be obtained from 100 cm^3 of the $AgNO_3$ solution?

In the precipitation of AgCl, equimolar amounts of Ag^+ and Cl^- are needed; therefore equal numbers of moles of $AgNO_3$ and NaCl must have been used.

$$n(AgNO_3) = n(NaCl) = (0.0360 \text{ L})(0.520 \text{ mol/L}) = 0.01872 \text{ mol}$$

Then 40.0 cm^3 of the $AgNO_3$ solution contains 0.01872 mol $AgNO_3$, or 0.01872 mol Ag, and so 100 cm^3 of solution contains

$$\left(\frac{100 \text{ cm}^3}{40.0 \text{ cm}^3}\right)(0.01872 \text{ mol Ag})(107.9 \text{ g Ag/mol Ag}) = 5.05 \text{ g Ag}$$

13.3. Exactly 40.0 cm^3 of 0.225 M $AgNO_3$ was required to react exactly with 25.0 cm^3 of a solution of NaCN, according to the following equation:

$$Ag^+ + 2CN^- \rightarrow Ag(CN)_2^-$$

What is the molar concentration of the NaCN solution?

$$n(AgNO_3) = (0.0400 \text{ L})(0.225 \text{ mol/L}) = 0.00900 \text{ mol}$$
$$n(NaCN) = 2 \times n(AgNO_3) = 0.0180 \text{ mol}$$

Then 25.0 cm^3 of the NaCN solution contains 0.0180 mol NaCN, so that

$$M = \frac{0.0180 \text{ mol}}{0.025 \text{ L}} = 0.72 \text{ M}$$

13.4. How many cubic centimeters of 6.0 N NaOH are required to neutralize 30 cm^3 of 4.0 N HCl?

$$(\text{Volume HCl}) \times (\text{normality HCl}) = (\text{volume NaOH}) \times (\text{normality NaOH})$$
$$(30 \text{ cm}^3)(4.0 \text{ N}) = (\text{volume NaOH})(6.0 \text{ N})$$
$$\text{volume NaOH} = \frac{(30 \text{ cm}^3)(4.0 \text{ N})}{6.0 \text{ N}} = 20 \text{ cm}^3$$

13.5. Determine the normality of an H_3PO_4 solution, 40 cm^3 of which neutralized 120 cm^3 of 0.531 N NaOH.

The solutions react exactly with each other. Therefore

$$(\text{Volume } H_3PO_4) \times (\text{normality } H_3PO_4) = (\text{volume NaOH}) \times (\text{normality NaOH})$$
$$(40 \text{ cm}^3)(\text{normality } H_3PO_4) = (120 \text{ cm}^3)(0.531 \text{ N})$$
$$\text{normality } H_3PO_4 = \frac{(120 \text{ cm}^3)(0.531 \text{ N})}{40 \text{ cm}^3} = 1.59 \text{ N}$$

Note: In this problem we do not have to know whether one, two, or three hydrogens of H_3PO_4 are replaceable. The normality was determined by reaction of the acid with a base of known concentration. Therefore the acid will have the same normality, 1.59 N, in reactions with any strong base under similar conditions. In order to know the molar concentration of the acid, however, it would be necessary to know the number of replaceable hydrogens in the reaction.

In a case like this, where a substance can have several equivalent weights, the normality determined by one type of reaction is not necessarily the normality in other types of reaction. If a weak base like NH_3 were used instead of a strong base for neutralizing the phosphoric acid, or if the method of detecting the point of neutralization were changed, the equivalent weight of phosphoric acid, and hence the normality of the above solution, might be different. In order to predict the number of replaceable hydrogens of the acid in each case, detailed information about the chemistry of the acid must be known.

13.6. (*a*) What volume of 5.00 N H_2SO_4 is required to neutralize a solution containing 2.50 g NaOH? (*b*) How many grams of pure H_2SO_4 are required?

(*a*)   One gram-equivalent H_2SO_4 reacts completely with one gram-equivalent NaOH. The equivalent weight of NaOH is the formula weight, 40.0.

$$\text{Number of gram-equivalents in 2.50 g NaOH} = \frac{2.50\text{ g}}{40.0\text{ g/g-eq}} = 0.062\,5\text{ g-eq NaOH}$$

$$\text{Number of liters} \times N = \text{number of g-eq}$$

$$\text{Number of liters} = \frac{\text{number of g-eq}}{N} = \frac{0.062\,5}{5.00}$$

$$= 0.012\,5\text{ L}\quad\text{or}\quad 12.5\text{ cm}^3\text{ of 5.00 N solution}$$

(*b*)   Equivalent weight of $H_2SO_4 = \frac{1}{2} \times$ (formula weight) $= \frac{1}{2}(98.08) = 49.04$

Mass of H_2SO_4 required $= (0.062\,5\text{ g-eq})(49.04\text{ g/g-eq}) = 3.07\text{ g}$

13.7. A 0.250-g sample of a solid acid was dissolved in water and exactly neutralized by 40.0 cm^3 of 0.125 N base. What is the equivalent weight of the acid?

$$\text{Number of milliequivalents of base} = (40.0\text{ cm}^3)(0.125\text{ meq/cm}^3) = 5.00\text{ meq}$$

$$\text{Number of milliequivalents of acid} = \text{number of meq of base} = 5.00\text{ meq}$$

$$\text{Equivalent weight of acid} = \frac{250\text{ mg}}{5.00\text{ meq}} = 50\text{ mg/meq} = 50\text{ g/g-eq}$$

13.8. Exactly 48.4 cm^3 of HCl solution is required to neutralize completely 1.240 g of pure $CaCO_3$. Calculate the normality of the acid.

Each CO_3^{2-} ion requires two H^+ for neutralization: $CO_3^{2-} + 2H^+ \rightarrow CO_2 + H_2O$. Thus, the equivalent weight of $CaCO_3$ is one-half the formula weight, or 50.05.

$$\text{Number of gram-equivalents in 1.240 g } CaCO_3 = \frac{1.240\text{ g}}{50.05\text{ g/g-eq}} = 0.024\,8\text{ g-eq } CaCO_3$$

Hence, 48.4 cm^3 of acid solution contains 0.024 8 g-eq HCl.

$$N = \frac{\text{number of g-eq}}{\text{number of liters}} = \frac{0.024\,8\text{ g-eq}}{0.048\,4\text{ L}} = 0.512\text{ N}$$

13.9. When 50.0 cm^3 of a certain Na_2CO_3 solution was titrated with 0.102 M HCl, 56.3 cm^3 were required for complete neutralization, according to the equation: $CO_3^{2-} + 2H^+ \rightarrow CO_2 + H_2O$. Calculate the number of grams of $CaCO_3$ that would be precipitated if an excess of $CaCl_2$ were added to a separate 50.0-cm^3 portion of the Na_2CO_3 solution.

Factor-Label Method.

$$\text{Amount of } CaCO_3 = \left(\frac{56.3\text{ cm}^3}{1\,000\text{ cm}^3/\text{L}}\right)\left(\frac{0.102\text{ mol HCl}}{1\text{ L}}\right)\left(\frac{1\text{ mol } Na_2CO_3}{2\text{ mol HCl}}\right)\left(\frac{1\text{ mol } CaCO_3}{1\text{ mol } Na_2CO_3}\right)\left(\frac{100.1\text{ g } CaCO_3}{1\text{ mol } CaCO_3}\right)$$

$$= 0.287\text{ g } CaCO_3$$

13.10. A 10.0-g sample of "gas liquor" is boiled with an excess of NaOH and the resulting ammonia is passed into 60 cm^3 of 0.90 N H_2SO_4. Exactly 10.0 cm^3 of 0.40 N NaOH is required to neutralize the excess sulfuric acid (not neutralized by the NH_3). Determine the percent ammonia in the "gas liquor" examined.

Number of meq NH_3 in 10.0 g of gas liquor = (number of meq H_2SO_4) − (number of meq NaOH)

$$= (60 \text{ cm}^3)(0.90 \text{ meq/cm}^3) - (10.0 \text{ cm}^3)(0.40 \text{ meq/cm}^3)$$

$$= 50 \text{ meq } NH_3$$

In neutralization experiments, the equivalent weight of NH_3 is the same as the molecular weight, 17.0, in accord with the equation

$$NH_3 + H^+ \rightarrow NH_4^+$$

Then the mass of NH_3 in the sample is (50 meq)(17.0 mg/meq) = 850 mg = 0.85 g, and

$$\text{Fraction } NH_3 \text{ in sample} = \frac{0.85 \text{ g}}{10.0 \text{ g}} = 0.085 = 8.5\%$$

13.11. A 40.8-cm^3 sample of an acid is equivalent to 50.0 cm^3 of a Na_2CO_3 solution, 25.0 cm^3 of which is equivalent to 23.8 cm^3 of a 0.102 N HCl. What is the normality of the first acid?

The volume of HCl that would have been required for 50.0 cm^3 of Na_2CO_3 solution is

$$\left(\frac{50.0}{25.0}\right)(23.8 \text{ cm}^3) = 47.6 \text{ cm}^3$$

$$\text{Volume}_1 \times \text{normality}_1 = \text{volume}_2 \times \text{normality}_2$$

$$(40.8 \text{ cm}^3)N_1 = (47.6 \text{ cm}^3)(0.102 \text{ N})$$

$$N_1 = 0.119 \text{ N}$$

13.12. Calculate the number of grams of $FeSO_4$ that will be oxidized by 24.0 cm^3 of 0.250 N $KMnO_4$ in a solution acidified with sulfuric acid. The unbalanced equation for the reaction is

$$MnO_4^- + Fe^{2+} + H^+ \rightarrow Fe^{3+} + Mn^{2+} + H_2O$$

and the normality of the $KMnO_4$ is with respect to this reaction.

It is not necessary to balance the complete equation. All that need be known is that Fe changes in oxidation state from +II in Fe^{2+} to +III in Fe^{3+}. Then

$$\text{Equivalent weight of } FeSO_4 = \frac{\text{formula weight}}{\text{oxidation state change}} = \frac{152}{1} = 152$$

The same result may be found from the balanced partial equation for the Fe^{2+}, $Fe^{2+} \rightarrow Fe^{3+} + e^-$.

$$\text{Equivalent weight of } FeSO_4 = \frac{\text{formula weight}}{\text{number of electrons transferred}} = \frac{152}{1} = 152$$

Let x = required mass of $FeSO_4$.

$$\text{Number of g-eq } KMnO_4 = \text{number of g-eq } FeSO_4$$

$$(\text{Volume } KMnO_4) \times (\text{normality } KMnO_4) = \frac{\text{mass of } FeSO_4}{\text{equivalent weight of } FeSO_4}$$

$$(0.0240 \text{ L})(0.250 \text{ g-eq/L}) = \frac{x}{152 \text{ g/g-eq}}$$

Solving, $x = 0.912$ g $FeSO_4$.

13.13. What volume of 0.100 0 N $FeSO_4$ is required to reduce 4.000 g $KMnO_4$ in a solution acidified with sulfuric acid?

The normality of the $FeSO_4$ is with respect to the oxidation-reduction reaction given in Problem 13.12. In this reaction the Mn changes in oxidation state from $+VII$ in MnO_4^- to $+II$ in Mn^{2+}. The net change is 5. Or, from the balanced partial equation

$$MnO_4^- + 8H^+ + 5e^- \rightarrow Mn^{2+} + 4H_2O$$

it can be seen that the electron transfer is 5 for each MnO_4^-. The equivalent weight of $KMnO_4$ in this reaction is then

$$\tfrac{1}{5} \times \text{(formula weight)} = \tfrac{1}{5}(158.0) = 31.6$$

$$\text{Number of g-eq } FeSO_4 = \text{number of g-eq } KMnO_4$$

$$(\text{Volume } FeSO_4) \times (0.100\,0\ \text{g-eq/L}) = \frac{4.000\ \text{g}}{31.6\ \text{g/g-eq}}$$

$$\text{Volume } FeSO_4 = 1.266\ \text{L}$$

13.14. A sample known to contain some As_2O_3 was brought into solution by a process which converted the arsenic to H_3AsO_3. This was titrated with a standard I_2 solution according to the equation

$$H_3AsO_3 + I_2 + H_2O \rightarrow H_3AsO_4 + 2I^- + 2H^+$$

Exactly 40.27 cm^3 of standard I_2 solution was required to reach the point of completion as indicated by the persistence of a faint I_2 color. The standard solution had been prepared by mixing 0.419 2 g of pure KIO_3 with excess KI and acid and diluting to exactly 250.0 cm^3. The I_2 is formed in the reaction:

$$IO_3^- + 5I^- + 6H^+ \rightarrow 3I_2 + 3H_2O$$

Calculate the mass of As_2O_3 in the sample.

First it is necessary to calculate the molar concentration of the I_2 solution. The *factor-label* method is used.

$$M = \left(\frac{0.419\,2\ \text{g } KIO_3}{214.0\ \text{g } KIO_3/\text{mol}}\right)\left(\frac{3\ \text{mol } I_2}{1\ \text{mol } KIO_3}\right)\left(\frac{1\,000\ \text{cm}^3/\text{L}}{250.0\ \text{cm}^3}\right) = 0.023\,51\ \text{mol/L} = 0.023\,51\ M$$

Then, by the same method:

$$(0.040\,27\ \text{L})(0.023\,51\ \text{mol/L})\left(\frac{1\ \text{mol } H_3AsO_3}{1\ \text{mol } I_2}\right)\left(\frac{1\ \text{mol } As_2O_3}{2\ \text{mol } H_3AsO_3}\right)(197.8\ \text{g/mol}) = 0.093\,63\ \text{g of } As_2O_3$$

Supplementary Problems

13.15. How many cubic centimeters of 0.25 M $BaCl_2$ are required to precipitate all the sulfate ion from 20 cm^3 of a solution containing 100 g of Na_2SO_4 per liter?

Ans.   56 cm^3

13.16. A 50.0-cm^3 sample of Na_2SO_4 solution is treated with an excess of $BaCl_2$. If the precipitated $BaSO_4$ is 1.756 g, what is the molar concentration of the Na_2SO_4 solution?

Ans.   0.150 5 M

13.17. What was the thorium content of a sample that required 35.0 cm^3 of 0.020 0 M $H_2C_2O_4$ to precipitate $Th(C_2O_4)_2$?

Ans. 81 mg

13.18. What is the molar concentration of a $K_4Fe(CN)_6$ solution if 40.0 cm^3 was required to titrate 150.0 mg Zn (dissolved) by forming $K_2Zn_3[Fe(CN)_6]_2$?

Ans. 0.038 2 M

13.19. A 50.0-cm^3 sample of NaOH solution requires 27.8 cm^3 of 0.100 N acid in titration. What is its normality? How many milligrams NaOH are in each cubic centimeter?

Ans. 0.055 6 N, 2.22 mg/cm^3

13.20. In standardizing HCl, 22.5 cm^3 was required to neutralize 25.0 cm^3 of 0.100 N Na_2CO_3 solution. What is the normality of the HCl solution? How much water must be added to 200 cm^3 of it to make it 0.100 N? Neglect volume changes.

Ans. 0.111 N, 22 cm^3

13.21. Exactly 21 cm^3 of 0.80 N acid was required to neutralize completely 1.12 g of an impure sample of calcium oxide. What is the purity of the CaO?

Ans. 42%

13.22. By the *Kjeldahl method*, the nitrogen contained in a foodstuff is converted into ammonia. If the ammonia from 5.0 g of a foodstuff is just sufficient to neutralize 20 cm^3 of 0.100 M nitric acid, calculate the percentage of nitrogen in the foodstuff.

Ans. 0.56%

13.23. What is the purity of concentrated H_2SO_4 (density 1.800 g/cm^3) if 5.00 cm^3 is neutralized by 84.6 cm^3 of 2.000 M NaOH?

Ans. 92.2%

13.24. A 10.0-cm^3 portion of $(NH_4)_2SO_4$ solution was treated with an excess of NaOH. The NH_3 gas evolved was absorbed in 50.0 cm^3 of 0.100 N HCl. To neutralize the remaining HCl, 21.5 cm^3 of 0.098 N NaOH was required. What is the molar concentration of the $(NH_4)_2SO_4$? How many grams of $(NH_4)_2SO_4$ are in a liter of solution?

Ans. 0.145 M, 19.1 g/L

13.25. Exactly 400 mL of an acid solution, when acted upon by an excess of zinc, evolved 2.430 L of H_2 gas measured over water at 21 °C and 747.5 torr. What is the normality of the acid? Vapor pressure of water at 21 °C is 18.6 torr.

Ans. 0.483 N

13.26. How many grams of copper will be replaced from 2 L of 0.150 M $CuSO_4$ solution by 2.7 g of aluminum?

Ans. 9.5 g

13.27. What volume of 1.50 M H_2SO_4 is needed to liberate 185 L of hydrogen gas at S.T.P. when treated with an excess of zinc?

Ans. 5.51 L

13.28. How many liters of hydrogen at S.T.P. would be replaced from 500 cm^3 of 3.78 M HCl by 125 g of zinc?

Ans. 21.2 L

13.29. Exactly 50.0 cm^3 of a solution of Na_2CO_3 was titrated with 65.8 cm^3 of 3.00 M HCl.

$$CO_3^{2-} + 2H^+ \rightarrow CO_2 + H_2O$$

If the density of the Na_2CO_3 solution is 1.25 g/cm^3, what percent Na_2CO_3 by weight does it contain?

Ans.   16.7%

13.30. What is the equivalent weight of an acid 1.243 g of which required 31.72 cm^3 of 0.192 3 N standard base for neutralization?

Ans.   203.8

13.31. The molecular weight of an organic acid was determined by the following study of its barium salt. 4.290 g of the salt was converted to the free acid by reaction with 21.64 cm^3 of 0.477 M H_2SO_4. The barium salt was known to contain 2 mol of water of hydration per mol Ba^{2+}, and the acid was known to be monoprotic (monobasic). What is the molecular weight of the anhydrous acid?

Ans.   122.1

13.32. A ferrous sulfate solution was standardized by titration. A 25.00-cm^3 portion of the solution required 42.08 cm^3 of 0.0800 N ceric sulfate for complete oxidation. What is the normality of the ferrous sulfate?

Ans.   0.134 7 N

13.33. How many cubic centimeters of 0.025 7 N KIO_3 would be needed to reach the end point in the oxidation of 34.2 cm^3 of 0.041 6 N hydrazine in hydrochloric acid solution?

Ans.   55.4 cm^3

13.34. How many grams of $FeCl_2$ will be oxidized by 28 cm^3 of 0.25 N $K_2Cr_2O_7$ in HCl solution? The unbalanced equation is

$$Fe^{2+} + Cr_2O_7^{2-} + H^+ \rightarrow Fe^{3+} + Cr^{3+} + H_2O$$

Ans.   0.89 g

13.35. What mass of MnO_2 is reduced by 35 cm^3 of 0.080 M oxalic acid, $H_2C_2O_4$, in sulfuric acid solution? The unbalanced equation is

$$MnO_2 + H^+ + H_2C_2O_4 \rightarrow CO_2 + H_2O + Mn^{2+}$$

Ans.   0.24 g

13.36. How many grams of $KMnO_4$ are required to oxidize 2.40 g of $FeSO_4$ in a solution acidified with sulfuric acid? What is the equivalent weight of $KMnO_4$ in this reaction?

Ans.   0.500 g, 31.6

13.37. Find the equivalent weight of $KMnO_4$ in the reaction

$$Mn^{2+} + MnO_4^- + H_2O \rightarrow MnO_2 + H^+ \qquad \text{(unbalanced)}$$

How many grams of $MnSO_4$ are oxidized by 1.25 g $KMnO_4$?

Ans.   52.7, 1.79 g

13.38. (*a*) What volume of 0.40 N $K_2Cr_2O_7$ is required to liberate the chlorine from 1.20 g of NaCl in a solution acidified with H_2SO_4?

$$Cr_2O_7^{2-} + Cl^- + H^+ \rightarrow Cr^{3+} + Cl_2 + H_2O \qquad \text{(unbalanced)}$$

(*b*) How many grams of $K_2Cr_2O_7$ are required?   (*c*) How many grams of chlorine are liberated?

Ans.   (*a*) 51 cm^3;   (*b*) 1.01 g;   (*c*) 0.73 g

13.39. If 25.0 cm^3 of an iodine solution is equivalent to 0.125 g of $K_2Cr_2O_7$, to what volume should 1 000 cm^3 be diluted to make the solution tenth normal?

Ans.   1 020 cm^3

13.40. How many grams of $KMnO_4$ should be taken to make up 250 cm^3 of a solution of such strength that 1 cm^3 is equivalent to 5.00 mg of iron in $FeSO_4$?

Ans.   0.707 g

13.41. How many grams of iodine are present in a solution which requires 40 cm^3 of 0.112 M $Na_2S_2O_3$ to react with it?

$$S_2O_3^{2-} + I_2 \to S_4O_6^{2-} + I^-   \text{(unbalanced)}$$

Ans.   0.57 g

13.42. To how many milligrams of iron (Fe^{2+}) is 1 cm^3 of 0.105 5 N $K_2Cr_2O_7$ equivalent?

Ans.   5.89 mg

13.43. Reducing sugars are sometimes characterized by a number R_{Cu}, which is defined as the number of milligrams of copper reduced by 1 gram of the sugar, in which the half-reaction for the copper is

$$Cu^{2+} + OH^- \to Cu_2O + H_2O   \text{(unbalanced)}$$

It is sometimes more convenient to determine the reducing power of a carbohydrate by an indirect method. In this method 43.2 mg of the carbohydrate was oxidized by an excess of $K_3Fe(CN)_6$. The $Fe(CN)_6^{4-}$ formed in this reaction required 5.29 cm^3 of 0.0345 N $Ce(SO_4)_2$ for reoxidation to $Fe(CN)_6^{3-}$ (the normality of the ceric sulfate solution is given with respect to the reduction of Ce^{4+} to Ce^{3+}). Determine the R_{Cu}-value for the sample. (*Hint*: The number of milliequivalents of Cu in a direct oxidation is the same as the number of milliequivalents of Ce^{4+} in the indirect method.)

Ans.   268

13.44. A volume of 12.53 cm^3 of 0.050 93 M selenium dioxide, SeO_2, reacted exactly with 25.52 cm^3 of 0.100 0 M $CrSO_4$. In the reaction, Cr^{2+} was oxidized to Cr^{3+}. To what oxidation state was the selenium converted by the reaction?

Ans.   0

13.45. An acid solution of a $KReO_4$ sample containing 26.83 mg of combined rhenium was reduced by passage through a column of granulated zinc. The effluent solution, including the washings from the column, was then titrated with 0.100 0 N $KMnO_4$; 11.45 cm^3 of the standard permanganate was required for the reoxidation of all the rhenium to the perrhenate ion, ReO_4^-. Assuming that rhenium was the only element reduced, what is the oxidation state to which rhenium was reduced by the zinc column?

Ans. $-I$

13.46. The iodide content of a solution was determined by titration with ceric sulfate in the presence of HCl, in which I^- is converted to ICl and Ce^{4+} to Ce^{3+}. A 250-cm^3 sample of the solution required 20.0 cm^3 of 0.050 M Ce^{4+} solution. What is the iodide concentration in the original solution, in mg/cm^3?

Ans.   0.25 mg/cm^3

13.47. A 0.518-g sample of limestone is dissolved, and then the calcium is precipitated as calcium oxalate, CaC_2O_4. After filtering and washing the precipitate, it requires 40.0 cm^3 of 0.0500 M $KMnO_4$ solution acidified with sulfuric acid to titrate it. What is the percent CaO in the limestone? The unbalanced equation for the titration is

$$MnO_4^- + CaC_2O_4 + (H^+)_2SO_4^{2-} \to CaSO_4 + Mn^{2+} + CO_2 + H_2O$$

Ans.   54.2%

Chapter 14

Properties of Solutions

INTRODUCTION

Just as dilute gases are characterized by more or less general adherence to a group of simple laws, so dilute solutions as a class have many properties that are determined by concentration alone, without reference to the particular nature of the dissolved materials. All solutions obey the laws described below when the concentration is sufficiently low. A few of the laws, specifically designated in the discussion below, are obeyed over the entire range of composition by certain pairs (or groups of more than two) of substances. Such pairs of substances are said to form *ideal solutions*. In an ideal solution the forces of interaction between solvent and solute molecules are the same as between the molecules in the separate components. In the formation of an ideal solution from the separate components, there are no volume changes and no enthalpy changes. Pairs of chemically similar substances, such as methanol (CH_3OH) and ethanol (C_2H_5OH), or benzene (C_6H_6) and toluene (C_7H_8), form ideal solutions. But dissimilar substances, such as C_2H_5OH and C_6H_6, form nonideal solutions.

VAPOR PRESSURE

The vapor pressures of all solutions of nonvolatile solutes in a solvent are *less than* that of the pure solvent. If we prepare solutions of different solutes in a given solvent by adding *equal numbers of solute molecules* to a fixed amount of solvent, as we do in preparing solutions of equal molality, we find that the *depression* of the vapor pressure is the same in every case in dilute solutions of nonvolatile nonelectrolytes.

Raoult's law states that in dilute solutions of nonvolatile nonelectrolytes *the depression is proportional to the mole fraction of the solute*, or *the solution vapor pressure is proportional to the mole fraction of the solvent.* In equation form:

Depression of solvent vapor pressure

$= \Delta P =$ (vapor pressure of pure solvent) $-$ (vapor pressure of solution)

$=$ (vapor pressure of pure solvent) $\times$ (mole fraction of solute)

or

Vapor pressure of solvent over solution

$=$ (vapor pressure of pure solvent) $\times$ (mole fraction of solvent)

In the second form, the vapor pressure of the solution has been identified with the vapor pressure of the solvent over the solution, since the solute is assumed nonvolatile.

In systems of liquids that mix with each other in all proportions to form ideal solutions, Raoult's law in the form of the second equation above applies to the partial pressure of each component separately.

Partial pressure of any component over solution

$=$ (vapor pressure of that pure component) $\times$ (mole fraction of component)

Raoult's law is explained by the hypothesis that solute molecules at the liquid surface interfere with the escape of solvent molecules into the vapor phase. Because of the vapor pressure lowering, the *boiling point of the solution is raised* and the *freezing point is lowered*, as compared with the pure solvent.

FREEZING-POINT LOWERING, ΔT_f

When most dilute solutions are cooled, pure solvent begins to crystallize before any solute crystallizes. The temperature at which the first crystals are in equilibrium with the solution is called the *freezing point of the solution*. The freezing point of such a solution is always *lower* than the freezing point of pure solvent.

In dilute solutions, the lowering of the freezing point is directly proportional to the number of solute molecules (or moles) in a given mass of solvent. Thus,

Lowering of freezing point = ΔT_f = (freezing point of solvent) − (freezing point of solution) = $K_f m$

where m is the molality (Chapter 12) of the solution. If this equation were valid up to a concentration of 1 molal, the lowering of the freezing point of a 1 molal solution of any nonelectrolyte dissolved in the solvent would be K_f, which is thus called the *molal freezing-point constant* of the solvent. The numerical value of K_f is a property of the solvent alone.

EXAMPLE 1   The molal freezing-point constant for water is 1.86 K · kg/mol or 1.86 °C · kg/mol. Thus if 1 mol of cane sugar (342 g sugar) is dissolved in 1 000 g of water, the solution will freeze at −1.86 °C.

BOILING-POINT ELEVATION, ΔT_b

The temperature at which a solution boils is *higher* than that of the pure solvent if the solute is relatively nonvolatile. In dilute solutions, the elevation of the boiling point is directly proportional to the number of solute molecules (or moles) in a given mass of solvent. Again, the molality scale is usually used, and the equation is

Elevation of boiling point = ΔT_b = (boiling point of solution) − (boiling point of solvent) = $K_b m$

K_b is called the *molal boiling-point constant* of the solvent. As with K_f, the numerical value of K_b is a property of the solvent alone and is independent of the nature of the solute, within the general requirements of nonvolatility and nondissociation into ions.

EXAMPLE 2   The molal boiling-point constant for water is 0.513 °C · kg/mol. Thus if 1 mol of cane sugar (342 g sugar) is dissolved in 1 000 g of water, the solution will boil at 100.513 °C, assuming a pressure of 1 atm. (If pure water boils at a temperature slightly different from 100 °C because the air pressure is not exactly one standard atmosphere, the magnitude of ΔT_b is still 0.513, and the boiling point of the solution is 0.513 °C higher than the actual boiling point of the water.) If a solution contains $\frac{1}{2}$ mol of sugar (171 g) and 1 000 g water, it will boil at 100.256 °C at 1 atm.

OSMOTIC PRESSURE

If a solution is separated from a sample of pure solvent by a porous sheet that allows solvent, but not solute molecules to pass through, solvent will move into the solution in an attempt to equalize the concentrations on the two sides of the sheet. Such a dividing sheet is called a *semipermeable membrane*. If the membrane is placed vertically and the vessel holding the solution can extend indefinitely in a horizontal direction to accommodate the incoming material, solvent will continue to flow until it is all used up or until the solution becomes so dilute that there is no more driving force due to the difference in solvent concentrations on the two sides. If, however, the solution vessel is closed on all sides except for an extension tube on top as shown in Fig. 14-1, the incoming solvent will force some of the solution up the extension tube. The weight of this solution in the tube will exert a downward pressure that will tend to oppose the inward penetration of more solvent through the membrane. Eventually the two forces will just balance each other and no more solvent will enter. These two forces are the weight of the hydrostatic head of solution in the tube and the driving force tending to equalize the concentrations on the two sides of the membrane. The hydrostatic head at this balance point is called the *osmotic pressure* of the solution. It can be measured in usual units of pressure, such as Pa, atm, or torr.

The osmotic pressure, π, of a dilute solution of a nonelectrolyte is given by an equation formally equivalent to the ideal gas law:

$$\pi = MRT$$

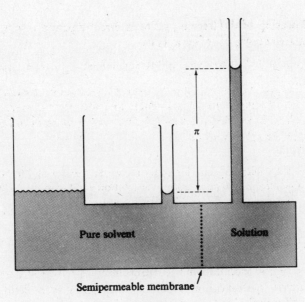

Fig. 14-1

With the molar concentration M in mol/L, the temperature T in K, and with

$$R = 0.082\,1\ \text{L} \cdot \text{atm} \cdot \text{K}^{-1} \cdot \text{mol}^{-1}$$

π will be obtained in atm.

DEVIATIONS FROM LAWS OF DILUTE SOLUTIONS

The above laws are valid only for dilute solutions of nonelectrolytes. For electrolyte solutions each ion contributes independently to the effective molality or molar concentration. On account of the electrical interactions between ions, however, none of the effects is as large as would be predicted on the basis of simple counting of ions.

EXAMPLE 3 A solution containing 0.100 mol of KCl per kilogram of water freezes at $-0.345\,°C$. The observed lowering is slightly smaller than

$$(1.86\ °C \cdot \text{Kg/mol})(0.2\ \text{mol/kg}) = 0.372\ °C$$

which would be predicted if each K^+ and Cl^- were truly independently effective (0.1 molal K^+ + 0.1 molal Cl^-). A solution containing 0.100 mol of $BaCl_2$ per kilogram of water freezes at $-0.470\,°C$, showing a lowering somewhat less than the $(1.86)(0.3) = 0.558\,°C$ predicted from the simple additivity of molalities (0.1 molal Ba^{2+} + 0.2 molal Cl^-).

For any solution not too concentrated, whether electrolyte or nonelectrolyte, the deviations from any one of the laws of the dilute solution are equal to the deviations from any of the others, on a fractional or percentage basis. That is,

$$\frac{\Delta T_f - (\Delta T_f)°}{(\Delta T_f)°} = \frac{\Delta T_b - (\Delta T_b)°}{(\Delta T_b)°} = \frac{\Delta P - (\Delta P)°}{(\Delta P)°} = \frac{\pi - \pi°}{\pi°}$$

in which a $°$ labels the magnitude of the effect that would be predicted by the laws of the dilute solution.

SOLUTIONS OF GASES IN LIQUIDS

At constant temperature, the concentration of a slightly soluble gas in a liquid (i.e., the mass or number of moles of gas dissolved in a given volume of liquid) is directly proportional to the partial pressure of the gas. This is known as *Henry's law*.

When a mixture of two gases is in contact with a solvent, the amount of each gas that is dissolved is the same as if it were present alone at a pressure equal to its own partial pressure in the gas mixture.

LAW OF DISTRIBUTION

A solute distributes itself between two immiscible solvents so that the ratio of its concentrations in dilute solutions in the two solvents is constant, regardless of the actual concentration in either solvent. Here, both concentrations are assumed to be on the same volumetric basis (e.g., mol/L).

EXAMPLE 4 For the distribution of iodine between ether and water at room temperature, the value of the constant is about 200. Thus

$$\frac{\text{Concentration of iodine in ether}}{\text{Concentration of iodine in water}} = K = 200$$

The value of this ratio of concentrations, called the *distribution ratio* or *distribution coefficient*, is equal to the ratio of the solubilities (per unit volume) in the two solvents if saturated solutions in these solvents are dilute enough for the law of distribution to apply.

Solved Problems

FREEZING-POINT LOWERING

14.1 The freezing point of pure camphor is 178.4 °C, and its molal freezing-point contant, K_f, is 40.0 °C · kg/mol. Find the freezing point of a solution containing 1.50 g of a compound of molecular weight 125 in 35.0 g of camphor.

The first step is to find the molality (m) of the solution.

$$m = \frac{\text{number of moles of solute}}{\text{number of kilograms of solvent}} = \frac{(1.50/125)\ \text{mol solute}}{\left(\frac{35}{1000}\right)\ \text{kg solvent}} = 0.343\ \text{mol/kg}$$

Lowering of freezing point $= \Delta T_f = K_f m = (40.0\ °\text{C} \cdot \text{kg/mol})(0.343\ \text{mol/kg}) = 13.7\ °\text{C}$

Freezing point of solution $=$ (freezing point of pure solvent) $- \Delta T_f$

$$= 178.4\ °\text{C} - 13.7\ °\text{C} = 164.7\ °\text{C}$$

14.2. A solution containing 4.50 g of a nonelectrolyte dissolved in 125 g of water freezes at $-0.372\ °\text{C}$. Calculate the approximate molecular weight of the solute.

First Method.

First compute the molality from the freezing-point equation.

$$\Delta T_f = K_f m$$

$$0.372\ °\text{C} = (1.86\ °\text{C} \cdot \text{kg/mol})m$$

$$m = \frac{0.372\ °\text{C}}{1.86\ °\text{C} \cdot \text{kg/mol}} = 0.200\ \text{mol/kg}$$

From the definition of molality compute the number of moles of solute in the sample.

$$n(\text{solute}) = (0.200 \text{ mol solute/kg solvent})(0.125 \text{ kg solvent}) = 0.025 \text{ mol solute}$$

Then
$$\text{Molecular weight} = \frac{4.50 \text{ g solute}}{0.025 \text{ mol solute}} = 180 \text{ g/mol}$$

Second Method.

Putting the definition of molality into the freeziing-point equation:

$$\Delta T_f = K_f m = K_f \times \frac{\text{amt. of solute in grams}/\mathcal{M} \text{ of solute}}{\text{amt. of solvent in kilograms}}$$

where $\mathcal{M}$ is the molecular weight of the solute. Solving for $\mathcal{M}$:

$$\mathcal{M} = \left(\frac{K_f}{\Delta T_f}\right)\left(\frac{\text{amt. of solute in grams}}{\text{amt. of solvent in kilograms}}\right) = \frac{(1.86)(4.50)}{(0.372)(0.125)} = 180 \text{ g/mol}$$

BOILING-POINT ELEVATION

14.3. The molecular weight of an organic compound is 58.0. Compute the boiling point of a solution containing 24.0 g of the solute and 600 g of water, when the barometric pressure is such that pure water boils at 99.725 °C.

$$\text{Molality} = m = \frac{n(\text{solute})}{\text{number of kg solvent}} = \frac{(24.0/58.0) \text{ mol solute}}{0.600 \text{ kg solvent}} = 0.690 \text{ mol/kg}$$

$$\text{Elevation of boiling point} = \Delta T_b = K_b m = (0.513 \text{ °C} \cdot \text{kg/mol})(0.690 \text{ mol/kg}) = 0.354 \text{ °C}$$

$$\text{Boiling point of solution} = (\text{boiling point of water}) + \Delta T_b = 99.725 \text{ °C} + 0.354 \text{ °C} = 100.079 \text{ °C}$$

14.4. A solution was made up by dissolving 3.75 g of a pure hydrocarbon in 95 g of acetone. The boiling point of pure acetone was observed to be 55.95 °C, and of the solution, 56.50 °C. If the molal boiling-point constant of acetone is 1.71 °C · kg/mol, what is the approximate molecular weight of the hydrocarbon?

First Method.

Compute the molality (m) from the boiling-point equation.

$$\Delta T_b = K_b m$$
$$56.50 - 55.95 = 1.71 m$$

Solving, $m = 0.322$ mol solute/kg solvent. Now find the number of moles of solute in the weighed sample.

$$n(\text{solute}) = \left(0.322 \, \frac{\text{mol solute}}{\text{kg solvent}}\right)(0.095 \text{ kg solvent}) = 0.0306 \text{ mol solute}$$

Then
$$\text{Molecular weight} = \frac{3.75 \text{ g solute}}{0.0306 \text{ mol solute}} = 123 \text{ g/mol}$$

Second Method.

As in Problem 14.2, we can develop the expression

$$\mathcal{M} = \left(\frac{K_b}{\Delta T_b}\right)\left(\frac{\text{amt. of solute in grams}}{\text{amt. of solvent in kilograms}}\right) = \frac{(1.71)(3.75)}{(0.55)(0.095)} = 123 \text{ g/mol}$$

VAPOR PRESSURE

14.5. The vapor pressure of water at 28 °C is 28.35 torr. Compute the vapor pressure at 28 °C of a solution containing 68 g of cane sugar, $C_2H_{22}O_{11}$, in 1 000 g of water.

$$\text{Number of moles of } C_{12}H_{22}O_{11} \text{ in 68 g} = \frac{68 \text{ g}}{342 \text{ g/mol}} = 0.20 \text{ mol } C_{12}H_{22}O_{11}$$

$$\text{Number of moles of } H_2O \text{ in 1 000 g} = \frac{1\,000 \text{ g}}{18.02 \text{ g/mol}} = 55.49 \text{ mol } H_2O$$

$$\text{Total number of moles} = 0.20 + 55.49 = 55.69 \text{ mol}$$

$$\text{Mole fraction } C_{12}H_{22}O_{11} = \frac{0.20}{55.69} = 0.003\,6 \qquad \text{Mole fraction } H_2O = \frac{55.49}{55.69} = 0.996\,4$$

First Method.

$$\text{Vapor pressure of solution} = (\text{v.p. of pure solvent}) \times (\text{mole fraction of solvent})$$
$$= (28.35 \text{ torr})(0.996\,4) = 28.25 \text{ torr}$$

Second Method.

$$\text{Vapor pressure depression} = \Delta P = (\text{v.p. of pure solvent}) \times (\text{mole fraction of solute})$$
$$= (28.35 \text{ torr})(0.003\,6) = 0.10 \text{ torr}$$
$$\text{Vapor pressure of solution} = (28.35 - 0.10) \text{ torr} = 28.25 \text{ torr}$$

14.6. At 30 °C, pure benzene (molecular weight 78.1) has a vapor pressure of 121.8 torr. Dissolving 15.0 g of a nonvolatile solute in 250 g of benzene produced a solution having a vapor pressure of 120.2 torr. Determine the approximate molecular weight of the solute.

Let $\mathcal{M}$ be the molecular weight of the solute.

$$\text{Number of moles of benzene in 250 g} = \frac{250 \text{ g}}{78.1 \text{ g/mol}} = 3.20 \text{ mol benzene}$$

$$\text{Number of moles of solute in 15.0 g} = \frac{15.0}{\mathcal{M}} \text{ mol solute}$$

Substituting in the relation v.p. solution = (v.p. pure solvent) × (mole fraction solvent),

$$120.2 \text{ torr} = (121.8 \text{ torr})\left[\frac{3.20 \text{ mol}}{(15.0/\mathcal{M}) \text{ mol} + 3.20 \text{ mol}}\right]$$

or

$$120.2 = (121.8)\left(\frac{3.20\mathcal{M}}{15.0 + 3.20\mathcal{M}}\right)$$

Solving, $\mathcal{M} = 350$. Note that the accuracy of the calculation is limited by the term $121.8 - 120.2$ that appears in the expansion. The answer is significant only to 1 part in 16.

14.7. At 20 °C the vapor pressure of methyl alcohol (CH_3OH) is 94 torr and the vapor pressure of ethyl alcohol (C_2H_5OH) is 44 torr. Being closely related, these compounds form a two-component system which adheres quite closely to Raoult's law throughout the entire range of concentrations. If 20 g of CH_3OH is mixed with 100 g of C_2H_5OH, determine the partial pressure exerted by each and the total pressure of the solution. Calculate the composition of the vapor above the solution by applying Dalton's law (Chapter 5).

In an ideal solution of two liquids, there is no distinction between solute and solvent, and Raoult's law holds for each component of such solutions. Hence, when two liquids are mixed to give an ideal solution, the partial pressure of each liquid is equal to its vapor pressure multiplied by its mole fraction in the solution. The molecular weights of CH_3OH and C_2H_5OH are 32 and 46, and so

$$\text{Partial pressure of } CH_3OH = (94 \text{ torr})\left(\frac{\frac{20}{32} \text{ mol } CH_3OH}{\frac{20}{32} \text{ mol } CH_3OH + \frac{100}{46} \text{ mol } C_2H_5OH}\right)$$

$$= (94 \text{ torr})(0.22) = 21 \text{ torr}$$

$$\text{Partial pressure of } C_2H_5OH = (44 \text{ torr})\left(\frac{\frac{100}{46} \text{ mol } C_2H_5OH}{\frac{20}{32} \text{ mol } CH_3OH + \frac{100}{46} \text{ mol } C_2H_5OH}\right)$$

$$= (44 \text{ torr})(0.78) = 34 \text{ torr}$$

The total pressure of the gaseous mixture is the sum of the partial pressures of all the components (Dalton's law): total pressure of solution = 21 torr + 34 torr = 55 torr. Dalton's law also indicates that the mole fraction of any component of a gaseous mixture is equal to its pressure fraction, i.e., its partial pressure divided by the total pressure.

$$\text{Mole fraction of } CH_3OH \text{ in vapor} = \frac{\text{partial pressure of } CH_3OH}{\text{total pressure}} = \frac{21 \text{ torr}}{55 \text{ torr}} = 0.38$$

$$\text{Mole fraction of } C_2H_5OH \text{ in vapor} = \frac{\text{partial pressure of } C_2H_5OH}{\text{total pressure}} = \frac{34 \text{ torr}}{55 \text{ torr}} = 0.62$$

Since the mole fraction of (ideal) gases is the same as the volume fraction, we may also say that the vapor consists of 38% CH_3OH by volume. Note that the vapor is relatively richer in the more volatile component, methyl alcohol (mole fraction 0.38), than is the liquid (mole fraction of CH_3OH, 0.22).

OSMOTIC PRESSURE

14.8. What would be the osmotic pressure at 17 °C of an aqueous solution containing 1.75 g of sucrose ($C_{12}H_{22}O_{11}$) per 150 cm^3 of solution?

$$\text{Molar concentration} = M = \frac{\text{number of moles of solute}}{\text{number of liters of solution}} = \frac{1.75 \text{ g}/(342 \text{ g/mol})}{0.150 \text{ L}} = 0.034\,1 \text{ mol/L}$$

$$\text{Osmotic pressure} = \pi = MRT = (0.034\,1 \text{ mol/L})(0.082\,1 \text{ L} \cdot \text{atm} \cdot \text{K}^{-1} \cdot \text{mol}^{-1})(290 \text{ K}) = 0.812 \text{ atm}$$

14.9. The osmotic pressure of a solution of a synthetic polyisobutylene in benzene was determined at 25 °C. A sample containing 0.20 g of solute per 100 cm^3 of solution developed a rise of 2.4 mm at osmotic equilibrium. The density of the solution was 0.88 g/cm^3. What is the molecular weight of the polyisobutylene?

The osmotic pressure is equal to that of a column of the solution 2.4 mm high. By the formula of Chapter 5,

$$\pi = \text{height} \times \text{density} \times g = (2.4 \times 10^{-3} \text{ m})(0.88 \times 10^3 \text{ kg/m}^3)(9.81 \text{ m/s}^2) = 20.7 \text{ Pa}$$

The molar concentration can now be determined from the osmotic pressure equation.

$$M = \frac{\pi}{RT} = \frac{20.7 \text{ N/m}^2}{(8.314 \text{ J} \cdot \text{K}^{-1} \cdot \text{mol}^{-1})(298 \text{ K})} = 8.3 \times 10^{-3} \text{ mol/m}^3 = 8.3 \times 10^{-6} \text{ mol/L}$$

The solution contained 0.20 g solute per 100 mL solution, or 2.0 g per liter, and has been found to contain 8.3×10^{-6} mol/L. Then

$$\text{Molecular weight} = \frac{2.0 \text{ g}}{8.3 \times 10^{-6} \text{ mol}} = 2.4 \times 10^5 \text{ g/mol}$$

14.10. An aqueous solution of urea had a freezing point of $-0.52\,°C$. Predict the osmotic pressure of the same solution at $37\,°C$. Assume that the molar concentration and the molality are numerically equal.

The concentration of the solution is not specified, but the effective molality may be inferred from the freezing-point lowering.

$$m = \frac{\Delta T_f}{K_f} = \frac{0.52\,°C}{1.86\,°C \cdot kg/mol} = 0.280\ mol/kg$$

The assumption that the molality and molar concentration are equal is not very bad for dilute aqueous solutions. (The relation found in Problem 12.58 shows that $M \approx m$ when $d \approx 1\ g/cm^3$ and $\mathcal{M} \ll 1\,000/m$. Urea has a molecular weight of 60.) Then $0.280\ mol/L$ may be used for the molar concentration in the osmotic pressure equation.

$$\pi = MRT = (0.280\ mol/L)(0.082\ L \cdot atm \cdot K^{-1} \cdot mol^{-1})(310\ K) = 7.1\ atm$$

SOLUTIONS OF GASES IN LIQUIDS

14.11. At $20\,°C$ and a total pressure of 760 torr, 1 L of water dissolves 0.043 g of pure oxygen or 0.019 g of pure nitrogen. Assuming that dry air is composed of 20% oxygen, and 80% nitrogen (by volume), determine the masses of oxygen and nitrogen dissolved by 1 L of water at $20\,°C$ exposed to air at a total pressure of 760 torr.

The *solubility* of a gas, i.e., the concentration of the dissolved gas, may be expressed as

$$\text{Solubility of } Y = k_H(Y) \times P(Y)$$

In words, the solubility of a gas dissolved from a gaseous mixture (air in the present case) is directly proportional to the partial pressure of the gas; the proportionality constant, k_H, is called the *Henry's law constant*. (Some authors define the Henry's law constant as the reciprocal of the k_H used here.) To evaluate k_H from the data, note that when pure oxygen is equilibrated with water at a total pressure of 760 torr,

$$P(O_2) = (760\ torr) - (\text{vapor pressure of water})$$

Then, from the data,

$$k_H(O_2) = \frac{\text{solubility of } O_2}{P(O_2)} = \frac{0.043\ g/L}{760\ torr - v.p.}$$

$$k_H(N_2) = \frac{\text{solubility of } N_2}{P(N_2)} = \frac{0.019\ g/L}{760\ torr - v.p.}$$

When water is exposed to air at a total pressure of 760 torr,

$$P(O_2) = (0.20)(760\ torr - v.p.) \qquad P(N_2) = (0.80)(760\ torr - v.p.)$$

Hence

$$\text{Solubility of } O_2 \text{ from air} = k_H(O_2) \times P(O_2)$$

$$= \left(\frac{0.043\ g/L}{760\ torr - v.p.}\right)(0.20)(760\ torr - v.p.) = 0.008\ 6\ g/L$$

Similarly, the solubility of N_2 from air is $(0.80)(0.019\ g/L) = 0.015\ g/L$.

14.12. A gaseous mixture of hydrogen and oxygen contains 70% hydrogen and 30% oxygen by volume. If the gas mixture at a pressure of 2.5 atm (excluding the vapor pressure of water) is allowed to saturate water at $20\,°C$, the water is found to contain 31.5 cm³ (S.T.P.) of hydrogen per liter. Find the solubility of hydrogen (reduced to S.T.P.) at $20\,°C$ and 1 atm partial pressure of hydrogen.

Since the volume of a gas at S.T.P. depends only on the mass, the volume of the dissolved gas (reduced to S.T.P.) is proportional to the partial pressure of the gas.

$$\text{Partial pressure of hydrogen} = (0.70)(2.5 \text{ atm}) = 1.75 \text{ atm}$$

$$\text{Solubility of } H_2 \text{ at } 20\,^\circ\text{C and 1 atm} = \left(\frac{1.00 \text{ atm}}{1.75 \text{ atm}}\right)(31.5 \text{ cm}^3/\text{L}) = 18.0 \text{ cm}^3 \text{ (S.T.P.)}/\text{L}$$

LAW OF DISTRIBUTION: EXTRACTION

14.13. (a) An aqueous solution of iodine, of volume 25 cm^3 and containing 2 mg of iodine, is shaken with 5 cm^3 of CCl$_4$, and the CCl$_4$ is allowed to separate. Given that the solubility of iodine per unit volume is 85 times greater in CCl$_4$ than in water at the temperature of the experiment and both saturated solutions may be considered to be "dilute," calculate the quantity of iodine remaining in the water layer.   (b) If a second extraction is made of the water layer using another 5 cm^3 of CCl$_4$, calculate the quantity of iodine remaining after the second extraction.

(a)   Let x = number of milligrams of iodine in H$_2$O layer at equilibrium

$2 - x$ = number of milligrams of iodine in CCl$_4$ layer at equilibrium

The concentration of iodine in the water layer will be $x/25$ (mg per cm^3 of water), and the concentration of iodine in the CCl$_4$ layer will be $(2 - x)/5$ (mg per cm^3 of CCl$_4$). Hence

$$\frac{\text{conc. I}_2 \text{ in CCl}_4}{\text{conc. I}_2 \text{ in H}_2\text{O}} = \frac{85}{1} \quad \text{or} \quad \frac{(2 - x)/5}{x/25} = \frac{85}{1} \quad \text{or} \quad \frac{2 - x}{x} = 17$$

Solving, $x = 0.11$ mg iodine.

Note: Any volumetric concentration units could have been used in this problem, so long as the same units were used in both numerator and denominator. The choice of mg/cm^3 was the most convenient for this particular case.

(b)   Let y = number of milligrams of iodine in H$_2$O layer after second extraction

$0.11 - y$ = number of milligrams of iodine in CCl$_4$ layer after second extraction

The concentration of iodine in the water layer will be $y/25$, and the concentration in the CCl$_4$ layer will be $(0.11 - y)/5$. Hence

$$\frac{\text{conc. I}_2 \text{ in CCl}_4}{\text{conc. I}_2 \text{ in H}_2\text{O}} = \frac{85}{1} \quad \text{or} \quad \frac{(0.11 - y)/5}{y/25} = \frac{85}{1} \quad \text{or} \quad \frac{0.11 - y}{y} = 17$$

Solving, $y = 0.006\,1$ mg iodine.

Supplementary Problems

14.14. A solution containing 6.35 g of a nonelectrolyte dissolved in 500 g of water freezes at $-0.465\,^\circ$C. Determine the molecular weight of the solute.

Ans.   50.8

14.15. A solution containing 3.24 g of a nonvolatile nonelectrolyte and 200 g of water boils at 100.130 °C at 1 atm. What is the molecular weight of the solute?

Ans.   63.9

14.16. Calculate the freezing point and the boiling point at 1 atm of a solution containing 30.0 g cane sugar (molecular weight 342) and 150 g water.

   Ans. $-1.09\ °C$, $100.300\ °C$

14.17. If glycerin, $C_3H_5(OH)_3$, and methyl alcohol, CH_3OH, sold at the same price per pound, which would be cheaper for preparing an antifreeze solution for the radiator of an automobile?

   Ans.   methyl alcohol

14.18. How much ethyl alcohol, C_2H_5OH, must be added to 1 L of water so that the solution will not freeze at $-4\ °F$?

   Ans.   495 g

14.19. If the radiator of an automobile contains 12 L of water, how much would the freezing point be lowered by the addition of 5 kg of Prestone [glycol, $C_2H_4(OH)_2$]? How many kilograms of Zerone (methyl alcohol, CH_3OH) would be required to produce the same result? Assume 100% purity.

   Ans. $12.5\ °C$, 2.6 kg

14.20. What is the freezing point of a 10% (by weight) solution of CH_3OH in water?

   Ans. $-6.5\ °C$

14.21. When 10.6 g of a nonvolatile substance is dissolved in 740 g of ether, its boiling point is raised $0.284\ °C$. What is the molecular weight of the substance? Molal boiling-point constant for ether is $2.11\ °C\cdot kg/mol$.

   Ans.   106

14.22. The freezing point of a sample of naphthalene was found to be $80.6\ °C$. When 0.512 g of a substance is dissolved in 7.03 g naphthalene, the solution has a freezing point of $75.2\ °C$. What is the molecular weight of the solute? The molal freezing-point constant of naphthalene is $6.80\ °C\cdot kg/mol$.

   Ans.   92

14.23. Pure benzene freezes at $5.45\ °C$. A solution containing 7.24 g of $C_2Cl_4H_2$ in 115.3 g of benzene was observed to freeze at $3.55\ °C$. What is the molal freezing-point constant of benzene?

   Ans. $5.08\ °C\cdot kg/mol$

14.24. What is the osmotic pressure at $0\ °C$ of an aqueous solution containing 46.0 g of glycerin ($C_3H_8O_3$) per liter?

   Ans.   11.2 atm

14.25. A solution of crab hemocyanin, a pigmented protein extracted from crabs, was prepared by dissolving 0.750 g in 125 cm³ of an aqueous medium. At $4\ °C$ an osmotic pressure rise of 2.6 mm of the solution was observed. The solution had a density of $1.00\ g/cm^{-3}$. Determine the molecular weight of the protein.

   Ans. 5.4×10^5

14.26. The osmotic pressure of blood is 7.65 atm at $37\ °C$. How much glucose, $C_6H_{12}O_6$, should be used per liter for an intravenous injection that is to have the same osmotic pressure as blood?

   Ans.   54.3 g/L

14.27. The vapor pressure of pure water at $26\ °C$ is 25.21 torr. What is the vapor pressure of a solution which contains 20.0 g glucose, $C_6H_{12}O_6$, in 70 g water?

   Ans.   24.51 torr

14.28. The vapor pressure of pure water at 25 °C is 23.76 torr. The vapor pressure of a solution containing 5.40 g of a nonvolatile substance in 90 g water is 23.32 torr. Compute the molecular weight of the solute.

   Ans.   57

14.29. Ethylene bromide, $C_2H_4Br_2$, and 1,2-dibromopropane, $C_3H_6Br_2$, form a series of ideal solutions over the whole range of composition. At 85 °C the vapor pressures of these two pure liquids are 173 torr and 127 torr, respectively. (*a*) If 10.0 g of ethylene bromide is dissolved in 80.0 g of 1,2-dibromopropane, calculate the partial pressure of each component and the total pressure of the solution at 85 °C. (*b*) Calculate the mole fraction of ethylene bromide in the vapor in equilibrium with the above solution. (*c*) What would be the mole fraction of ethylene bromide in a solution at 85 °C equilibrated with a 50:50 mole mixture in the vapor?

   Ans.   (*a*) ethylene bromide, 20.5 torr; 1,2-dibromopropane, 112 torr; total, 132 torr;   (*b*) 0.155;   (*c*) 0.42

14.30. At 40 °C the vapor pressure, in torr, of methyl alcohol–ethyl alcohol solutions is represented by the equation

$$P = 119x(CH_3OH) + 135$$

where $x(CH_3OH)$ is the mole fraction of methyl alcohol. What are the vapor pressures of the pure components at this temperature?

   Ans.   methyl alcohol, 254 torr; ethyl alcohol, 135 torr

14.31. A 0.100 molal solution of $NaClO_3$ freezes at $-0.343\,3$ °C. What would you predict for the boiling point of this aqueous solution at 1 atm pressure? At 0.001 molal concentration of this same salt, the electrical interferences between the ions no longer exist, because the ions are, on the average, too far apart from each other. Predict the freezing point of this more dilute solution.

   Ans.   100.095 °C, $-0.003\,7$ °C

14.32. The molecular weight of a newly synthesized organic compound was determined by the method of *isothermal distillation*. In this procedure two solutions, each in an open calibrated vial, are placed side by side in a closed chamber. One of the solutions contained 9.3 mg of the new compound, the other 13.2 mg of azobenzene (molecular weight 182). Both were dissolved in portions of the same solvent. During a period of three days of equilibration, solvent distilled from one vial into the other until the same partial pressure of solvent was reached in the two vials. At this point the distillation of solvent stopped. Neither of the solutes distilled at all. The volumes of the two solutions at equilibrium were then read on the calibration marks of the vials. The solution containing the new compound occupied 1.72 cm³ and the azobenzene solution occupied 1.02 cm³. What is the molecular weight of the new compound? The mass of solvent in solution may be assumed to be proportional to the volume of the solution.

   Ans.   76

14.33. If 29 mg of N_2 dissolves in 1 L of water at 0 °C and 760 torr N_2 pressure, how much N_2 will dissolve in 1 L of water at 0 °C and 5.00 atm N_2 pressure?

   Ans.   145 mg

14.34. At 20 °C and 1.00 atm partial pressure of hydrogen, 18 cm³ of hydrogen, measured at S.T.P., dissolves in 1 L of water. If water at 20 °C is exposed to a gaseous mixture having a total pressure of 1 400 torr (excluding the vapor pressure of water) and containing 68.5 % H_2 by volume, find the volume of H_2, measured at S.T.P., which will dissolve in 1 L of water.

   Ans.   23 cm³

14.35. A liter of CO_2 gas at 15 °C and 1.00 atm dissolves in a liter of water at the same temperature when the pressure of CO_2 is 1.00 atm. Compute the molar concentration of CO_2 in a solution over which the partial pressure of CO_2 is 150 torr at this temperature.

   Ans.   0.008 3 M

14.36. (a) The solubility of iodine per unit volume is 200 times greater in ether than in water at a particular temperature. If an aqueous solution of iodine, 30 cm^3 in volume and containing 2.0 mg of iodine, is shaken with 30 cm^3 of ether and the ether is allowed to separate, what quantity of iodine remains in the water layer? (b) What quantity of iodine remains in the water layer if only 3 cm^3 of ether is used? (c) How much iodine is left in the water layer if the extraction in (b) is followed by a second extraction, again using 3 cm^3 of ether? (d) Which method is more efficient, a single large washing or repeated small washings?

   Ans.   (a) 0.010 mg;   (b) 0.095 mg;   (c) 0.0045 mg

14.37. The ratio of the solubility of stearic acid per unit volume of *n*-heptane to that in 97.5% acetic acid is 4.95. How many extractions of a 10-cm^3 solution of stearic acid in 97.5% acetic acid with successive 10-cm^3 portions of *n*-heptane are needed to reduce the residual stearic acid content of the acetic acid layer to less than 0.5% of its original value?

   Ans.   3

14.38. One method of purifying penicillin is by extraction. The distribution coefficient for penicillin G between isopropyl ether and an aqueous phosphate medium is 0.34 (lower solubility in the ether). The corresponding ratio for penicillin F is 0.68. A preparation of penicillin G has penicillin F as a 10.0% impurity. (a) If an aqueous phosphate solution of this preparation is extracted with an equal volume of isopropyl ether, what will be the % recovery of the initial G in the residual aqueous-phase product after one extraction, and what will be the % impurity in this product? (b) Compute the same two quantities for the product remaining in the aqueous phase after a second extraction with an equal volume of ether.

   Ans.   (a) 75% recovery, 8.1% impurity;   (b) 56% recovery, 6.6% impurity

Chapter 15

Organic and Biochemistry

INTRODUCTION

Compounds of the element carbon, with a few exceptions, are called *organic* compounds; others are called *inorganic*. Originally, the distinction was based on the mistaken notion that living matter, organic, was chemically different from nonliving matter, inorganic. We now know that organic compounds, which constitute most of all known compounds, can be synthesized not only by the natural processes of living organisms but also in the laboratory by processes that do not require the intervention of biological events.

The principles of chemistry as presented in this text apply equally of course to inorganic, organic, and biochemistry. This chapter is considered necessary, however, because the great variety of organic compounds demands, in addition to a consideration of their most important chemical reactions, a special nomenclature and special attention to the details of isomerism. Biochemistry will be introduced by way of examples.

NOMENCLATURE

Because carbon atoms readily bond to each other, organic compounds typically have a large number of atoms per molecule and their names must indicate not only the number of atoms of each kind but the pattern of connections. We begin with compounds of carbon and hydrogen, called *hydrocarbons*. If there are no multiple bonds these are termed *alkanes*, and if there are no rings they must all have the empirical formula C_nH_{n+2} in order to satisfy the rules for bonding as in Chapter 9. The name of an organic compound always provides the number of carbon atoms explicitly. The root of the name indicates the number of carbon atoms, *meth* meaning one, *eth* two, *prop* three, *but* four, *pent* five, *hex* six, *hept* seven, and so forth. The number of atoms of other elements is often implied. For instance, in cyclic compounds each ring of carbon atoms reduces the number of hydrogens by 2; each double bond (in compounds called *olefins* or *alkenes*) likewise by 2; and each triple bond (in *alkynes*) by 4.

In older less systematic nomenclature, still encountered, the root counted all the carbons. Thus "butane" meant C_4H_{10}, the suffix *ane* indicating an alkane. To indicate that the structure was branched rather than straight the prefix *iso* was used. Clearly such a scheme is inadequate for five or more carbons since various branched structures are possible.

Modern organic nomenclature, as prescribed by the International Union of Pure and Applied Chemistry (IUPAC), avoids such ambiguities. For noncyclic alkanes the root gives the number of carbons in the longest chain. Each branch is described by a prefix indicating the number of carbons in the branch, and a numeral indicating the point of connection. (Numbering starts at the end of the chain which produces the smallest numerals.)

EXAMPLE 1 Before the IUPAC rules were developed, both of the following would have been called "hexane." Give the IUPAC names.

2-Methylpentane   2,3-Dimethylbutane

The system is extended to include alkenes by changing the suffix from -ane to -ene, and to alkynes by the suffix -yne. A numeral preceding the root of the name gives the location of the multiple bond, which is always in the lower numbered end of the molecule.

EXAMPLE 2   Give the IUPAC names of the hydrocarbons below. (Notice that the bonds to the individual H atoms have been omitted for simplicity.)

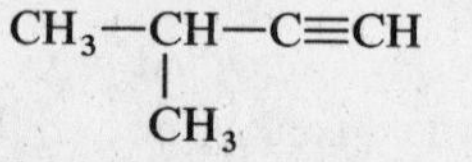

$$CH_3—CH—C≡CH \qquad\qquad CH_2=CH—CH=CH_2$$
$$\quad\;\; |$$
$$\quad\;\; CH_3$$

3-Methyl-1-butyne   1,3-Butadiene

The prefix cyclo- appears before the root of the name in ring compounds, except for those involving the benzene ring (see Problem 9.8), which form a special category of compounds called *aromatic*. The opposite of *aromatic* is *aliphatic*. Further modifications of the system when elements other than carbon and hydrogen are involved will be noted below when *functional groups* are discussed.

ISOMERISM

Recalling the principles and definitions in Chapter 9, isomers are compounds with the same number of toms of each kind per molecule, but that are different substances because of differences in molecular structure. There are three classes of isomers; (*a*) structural, (*b*) geometric, and (*c*) optical.

In (*a*) structural isomerism, molecules differ in the sequence of atomic attachments, i.e., in the structural skeleton of the molecule. For example, the compounds in Example 1 are *structural isomers*. The IUPAC names clearly distinguish between structural isomers.

Geometric isomers differ in the three-dimensional shape of the molecule, which is rigidly fixed because of the presence of a double bond or a ring. In simple cases geometrical isomers can be distinguished in their names by the prefixes *cis-* or *trans-* according to whether two groups are close (same side of the double bond) or far apart (opposite sides of the double bond).

Optical isomers are nonsuperimposable mirror images of one another. In most cases they arise from the presence in the molecule of a tetrahedral carbon atom with four different attachments. Such an atom is termed *asymmetric* or *chiral*. Two optical isomers can be distinguished in their names by the prefixes *dextro-* or *levo-* according to whether a solution of the compound rotates a beam of polarized light to the right or to the left. The abbreviations D- and L- are commonly used for dextro- and levo-, and in a newer system the abbreviations are *R*- (for rectus) and *S*- (for sinister).

EXAMPLE 3   (*a*) Draw the geometric isomers of 2-butene.   (*b*) Draw the optical isomers of 1-bromo 1-chloro-ethane.

(*a*)   Because of the double bond all atoms except the six end-hydrogens are constrained to one plane, which is chosen as the plane of the paper.

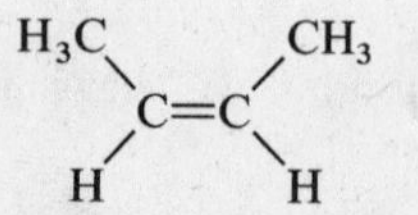

cis 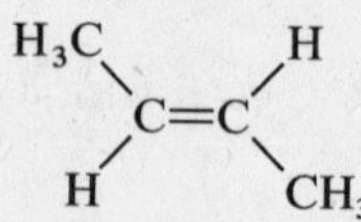   trans

(*b*)   If both carbons and the bromine are placed in the plane of the paper the chlorine and hydrogen atoms are on opposite sides of the plane and their bonds to carbon form 109°28′ angles to the other bonds of that carbon atom. The chlorine is shown above and the hydrogen below the plane of the paper.

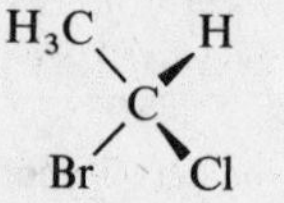

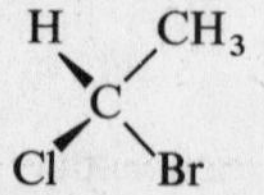

FUNCTIONAL GROUPS

Alkanes are comparatively unreactive and chemically uninteresting. But if a multiple bond or some other kind of atom is introduced, characteristic properties and reactions result. These properties depend rather little on the number or arrangement of the carbon atoms in the rest of the molecule. The group of atoms that determine the characteristic properties of the compound is called a *functional group*. Learning the properties and reactions of the functional groups greatly simplifies the study of organic chemistry. Table 15-1 lists some of the more common functional groups and illustrates how the IUPAC name is constructed, usually by using a characteristic suffix. In the case of acids the extra word "acid" is added in addition to the suffix "-oic." The word "ether" also appears separate from the rest of the name. Esters have two-part names indicating the structures of the two parts of the molecule joined by the ester group.

In naming aldehydes, acids, and esters the roots for one and two carbons are usually "form-" and "acet-," as in "formaldehyde" and "acetic acid," these historical names being more commonly used than the strict IUPAC "methanal" and "ethanoic acid." Departures from IUPAC nomenclature are common

Table 15-1

Structure	Group Name	Example	Name of Example
—C=C—	Alkene, double bond	$CH_3CH=CHCH_3$	2-Butene
—C≡C—	Alkyne, triple bond	$CH_3—C≡CH$	1-Propyne
—C—X (X = F, Cl, Br, I)	Halide	CH_3CHCH_3 with F	2-Fluoropropane
—C—OH	Alcohol, hydroxyl	CH_3OH	Methanol
—C(=O)—H	Aldehyde	CH_3CHO	Ethanal
—C(=O)—OH	Carboxylic acid	$CH_3CHCOOH$ with CH_3	2-Methylpropanoic acid
—C—C(=O)—C—	Ketone	$CH_3CH_2CH_2CCH_3$ (=O)	2-Pentanone
—C—O—C—	Ether	$CH_3OCH_2CH_3$	Methyl ethyl ether
—C—N(H)—H	Amine	$CH_3CH_2NH_2$	Ethylamine
—C(=O)—NH_2	Amide	$CH_3CH_2CONH_2$	Propanamide
—C—C(=O)—O—C—	Ester	$CH_3CH_2CH_2C(=O)—O—CH_2CH_3$	Ethyl butanoate

for very common substances, and fortunately they rarely can be misunderstood, such as "ethyl alcohol" for "ethanol."

PROPERTIES AND REACTIONS

Alkanes

Alkanes are colorless, water-insoluble compounds the simplest of which have rather low boiling points and are commonly used as fuels. In combustion reactions the products are CO_2 and H_2O along with some CO if the amount of oxygen is insufficient. The reaction of alkanes with halogens can be controlled, resulting in the *substitution* of a halogen for a hydrogen without disruption of the carbon skeleton.

$$CH_3CH_2CH_3 + Cl_2 \rightarrow CH_3CH_2CH_2Cl + HCl$$

Alkenes

Alkenes are physically similar to alkanes and are also easily combusted. However, they are much more likely to undergo *addition* reactions with halogens rather than *substitution*. Addition reactions occur with a wide variety of reagents.

$$CH_3CH{=}CH_2 + Cl_2 \rightarrow CH_3\underset{|}{C}H\underset{|}{C}H_2$$
$$\qquad\qquad\qquad\qquad\quad Cl\ \ Cl$$

$$CH_3CH{=}CH_2 + HCN \rightarrow CH_3\underset{|}{C}HCH_3$$
$$\qquad\qquad\qquad\qquad\qquad CN$$

A very important reaction, especially in the plastics industry, is *addition polymerization*, in which alkene molecules add to themselves to form long chains, for example, from $CH_3CH{=}CH_2$:

$$\text{etc.}\quad {-}\underset{|}{C}HCH_2\underset{|}{C}HCH_2\underset{|}{C}HCH_2\underset{|}{C}HCH_2{-}\quad \text{etc.}$$
$$\qquad\quad CH_3\quad\ CH_3\quad\ CH_3\quad\ CH_3$$

Halides

Halogenated compounds are water-insoluble and have much higher boiling points than the corresponding hydrocarbons. They find wide use as solvents, cleaning fluids, insecticides, and (when polymerized) plastics. They are also important starting materials for the synthesis of other compounds since the halogen can be *replaced* with OH (using NaOH) and other groups.

Alcohols

Alcohols are pleasant-smelling liquids, owing both their water solubility and high boiling point to extensive hydrogen bonding. Two very important *condensation* reactions are (a) with another alcohol to form an ether, and (b) with an acid to form an ester. A *condensation* reaction in general is one in which two molecules become joined along with the elimination of water or some other small molecule. The reaction of an ester with water to produce an alcohol and an acid, i.e., the reverse of esterification (the ester-forming condensation), is called *hydrolysis*. If the ester happens to be a natural fat, the process is called *saponification* since the sodium salt of the acid formed is a *soap*. The alcohol listed in Table 15-1, methanol, is also called methyl alcohol. The CH_3 grouping is called the *methyl* group. Similarly other groups derived from any alkane by the loss of one hydrogen atom are called *alkyl* groups. An alkyl group is not a compound by itself but retains its identity in most types of compounds, including halides, alcohols, and ethers.

Aldehydes and ketones

Aldehydes and ketones have much lower boiling points and are less water-soluble than the corresponding alcohols. In all cases, whether alcohols, aldehydes, ketones, or acids, water solubility gradually decreases as the number of carbon atoms increases.

An alcohol can be considered the product of the first step in the oxidation of a hydrocarbon, in the sense that the overall result can be considered to be the insertion of one O atom between C and H. If another O atom is inserted on the same C, and H_2O eliminated, the result is an aldehyde (if the C was at the end of chain) or a ketone (if in the middle). The third step, inserting the last possible O atom on the C of an aldehyde, results in an acid.

Acids

Carboxylic acids are quite soluble in water, in which they ionize slightly (typically a few percent, depending on concentration), and undergo typical acid-base reactions. Beside the condensation reaction with alcohols, acids condense with ammonia or with amines to form amides.

$$CH_3C{-}OH + CH_3CH_2CH_2NH_2 \longrightarrow H_2O + CH_3C{-}N{-}CH_2CH_2CH_3$$

Acetic acid   Propylamine   n-Propylacetamide

Salts of carboxylic acids are fully ionized in the solid state and in aqueous solution. Salts of fatty acids (from the hydrolysis of fats) are soaps.

Amines and amides

Amines can be considered derivatives of ammonia, NH_3, in which one or more H atoms is replaced by an alkyl group. The example in Table 15-1 is called a primary amine; replacement of two H atoms gives a secondary amine, and all three a tertiary amine. Amines are somewhat soluble in water in which, like ammonia, they abstract a proton leaving the solution basic. They dissolve readily in strong acid to form salts analogous to ammonium salts.

The condensation of an acid with ammonia or an amine gives an amide. The example in Table 15-1 resulted from the condensation of propanoic acid with ammonia; the example in the preceding section involved the condensation of the same acid with a primary amine. A very important class of biochemical molecules are the amino acids which join to form protein molecules by the condensation of the amine group of one molecule to the acid group of the next. The amide linkage in this case has the special name "peptide."

Ethers and Esters

When an alcohol is heated under proper conditions with concentrated sulfuric acid, water is eliminated by the condensation of two molecules of alcohol and the two alkyl groups are joined through an oxygen atom to form an ether. The product formed from ethanol, diethyl ether, is the familar "ether" known for its use as a total anesthetic. Mixed ethers, such as methylethyl ether, may be formed by a condensation reaction of two different alcohols. Ethers are good organic solvents, rather insoluble in water.

As discussed under alcohols, esters are the condensation products of alcohols and acids. They are also good organic solvents, insoluble in water. Many found in nature are flavorful and sweet smelling; amyl acetate is the fragrant component of "banana oil" ("amyl" is a traditional name for the 5-carbon group). Natural fats and oils are esters of the trifunctional alcohol, gycerol, and long-chain acids called "fatty acids."

Aromatic compounds

Benzene and compounds containing the benzene ring are termed "aromatic." Aromatic compounds have somewhat different chemical properties than their aliphatic counterparts. For instance the hydroxyl group attached to a benzene ring is a weak acid (though much weaker than a carboxylic acid). The three alternating double bonds of the ring are not readily saturated. If benzene is treated with chlorine (in the dark) substitution rather than addition will occur.

$$\text{(benzene)} + Cl_2 \rightarrow \text{(chlorobenzene)}-Cl + HCl$$

Functional groups on aliphatic side chains attached to aromatic rings behave in a manner typical of aliphatic compounds.

Multifunctional molecules

Many molecules contain more than one functional group (either the same or different). In such cases condensation reactions involving two or more groups per molecule can lead to the formation of polymers, as illustrated by the formation of proteins from amino acids. Another example is the synthesis of polyester fiber.

$$HO—CH_2CH_2—OH + HO—\overset{\overset{O}{\|}}{C}\!\!-\!\!\bigcirc\!\!-\!\!\overset{\overset{O}{\|}}{C}—OH \rightarrow H_2O +$$

1,2-Ethanediol   Terephthalic acid

$$\text{etc.} \quad —O—CH_2CH_2—O—\overset{\overset{O}{\|}}{C}\!\!-\!\!\bigcirc\!\!-\!\!\overset{\overset{O}{\|}}{C}—O—CH_2CH_2—O—\overset{\overset{O}{\|}}{C}\!\!-\!\!\bigcirc\!\!-\!\!\overset{\overset{O}{\|}}{C}— \quad \text{etc.}$$

Dacron polyester

(The above equation is not balanced. Thousands of molecules of each reagent combine, eliminating a like number of water molecules, one for each ester linkage.)

BIOCHEMISTRY

Certain aspects of biochemistry which distinguish it from nonliving chemistry are listed below.

1. Molecules may be very large, but unlike a synthetic polymer, a biopolymer has a fixed number and sequence of monomer units, and often a fixed three-dimensional shape as well.
2. Classes of molecules are based primarily on structure, but also on their function in the living cell.
3. Reactions occur at a modest temperature; extremely sensitive catalysts, called enzymes, are involved; and the energy to drive one reaction may come from the energy released in some coupled reaction.
4. Throughout the animal and plant kingdom there is little or no difference in the structures of the molecules chosen by nature for a given biological function.

Some of the principal classes of compounds are (a) proteins, (b) carbohydrates, (c) fats, and (d) nucleic acids. A brief account of their structures and functions follows.

(a) Proteins are polymers of amino acids joined by peptide linkages as discussed earlier. They form the bulk of living tissue. Enzymes, those very selective and powerful catalysts, are also proteins, which may contain a prosthetic group based on some metal atom. Some proteins serve special functions, such as hemoglobin and myoglobin which are oxygen carriers.

(b) The simplest carbohydrates are the sugars, relatively small molecules typically of four, five or six carbon atoms which have hydroxyl and aldehyde (or ketone) functional groups. They function as fuels or as building blocks for polymeric materials. Plants contain polymers of the sugars called starch and cellulose formed by ether linkages. They serve to store energy and to provide structural support.

(c) Fats are esters of glycerol and the fatty acids. In plants many of the fats are unsaturated (containing double bonds) and liquid, and thus are called oils. Fats also function as fuels, but for long-term energy storage, rather than the short-term delivery of energy provided by sugars. Fats are part of a class of substances called lipids based on their insolubility in water and solubility in ether. Among other lipids are cholesterol and cell-membrane materials.

(d) Nucleic acids are very high polymers joined by ester linkages between phosphoric acid and sugar monomer units. These sugars carry nitrogenous side chains called purine or pyrimidine bases. The sequence of bases constitutes the genetic code of the individual living organism. The nucleic acid molecule directs the sequencing of amino acids in the synthesis of protein, as well as its own reproduction.

Solved Problems

NOMENCLATURE

15.1. Name each of the following compounds according to the IUPAC rules:

(a) $CH_3CHCH_2CHCH_3$
        | |
       CH_3 CH_2CH_3

(b) [cyclohexadiene ring structure with CH_2, H_2C, CH, HC, CH, CH]

(c) $ClCH_2CH_2\overset{O}{\overset{\|}{C}}CH_2CH_3$

(d) $CH_3CHCH_2—O—\overset{O}{\overset{\|}{C}}—CH_3$
         |
        CH_3

(e) $CH_3NHCH_2CH_2CH_3$

(f) $CH_2CH_2CH_2—O—\overset{}{C}—C_{17}H_{35}$
     | | $\overset{\|}{O}$
    OH   OH

(a) Don't be deceived by the manner in which the structure is drawn. The longest chain of linearly connected carbon atoms is six carbon atoms and numbering must start at the left to minimize the numbers of the points of substitution: 2,4-dimethylhexane.

(b) The ring carbons are numbered so as to minimize the sum of the numerals: 1,3-cyclohexadiene.

(c) The chloro group is treated as a prefix but the carbonyl oxygen is a suffix: 1-chloro-3-pentanone.

(d) The compound is an ester. Contrary to the example in Table 15-1 the alcohol residue is to the left. In either case the alcohol part is always named first: 2-methylpropyl ethanoate, or 2-methylpropyl acetate.

(e) The compound is a secondary amine: methylpropylamine.

(f) The fatty acids have special names. The saturated C_{18} acid is called "stearic": 2,3-dihydroxypropyl stearate. The alcohol residue is glycerol. Since only one hydroxyl is esterified this type of ester is called a monoglyceride and the compound may be named monoglyceryl stearate. It is valuable in food processing as an emulsifier.

ISOMERISM

15.2. Identify all the isomers with the formula C_5H_{10}, and give their IUPAC names.

Consider first all the possible structural isomers. The cyclic structures will be saturated; the open-chain structures will have one double bond.

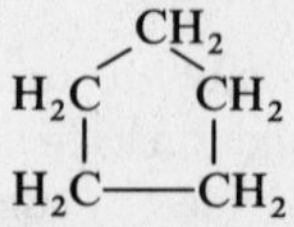

(a) Cyclopentane 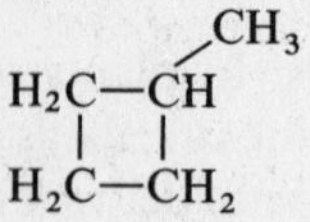   (b) Methylcyclobutane 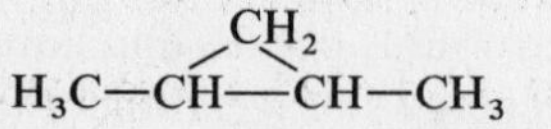   (c) 1,2-Dimethylcyclopropane

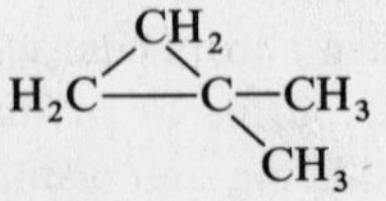

(d) 1,1-Dimethylcyclopropane 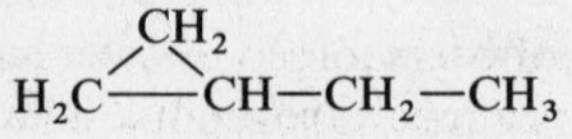   (e) Ethylcyclopropane

$$CH_2{=}CH{-}CH_2{-}CH_2{-}CH_3 \qquad CH_3{-}CH{=}CH{-}CH_2{-}CH_3$$

(f) 1-Pentene   (g) 2-Pentene

$$CH_2{=}C{-}CH_2{-}CH_3 \qquad CH_3{-}C{=}CH{-}CH_3 \qquad CH_2{=}CH{-}CH{-}CH_3$$
$$\quad\ \ |\qquad\qquad\qquad\qquad |\qquad\qquad\qquad\qquad\qquad\quad |$$
$$\quad CH_3 \qquad\qquad\qquad\qquad CH_3 \qquad\qquad\qquad\qquad\qquad CH_3$$

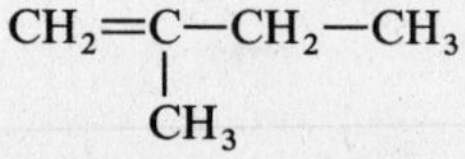

(h) 2-Methyl-1-butene 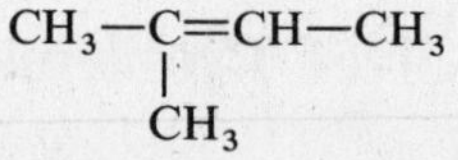   (i) 2-Methyl-2-butene 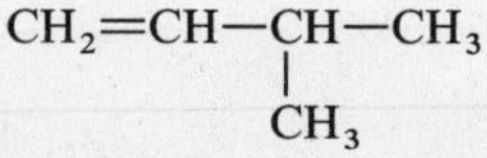   (j) 3-Methyl-1-butene

Next consider *cis-trans* isomerism. The only case among the alkenes is (g). The two structures are shown below.

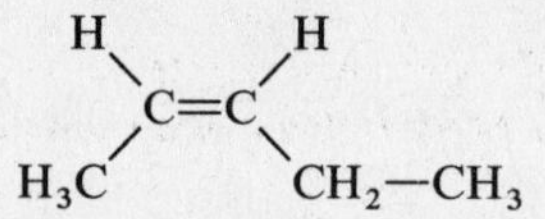

cis 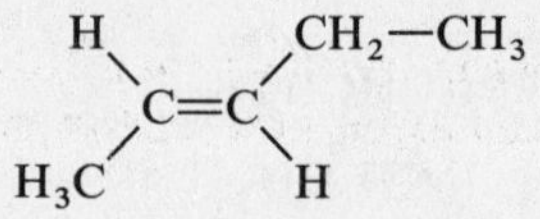   *trans*

However, structure (c) also shows *cis-trans* isomerism. The three carbons in the ring define a plane (the plane of the paper). Either both hydrogens of the CH groups are on the same side of this plane or one is above and the other below.

Structure (c) is also the only one which has a chiral carbon atom (the carbons of the two CH groups). However the *cis* isomer has a plane of symmetry and thus cannot have an optical isomer. Only the *trans* isomer has an optical isomer.

15.3. Draw the structures of all the structural isomers with the formula C_3H_6O. Point out any that have geometrical or optical isomers.

$$CH_3{-}CH{=}CH{-}OH \qquad CH_2{=}CH{-}CH_2{-}OH \qquad CH_3{-}\overset{\overset{\textstyle OH}{|}}{C}{=}CH_2$$

cis-trans forms

$$CH_3{-}\overset{\overset{\textstyle O}{\|}}{C}{-}CH_3 \qquad CH_3{-}CH_2{-}\overset{\overset{\textstyle O}{\|}}{C}{-}H \qquad CH_2{=}CH{-}O{-}CH_3$$

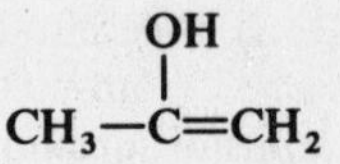

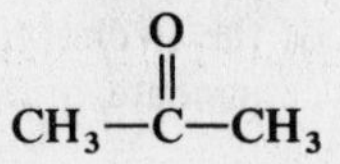

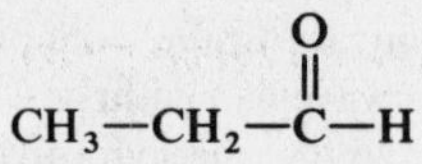

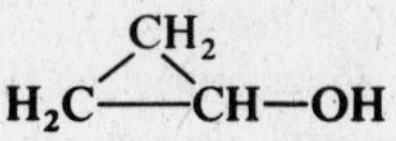

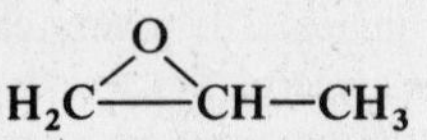

CH carbon is chiral.

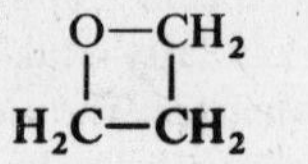

FUNCTIONAL GROUPS AND REACTIONS

15.4. Draw the structure of the organic product resulting from each of the reactions below.
   (a)   1-Propanol in dehydrating medium
   (b)   1-Propanol in mildly oxidizing medium
   (c)   2-Propanol in mildly oxidizing medium
   (d)   2-Propanol plus butanoic acid in dehydrating medium
   (e)   1-Propanol plus metallic sodium

   (a)   A condensation results in elimination of water and formation of an ether:

$$CH_3CH_2CH_2-O-CH_2CH_2CH_3$$

   (b)   One stage of oxidation of a hydroxyl group produces a carbonyl group ($-C=O$). When the alcohol is at the end of the chain (primary alcohol) the product is an aldehyde:

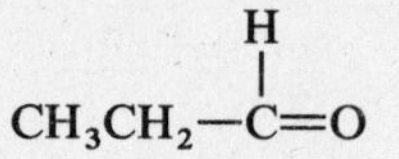

$$CH_3CH_2-\overset{\overset{\displaystyle H}{|}}{C}=O$$

   (c)   When the alcohol is not at the end (secondary alcohol in this case) the product is a ketone:

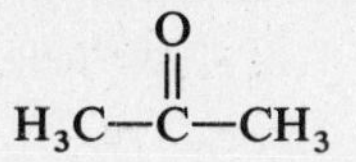

$$H_3C-\overset{\overset{\displaystyle O}{\|}}{C}-CH_3$$

   (d)   A condensation results in elimination of water and formation of an ester:

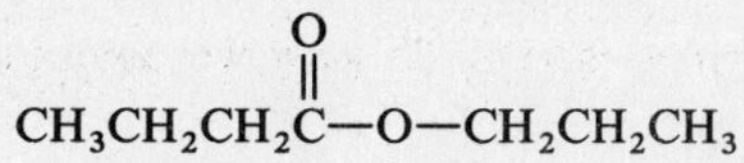

$$CH_3CH_2CH_2\overset{\overset{\displaystyle O}{\|}}{C}-O-CH_2CH_2CH_3$$

   (e)   Considering the alcohol to be a derivative of water, one expects the sodium to displace the hydrogen of the hydroxyl group. One product is H_2; the other is the sodium salt of the alcohol:

$$CH_3CH_2CH_2O^-\ Na^+$$

15.5. Describe the products of the following reactions, using structural formulas:
   (a)   Vinyl chloride (alternate name: chloroethene) is treated with a polymerization catalyst.
   (b)   The amino acid alanine (2-aminopropanoic acid) is treated with the appropriate polymerase enzyme and energy source molecules to cause it to polymerize.

   (a)   The product is a saturated high polymer, the double bonds having been used up in joining the monomer molecules together:

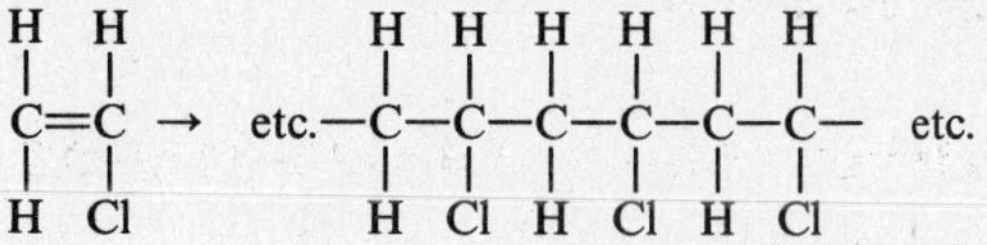

$$\underset{\overset{|}{H}\ \ \overset{|}{Cl}}{\overset{\overset{H}{|}\ \ \overset{H}{|}}{C=C}} \rightarrow etc.-\underset{\overset{|}{H}\ \overset{|}{Cl}\ \overset{|}{H}\ \overset{|}{Cl}\ \overset{|}{H}\ \overset{|}{Cl}}{\overset{\overset{H}{|}\ \overset{H}{|}\ \overset{H}{|}\ \overset{H}{|}\ \overset{H}{|}\ \overset{H}{|}}{C-C-C-C-C-C}}-\ etc.$$

   (b)   Water is condensed out, H from the amino group of one molecule and OH from the acid of the next:

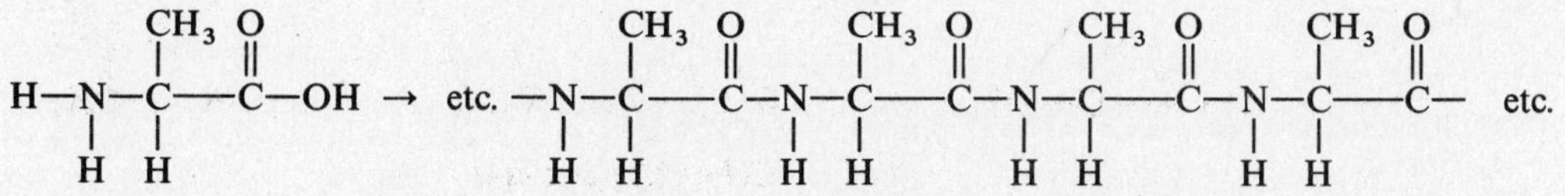

$$H-\underset{\overset{|}{H}}{\overset{\overset{CH_3}{|}}{N}}-\underset{\overset{|}{H}}{\overset{|}{C}}\overset{\overset{O}{\|}}{C}-OH \rightarrow etc.\ -\overset{\overset{CH_3}{|}}{N}-\overset{|}{C}\overset{\overset{O}{\|}}{C}-\overset{\overset{CH_3}{|}}{N}-\overset{|}{C}\overset{\overset{O}{\|}}{C}-\overset{\overset{CH_3}{|}}{N}-\overset{|}{C}\overset{\overset{O}{\|}}{C}-\overset{\overset{CH_3}{|}}{N}-\overset{|}{C}\overset{\overset{O}{\|}}{C}-\ etc.$$

Supplementary Problems

NOMENCLATURE

15.6. Give the IUPAC name of each of the following compounds:

(a) $CH_3-CH_2-\underset{\underset{\displaystyle CH_3}{|}}{CH}-CH_3$

(b) $CH_3-CH_2-\underset{\underset{\displaystyle H_3C-CH-CH_3}{|}}{CH}-CH_3$

(c)
$$\underset{\underset{\displaystyle CH_2}{|}}{\overset{\overset{\displaystyle CH_2}{}}{H_2C}}\underset{\underset{\displaystyle CH_2}{|}}{\overset{\overset{\displaystyle CH-CH_3}{}}{H_2C}}$$

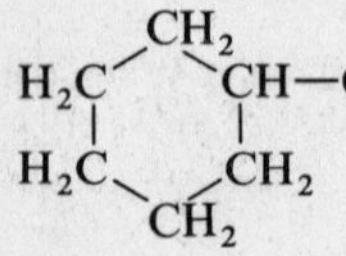

(d) $CH_3-CH=CH-CH_2-CH_3$

(e) $CH_2=CH-CH_2-\underset{\underset{\displaystyle CH_3}{|}}{CH}-CH=CH_2$

(f) $H_3C-C\equiv C-CH_3$

Ans. (a) 2-methylbutane; (b) 2,3-dimethylpentane; (c) methylcyclohexane; (d) 2-pentene; (e) 3-methyl-1,5-hexadiene; (f) 2-butyne

15.7. Name each of the following:

(a) $CH_3-O-CH_2-CH_3$

(b) $CH_3-\underset{\underset{\displaystyle Cl}{|}}{CH}-CH_2-CH_3$

(c) $\overset{\overset{\displaystyle CH_2}{}}{H_2C}\!\!-\!\!CH-O-\overset{\overset{\displaystyle O}{\|}}{C}-CH_3$

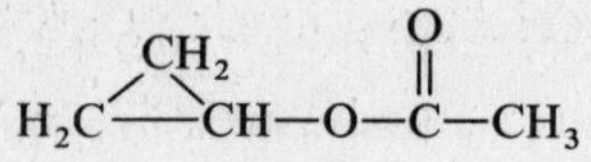

(d) $CH_3-\underset{\underset{\displaystyle CH_2-CH_3}{|}}{CH}-CH_2-CH_2-OH$

(e) $CH_3-CH_2-\underset{\underset{\displaystyle CH_2-CH_3}{|}}{N}-CH_2-CH_3$

(f) $CH_3-\overset{\overset{\displaystyle O}{\|}}{C}-CH_2-CH_2-CH_3$

Ans. (a) methyl ethyl ether, (b) 2-chlorobutane, (c) cyclopropyl acetate (or ethanoate), (d) 3-methyl-1-pentanol, (e) triethylamine, (f) 2-pentanone

15.8. Name each of the following:

(a) $CH_3-CH_2-\overset{\overset{\displaystyle O}{\|}}{C}-H$

(b) $H-\overset{\overset{\displaystyle O}{\|}}{C}-NH_2$

(c) $CH_3-CH_2-\underset{\underset{\displaystyle CH_3-CH_2}{|}}{CH}-CH_2-\overset{\overset{\displaystyle O}{\|}}{C}-OH$

(d) $CH_3-CH_2-\overset{\overset{\displaystyle O}{\|}}{C}-\overset{\overset{\displaystyle H}{|}}{N}-CH_2-CH_3$

(e) (benzene ring)

(f) (benzene ring with Cl at two adjacent positions)

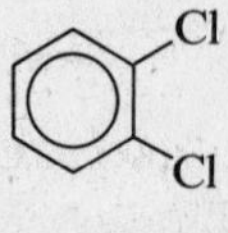

Ans. (a) propanal; (b) formamide (methanamide); (c) 3-ethylpentanoic acid; (d) N-ethylpropanamide; (e) benzene; (f) 1,2-dichlorobenzene

15.9. Draw the structures of the following compounds: (*a*) 3-ethyl-4-methyl-2-hexanone, (*b*) 2-chloro-butyl acetate, (*c*) ethylbenzene, (*d*) 3-ethylcyclohexene, (*e*) 2-methyl-3-pentanol, (*f*) 2-methylpropyl methyl ether.

Ans. (*a*)

$$CH_3-\overset{\overset{\displaystyle O}{\|}}{C}-CH-\overset{\overset{\displaystyle CH_3}{|}}{CH}-CH_2-CH_3$$
$$|$$
$$CH_2-CH_3$$

(*b*)

$$CH_3-CH_2-\overset{\overset{\displaystyle Cl}{|}}{CH}-CH_2-O-\overset{\overset{\displaystyle O}{\|}}{C}-CH_3$$

(*c*)

$-CH_2-CH_3$

(*d*)

(*e*)

$$CH_3-CH-CH-CH_2-CH_3$$
$$|\quad\,\,|$$
$$CH_3\,\,OH$$

(*f*)

$$CH_3-CH-CH_2-O-CH_3$$
$$|$$
$$CH_3$$

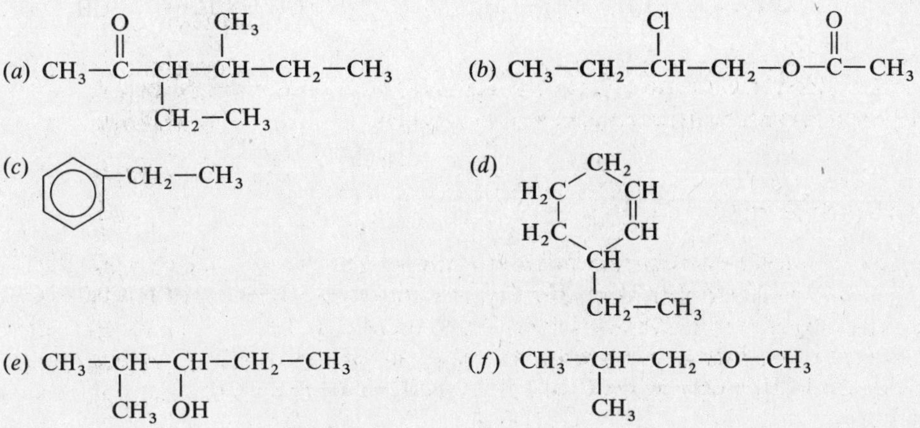

15.10. Explain what is wrong with each of the following names: (*a*) 3-methyl-2-propanol; (*b*) 3,3-dimethyl-2-pentene; (*c*) 1,4-dichlorocyclobutane; (*d*) 2-propanal; (*e*) 2-methyl-1-butyne; (*f*) pentanoicacid.

Ans. (*a*) The longest chain is four carbons. The correct name is 2-butanol. (*b*) Such a compound is impossible because it would require five bonds on the third carbon. (*c*) Positions 1 and 4 are equivalent to 1 and 2. The correct name is 1,2-dichlorocyclobutane. (*d*) Such a compound is impossible because the aldehyde carbon must be at the end of the chain. (*e*) Such a compound is impossible because it would require five bonds on the second carbon. (*f*) The suffix "acid" should be a separate word. The correct name is pentanoic acid.

15.11. Draw the structures of all the structural isomers having the formulas: (*a*) C_4H_9Br, (*b*) C_4H_8, (*c*) $C_2H_4O_2$ (omit any peroxides, $-O-O-$), (*d*) C_6H_{14}, (*e*) $C_4H_8Cl_2$.

Ans. (*a*) $CH_3-CH_2-CH_2-CH_2-Br$

$$CH_3-CH_2-CH-CH_3$$
$$|$$
$$Br$$

$$CH_3-CH-CH_3$$
$$|$$
$$CH_2Br$$

$$CH_3-\overset{\overset{\displaystyle Br}{|}}{C}-CH_3$$
$$|$$
$$CH_3$$

(*b*) $CH_3-CH_2-CH=CH_2$ $CH_3-CH=CH-CH_3$

$$CH_3-C=CH_2$$
$$|$$
$$CH_3$$

(*c*)

$$CH_3-\overset{\overset{\displaystyle O}{\|}}{C}-OH$$

$$H-\overset{\overset{\displaystyle O}{\|}}{C}-O-CH_3$$

$$HO-CH_2-\overset{\overset{\displaystyle O}{\|}}{C}-H$$

$$HO-CH=CH-OH$$

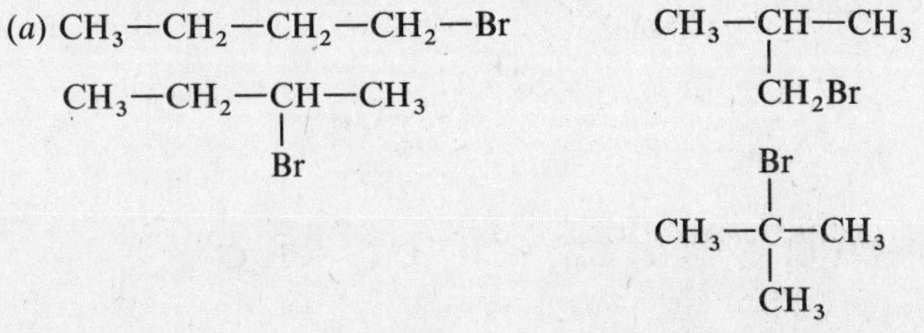

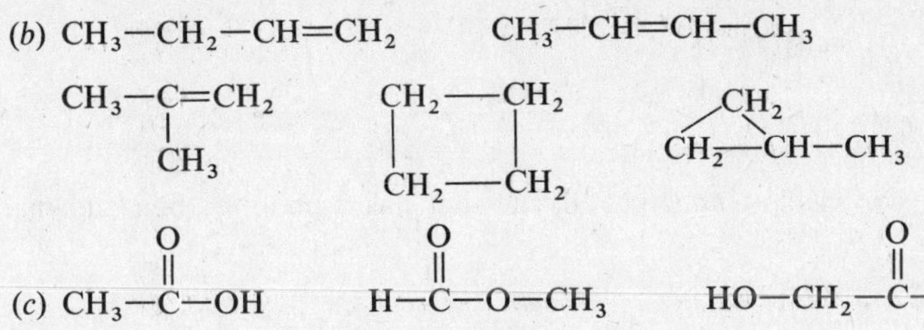

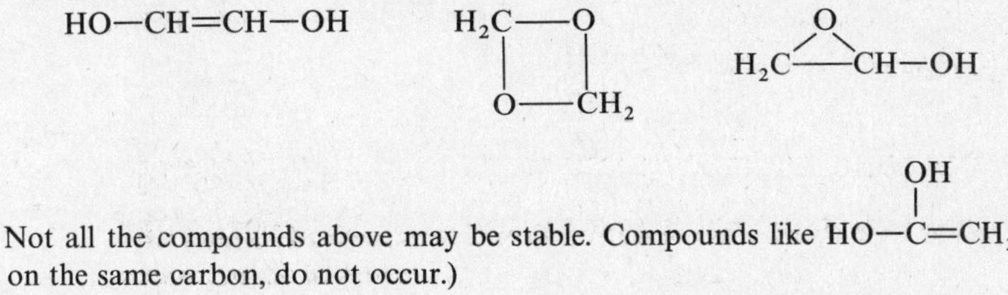

(*Note*: Not all the compounds above may be stable. Compounds like $HO-\overset{\overset{\displaystyle OH}{|}}{C}=CH_2$, with two hydroxyl groups on the same carbon, do not occur.)

(d) $CH_3-CH_2-CH_2-CH_2-CH_2-CH_3$ $CH_3-\underset{\underset{CH_3}{|}}{CH}-CH_2-CH_2-CH_3$

$CH_3-CH_2-\underset{\underset{CH_3}{|}}{CH}-CH_2-CH_3$ $CH_3-\underset{\underset{CH_3}{|}}{CH}-\underset{\underset{CH_3}{|}}{CH}-CH_3$

$CH_3-\overset{\overset{CH_3}{|}}{\underset{\underset{CH_3}{|}}{C}}-CH_2-CH_3$

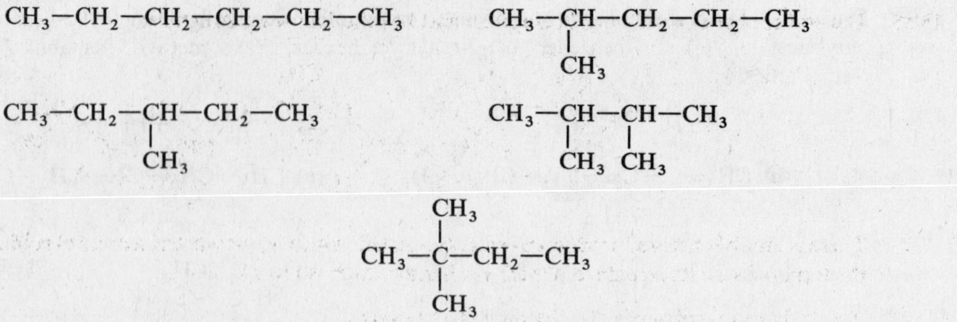

(e) $CH_3-CH_2-CH_2-\overset{\overset{Cl}{|}}{CH}-Cl$ $Cl-CH_2-CH_2-CH_2-CH_2Cl$ $CH_3-\underset{\underset{CH_3}{|}}{CH}-CHCl_2$

$CH_3-CH_2-\underset{\underset{Cl}{|}}{CH}-CH_2Cl$ $CH_3-\underset{\underset{Cl}{|}}{CH}-\underset{\underset{Cl}{|}}{CH}-CH_3$ $CH_3-\underset{\underset{CH_2Cl}{|}}{CH}-CH_2Cl$

$CH_3-\underset{\underset{Cl}{|}}{CH}-CH_2-CH_2Cl$ $CH_3-CH_2-\overset{\overset{Cl}{|}}{\underset{\underset{Cl}{|}}{C}}-CH_3$ $CH_3-\overset{\overset{Cl}{|}}{\underset{\underset{CH_3}{|}}{C}}-CH_2Cl$

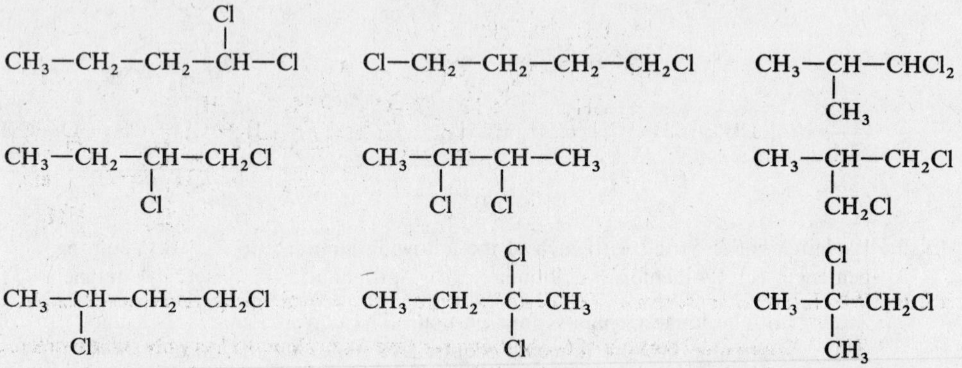

15.12. Examine all the answers to Problem 15.11 and select all that exhibit geometric isomerism, pointing out the two carbon atoms involved in each.

Ans.   (b) $CH_3-\overset{*}{C}H=\overset{*}{C}H-CH_3$,   (c) $HO-\overset{*}{C}H=\overset{*}{C}H-OH$

15.13. Examine all the answers to Problem 15.11 and select all that exhibit optical isomerism, pointing out the chiral carbon (or carbons) involved in each.

Ans.   (a) $CH_3-CH_2-\underset{\underset{Br}{|}}{\overset{*}{C}H}-CH_3$ (c) $H_2C\overset{\overset{O}{\diagup\diagdown}}{\underline{\quad\quad}}\overset{*}{C}H-OH$

(e) $CH_3-CH_2-\underset{\underset{Cl}{|}}{\overset{*}{C}H}-CH_2Cl$ (e) $CH_3-\underset{\underset{Cl}{|}}{\overset{*}{C}H}-CH_2-CH_2Cl$

(e) $CH_3-\underset{\underset{Cl}{|}}{\overset{*}{C}H}-\underset{\underset{Cl}{|}}{\overset{*}{C}H}-CH_3$

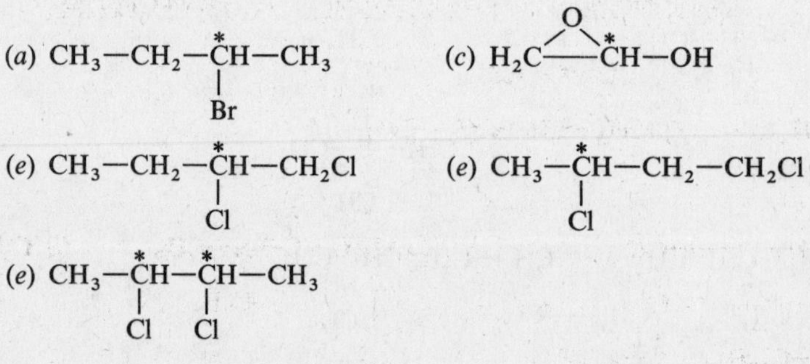

15.14. Draw the structure of the smallest noncyclic alkane that has a chiral carbon and mark it. Name the compound.

Ans.

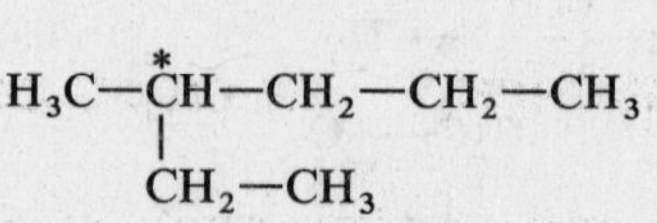

$H_3C-\underset{\underset{CH_2-CH_3}{|}}{\overset{*}{C}H}-CH_2-CH_2-CH_3$ 3-methylhexane

Also

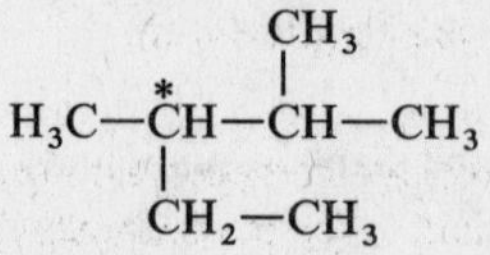

$H_3C-\underset{\underset{CH_2-CH_3}{|}}{\overset{*}{C}H}-\overset{\overset{CH_3}{|}}{CH}-CH_3$ 2,3-dimethylpentane

15.15. Natural rubber is a polymer of isoprene and has the following structure:

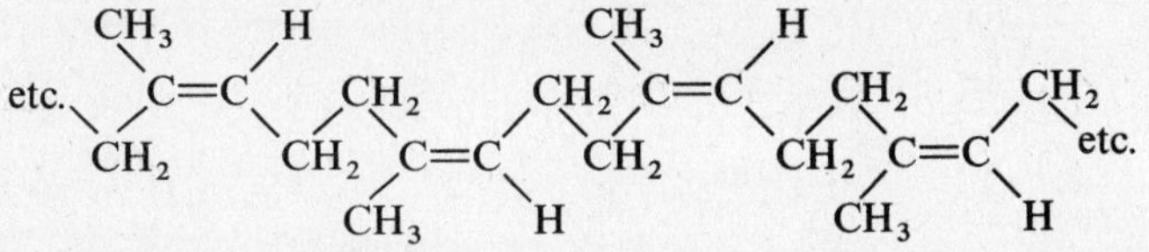

There is another natural product called gutta-percha which is a geometric isomer of rubber but is useless as an elastomer. Draw its structure. Which is *cis* and which is *trans*?

Ans.   Natural rubber is *cis*; gutta-percha is *trans*.

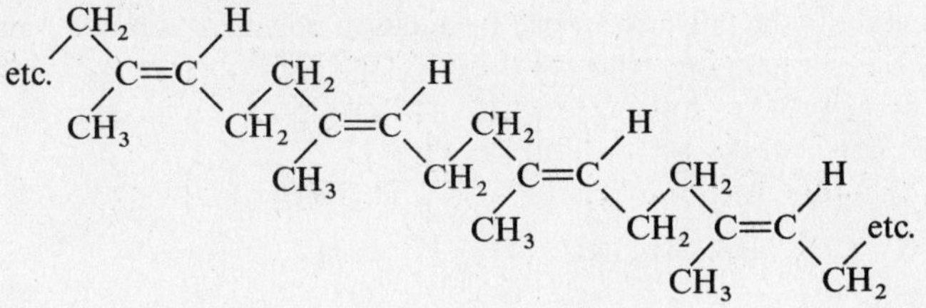

15.16. Which has more isomers, (*a*) dimethylbenzene or (*b*) dimethylcyclohexene? Explain.

Ans.   Because all corners of the benzene ring are equivalent (*a*) has only three isomers (structural) with the methyl groups in positions 1,2; 1,3; or 1,4. However (*b*) can have many more structural isomers; for example, 1,2 is different from 1,6 and 1,3 is different from 2,4. There are also many geometric isomers; for example, the 3, 4 structural isomer must have *cis* and *trans* forms. Optical isomerism also occurs; for instance, both carbons 3 and 4 of the 3, 4 isomer are chiral.

15.17. Name the functional group in each of the following molecules:

(*a*) $CH_3CH_2CH_2OH$ (*b*) $(CH_3)_3N$ (*c*) CH_3OCCH_3 with O double bond   (*d*) CH_3CHCH_3 with Cl

(*e*) $CH_3CH_2CH_2CCH_3$ with O double bond   (*f*) $CH_3-CH-CH_2-CH_3$ with $O=C-OH$

(*g*) triglyceride structure   (*h*) $C_2H_5-C(=O)-N(CH_3)-C_2H_5$ (*i*) $C_4H_9OC_4H_9$

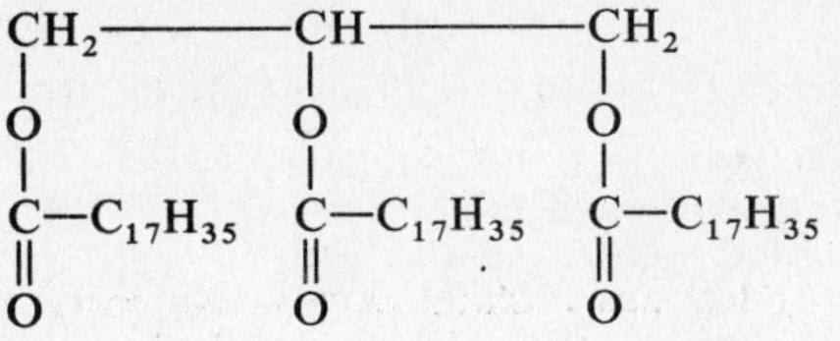

(*j*) C_2H_5CH with O double bond   (*k*) $Br-CH_2CH_2CHCH_3$ with Br (*l*) $CH_3CH_2C\equiv CCH_3$

Ans.   (*a*) alcohol, (hydroxyl); (*b*) amine, (tertiary amine); (*c*) ester; (*d*) chloride; (*e*) ketone; (*f*) acid, (carboxylic acid) (*g*) ester, (triglyceride); (*h*) amide; (*i*) ether; (*j*) aldehyde; (*k*) bromide, (dibromide); (*l*) alkyne, (triple bond).

15.18. What are the structures of the following compounds: (*a*) butanamide; (*b*) methylpropylamine; (*c*) diethyl ether; (*d*) 2,3-dimethyl-1-hexanol; (*e*) ethyl 2-methylpropionate; (*f*) 3-iodo-2-pentanone.

Ans. (a) $CH_3CH_2CH_2\overset{\displaystyle O}{\overset{\displaystyle \|}{C}}NH_2$ (b) $CH_3-\overset{\displaystyle H}{\overset{\displaystyle |}{N}}-CH_2CH_2CH_3$

(c) $C_2H_5OC_2H_5$ (d) $HO-CH_2-\underset{\overset{\displaystyle |}{CH_3}}{CH}-\underset{\overset{\displaystyle |}{CH_3}}{CH}-CH_2-CH_2-CH_3$

(e) $CH_3-\underset{\overset{\displaystyle |}{CH_3}}{CH}-\overset{\displaystyle O}{\overset{\displaystyle \|}{C}}-O-C_2H_5$ (f) $CH_3-\overset{\displaystyle O}{\overset{\displaystyle \|}{C}}-\underset{\overset{\displaystyle |}{I}}{CH}-CH_2-CH_3$

15.19. Draw the structure of the polymer formed by addition polymerization of 2-methyl-1-propene.

Ans. etc.$-\underset{\overset{\displaystyle |}{H}}{\overset{\overset{\displaystyle H}{\displaystyle |}}{C}}-\underset{\overset{\displaystyle |}{CH_3}}{\overset{\overset{\displaystyle CH_3}{\displaystyle |}}{C}}-\underset{\overset{\displaystyle |}{H}}{\overset{\overset{\displaystyle H}{\displaystyle |}}{C}}-\underset{\overset{\displaystyle |}{CH_3}}{\overset{\overset{\displaystyle CH_3}{\displaystyle |}}{C}}-\underset{\overset{\displaystyle |}{H}}{\overset{\overset{\displaystyle H}{\displaystyle |}}{C}}-\underset{\overset{\displaystyle |}{CH_3}}{\overset{\overset{\displaystyle CH_3}{\displaystyle |}}{C}}-\underset{\overset{\displaystyle |}{H}}{\overset{\overset{\displaystyle H}{\displaystyle |}}{C}}-\underset{\overset{\displaystyle |}{CH_3}}{\overset{\overset{\displaystyle CH_3}{\displaystyle |}}{C}}-$etc.

15.20. What are the products of the following reactions? Give the structures of the organic products.

(a) $C_2H_5\overset{\displaystyle O}{\overset{\displaystyle \|}{C}}-NH_2 + H_2O \xrightarrow[\text{catalyst}]{\text{hydrolysis}}$ (b) $C_3H_8 + O_2 \xrightarrow[\text{temp}]{\text{high}}$

(c) $C_2H_5OH \xrightarrow[\text{medium}]{\text{dehydration}}$ (d) $C_2H_5OH \xrightarrow[\text{oxidizer}]{\text{mild}}$

(e) $CH_3\overset{\displaystyle O}{\overset{\displaystyle \|}{C}}-OH + CH_3-\underset{\overset{\displaystyle |}{OH}}{\overset{\overset{\displaystyle OH}{\displaystyle |}}{CH}}-C_2H_5 \xrightarrow[\text{medium}]{\text{dehydration}}$ (f) $CH_3CH_2CH_3 + Cl_2 \rightarrow$ (first stage)

(g) $CH_2{=}CHCH_3 + Cl_2 \rightarrow$ (first stage) (h) $CH_3\underset{\overset{\displaystyle |}{OH}}{\overset{\overset{\displaystyle OH}{\displaystyle |}}{CH}}CH_2CH_3 \xrightarrow[\text{oxidizer}]{\text{mild}}$

(i) $CH_3CH_2CH_2\overset{\displaystyle O}{\overset{\displaystyle \|}{C}}-H \xrightarrow[\text{oxidizer}]{\text{mild}}$ (j) $CH_3-\underset{\overset{\displaystyle |}{OH}}{\overset{\overset{\displaystyle OH}{\displaystyle |}}{CH}}CH_2CH_3 + K \rightarrow$

Ans. (a) NH_3 and $C_2H_5\overset{\displaystyle O}{\overset{\displaystyle \|}{C}}-OH$ (b) CO_2 and H_2O (c) H_2O and $C_2H_5OC_2H_5$

(d) H_2O and $CH_3\overset{\displaystyle O}{\overset{\displaystyle \|}{C}}-H$ (e) H_2O and $CH_3\overset{\displaystyle O}{\overset{\displaystyle \|}{C}}-O-\underset{\overset{\displaystyle |}{CH_3}}{CH}-C_2H_5$

(f) HCl plus $CH_3CH_2CH_2-Cl$ and $CH_3\underset{\overset{\displaystyle |}{Cl}}{CH}CH_3$

(g) $\underset{\overset{\displaystyle |}{Cl}}{CH_2}\underset{\overset{\displaystyle |}{Cl}}{CH}CH_3$ (h) H_2O and $CH_3\overset{\displaystyle O}{\overset{\displaystyle \|}{C}}CH_2CH_3$

(i) $CH_3CH_2CH_2\overset{\displaystyle O}{\overset{\displaystyle \|}{C}}-OH$ (j) H_2 and $CH_3\underset{\overset{\displaystyle |}{O^-K^+}}{CH}CH_2CH_3$

Chapter 16

Thermodynamics and Chemical Equilibrium

THE FIRST LAW

The principle of conservation of energy, referred to in Chapter 7, constitutes the first law of thermodynamics. The quantities E and H are properties of a system that along with others define the state of the system. Such properties are termed *state functions*. If any of these properties changes, the system is said to have undergone a change in state. Any change in E must be equal to the amount of heat absorbed by the system plus the amount of work performed on the system.

$$\Delta E = q + w \tag{16-1}$$

In this equation, w is defined as the work done on the system of interest. If a chemical reaction occurs within a system work may be done upon it if gases are consumed and its volume is thereby decreased, or the system may perform work if gases are produced or if the system delivers an electric current to an external circuit. [In some books, w is defined as the negative of the above; that is, the work done by the system upon its surroundings. With the changed convention, (16-1) is rewritten with $-w$ instead of $+w$.]

THE SECOND LAW

The existence of an energy balance does not suffice to answer all questions about a chemical reaction. Does a given reaction take place at all? If so, to what extent does it proceed? Such questions require the introduction of some new thermodynamic functions, which like E and H are properties of the state of the system. These new functions are the *entropy*, S, and the *Gibbs free energy*, G. One mathematical statement of the second law of thermodynamics is

$$\Delta S \geq \frac{q}{T} \tag{16-2}$$

In words: the increase of entropy in a process is equal to or greater than the heat absorbed in the process divided by the temperature. The equality, which provides a definition of entropy increment, applies to any *reversible process*, defined as one that can be made to go in the reverse direction by a very small change in one of the variables, such as temperature, pressure, or the concentration of a chemical substance. The inequality refers to a spontaneous, or *irreversible*, *process*, defined as one which cannot be made to go in the reverse direction by a very small change in one of the variables.

EXAMPLE 1 The freezing of a liquid *at* its freezing point is a reversible process, because the elevation of the temperature by the slightest amount, e.g., 0.01 K, would cause the frozen material to begin to melt. An example of an irreversible process is the freezing of a supercooled liquid at a temperature significantly *below* its freezing point because melting would require that the temperature be raised *above* the freezing point from a value well below the freezing point.

The above statement of the second law implies the differentiation between reactions that can occur spontaneously (for which $\Delta S > q/T$) and those that cannot (for which $\Delta S < q/T$). For chemical processes, a more convenient differentiation between reactions that can occur and those that cannot is made in terms of G, the free energy, which is defined as

$$G = H - TS \tag{16-3}$$

Equation (16-2) leads, by a complex argument, to the following free-energy principle:

$$\Delta G_{T,P} \leq 0 \tag{16-4}$$

223

The change of free energy at constant temperature and pressure, during processes in which the only form of work is expansion against the surroundings (or the reverse, contraction), can be negative (for spontaneous irreversible processes) or zero (for reversible processes); it can be positive only with the application of work from some external source (such as electrolytic decomposition by the application of an electrical potential). Another formulation of the free-energy principle is that the maximum amount of work which a system can perform at constant temperature and constant pressure in forms other than expansive or contractive work is equal to the decrease in free energy of the system.

Entropy

Via (16-2), entropy changes can be calculated from gross thermal measurements of reversible processes. However, they may also be related to the molecular properties of matter. Entropy is positively correlated with the number of different collections of molecular states consistent with the observed state of the bulk system. Some generally observed examples are given below.

1. Liquids have more entropy than their corresponding crystalline forms. (Every atom or molecule in a crystal is in a prescribed position in the lattice; in a liquid, positions of the particles may take on many different values consistent with the overall liquid nature.)
2. Gases have more entropy than their corresponding liquids. (Although the molecules in a liquid are free to occupy a variety of positions, they are constrained to be in close contact with their nearest neighbors; in a gas, the number of positional possibilities for the molecules is much greater because there is much more free space available per molecule.
3. Gases at low pressure have higher entropy than at high pressure. (The argument is an extension of that given above in 2.)
4. A large molecule has a greater entropy than any of its submolecular fragments existing in the same phase of matter. (The internal vibrations and rotations of atoms within molecules give many possibilities for the distribution of intramolecular motions.)
5. The entropy of a substance always increases when its temperature is raised. [The temperature measures the average energy per molecule and, therefore, the total energy. The higher the temperature, the greater the total energy; and the greater the total energy, the greater the number of ways of apportioning this energy among the fixed number of molecules. Thus there are more different collections of molecular (energy) states corresponding to a higher temperature than to a lower temperature.]
6. If a chemical reaction is accompanied by a change in the number of gas molecules, ΔS is positive in the reaction direction leading to an increase in the number of gas molecules.
7. When one substance dissolves in another, ΔS is positive. (The number of possible configurations of randomly arranged, unlike particles is greater than the number of configurations of separately packaged, like particles.)

THE THIRD LAW

Equation (16-2) allows the calculation of *changes* in the entropy of a substance, specifically by measuring the heat capacities at different temperatures and enthalpies of phase changes. If the *absolute value* of the entropy were known at any one temperature, the measurements of *changes* in entropy in proceeding from that one temperature to any other temperature would allow the determination of the absolute value of the entropy at the other temperature. The third law of thermodynamics provides the basis for establishing absolute entropies; it states that the entropy of any perfect crystal is zero at the absolute zero of temperature. This is understandable in terms of the molecular interpretation of entropy. In a perfect crystal every atom is fixed in position; and at the absolute zero of temperature every form of internal energy (such as atomic vibrations) has its lowest possible value, so that there is only one way of distributing the energy among the available atoms or molecules.

STANDARD STATES AND REFERENCE TABLES

Although enthalpies of substances are relatively independent of pressure (for gases) and of concentration (for dissolved species), their entropies, and thus the free energies as well, depend markedly on these variables. Tabulated value for S and G usually refer to the idealized state of 1 bar or 1 atm pressure for gases, 1 M concentration for solutes, and to the pure substances for liquids and solids. Table 16-1 compiles some data for $S°$, the molar entropy, and $\Delta G_f°$, the free energy of formation from the elements, both at 25 °C. $\Delta G_f°$ is defined, analogously to $\Delta H_f°$, as the free-energy change associated with the reaction in which 1 mol of the substance in its standard state is formed from the stoichiometrically required numbers of moles of the constituent elements in their respective standard states. Note that the third law allows the determination of the entropy of a substance without reference to its constituent elements; note also that there are no zero values for the entropies of the elements at temperatures above

Table 16-1. Standard Entropies and Free Energies of Formation at 25 °C and 1 atm

Substance	$S°/\mathrm{J \cdot K^{-1} \cdot mol^{-1}}$	$\Delta G_f°/\mathrm{kJ \cdot mol^{-1}}$
Ag_2O (s)	121.3	−11.21
Br_2 (l)	152.23	0
Br_2 (g)	245.35	3.14
C (s, graphite)	5.74	0
CH_3OH (l)	126.8	−166.36
CH_3OH (g)		−162.00
C_2H_5OH (g)	282.6	−168.57
CO (g)	197.56	−137.15
CO_2 (g)	213.68	−394.37
Cl_2 (g)	222.96	0
Cl_2O (g)	266.10	97.9
H_2 (g)	130.57	0
H_2O (l)	69.95	−237.19
H_2O (g)	188.72	−228.59
N_2 (g)	191.50	0
NO_2 (g)	239.95	51.30
N_2O_4 (g)	304.18	97.82
O_2 (g)	205.03	0
PCl_3 (l)	217.1	−272.4
PCl_3 (g)	311.7	−267.8
PCl_5 (g)		−305.0
SO_3 (s)	52.3	−369.0
SO_3 (l)	95.6	−368.4
Sn (s, white)	51.5	0
Sn (s, gray)	44.1	0.12

0 K as there are for ΔH_f° and ΔG_f°. Unless otherwise stated, calculations in this text will be referred to 1 atm standard state, still the common practice in the United States. However, use of the bar as standard state will be illustrated.

In general, for the reaction

$$a\text{A} + b\text{B} \rightleftharpoons c\text{C} + d\text{D}$$

the concentration dependence of the free-energy change may be expressed as

$$\Delta G = \Delta G^\circ + RT \ln \frac{[\text{C}]^c[\text{D}]^d}{[\text{A}]^a[\text{B}]^b} \qquad (16\text{-}5)$$

In this equation, [X] denotes the relative concentration of X, i.e., the ratio of the concentration to the concentration at the standard state. For a substance whose standard state is 1 mol/L, [X] is the *numerical* value of the molar concentration. Also, ln is the symbol for the natural logarithm (see Appendix C); ΔG is the free-energy change at the given concentrations; ΔG° is the free-energy change for the hypothetical reaction in which all reactants are in their standard states and all products are in their standard states; it is called the *standard* free energy change. Note that if the concentration of every species is 1, the logarithmic term in (16-5) becomes 0, and $\Delta G = \Delta G^\circ$; this is as it should be, because unit concentration implies the standard state.

In (16-5) if ΔG is to be expressed in joules, the universal gas constant must be taken as

$$R = 8.314 \text{ J} \cdot \text{K}^{-1} \cdot \text{mol}^{-1}$$

On the right side of (16-5) the term ΔG° lacks the unit mol^{-1}, which must be removed from the second term as well, having been absorbed into the logarithmic term by mathematical manipulation. The value of the second term applies to the number of moles indicated by the coefficients a, b, c, d. This same value for R *must be used in all calculations in this chapter*, except those directly involving the ideal gas law.

The generalization of (16-5) to any number of reactants and products is

$$\Delta G = \Delta G^\circ + RT \ln Q \qquad (16\text{-}6)$$

where Q, the *reaction quotient*, is defined as the product of all reaction-product relative concentrations, each raised to a power equal to the coefficient of that particular product in the balanced equation, divided by the product of all reactant relative concentrations, each raised to a power equal to the coefficient of that reactant in the balanced equation.

CHEMICAL EQUILIBRIUM

Theoretically, any chemical reaction could take place in the reverse direction to some extent. Often the *driving force* of a reaction favors one direction so greatly that the extent of the reverse reaction is infinitesimal and impossible to measure. The driving force of a chemical reaction, or the *change in free energy* accompanying the reaction, is an exact measure of the tendency of the reaction to go to completion. When the magnitude of ΔG° is very large and the sign negative, the reaction may go practically to completion in the forward direction; if ΔG° is only slightly negative, the reaction may proceed to a small extent until it reaches a point where ΔG (as distinct from ΔG°) for any further reaction would be zero, and the reaction could be reversed with a slight change in concentrations. In the latter case, the reaction is said to be thermodynamically reversible. Since many organic and metallurgical reactions are of this reversible type, it is necessary to learn how conditions should be altered to obtain economical yields, hasten desirable reactions, and minimize undesirable reactions.

A chemical system which has reached a state of thermodynamic reversibility shows no further *net* reaction in either direction, since $\Delta G = 0$. Such a system is said to be in a state of *equilibrium*, from which it could be displaced either forwards or backwards by an appropriate very small change in one of the variables. Although a system at equilibrium shows no *net* changes in the concentrations of any of the chemical species, the actual molecular or ionic rearrangements constituting both forward and reverse reactions may still be occurring, but the opposing forward and reverse reactions proceed at equal speeds. Consequently the composition of the mixture remains constant and does not change with time. This condition of *dynamic* equilibrium, involving balanced reactions in both directions, contrasts with the *static* equilibrium of pulleys, levers, and springs.

THE EQUILIBRIUM CONSTANT

For a reversible reaction at equilibrium, $\Delta G = 0$, and (*16-6*) becomes

$$0 = \Delta G^\circ + RT \ln Q_{eq}$$

or

$$\Delta G^\circ = -RT \ln Q_{eq} \qquad (16\text{-}7)$$

Equation (*16-7*) is a remarkable statement. It implies that Q_{eq}, the value of the reaction quotient under equilibrium conditions, *depends only on thermodynamic quantities that are constant in the reaction* (the temperature, and the *standard* free-energy change for the reaction at that temperature), *and is independent of the actual starting concentrations of reactants or products.* For this reason, Q_{eq} is usually denoted the *equilibrium constant*, K, and (*16-7*) is rewritten as

$$\Delta G^\circ = -RT \ln K \qquad (16\text{-}8)$$

EXAMPLE 2   Consider the reversible gaseous reaction system

$$H_2 + I_2 \rightleftharpoons 2HI$$

A reaction mixture could be made up by starting with H_2 and I_2 alone, HI plus I_2, HI plus H_2, HI alone, or a mixture of all three substances. In each case a net reaction would occur in one direction or the other until the system comes to an eventual state of no further net change, i.e., equilibrium, which could be described by specifying the concentration of the three substances. Because of the variety of ways of making up the initial mixture, differing in the relative amounts of the various substances used, there are an infinite number of equilibrium states, each describable by a set of the concentrations of the three participating substances. Yet, for each of this infinite number of states, all at the same temperature,

$$\frac{[HI]^2}{[H_2][I_2]} = K$$

That is, the particular function of the three concentrations defined by Q is always the same, even though any individual concentration may vary by as much as 10 orders of magnitude. This unifying principle allows the calculation of conditions at equilibrium under as many sets of starting conditions as may be desired.

Experimental measurements show that molecules in highly compressed gases or highly concentrated solutions, especially if electrically charged, abnormally affect each other. In such cases the true *activity* or *effective concentration* may be greater or less than the measured concentration. Hence when the molecules involved in equilibrium are relatively close together, the concentration should be multiplied by an *activity coefficient* (determined experimentally). At moderate pressures and dilutions, the activity coefficient for nonionic compounds is close to unity. In any event, the activity coefficient correction will not be made in the problems in this book.

The equilibrium constant K is a pure number whose magnitude depends not only on the temperature but also, generally, on the standard-state concentration to which all concentrations are referred. In this chapter, the standard state for dissolved substances will be taken as a 1 M concentration, unless a statement is made to the contrary. The magnitude of K is independent of the choice of standard concentration in the special case where the sum of the concentration exponents in the numerator equals the sum of the exponents in the denominator.

The concentration of a gas is proportional to its partial pressure ($n/V = P/RT$). Hence, when all substances in the reversible reaction

$$aA + bB \rightleftharpoons cC + dD$$

are gases, the equilibrium constant may be written as

$$\frac{P^c(C)P^d(D)}{P^a(A)P^b(B)} = K_p$$

K_p can replace K in equation (16-8). Then if $\Delta G°$ is obtained from a table based on the standard state of 1 atm, K_p will normally be correct only for P in atmospheres; and likewise for a standard state of 1 bar, K_p will normally apply only to P in bars. When the equation shows no change in the total number of moles of gas as the reaction proceeds (e.g., $N_2 + O_2 \rightleftharpoons 2NO$), K_p will be the same regardless of the pressure units used and will be identical with the K expressed in molar concentrations.

Terms representing the concentrations of undissolved *solid* reactants or products are conventionally omitted from the expression for K because their concentrations cannot be varied. Such substances are always in their standard states and would merely contribute a $[X] = 1$ in the logarithmic terms of (16-5) through (16-8). Another exception occurs when the solvent is one of the products or reactants, as in the hydrolysis of urea in aqueous solution:

$$CO(NH_2)_2 + H_2O \rightleftharpoons CO_2 + 2NH_3 \qquad K = \frac{[CO_2][NH_3]^2}{[CO(NH_2)_2]}$$

So long as the solution is moderately dilute, the concentration of water cannot be much less than in pure water itself, 55.5 mol/L. It is therefore customary to remove $[H_2O]$ from the expression for K by choosing 55.5 mol/L as the standard state for water, while retaining 1 mol/L as the standard state for the other substances.

LE CHATELIER'S PRINCIPLE

If some stress (such as a change in temperature, pressure, or concentration) is brought to bear upon a system in equilibrium, a reaction occurs in the direction which tends to relieve the stress. This generalization is extremely useful in predicting the effects of changes in temperature, pressure, or concentration upon a system at equilibrium.

Effect of changes in temperature

The effect will first be discussed in qualitative terms. When the temperature of a system at equilibrium is raised, the reaction occurs in the direction which absorbs heat.

EXAMPLE 3   In the thermochemical equation for the synthetic methanol process (all substances are in the gaseous state)

$$CO + 2H_2 \rightleftharpoons CH_3OH \qquad \Delta H = -22 \text{ kcal}$$

the forward reaction liberates heat (exothermic), while the reverse reaction absorbs heat (endothermic). If the temperature of the system is raised, the reverse reaction (from right to left) occurs in an attempt to absorb the heat being supplied. Eventually a new state of equilibrium will be achieved at the higher temperature, containing more reactants (CO and H_2) and less product than at the initial temperature. Conversely, the equilibrium yield of methanol is increased by lowering the temperature of this system.

The same effect can be explained in quantitative terms. For any chemical process, an equation like (16-3) can be written for the free energy of each substance appearing as a reactant or a product. By adding up all such equations for the products and subtracting all the equations for the reactants, and noting that the temperature is the same for all substances, one obtains (16-9), which represents ΔG, the *change* in free energy for the process.

$$\Delta G = \Delta H - T \Delta S \qquad\qquad (16-9)$$

If all substances are in their standard states, (16-9) becomes

$$\Delta G° = \Delta H° - T\,\Delta S°$$ (16-10)

Combination of (16-8) and (16-10) gives

$$\ln K = -\frac{\Delta H°}{RT} + \frac{\Delta S°}{R}$$ (16-11)

If $\Delta H°$ and $\Delta S°$ are fairly independent of temperature, as they are for most reactions, (16-11) shows that $\ln K$ is a decreasing function of T for $\Delta H° < 0$ (exothermic reaction). Then K itself is a decreasing function of T. A decrease in K means a shift of the equilibrium to favor the formation of the substances whose concentrations appear in the denominator of the K expression, namely the reactants, at the expense of the products of the reaction, whose concentrations appear in the numerator of the K expression.

Conversely, if $\Delta H° > 0$ (endothermic reaction), K increases with T, and the equilibrium shifts to favor the formation of the reaction products.

Effect of changes in pressure

When the pressure of a system in equilibrium is increased (for example, as shown in Fig. 16-1), the reaction occurs in the direction that lowers the pressure by reducing the volume of gas.

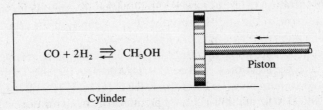

Fig. 16-1

EXAMPLE 4   In the synthetic methanol process (all substances are in the gaseous state)

$$CO + 2H_2 \quad \rightleftharpoons \quad CH_3OH$$

$$\begin{pmatrix} 3 \text{ molecules of gas} \\ 3 \text{ volumes of gas} \end{pmatrix} \quad \begin{pmatrix} 1 \text{ molecule of gas} \\ 1 \text{ volume of gas} \end{pmatrix}$$

the forward reaction is accompanied by a decrease in volume. Then an increase in pressure will increase the equilibrium yield of CH_3OH. (This increase in yield occurs even though the value of K, depending only on temperature, does not change.) When the new equilibrium state is finally established the total pressure will exceed the original pressure, but will be much less than it would have been if no reaction had occurred.

A pressure change will *not* affect the relative amounts of the substances at equilibrium in any gaseous system where the number of molecules reacted equals the number produced. For example, $H_2 + CO_2 \rightleftharpoons CO + H_2O$ (*steam*) and $H_2 + I_2 \rightleftharpoons 2HI$ (all substances in the gaseous state).

The effect of pressure on equilibrium systems involving gases and liquids or solids is usually due to a change in the number of gaseous molecules, since molar volumes of gases are so much larger than of liquids or solids. In the case of equilibrium reactions of solids or liquids, with no gases, the pressure effect is trivially small except at thousands of atmospheres.

Effect of changes in amount of solvent

For reactions that take place in solution, increasing the amount of solvent (dilution) will displace the equilibrium in the direction of forming the larger number of dissolved particles. This is analogous to decreasing the pressure in a gas reaction.

EXAMPLE 5   Consider the dimerization of acetic acid in benzene solutions.

$$2HC_2H_3O_2 \ (in \ solution) \rightleftharpoons (HC_2H_3O_2)_2 \ (in \ solution) \qquad K = \frac{[(HC_2H_3O_2)_2]}{[HC_2H_3O_2]^2}$$

$$\text{(2 dissolved particles)} \qquad\qquad\qquad \text{(1 dissolved particle)}$$

Let us imagine an equilibrium benzene solution of acetic acid. If the solution is suddenly diluted to twice its original volume, and if there were no change in the relative amounts of the two forms of acetic acid, the concentration of each form would be just one-half of what it had been before dilution. In the equilibrium constant equation, the numerator would be $\frac{1}{2}$ its former value and the denominator would be $\frac{1}{4}$ its former value ($\frac{1}{2}$ squared). The ratio of numerator to denominator would be 2 times its original value ($\frac{1}{2}$ divided by $\frac{1}{4}$). But this ratio must return to its original value of K. It can do this if the numerator becomes smaller and the denominator larger. In other words, some of the dimer, $(HC_2H_3O_2)_2$, must react backwards to form $2HC_2H_3O_2$.

Changes in the amount of solvent will not affect the equilibrium in any system where the number of dissolved particles reacted equals the number produced. For example, the esterification of methyl alcohol with formic acid in an inert solvent:

$$CH_3OH + HCO_2H \rightleftharpoons HCO_2CH_3 + H_2O$$

Effect of varying the concentration

Increasing the concentration of any component of a system in equilibrium will promote the reaction which tends to consume some of the added substance. For example, in the reaction $H_2 + I_2 \rightleftharpoons 2HI$, the consumption of iodine is improved by adding excess hydrogen.

This effect, too, can be explained in terms of the equilibrium constant. Consider a system at equilibrium with respect to the reaction $A + B \rightleftharpoons C + D$. Allow an increase in the concentration of A by introducing an additional amount of A. To bring the value of [C] [D]/[A] [B] back to K, further reaction takes place between A and B to form more of C and D, thus increasing the concentrations of C and D (the numerator) while reducing the resulting concentration of B (the denominator), until the fraction attains the value K. At this point equilibrium is again established. The resulting concentration of A is greater than before the supplementary addition of A but less than it would have been had there been no net chemical reaction of adjustment following the addition.

Effect of catalysts

Catalysts accelerate both forward and backward reaction rates equally. They speed up the approach to equilibrium, but do not alter the equilibrium concentrations.

Solved Problems

THERMODYNAMICS

16.1.   Without consulting entropy tables, predict the sign of ΔS for each of the following processes:

       (a) $O_2 \ (g) \rightarrow 2O(g)$ (f)   Desalination of seawater.

       (b) $N_2 \ (g) + 3H_2 \ (g) \rightarrow 2NH_3 \ (g)$ (g)   Devitrification of glass.

       (c) $C \ (s) + H_2O \ (g) \rightarrow CO \ (g) + H_2 \ (g)$ (h)   Hard-boiling of an egg.

       (d) $Br_2(l) \rightarrow Br_2 \ (g)$ (i) $C \ (s, \ graphite) \rightarrow C \ (s, \ diamond)$

       (e) $N_2 \ (g, \ 10 \ atm) \rightarrow N_2 \ (g, \ 1 \ atm)$

(a) Positive. There is an increase in the number of gas molecules.

(b) Negative. There is a decrease in the number of gas molecules.

(c) Positive. There is an increase in the number of gas molecules.

(d) Positive. S is always greater for a gas than for its corresponding liquid.

(e) Positive. Entropy increases on expansion.

(f) Negative. Desalination is the opposite of solution; a solute must be removed from a solution.

(g) Negative. Devitrification is the onset of crystallization in a supercooled liquid.

(h) Positive. The fundamental process in the "boiling" of an egg is not a literal boiling, in the sense of vaporization, but a denaturation of the egg protein. A protein is a large molecule which exists in a particular configuration in the so-called native state but may occupy a large number of almost random configurations in the denatured state, resulting from rotations around the bonds. The increase in the number of possible configurations is analogous to the melting process.

(i) Negative. Diamond, being a harder solid, would be expected to have more restricted atomic motions within the crystal. Diamond is denser and has less entropy on that account also.

16.2. Calculate ΔS for the following phase transitions: (a) melting of ice at 0 °C,   (b) vaporization of water at 100 °C. Use data from Chapter 7.

(a) ΔH of fusion of ice $= 1.44$ kcal/mol $= 6.02$ kJ/mol

Since the melting of ice at 0 °C is a reversible process, (16-2) may be used with the equals sign. (Recall that $q = \Delta H$ at constant pressure.)

$$\Delta S = \frac{q_{rev}}{T} = \frac{6.02 \times 10^3 \text{ J/mol}}{273.1 \text{ K}} = 22.1 \text{ J} \cdot \text{K}^{-1} \cdot \text{mol}^{-1}$$

(b) ΔH of vaporization of water $= 9.72$ kcal/mol $= 40.7$ kJ/mol

Since the vaporization at 100 °C is reversible, (16-2) may be used with the equals sign.

$$\Delta S = \frac{q_{rev}}{T} = \frac{4.07 \times 10^4 \text{ J/mol}}{373.1 \text{ K}} = 109.1 \text{ J} \cdot \text{K}^{-1} \cdot \text{mol}^{-1}$$

16.3. After comparing data in Table 7-1 and the answer to Problem 16.1(i), how do you account for the fact that ΔH and ΔS for the phase transition from diamond to graphite are not related by the same equation that applied in Problem 16.2?

From Table 7-1, the formation of diamond from graphite (the standard state of carbon) is accompanied by a *positive* ΔH of 1.88 kJ/mol at 25 °C. From Problem 16.1(i), ΔS for the same process is *negative*. Since 25 °C is not the transition temperature, the process is not a reversible one. In fact, it is not even a spontaneous irreversible process, and (16-2) does not apply with the inequality sign. On the contrary, the opposite process, the conversion of diamond to graphite, is thermodynamically spontaneous (although it might require billions of years in the absence of a suitable catalyst), and ΔS for *this* process would obey (16-2) with the inequality sign.

16.4. Calculate ΔH_f° for C_2H_5OH (g).

For the special process in which a substance in its standard state is formed from its elements in their standard states, (16-10) gives

$$\Delta G_f^\circ = \Delta H_f^\circ - T \Delta S_f^\circ \tag{16-12}$$

ΔH_f° can be computed from (16-12) with the use of data from Table 16-1.

Write the balanced formation equation for C_2H_5OH (g), and write the (nS°)-value under each substance.

$$2C \text{ (s)} + 3H_2 \text{ (g)} + \tfrac{1}{2}O_2 \text{ (g)} \rightarrow C_2H_5OH \text{ (g)}$$

$(nS^\circ)/\text{J} \cdot \text{K}^{-1}$:   2(5.74)   3(130.57) $\tfrac{1}{2}$(205.03)   282.6

For the process, then,

$$\Delta S^\circ = 282.6 - 2(5.74) - 3(130.57) - \tfrac{1}{2}(205.03) = -223.11 \text{ J} \cdot \text{K}^{-1}$$

whence

$$\Delta S_f^\circ = \frac{-223.11 \text{ J} \cdot \text{K}^{-1}}{1 \text{ mol C}_2\text{H}_5\text{OH}} = -223.11 \text{ J} \cdot \text{K}^{-1} \cdot \text{mol}^{-1}$$

Now, from (16-12),

$$\Delta H_f^\circ = \Delta G_f^\circ + T \Delta S_f^\circ$$
$$= -168.57 \text{ kJ} \cdot \text{mol}^{-1} + (298.1 \text{ K})(-0.223\,11 \text{ kJ} \cdot \text{K}^{-1} \cdot \text{mol}^{-1}) = -235.09 \text{ kJ} \cdot \text{mol}^{-1}$$

Note again that ΔG_f° for a substance is listed as a single entry in Table 16-1, but ΔS_f° must be computed by taking the difference of the tabulated absolute entropies of the substance and its constituent elements.

16.5. (a)   What is ΔG° at 25 °C for the following reaction?

$$\text{H}_2 (g) + \text{CO}_2 (g) \rightleftharpoons \text{H}_2\text{O} (g) + \text{CO} (g)$$

(b)   What is ΔG at 25 °C under conditions where the partial pressures of H_2, CO_2, H_2O, and CO are 10, 20, 0.02, and 0.01 atm, respectively?

First tabulate $(n \Delta G_f^\circ)$ values under each substance (in this case, $n = 1$ mol for each substance).

$$\text{H}_2 (g) + \text{CO}_2 (g) \rightleftharpoons \text{H}_2\text{O} (g) + \text{CO} (g)$$

$n \Delta G_f^\circ/\text{kJ}$:   0 -394.37 -228.59 -137.15

(a)   The computation of ΔG° is analogous to that of ΔH° (see Problem 7.12).

$$\Delta G^\circ = (-228.59 - 137.15) - (0 - 394.37) = 28.63 \text{ kJ}$$

(b) $$\Delta G = \Delta G^\circ + RT \ln Q$$

$$= (28.63 \text{ kJ}) + (8.314 \times 10^{-3} \text{ kJ/K})(298.1 \text{ K})\left[2.303 \log \frac{P(\text{H}_2\text{O})P(\text{CO})}{P(\text{H}_2)P(\text{CO}_2)} \right]$$

$$= \left[28.63 + 5.708 \log \frac{(0.02)(0.01)}{(10)(20)} \right] \text{kJ} = (28.63 - 34.25) \text{ kJ} = -5.62 \text{ kJ}$$

Note that the reaction, although not possible under standard conditions, becomes possible ($\Delta G < 0$) under this set of experimental conditions. Note the conversion from natural to common logarithms by use of the factor 2.303.

16.6.   Calculate the absolute entropy of CH_3OH (g) at 25 °C.

Although there is no S°-entry in Table 16-1 for CH_3OH (g), the ΔG_f°-value for this substance listed in Table 16-1, the ΔH_f°-value listed in Table 7-1 and the S°-values for the constituent elements may be combined to yield the desired value. From (16-12),

$$\Delta S_f^\circ = \frac{\Delta H_f^\circ - \Delta G_f^\circ}{T} = \frac{(-200.7 + 162.0) \text{ kJ} \cdot \text{mol}^{-1}}{298.1 \text{ K}} = -129.8 \text{ J} \cdot \text{K}^{-1} \cdot \text{mol}^{-1}$$

From the equation for the formation of 1 mol of CH_3OH (g) under standard conditions,

$$\text{C}(s) + 2\text{H}_2 (g) + \tfrac{1}{2}\text{O}_2 (g) \rightarrow \text{CH}_3\text{OH} (g)$$

we can write

$$-129.8 \text{ J} \cdot \text{K}^{-1} = (1 \text{ mol})[S^\circ(\text{CH}_3\text{OH})] - (1 \text{ mol}) [S^\circ(\text{C})] - (2 \text{ mol})[S^\circ(\text{H}_2)] - (\tfrac{1}{2} \text{ mol}) [S^\circ(\text{O}_2)]$$
$$= (1 \text{ mol})[S^\circ(\text{CH}_3\text{OH})] - [5.7 + 2(130.6) + \tfrac{1}{2}(205.0)] \text{ J} \cdot \text{K}^{-1}$$

Solving, $S^\circ(\text{CH}_3\text{OH}) = 239.6 \text{ J} \cdot \text{K}^{-1} \cdot \text{mol}^{-1}$.

16.7. Estimate the boiling point of PCl_3.

The boiling point is the temperature at which $\Delta G°$ for the following reaction is zero:

$$PCl_3 \ (l) \rightleftharpoons PCl_3 \ (g)$$

The reaction is not spontaneous at 25 °C, where, according to Table 16-1,

$$\Delta G° = (1 \ \text{mol})[\Delta G_f°(PCl_3, g)] - (1 \ \text{mol})[\Delta G_f°(PCl_3, l)] = -267.8 + 272.4 = 4.6 \ \text{kJ}$$

If we assume that $\Delta H°$ and $\Delta S°$ are both independent of temperature, then the temperature dependence of $\Delta G°$ is given by the factor T in (16-10). If $\Delta H°$ and $\Delta S°$ are known from the data at 25 °C, T can be calculated to satisfy the condition that $\Delta G°$ equals zero.

$$\Delta G° = \Delta H° - T \ \Delta S° = 0 \qquad \text{or} \qquad T(\text{b.p.}) = \frac{\Delta H°}{\Delta S°}$$

Now

$$\Delta S° = (1 \ \text{mol})[S°(g)] - (1 \ \text{mol})[S°(l)] = 311.7 - 217.1 = 94.6 \ \text{J/K}$$

and, from Table 7-1,

$$\Delta H° = (1 \ \text{mol})[\Delta H_f° \ (g)] - (1 \ \text{mol})[\Delta H_f°(l)] = -287.0 - (-319.7) = 32.7 \ \text{kJ}$$

Then

$$T(\text{b.p.}) = \frac{\Delta H°}{\Delta S°} = \frac{32.7 \times 10^3 \ \text{J}}{94.6 \ \text{J/K}} = 346 \ \text{K}$$

or 73 °C. The observed boiling point, 75 °C, is very close to the estimated 73 °C.

The validity of this type of estimate is no better than the assumption of constancy of $\Delta S°$ and $\Delta H°$. In general, the smaller the temperature range over which the extrapolation must be made, the greater the accuracy of the prediction.

16.8. For the gaseous equilibrium

$$PCl_5 \ (g) \rightleftharpoons PCl_3 \ (g) + Cl_2 \ (g)$$

explain the effect upon the composition of   (a) increased temperature,   (b) increased pressure, (c) higher concentration of Cl_2 (d) higher concentration of PCl_5,   (e) presence of a catalyst.

(a) When the temperature of a system in equilibrium is raised, the equilibrium point is displaced in the direction which absorbs heat. Table 7-1 can be used to determine that the *forward* reaction as written is endothermic.

$$\Delta H = (1 \ \text{mol})[\Delta H_f(PCl_3)] + (1 \ \text{mol})[\Delta H_f(Cl_2)] - (1 \ \text{mol})[\Delta H_f(PCl_5)]$$
$$= -287.0 + 0 - (-374.9) = 87.9 \ \text{kJ}$$

Hence increasing the temperature will cause more PCl_5 to dissociate.

(b) When the pressure of a system in equilibrium is increased, the equilibrium point is displaced in the direction of the smaller volume. One volume each of PCl_3 and Cl_2, a total of 2 gas volumes, forms 1 volume of PCl_5. Hence a pressure increase will promote the reaction between PCl_3 and Cl_2 to form more PCl_5.

(c) Increasing the concentration of any component will displace the equilibrium in the direction that tends to lower the concentration of the component added. Increasing the concentration of Cl_2 will result in the consumption of more PCl_3 and the formation of more PCl_5, and this action will partially offset the increased concentration of Cl_2.

(d) Increasing the concentration of PCl_5 will result in the formation of more PCl_3 and Cl_2.

(e) A catalyst accelerates both forward and backward reactions equally. It speeds up the approach to equilibrium, but does not favor reaction in either direction over the other.

16.9. What conditions would you suggest for the manufacture of ammonia by the Haber process?

$$N_2\,(g) + 3H_2\,(g) \rightleftharpoons 2NH_3(g) \qquad \Delta H = -22 \text{ kcal}$$

From the sign of ΔH, we see that the forward reaction gives off heat (exothermic). Therefore the equilibrium formation of NH_3 is favored by as low a temperature as practicable. In a reaction like this, the choice of temperature requires a compromise between equilibrium considerations and rate considerations. The equilibrium yield of NH_3 is higher, the lower the temperature. On the other hand, the rate at which the system will reach equilibrium is lower, the lower the temperature.

The forward reaction is accompanied by a decrease in volume, since 4 volumes of initial gases yield 2 volumes of NH_3. Therefore increased pressure will give a larger proportion of NH_3 in the equilibrium mixture.

A catalyst should be employed to speed up the approach to equilibrium.

16.10. Consider the gaseous reaction

$$2NO_2 \rightleftharpoons N_2O_4$$

(a) Calculate ΔG° and K_p for this reaction at 25 °C. (b) Calculate ΔG° and K_p for the reverse reaction:

$$N_2O_4 \rightleftharpoons 2NO_2$$

(c) Calculate ΔG° and K_p for the forward reaction written with different balancing coefficients:

$$NO_2 \rightleftharpoons \tfrac{1}{2}N_2O_4$$

(d) Repeat calculation (a) for a standard state pressure of 1 bar. For NO_2 (g) ΔG_f° based on 1 bar is 51.32 kJ mol^{-1}, and for N_2O_4 (g) 97.89 kJ mol^{-1}.

(a)
$$\Delta G^\circ = (1 \text{ mol})[\Delta G_f^\circ(N_2O_4)] - (2 \text{ mol})[\Delta G_f^\circ(NO_2)]$$

$$= 97.82 - 2(51.30) = -4.78 \text{ kJ} = -RT \ln K_p$$

$$\log K_p = \frac{-\Delta G^\circ}{2.303RT} = \frac{4.78 \times 10^3 \text{ J}}{(2.303)(8.314 \text{ J/K})(298.1 \text{ K})} = 0.837$$

$$K_p = 6.87$$

(b)
$$\Delta G^\circ = (2 \text{ mol})[\Delta G_f^\circ(NO_2)] - (1 \text{ mol})[\Delta G_f^\circ(N_2O_4)] = 2(51.30) - 97.82 = 4.78 \text{ kJ}$$

$$\log K_p = \frac{-\Delta G^\circ}{2.303RT} = \frac{-4.78 \times 10^3}{(2.303)(8.314)(298.1)} = -0.837$$

$$K_p = 0.146$$

Parts (a) and (b) illustrate the general requirement that ΔG of a reverse reaction is the negative of ΔG for the forward reaction and K for a reverse reaction is the reciprocal of K for the forward reaction.

(c)
$$\Delta G^\circ = (\tfrac{1}{2} \text{ mol})[\Delta G_f^\circ(N_2O_4)] - (1 \text{ mol})[\Delta G_f^\circ(NO_2)]$$

$$= \tfrac{1}{2}(97.82) - 51.30 = -2.39 \text{ kJ}$$

$$\log K_p = \frac{-\Delta G^\circ}{2.303RT} = \frac{2.39 \times 10^3}{(2.303)(8.314)(298.1)} = 0.418$$

$$K_p = 2.62$$

Parts (a) and (c) illustrate the general result that ΔG for a reaction with halved coefficients is half the ΔG-value for the standard coefficients and that K for the reaction with halved coefficients is the one-half power of the K for the standard reaction. Note that the value of the ratio

$$\frac{P(N_2O_4)}{P^2(NO_2)}$$

at equilibrium must be independent of the way we write the balanced equation. Any equilibrium constant for the reaction must involve the two partial pressures in exactly this way. For part (a), K is equal to this ratio; for part (c), K is equal to the square-root of this ratio.

(d)
$$\Delta G^\circ = (1 \text{ mol})[\Delta G_f^\circ(N_2O_4)] - (2 \text{ mol})[\Delta G_f^\circ(NO_2)]$$
$$= 97.89 - 2(51.32) = -4.75 \text{ kJ} = -RT \ln K_p$$

$$\ln K_p = \frac{4.75 \times 10^3 \text{ J}}{(8.314 \text{ J/K})(298.1 \text{ K})} = 1.917 \qquad K_p = 6.80$$

Since 1 atm and 1 bar are close　(1 atm = 1.013 25 bar),　the differences are small, but not insignificant. K_p (bar) can also be calculated from K_p (atm) by simply converting units.

$$K_p(\text{bar}) = \frac{P(\text{atm})(N_2O_4) \times 1.013\,25}{[P(\text{atm})(NO_2) \times 1.013\,25]^2} = \frac{K_p(\text{atm})}{1.013\,25}$$

$$= \frac{6.87}{1.013\,25} = 6.78$$

16.11. A quantity of PCl_5 was heated in a 12-L vessel at 250 °C.

$$PCl_5\,(g) \rightleftharpoons PCl_3\,(g) + Cl_2\,(g)$$

At equilibrium the vessel contains 0.21 mol PCl_5, 0.32 mol PCl_3, and 0.32 mol Cl_2.　(a) Compute the equilibrium constant K_p for the dissociation of PCl_5 at 250 °C when pressures are referred to the standard state of 1 atm.　(b) What is ΔG° for the reaction?　(c) Estimate ΔG° from the data in Tables 7-1 and 16-1, assuming constancy of ΔH° and ΔS°.　(d) Calculate K_p from the original data using SI units and a standard state of 1 bar.

(a)
$$P(PCl_5) = \frac{nRT}{V} = \frac{(0.21 \text{ mol})[0.082\,1 \text{ L} \cdot \text{atm}/(\text{mol} \cdot \text{K})](523 \text{ K})}{12 \text{ L}} = 0.751 \text{ atm}$$

$$P(Cl_2) = P(PCl_3) = \frac{(0.32)(0.082\,1)(523)}{12} = 1.145 \text{ atm}$$

$$K_p = \frac{P(PCl_3)P(Cl_2)}{P(PCl_5)} = \frac{(1.145)(1.145)}{0.751} = 1.75$$

(b)
$$\Delta G^\circ = -RT \ln K = -(8.314 \text{ J/K})(523 \text{ K}) \ln 1.75 = -2.4 \text{ kJ}$$

(c)
$$\Delta G_{298}^\circ = -267.8 + 305.0 = 37.2 \text{ kJ}$$
$$\Delta H_{298}^\circ = -287.0 + 374.9 = 87.9 \text{ kJ}$$

$$\Delta G^\circ = \Delta H^\circ - T\,\Delta S^\circ \qquad \Delta S^\circ = \frac{\Delta H^\circ - \Delta G^\circ}{T} = \frac{(87.9 - 37.2)(1000)}{298.1} = 170.1 \text{ J/K}$$

At 523 K, $\Delta G_{523}^\circ = 87.9 - \dfrac{(523)(170.1)}{1\,000} = 87.9 - 89.0 = -1.1 \text{ kJ}$

The estimate is close to the experimental value based on the equilibrium measurement, despite the large temperature range over which ΔH° and ΔS° were assumed constant.

(d)
$$P(PCl_5) = \frac{nRT}{V} = \frac{(0.21 \text{ mol})[8.314 \text{ m}^3 \cdot \text{Pa}/(\text{mol} \cdot \text{K})](523 \text{ K})(1 \text{ bar}/10^5 \text{Pa})}{(12 \text{ L})(1 \text{ m}^3/10^3 \text{ L})} = 0.761 \text{ bar}$$

$$P(Cl_2) = P(PCl_3) = \frac{(0.32)(8.314)(523)(10^3)}{12 \times 10^5} = 1.160 \text{ bar}$$

$$K_p(\text{bar}) = \frac{(1.160)(1.160)}{0.761} = 1.77$$

16.12. When 1 mol of pure ethyl alcohol is mixed with 1 mol of acetic acid at room temperature, the equilibrium mixture contains $\frac{2}{3}$ mol each of ester and water. (a) What is the equilibrium constant? (b) What is $\Delta G°$ for the reaction? (c) How many moles of ester are formed at equilibrium when 3 mol of alcohol is mixed with 1 mol of acid? All substances are liquids at the reaction temperature.

(a)

	alcohol	**acid**	**ester**	**water**
	C_2H_5OH (l) +	CH_3COOH (l) $\rightleftharpoons$	$CH_3COOC_2H_5$ (l) +	H_2O (l)
(1) n at start:	1	1	0	0
(2) Change by reaction:	$-\frac{2}{3}$	$-\frac{2}{3}$	$+\frac{2}{3}$	$+\frac{2}{3}$
(3) n at equilibrium:	$1-\frac{2}{3}=\frac{1}{3}$	$1-\frac{2}{3}=\frac{1}{3}$	$\frac{2}{3}$	$\frac{2}{3}$

This tabular representation is a convenient way of doing the bookkeeping for equilibrium problems. Under each substance in the balanced equation an entry is made on three lines: (1) the amount of starting material; (2) the change (plus or minus) in the number of moles due to the attainment of equilibrium; and (3) the equilibrium amount, which is the algebraic sum of entries (1) and (2). The entries in line (2) must be in the same ratio to each other as the coefficients in the balanced chemical equation. The equilibrium constant can be found from the entries in line (3).

Let v = number of liters of mixture and choose, as usual, 1 mol/L as the standard-state concentration. Then

$$K = \frac{[\text{ester}][\text{water}]}{[\text{alcohol}][\text{acid}]} = \frac{\left(\frac{2}{3}\middle/v\right)\left(\frac{2}{3}\middle/v\right)}{\left(\frac{1}{3}\middle/v\right)\left(\frac{1}{3}\middle/v\right)} = 4$$

Note that the concentration of water is not so large compared with the other reaction components that it remains constant under all reaction conditions; the concentration of water must therefore appear in the expression for K, along with the concentrations of the reactants and of the other product.

(b)
$$\Delta G° = -RT \ln K = -(8.314 \text{ J/K})(298.1 \text{ K})[(2.303)(\log 4)]$$

$$= -3440 \text{ J} = -3.44 \text{ kJ}$$

Because K is independent of the choice of standard-state concentration (the sum of the exponents in the numerator equals the sum of the exponents in the denominator), so also is $\Delta G°$.

(c)　Let x = number of moles of alcohol reacting.

	C_2H_5OH +	CH_3COOH $\rightleftharpoons$	$CH_3COOC_2H_5$ +	H_2O
n at start:	3	1	0	0
Change by reaction:	$-x$	$-x$	$+x$	$+x$
n at equilibrium:	$3-x$	$1-x$	x	x

$$K = 4 = \frac{\left(\frac{x}{v}\right)\left(\frac{x}{v}\right)}{\left(\frac{3-x}{v}\right)\left(\frac{1-x}{v}\right)} = \frac{x^2}{3-4x+x^2}$$

Then $x^2 = 4(3-4x+x^2)$ or $3x^2 - 16x + 12 = 0$. Solving by the quadratic formula

$$x = \frac{-b \pm \sqrt{b^2-4ac}}{2a} = \frac{+16 \pm \sqrt{(16)^2-4(3)(12)}}{2(3)} = \frac{16 \pm 10.6}{6} = 4.4 \text{ or } 0.9$$

Of the two roots to the quadratic equation, only one has physical meaning. The applicable root can generally be selected very easily. In this problem, we started with 3 mol of alcohol and 1 mol of acid. We can see from the reaction equation that we cannot form more than 1 mol of ester, even if we use up all the acid. Therefore the correct root of the quadratic is 0.9. Consequently, 0.9 mol of ester is formed at equilibrium.

It is wise to confirm the numerical result, especially if the calculation is complex, by substituting back in the expression for K.

$$K = \frac{(0.9)(0.9)}{(2.1)(0.1)} = 3.86 = 4 \qquad \text{(within the limits of precision of the calculation)}$$

Note that the number of moles of ester formed is greater than the number of moles of ester in equilibrium in (a) (0.9 compared with 2/3). This result was to be expected because of the increased concentration of alcohol, one of the reactants. A further addition of alcohol would lead to an even greater yield of ester, but in no case could the amount of ester formed exceed 1 mol, since 1 mol would represent 100 percent conversion of the acid. In practice, the actual selection of the excess concentration to be used may depend on economic factors. If alcohol is cheap compared to acid and ester, a large excess of alcohol might be used to assure a high percentage conversion of the acid. If, on the other hand, alcohol costs more per mole than acid, it would be more sensible to use an excess of acid and aim for a high percentage conversion of the alcohol.

16.13. In a 10-L evacuated chamber, 0.5 mol H_2 and 0.5 mol I_2 are reacted at 448 °C.

$$H_2\,(g) + I_2\,(g) \rightleftharpoons 2HI\,(g)$$

At the given temperature, and for a standard state of 1 mol/L, $K = 50$ for the reaction. (a) What is the total pressure in the chamber? (b) How many moles of the iodine remain unreacted at equilibrium? (c) What is the partial pressure of each component in the equilibrium mixture?

(a) Before the reaction sets in, the total number of gas moles is $0.5 + 0.5 = 1$. During the reaction, there is no change in the total number of moles. The total pressure can be computed from the ideal gas law:

$$P(\text{total}) = \frac{n(\text{total})RT}{V} = \frac{(1\ \text{mol})(0.082\ 1\ \text{L}\cdot\text{atm}\cdot\text{K}^{-1}\cdot\text{mol}^{-1})(721\ \text{K})}{10\ \text{L}} = 5.9\ \text{atm}$$

(b) Let x = number of moles of iodine reacting.

	H_2	$+$	I_2	$\rightleftharpoons$	$2HI$
n at start:	0.5		0.5		0
Change by reaction:	$-x$		$-x$		$2x$
n at equilibrium:	$0.5 - x$		$0.5 - x$		$2x$

Note the 2:1 ratio of moles of HI formed to moles of H_2 reacted. This ratio is demanded by the coefficients in the balanced chemical equation. Regardless of how complete or incomplete the reaction may be, 2 mol of HI is always formed for every 1 mol of H_2 that reacts.

Since the number of moles of reactant gas equals the number of moles of product gas, moles may be used in place of concentrations (as in Prob. 16-12), and $K_p = K$.

$$K_p = 50 = \frac{(2x)^2}{(0.5 - x)(0.5 - x)} \qquad \text{or} \qquad \sqrt{50} = 7.1 = \frac{2x}{0.5 - x}$$

Then $2x = 7.1(0.5 - x)$. Solving, $x = 0.39$. Thus, 0.39 mol I_2 reacts, leaving

$$0.5 - 0.39 = 0.11\ \text{mol}\ I_2$$

unreacted at equilibrium. Note that if the negative square root were chosen the value of x would have been 0.7 mol, an impossibly high result, because no more I_2 can react than the amount initially added. Checking the result:

$$K = \frac{(2 \times 0.39)^2}{(0.11)^2} = 50.3 = 50 \qquad \text{(within the limits of precision of the calculation)}$$

(c)
$$P(I_2) = \frac{n(I_2)}{n(\text{total})} \times (\text{total pressure}) = \left(\frac{0.11}{1}\right)(5.9\ \text{atm}) = 0.65\ \text{atm}$$

$$P(H_2) = P(I_2) = 0.65\ \text{atm}$$

$$P(HI) = (\text{total pressure}) - [P(H_2) + P(I_2)] = 5.9 - 1.3 = 4.6\ \text{atm}$$

or
$$P(HI) = \frac{n(HI)}{n(\text{total})} \times (\text{total pressure}) = \left(\frac{0.78}{1}\right)(5.9\ \text{atm}) = 4.6\ \text{atm}$$

16.14. Sulfide ion in alkaline solution reacts with solid sulfur to form polysulfide ions having formulas S_2^{2-}, S_3^{2-}, S_4^{2-}, and so on. The equilibrium constant for the formation of S_2^{2-} is 12 and for the formation of S_3^{2-} is 130, both from S and S^{2-}. What is the equilibrium constant for the formation of S_3^{2-} from S_2^{2-} and S?

To avoid confusion, let us designate the equilibrium constants for the various reactions by subscripts. Also, we note that only the ion concentrations appear in the equilibrium constant equations, because solid sulfur, S, is always in its standard state.

$$S + S^{2-} \rightleftharpoons S_2^{2-} \qquad K_1 = [S_2^{2-}]/[S^{2-}] = 12$$
$$2S + S^{2-} \rightleftharpoons S_3^{2-} \qquad K_2 = [S_3^{2-}]/[S^{2-}] = 130$$
$$S + S_2^{2-} \rightleftharpoons S_3^{2-} \qquad K_3 = [S_3^{2-}]/[S_2^{2-}]$$

The desired constant, K_3, expresses the equilibrium ratio of S_3^{2-} to S_2^{2-} concentrations in a solution equilibrated with solid sulfur. Such a solution must also contain sulfide ion, S^{2-}, resulting from the dissociation of S_2^{2-} by the reverse of the first equation. Since all four species, S, S^{2-}, S_2^{2-} and S_3^{2-}, are present, all the equilibria represented above must be satisfied. The three equilibrium ratios are not all independent, because

$$\frac{[S_3^{2-}]}{[S_2^{2-}]} = \frac{[S_3^{2-}]/[S^{2-}]}{[S_2^{2-}]/[S^{2-}]} \qquad \text{or} \qquad K_3 = \frac{K_2}{K_1} = \frac{130}{12} = 11$$

The result, $K_2 = K_1 K_3$, is a general one for any case where one chemical equation (the second in this case) can be written as the sum of two other equations (the first and third in this case).

16.15. At 27 °C and 1 atm, N_2O_4 is 20 percent dissociated into NO_2. Find   (a) K_p and   (b) the percent dissociation at 27 °C and a total pressure of 0.10 atm.   (c) What is the extent of dissociation in a 69-g sample of N_2O_4 confined in a 20-L vessel at 27 °C?

(a)   When 1 mol of N_2O_4 gas dissociates completely, 2 mol of NO_2 gas is formed. Since this problem does not specify a particular size of reaction vessel or a particular weight of sample, we are free to choose 1 mol (92 g) as the starting amount of N_2O_4. For the given total pressure of 1 atm, the tabular computation of the partial pressures proceeds as follows:

	N_2O_4 (g)	$\rightleftharpoons$	$2NO_2$ (g)
n at start:	1		0
Change by reaction:	-0.2		$+0.4$
n at equilibrium:	$1 - 0.2 = 0.8$		0.4
Mole fraction:	$\dfrac{0.8}{0.8 + 0.4} = 0.67$		$\dfrac{0.4}{0.8 + 0.4} = 0.33$
p.p. = (mole frac.) × (1 atm):	0.67 atm		0.33 atm

Then, in terms of the numerical values of the partial pressures,

$$K_p = \frac{P(NO_2)^2}{P(N_2O_4)} = \frac{(0.33)^2}{0.67} = 0.17$$

(b)   Let α = fraction of N_2O_4 dissociated at equilibrium at 0.1 atm total pressure.

	N_2O_4	$\rightleftharpoons$	$2NO_2$
n at start:	1		0
Change by reaction:	$-\alpha$		2α
n at equilibrium:	$1 - \alpha$		2α
Mole fraction:	$\dfrac{1-\alpha}{1+\alpha}$		$\dfrac{2\alpha}{1+\alpha}$
Partial pressure:	$\left(\dfrac{1-\alpha}{1+\alpha} \times 0.1\right)$ atm		$\left(\dfrac{2\alpha}{1+\alpha} \times 0.1\right)$ atm

From (a), $K_p = 0.17$. Thus,

$$0.17 = K_p = \frac{P(NO_2)^2}{P(N_2O_4)} = \frac{\left(\dfrac{2\alpha}{1+\alpha} \times 0.1\right)^2}{\dfrac{1-\alpha}{1+\alpha} \times 0.1} = \frac{0.4\alpha^2}{1-\alpha^2}$$

or $0.4\alpha^2 = 0.17(1 - \alpha^2)$. Solving, $\alpha = 0.55 = 55\%$ dissociated at 27 °C and 0.1 atm.

Note that a larger fraction of the N_2O_4 is dissociated at 0.1 atm than at 1 atm. This is in agreement with Le Chatelier's principle: Decreasing the pressure should favour the side with the greater volume (NO_2).

(c)   If the sample were all N_2O_4, it would contain

$$\frac{69 \text{ g}}{92 \text{ g/mol}} = 0.75 \text{ mol}$$

Let α be the fractional dissociation. The tabular analysis follows:

	N_2O_4	$\rightleftharpoons$	$2NO_2$
n at start:	0.75		0
Change by reaction:	-0.75α		$+2(0.75\alpha)$
n at equilibrium:	$0.75(1-\alpha)$		1.50α

Because the total gas pressure is unknown, it is simplest to obtain the partial pressures directly from Dalton's law (Chapter 5).

$$P(N_2O_4) = \frac{n(N_2O_4)RT}{V} = \frac{[0.75(1-\alpha) \text{ mol}](0.082 \text{ L} \cdot \text{atm} \cdot \text{K}^{-1} \cdot \text{mol}^{-1})(300 \text{ K})}{20 \text{ L}}$$

$$= 0.92(1-\alpha) \text{ atm}$$

$$P(NO_2) = \frac{n(NO_2)RT}{V} = \frac{1.50\alpha}{0.75(1-\alpha)}[0.92(1-\alpha) \text{ atm}] = 1.84\alpha \text{ atm}$$

Then

$$0.17 = K_p = \frac{P(NO_2)^2}{P(N_2O_4)} = \frac{(1.84\alpha)^2}{0.92(1-\alpha)} = \frac{3.68\alpha^2}{1-\alpha}$$

or

$$3.68\alpha^2 + 0.17\alpha - 0.17 = 0$$

Solving,

$$\alpha = \frac{-0.17 \pm \sqrt{(0.17)^2 + 4(0.17)(3.68)}}{2(3.68)} = \frac{-0.17 \pm 1.59}{7.36} = -0.24 \text{ or } +0.19$$

The negative root must obviously be discarded. The extent of dissociation is 19%.

16.16.   Under what conditions will $CuSO_4 \cdot 5H_2O$ be *efflorescent* at 25 °C? How good a drying agent is $CuSO_4 \cdot 3H_2O$ at the same temperature? For the reaction

$$CuSO_4 \cdot 5H_2O \text{ (solid)} \rightleftharpoons CuSO_4 \cdot 3H_2O \text{ (solid)} + 2H_2O \text{ (gas)}$$

K_p at 25 °C is 1.086×10^{-4}; the vapor pressure of water at 25 °C is 23.8 torr.

An efflorescent salt is one that loses water to the atmosphere. This will occur if the water vapor pressure in equilibrium with the salt is greater than the water vapor pressure in the atmosphere. The mechanism by which $CuSO_4 \cdot 5H_2O$ could be efflorescent is that the salt would lose 2 molecules of H_2O and simultaneously form 1 formula unit of $CuSO_4 \cdot 3H_2O$ for each unit of the original salt that dissociates. Then the above equilibrium equation would apply, since all three components would be present.

Since $CuSO_4 \cdot 5H_2O$ and $CuSO_4 \cdot 3H_2O$ are both solids,

$$K_p = P(H_2O)^2 = 1.086 \times 10^{-4}$$

where $P(H_2O)$ is the partial pressure of water vapor (relative to the standard state pressure of 1 atm) in equilibrium with the two solids. Solving,

$$P(H_2O) = 1.042 \times 10^{-2} \text{ atm} = (1.042 \times 10^{-2} \text{ atm})(760 \text{ torr/atm}) = 7.92 \text{ torr}$$

Since $P(H_2O)$ is less than the vapor pressure of water at the same temperature, $CuSO_4 \cdot 5H_2O$ will not always effloresce. It will effloresce only on a dry day, when the partial pressure of moisture in the air is less than 7.92 torr. This will occur when the relative humidity is less than

$$\frac{7.92 \text{ torr}}{23.8 \text{ torr}} = 0.333 = 33.3\%$$

$CuSO_4 \cdot 3H_2O$ could act as a drying agent by reacting with 2 molecules of H_2O to form $CuSO_4 \cdot 5H_2O$. The same equilibrium would be set up as above, and the vapor pressure of water would be fixed at 7.92 torr. In other words, $CuSO_4 \cdot 3H_2O$ can reduce the moisture content of any confined volume of gas to 7.92 torr. It cannot reduce the moisture content below this value.

To find out the conditions under which $CuSO_4 \cdot 3H_2O$ would be efflorescent, we would have to know the equilibrium constant for another reaction, which shows the product of dehydration of $CuSO_4 \cdot 3H_2O$. This reaction is

$$CuSO_4 \cdot 3H_2O \ (s) \rightleftharpoons CuSO_4 \cdot H_2O \ (s) + 2H_2O \ (g)$$

Supplementary Problems

Use may be made of data in Tables 7-1 and 16-1 for the solution of the following problems. These data are based on experimental measurements and are subject to reevaluation. Thus, tables in different books may not always agree. The data posted in this book are part of an internally consistent set, and the answers are based on them.

THERMODYNAMICS

16.17. Calculate $S°$ at 25 °C for PCl_5 (g).

Ans.   364.3 J/K · mol

16.18. Calculate $\Delta H_f°$ for Cl_2O (g) at 25 °C.

Ans.   80.2 kJ/mol

16.19. Predict the phase-transition temperature for the conversion of gray to white tin, using the tabulated thermodynamic data.

Ans.   9 °C (the observed value is 13 °C)

16.20. Consider the production of water gas:

$$C \ (s, \ graphite) + H_2O \ (g) \rightleftharpoons CO \ (g) + H_2 \ (g)$$

(a) What is $\Delta G°$ for this reaction at 25 °C?   (b) Estimate the temperature at which $\Delta G°$ for the reaction becomes zero.

Ans.   (a) 91.44 kJ;   (b) 982 K. The extrapolation for this estimate is extended over such a large temperature range that an appreciable error might be expected. The actual experiment value, 947 K, is not far from the estimate.

16.21. ΔG_f° for the formation of HI (g) from its gaseous elements is -10.10 kJ/mol at 500 K. When the partial pressure of HI is 10 atm, and of I_2 0.001 atm, what must the partial pressure of hydrogen be at this temperature to reduce the magnitude of ΔG for the reaction to zero?

Ans.   776 atm

16.22. Under what conditions could the decomposition of Ag_2O (s) into Ag (s) and O_2 (g) proceed spontaneously at 25 °C?

Ans.   The partial pressure of O_2 must be kept below 0.089 6 torr.

16.23. The effect of changing the standard state from 1 atm to 1 bar is to raise the molar entropies, S°, of all gaseous substances by 0.109 J/(K · mol) at 298.1 K. Convert ΔG_f° of   *(a)* $CH_3OH(l)$,   *(b)* $CH_3OH(g)$,   *(c)* $H_2O(l)$, and   *(d)* Sn $(s,$ gray$)$ to the standard state of 1 bar. Assume that within the precision of the tables in this text ΔH_f° of all substances is unchanged by this small change in standard state.

Ans.   *(a)* -166.28 kJ/mol;   *(b)* -161.95 kJ/mol;   *(c)* -237.14 kJ/mol;   *(d)* 0.12 kJ/mol

16.24. $N_2 + O_2 \rightleftharpoons 2NO$. State the effect upon the reaction equilibrium of   *(a)* increased temperature,   *(b)* decreased pressure,   *(c)* higher concentration of O_2,   *(d)* lower concentration of N_2,   *(e)* higher concentration of NO,   *(f)* presence of a catalyst.

Ans.   Favors   *(a)* forward reaction,   *(b)* neither reaction,   *(c)* forward reaction,   *(d)* backward reaction,   *(e)* backward reaction,   *(f)* neither reaction.

16.25. Predict the effect upon the following reaction equilibria of: *(a)* increased temperature,   *(b)* increased pressure.
   1. CO (g) + H_2O (g) $\rightleftharpoons$ CO_2 (g) + H_2 (g)
   2. $2SO_2$ (g) + O_2 (g) $\rightleftharpoons$ $2SO_3$ (g)
   3. N_2O_4 (g) $\rightleftharpoons$ $2NO_2$ (g)
   4. H_2O (g) $\rightleftharpoons$ H_2 (g) + $\frac{1}{2}O_2$ (g)
   5. $2O_3$ (g) $\rightleftharpoons$ $3O_2$ (g)
   6. CO (g) + $2H_2$ (g) $\rightleftharpoons$ CH_3OH (g)
   7. $CaCO_3$ (s) $\rightleftharpoons$ CaO (s) + CO_2 (g)
   8. C (s) + H_2O (g) $\rightleftharpoons$ H_2 (g) + CO (g)
   9. 4HCl (g) + O_2 (g) $\rightleftharpoons$ $2H_2O$ (g) + $2Cl_2$ (g)
   10. C$(s,$ *diamond*$)$ $\rightleftharpoons$ C$(s,$ *graphite*$)$ (This equilibrium can exist only under very special conditions.) Densities of diamond and graphite are 3.5 and 2.3 g/cm^3, respectively.

Ans.   F = favors forward reaction, B = favors backward reaction.

1. *(a)* B; *(b)* neither	*4.* *(a)* F; *(b)* B	*7.* *(a)* F; *(b)* B	*10.* *(a)* B; *(b)* B
2. *(a)* B; *(b)* F	*5.* *(a)* B; *(b)* B	*8.* *(a)* F; *(b)* B	
3. *(a)* F; *(b)* B	*6.* *(a)* B; *(b)* F	*9.* *(a)* B; *(b)* F	

EQUILIBRIUM CONSTANTS

16.26. When α-D-glucose is dissolved in water it undergoes a partial conversion to β-D-glucose, a sugar of the same molecular weight but of slightly different physical properties. This conversion, called *mutarotation*, stops when 63.6 percent of the glucose is in the β-form. Assuming that equilibrium has been attained, calculate K and $\Delta G°$ for the reaction α-D-glucose $\rightleftharpoons$ β-D-glucose at the experimental temperature.

Ans.   1.75, -1.38 kJ

16.27. The equilibrium constant for the reaction H_3BO_3 + glycerin $\rightleftharpoons$ (H_3BO_3-glycerin) is 0.9. How much glycerin should be added per liter of 0.10 molar H_3BO_3 solution so that 60 percent of the H_3BO_3 is converted to the boric acid–glycerin complex?

Ans.   1.7 mol

16.28. The equilibrium

$$p\text{-xyloquinone} + \text{methylene white} \rightleftharpoons p\text{-xylohydroquinone} + \text{methylene blue}$$

may be studied conveniently by observing the difference in color between methylene blue and methylene white. One millimole of methylene blue was added to a liter of solution that was 0.24 M in p-xylohydroquinone and 0.0120 M in p-xyloquinone. It was then found that 4.0 percent of the added methylene blue was reduced to methylene white. What is the equilibrium constant for the above reaction? The equation is balanced with one molecule of each of the four substances.

Ans. 4.8×10^2

16.29. A saturated solution of iodine in water contains 0.33 g I_2 per liter. More than this can dissolve in a KI solution because of the following equilibrium:

$$I_2\,(aq) + I^- \rightleftharpoons I_3^-$$

A 0.10 M KI solution (0.10 M I^-) actually dissolves 12.5 g of iodine per liter, most of which is converted to I_3^-. Assuming that the concentration of I_2 in all saturated solutions is the same, calculate the equilibrium constant for the above reaction. What is the effect of adding water to a clear saturated solution of I_2 in the KI solution?

Ans. 710; backward reaction is favored.

16.30. $H_2\,(g) + I_2\,(g) \rightleftharpoons 2HI\,(g)$. When 46 g of I_2 and 1.0 g of H_2 are heated to equilibrium at 470 °C, the equilibrium mixture contains 1.9 g I_2. (*a*) How many moles of each gas are present in the equilibrium mixture? (*b*) Compute the equilibrium constant.

Ans. (*a*) 0.0075 mol I_2, 0.32 mol H_2, 0.35 mol HI; (*b*) $K = 50$

16.31. Exactly 1 mol each of H_2 and I_2 are heated in a 30-L evacuated chamber to 470 °C. Using the value of K from Problem 16.30, determine (*a*) how many moles of I_2 remain unreacted when equilibrium is established, (*b*) the total pressure in the chamber, (*c*) the partial pressures of I_2 and of HI in the equilibrium mixture. (*d*) Now if one additional mole of H_2 is introduced into this equilibrium system, how many moles of the original iodine will remain unreacted?

Ans. (*a*) 0.22 mol; (*b*) 4.1 atm; (*c*) $P(H_2) = P(I_2) = 0.45$ atm, $P(HI) = 3.2$ atm; (*d*) 0.065 mol

16.32. $PCl_5\,(g) \rightleftharpoons PCl_3\,(g) + Cl_2\,(g)$. Calculate the number of moles of Cl_2 produced at equilibrium when 1 mol of PCl_5 is heated at 250 °C in a vessel having a capacity of 10 L. At 250 °C, $K = 0.041$ for this dissociation based on the 1 mol/L standard state.

Ans. 0.47 mol

16.33. Pure PCl_5 is introduced into an evacuated chamber and comes to equilibrium (see Problem 16.32) at 250 °C and 2.00 atm. The equilibrium gas contains 40.7% chlorine by volume.
(*a*) 1. What are the partial pressures of the gaseous components at equilibrium?
   2. Calculate K_p at 250 °C, based on the 1 atm standard state, for the reaction as written in Problem 16.32.
(*b*) If the gas mixture is expanded to 0.200 atm at 250 °C, calculate:
   1. The percent of PCl_5 that would be dissociated at equilibrium
   2. The percent by volume of Cl_2 in the equilibrium gas mixture
   3. The partial pressure of Cl_2 in the equilibrium mixture

Ans. (*a*) 1. $P(Cl_2) = P(PCl_3) = 0.814$ atm, $P(PCl_5) = 0.372$ atm; 2. 1.78
(*b*) 1. 94.8%; 2. 48.7%; 3. 0.0974 atm

16.34. At 46 °C, K_p for the reaction

$$N_2O_4\,(g) \rightleftharpoons 2NO_2\,(g)$$

is 0.67, based on the 1 bar standard state. Compute the percent dissociation of N_2O_4 at 46 °C and a total pressure of 0.507 bar. What are the partial pressures of N_2O_4 and NO_2 at equilibrium?

Ans. 50%; $P(N_2O_4) = 0.17$ bar, $P(NO_2) = 0.34$ bar

16.35. $2NOBr\ (g) \rightleftharpoons 2NO\ (g) + Br_2\ (g)$. If nitrosyl bromide (NOBr) is 34 percent dissociated at 25 °C and a total pressure of 0.25 bar, calculate K_p for the dissociation at this temperature, based on the 1 bar standard state.

Ans. 1.0×10^{-2}

16.36. The equilibrium constant for the reaction

$$CO\ (g) + H_2O\ (g) \rightleftharpoons CO_2\ (g) + H_2\ (g)$$

at 986 °C is 0.63. A mixture of 1 mol of water vapor and 3 mol of CO is allowed to come to equilibrium at a total pressure of 2 atm.   (a) How many moles of H_2 are present at equilibrium?   (b) What are the partial pressures of the gases in the equilibrium mixture?

Ans.   (a) 0.68 mol;   (b) $P(CO) = 1.16$ atm, $P(H_2O) = 0.16$ atm, $P(CO_2) = P(H_2) = 0.34$ atm

16.37. For the reaction

$$SnO_2\ (solid) + 2H_2\ (gas) \rightleftharpoons 2H_2O\ (steam) + Sn\ (molten)$$

calculate K_p (a) at 900 K, where the equilibrium steam-hydrogen mixture was 45% H_2 by volume;   (b) at 1100 K, where the equilibrium steam-hydrogen mixture was 24% H_2 by volume.   (c) Would you recommend higher or lower temperatures for more efficient reduction of tin?

Ans.   (a) 1.5;   (b) 10;   (c) higher

16.38. In the preparation of quicklime from limestone, the reaction is

$$CaCO_3\ (s) \rightleftharpoons CaO\ (s) + CO_2\ (g)$$

Experiments carried out between 850 °C and 950 °C led to a set of K_p-values fitting an empirical equation

$$\log K_p = 7.282 - \frac{8\,500}{T}$$

where T is the absolute temperature. If the reaction is carried out in quiet air, what temperature would be predicted from this equation for the complete decomposition of the limestone? In quiet air assume it is necessary to build up the CO_2 pressure to 1 atm to ensure continual removal of the product.

Ans.   894 °C

16.39. The moisture content of a gas is often expressed in terms of the *dew point*. The dew point is the temperature to which the gas must be cooled before the gas becomes saturated with water vapor. At this temperature, water or ice (depending on the temperature) will be deposited on a solid surface.

     The efficiency of $CaCl_2$ as a drying agent was measured by a dew point experiment. Air at 0 °C was allowed to pass slowly over large trays containing $CaCl_2$. The air was then passed through a glass vessel through which a copper rod was sealed. The rod was cooled by immersing the emergent part of it in a dry ice bath. The temperature of the rod inside the glass vessel was measured by a thermocouple. As the rod was cooled slowly the temperature at which the first crystals of frost were deposited was observed to be −43 °C. The vapor pressure of ice at this temperature is 0.07 torr. Assuming that the $CaCl_2$ owes its desiccating properties to the formation of $CaCl_2 \cdot 2H_2O$, calculate K_p at 0 °C for the reaction

$$CaCl_2 \cdot 2H_2O\ (s) \rightleftharpoons CaCl_2\ (s) + 2H_2O\ (g)$$

Ans. 8×10^{-9}

16.40. At high temperatures the following equilibria prevail in a mixture of carbon, oxygen, and their compounds:

$$C\ (s) + O_2\ (g) \rightleftharpoons CO_2\ (g) \qquad K_1$$
$$2C\ (s) + O_2\ (g) \rightleftharpoons 2CO\ (g) \qquad K_2$$
$$C\ (s) + CO_2\ (g) \rightleftharpoons 2CO\ (g) \qquad K_3$$
$$2CO\ (g) + O_2\ (g) \rightleftharpoons 2CO_2\ (g) \qquad K_4$$

If it were possible to measure K_1 and K_2 independently, how could K_3 and K_4 be calculated?

Ans. $K_3 = K_2/K_1$; $K_4 = K_1/K_3 = K_1^2/K_2$

Chapter 17

Acids and Bases

The general principles of chemical equilibrium apply equally to reactions of neutral molecules and to reactions of ions. Because of the special interest which chemists have in ionic equilibria in aqueous solutions and because of some common methods of treating some of these problems, this chapter and the following are devoted to the specific applications of chemical equilibrium to ionic reactions. As in Chapter 16, concentrations will always be expressed in mol/L, and [X] will represent the numerical value of the concentration of X.

ACIDS AND BASES

Arrhenius concept

According to the classical definition as formulated by Arrhenius, an *acid* is a substance which can yield H^+ in solution. $HClO_4$ and HNO_3, which are completely ionized in water into H^+ and ClO_4^- and into H^+ and NO_3^-, respectively, are called *strong acids*. $HC_2H_3O_2$, acetic acid, and HNO_2, nitrous acid, are only partly ionized into H^+ and $C_2H_3O_2^-$ and into H^+ and NO_2^-, and these substances are called *weak acids*. The dissociation of a *weak acid* is reversible in aqueous solutions and may be described by an equilibrium constant, usually designated as K_a. Thus

$$HC_2H_3O_2 \rightleftharpoons H^+ + C_2H_3O_2^- \qquad K_a = \frac{[H^+][C_2H_3O_2^-]}{[HC_2H_3O_2]} \qquad (17\text{-}1)$$

Similarly, a *base* is a substance which can yield OH^-. NaOH, a *strong base*, is completely ionized in water into Na^+ and OH^-; even relatively insoluble hydroxides, such as $Ca(OH)_2$, give solutions within their limited solubility range which are completely ionized. A solution of ammonia, NH_3, in water also produces hydroxide ions, and so at one time was thought to consist of NH_4OH in solution. Since the OH^- concentration is only a few percent of the ammonia concentration, ammonia is considered a *weak base*.

Brönsted-Lowry concept

The Arrhenius concept of acids and bases has been modified and generalized by the Brönsted-Lowry picture, which allows for the specific role of the solvent. The proton is the essential ingredient in the definition of both acids and bases in the Brönsted-Lowry scheme, according to which an *acid* is a substance which can transfer a proton to another substance, which could be the solvent itself, and a *base* is a substance, which might be the solvent, which accepts a proton from an acid. The species remaining after an acid has transferred a proton is also a base, since in principle it could regain the proton to form the original acid. Such a grouping of an acid and its deprotonated form is said to constitute a *conjugate acid-base pair*. The generalized proton-transfer reaction is then represented as

$$HA + B \rightleftharpoons A^- + BH^+ \qquad (17\text{-}2)$$

In this equation, HA-A^- and B-BH^+ constitute conjugate acid-base pairs. Neither HA nor B is necessarily a neutral species, but in any event the charge on the base conjugate to HA is algebraically one unit less than the charge on HA, and the charge on the acid conjugate to B is one positive unit greater than on B. In this notation, (17-1) would be written as follows:

$$HC_2H_3O_2 + H_2O \rightleftharpoons C_2H_3O_2^- + H_3O^+ \qquad K_a = \frac{[H_3O^+][C_2H_3O_2^-]}{[HC_2H_3O_2]} \qquad (17\text{-}3)$$

In a similar fashion the ionization of the weak base, ammonia, is written:

$$H_2O + NH_3 \rightleftharpoons OH^- + NH_4^+ \qquad K_b = \frac{[NH_4^+][OH^-]}{[NH_3]} \qquad (17\text{-}4)$$

H_2O serves as a base in (17-3) and as an acid in (17-4). Note that the bare H^+ of (17-1) becomes the hydrated proton or hydronium ion, H_3O^+, of (17-3). In the formulation of equilibrium constants, $[H^+]$ and $[H_3O^+]$ are always equivalent to each other; the two forms are used interchangeably in most contexts and will be so used in all ionic equilibrium problems in this book. Although the proton is indeed hydrated in aqueous solutions, the notation H^+ is often used instead of H_3O^+ because the reader understands the fact of hydration, because he need not worry about specifying the exact extent of hydration (which exceeds one H_2O per proton), and because the specific use of the hydrated formula for the proton might obscure the important fact that *all* ions in water are extensively hydrated. Note also that the denominator in the K_a expression in (17-3) is identical with that in (17-1); since applications of these equilibria are intended for dilute solutions, H_2O is always taken to be in its standard state and therefore need not be represented by a term in the K_a expression. Equation (17-4) avoids describing aqueous ammonia as NH_4OH, a species that probably does not exist at ambient temperatures.

The strengths of acids can be compared in terms of their K_a-values; the stronger the acid, the larger its K_a. The strong acids, which are completely ionized in water, cannot be described by the K_a formalism applicable to dilute aqueous solutions, but other solvents not as basic as water (i.e., not as effective in removing protons) might allow only partial and reversible ionization and thus permit distinctions among the strong acids to be made. For example, HNO_3 becomes a weak acid with a measurable K_a in ethanol solvent, while $HClO_4$ and HCl remain strong. In acetone solvent, HCl becomes measurably weak and $HClO_4$ remains strong.

Lewis concept

The Lewis categorization of acids and bases uses an even more general picture. In this view an *acid* is a structure which has an affinity for electron pairs contributed by *bases*, defined as substances with unshared pairs of electrons. Examples of acids include not only the proton, which can react with the unshared pairs of such bases as H_2O, OH^-, and $C_2H_3O_2^-$, but also transition-metal ions, which can react with ligands (bases) to form complexes, and electron-deficient substances such as BF_3, which can react with a base like NH_3 to form the compound

$$\begin{array}{ccc} H & & F \\ | & & | \\ H-N&-B&-F \\ | & & | \\ H & & F \end{array}$$

IONIZATION OF WATER

Since water is amphiprotic, that is, it can act both as an acid and as a base, every aqueous solution is characterized by the auto-ionization process, in which one H_2O molecule transfers a proton to another H_2O molecule. The water auto-ionization equilibrium must always be satisfied, whether other acids or bases are present in the solution or not.

$$2H_2O \rightleftharpoons H_3O^+ + OH^- \qquad K_w = [H^+][OH^-] \qquad (17\text{-}5)$$

The unhydrated form of the proton is written in the K_w expression, but the H_3O^+ notation is of course acceptable. Although the number of H_2O molecules in the balanced chemical equation depends on whether the proton or hydronium ion is written, $[H_2O]$ does not appear in the K_w expression since applications are usually limited to dilute solutions where H_2O remains in its standard state. In the problems of this book, total concentrations of nonelectrolytes do not exceed 1 M, and concentrations of electrolytes normally do not exceed 0.1 M. For solutions containing larger ionic concentrations, the same laws of equilibrium apply if proper correction is made for the electrical interactions between ions. Such corrections will not be made in this book. Instead, examples will be chosen for which the numerical solutions should be correct to within 10 percent even without corrections for these electrostatic interactions. In general, 10 percent is about the limit of accuracy expected for a calculated concentration in this and in the following chapter.

At 25 °C, $K_w = [H^+][OH^-] = 1.00 \times 10^{-14}$; this value should be memorized. In pure water containing no other acid or base, the concentrations of H^+ and OH^- must equal each other. Therefore, at 25 °C,

$$[H^+] = [OH^-] = \sqrt{1.00 \times 10^{-14}} = 1.00 \times 10^{-7}$$

Hence a *neutral solution* may be defined as one in which $[H^+] = [OH^-] = \sqrt{K_w}$. (The value of K_w depends on temperature. Thus, at 0 °C in pure water, $[H^+] = [OH^-] = 0.34 \times 10^{-7}$.) At 25 °C, an *acid solution* is one in which the $[H^+]$ is greater than 10^{-7}, or one in which the $[OH^-]$ is less than 10^{-7}; a *basic solution* is one in which the $[H^+]$ is less than 10^{-7}, or one in which the $[OH^-]$ is greater than 10^{-7}.

The acidity of alkalinity of a solution is often expressed by its *pH*. By definition,

$$pH = \log \frac{1}{[H^+]} = -\log [H^+] \quad \text{or} \quad [H^+] = 10^{-pH}$$

Thus, if the $[H^+]$ of a solution is expressed as a simple power of 10, the pH of the solution is equal to the negative of the exponent. The smaller the pH, the greater the acidity. Similarly, we define

$$pOH = -\log [OH^-]$$

The smaller the pOH, the greater the alkalinity.

The two measures are related by

$$pH + pOH = -\log K_w = 14.00 \quad \text{(at 25 °C)} \quad (17\text{-}6)$$

Table 17-1 summarizes the pH and pOH scales.

Table 17-1

[H$^+$]		[OH$^-$]	pH	pOH	
1	$= 10^0$	10^{-14}	0	14	Strongly acidic
0.1	$= 10^{-1}$	10^{-13}	1	13	
0.001	$= 10^{-3}$	10^{-11}	3	11	
0.000 01	$= 10^{-5}$	10^{-9}	5	9	Weakly acidic
0.000 000 1	$= 10^{-7}$	10^{-7}	7	7	Neutral
0.000 000 001	$= 10^{-9}$	10^{-5}	9	5	Weakly basic
0.000 000 000 01	$= 10^{-11}$	10^{-3}	11	3	
0.000 000 000 000 1	$= 10^{-13}$	10^{-1}	13	1	
0.000 000 000 000 01	$= 10^{-14}$	1	14	0	Strongly basic

An uncertainty of 10 percent in $[H^+]$, the upper limit for problems in this book, corresponds to an uncertainty of 0.04 pH unit.

For convenience, pK_a, defined as $-\log K_a$, is often used to express the strength of an acid. Thus an acid whose ionization constant is 10^{-4} has a pK_a of 4. Similarly, $pK_b = -\log K_b$; and for any equilibrium constant, $pK = -\log K$.

HYDROLYSIS

A salt containing at least one ion which is conjugate to a weak acid or base undergoes a reaction with water of an acid-base nature. For example, sodium acetate, $NaC_2H_3O_2$, has an ion which is the conjugate base to acetic acid, a weak acid. The acetate ion is a base and can accept protons from acids, or from the solvent, water:

$$C_2H_3O_2^- + H_2O \rightleftharpoons HC_2H_3O_2 + OH^- \qquad K_b = \frac{[HC_2H_3O_2][OH^-]}{[C_2H_3O_2^-]} \qquad (17\text{-}7)$$

The above reaction is called *hydrolysis*. As a result of this process, a solution of sodium acetate in water is basic because an excess of OH^- is produced. Note that the reaction (17-5) is the algebraic sum of the reactions (17-3) and (17-7). The equilibrium constants of these three reactions must then be related as follows (see Problem 16.14):

$$K_w = K_a K_b \qquad \text{or} \qquad K_b = \frac{K_w}{K_a} \qquad (17\text{-}8)$$

An equation of the type (17-8) applies to the hydrolysis of any species which is the conjugate base to an acid having an ionization constant K_a. Some examples are CN^-, HS^-, SCN^-, NO_2^-. Since hydrolysis involves a reverse of acid dissociation, the tendency toward hydrolysis runs counter to the tendency of the conjugate acid toward ionization. The weaker the acid, the greater the difficulty of removing a proton from the acid, and the easier for its anion, or conjugate base, to attach a proton from water (that is, hydrolyze). This relationship appears mathematically as the inverse proportionality between K_a of the acid and K_b of the conjugate base. Acetic acid is a *moderately weak* acid, and the acetate ion hydrolyzes to a *slight* extent. HCN is a *very weak* acid and the cyanide ion, CN^-, hydrolyzes to a *great* extent. Chloride ion, on the other hand, does not undergo the hydrolysis reaction at all, because its conjugate acid, HCl, is a *strong* acid and cannot form to any appreciable extent in dilute aqueous solutions.

An analogous reaction will be shown by a positive ion which can act as an acid in water solution to form the weak conjugate base. Thus, a solution of ammonium chloride would be acidic because of the reaction

$$NH_4^+ + H_2O \rightleftharpoons NH_3 + H_3O^+$$

K_a for NH_4^+ can be obtained from K_b for NH_3, its conjugate base, by means of Equation (17-8)

$$K_a = \frac{K_w}{K_b}$$

Many heavy-metal cations hydrolyze to some extent in aqueous solution, a reaction characterized by an equilibrium constant, K_a. For example,

$$Fe^{3+} + H_2O \rightleftharpoons Fe(OH)^{2+} + H^+ \qquad K_a = \frac{[Fe(OH)^{2+}][H^+]}{[Fe^{3+}]} \qquad (17\text{-}9)$$

The above reaction is sometimes written with the hydrated forms of ions to show that ferric ion, like the neutral acids, demonstrates its acidity by loss of a proton.

$$Fe(H_2O)_6^{3+} \rightleftharpoons Fe(H_2O)_5(OH)^{2+} + H^+$$

The two formulations of the equation are equivalent.

BUFFER SOLUTIONS AND INDICATORS

If the $[H^+]$ (or pH) of a solution is not appreciably affected by the addition of small amounts of acids and bases, the solution is said to be *buffered*. A solution will have this property if it contains relatively large amounts of both a weak acid and a weak base. If a small amount of a strong acid is added to this solution, most of the added H^+ will combine with an equivalent amount of the weak base of the buffer to form the conjugate acid of that weak base; thus the $[H^+]$ of the solution remains almost constant. If a small amount of a strong base is added to the buffer solution, most of the OH^- will combine with an equivalent amount of the weak acid of the buffer to form the conjugate base of that weak acid. In this way, the $[H^+]$ of the buffer solution is not appreciably affected by the addition of small amounts of acid or base.

Any pair of weak acid and base can be used to form a buffer solution, as long as each can form its conjugate base or acid in water solution.

EXAMPLE 1   A particularly simple and common case of the buffer solution is one in which the weak acid and weak base are conjugates of each other. Thus acetic acid may be chosen as the weak acid and acetate ion as the weak base. Since relatively large amounts of each are needed, it would not be satisfactory to use just a solution of acetic acid in water, in which the concentration of the acetate ion is relatively small. An acetic acid–acetate buffer can be made in any of the following three ways:

1.   Dissolve relatively large amounts of acetic acid and an acetate salt (sodium or potassium acetate, for example) in water.
2.   Dissolve a relatively large amount of acetic acid in water. *Partially* neutralize the acid by adding some strong base, like sodium hydroxide. The amount of acetate formed will be equivalent to the amount of strong base added. The amount of acetic acid left in solution will be the starting amount minus the amount converted to acetate.
3.   Dissolve a relatively large amount of an acetate salt in water. *Partially* neutralize the acetate by adding some strong acid, like HCl. The amount of acetic acid formed will be equivalent to the amount of strong acid added. The amount of acetate ion left in solution will be the starting amount minus the amount converted to acetic acid.

The ratio of acetic acid to acetate ion in solution can be chosen so as to give a desired $[H^+]$ or pH for the buffer solution. Usually the ratio is kept within the limits 10 and 0.1. According to a rearranged form of the ionization equilibrium equation for acetic acid:

$$[H^+] = K_a \times \frac{[HC_2H_3O_2]}{[C_2H_3O_2^-]}$$

More generally,

$$[H^+] = K_a \times \frac{[\text{acid}]}{[\text{conjugate base}]} \qquad (17\text{-}10)$$

Taking logarithms of (*17-10*) and reversing the signs yields the convenient form:

$$pH = pK_a + \log \frac{[\text{base}]}{[\text{acid}]}$$

An *indicator* is used to give a visual indication of the pH of a solution. Like a buffer it is a conjugate acid-base pair, but contrary to a buffer it is present in such a small concentration that it does not affect the overall pH. Rather the pH determines the ratio of acid form to conjugate base form. Since at least one form, and usually both, are intensely colored (but different in color), the pH determines the color of the solution. If the acid dissociation of the indicator is written,

$$HIn \rightleftharpoons H^+ + In^- \qquad (17\text{-}11)$$

then
$$[H^+] = K_{a(\text{indicator})} \times \frac{[HIn]}{[In^-]}$$

If the ratio exceeds about 10 the color will appear to be purely that of HIn, the acid form, and if it is less than about 0.1 it will be purely that of the base form In^-. Hence there will be a gradual change in color over about a 100-fold range in $[H^+]$ or about a range of 2 pH units, with pK_a near the center of the range.

(The greater the difference in color intensity between the acid and base forms, the further off-center will be pK_a.) A variety of indicators is known, each with its own pK_a and range.

WEAK POLYPROTIC (POLYBASIC) ACIDS

If multiple ionization is possible, as in H_2S and H_2CO_3, each stage of ionization has its own equilibrium constant. Subscripts are usually used to distinguish the various constants.

$$\text{Primary ionization:} \qquad H_2S \rightleftharpoons H^+ + HS^- \qquad K_1 = \frac{[H^+][HS^-]}{[H_2S]} \qquad (17\text{-}12)$$

$$\text{Secondary ionization:} \qquad HS^- \rightleftharpoons H^+ + S^{2-} \qquad K_2 = \frac{[H^+][S^{2-}]}{[HS^-]} \qquad (17\text{-}13)$$

The secondary ionization constant of polyprotic acids is always smaller than the primary ($K_2 < K_1$); the tertiary, if there is one, is smaller than the secondary; and so on.

It should be clear that $[H^+]$ means the actual concentration of hydrogen ions *in solution*. In an aqueous mixture containing several acids, the different acids all contribute to the hydrogen ion concentration in solution, but there is only one value of $[H^+]$ in any given solution and this value must simultaneously satisfy the equilibrium conditions for all the different acids. Although it might seem very complex to solve a problem in which several equilibria are involved, simplifications can be made when all sources but one make only a negligible contribution (less than 10 percent, for the purposes of this book) to the total concentration of a particular ion.

In the case of polyprotic acids, K_1 is often so much greater than K_2 that only the K_1 equilibrium need be considered to compute $[H^+]$ in a solution of the acid. Examples where this assumption may and may not be made will be given in specific problems.

Another problem of interest is the calculation of the concentration of the divalent ion in a solution of a weak polyprotic acid when the total $[H^+]$ is essentially that due to a stronger acid present in the solution or to a buffer. In such a case the concentration of the divalent ion can best be calculated by multiplying the expressions for K_1 and K_2. Again illustrating with H_2S, we find

$$K_1K_2 = \frac{[H^+][HS^-]}{[H_2S]} \times \frac{[H^+][S^{2-}]}{[HS^-]} = \frac{[H^+]^2[S^{2-}]}{[H_2S]} \qquad (17\text{-}14)$$

APPARENT EQUILIBRIUM CONSTANTS

Simplifications of multiple equilibria systems are often used in biochemical and other organic reactions involving acids or bases. Consider the hydrolysis of acetyl phosphate as an example. (This is not the same type of hydrolysis as was discussed above.)

$$CH_3CO_2PO_3^{2-} + H_2O \rightleftharpoons CH_3CO_2^- + HPO_4^{2-} + H^+ \qquad (17\text{-}15)$$

In addition to the above fundamental hydrolysis reaction, separate acid-base equilibria exist for the acetyl phosphate, acetate, and hydrogen phosphate ions. The biochemist may want to concentrate on the hydrolysis itself and not be concerned with the distribution of $CH_3CO_2PO_3^{2-}$ and its conjugate acid $CH_3CO_2PO_3H^-$ or of HPO_4^{2-} and its conjugate acid $H_2PO_4^-$. In fact, analytical methods do not recognize the separate stages of ionization of a weak acid but monitor only the stoichiometric concentrations, i.e., the sum of the concentrations of all ionization species of each substance. In order to avoid confusion with a true, well-characterized ionization species, the stoichiometric concentration is usually designated by a verbal notation or abbreviation instead of a formula. Thus [acetyl phosphate] or [AcP] might be used to designate the sum of the concentrations of all the ionized or un-ionized forms of acetyl phosphoric acid, and [phosphate] or [P_i] to designate the sum of the concentrations of all the inorganic forms of phosphoric acid and its successive conjugate bases. The convention has been adopted to define K', an *apparent equilibrium constant*, in which stoichiometric concentrations appear rather than concentrations of species in particular states of ionization.

$$K' = \frac{[Ac][P_i]}{[AcP]} \tag{17-16}$$

Calculations involving K' are no different from those of Chapter 16, but in giving the value of K' the pH as well as the temperature must be specified.

TITRATION

When a base is added in small increments to a solution of acid, the pH of the solution rises on each addition of base. When the pH is plotted against the amount of base added, the steepest rise occurs at the equivalence point, when the acid is exactly neutralized. This region of steepest rise is called the *end point*, and the whole process of base addition and determination of the end point is called *titration*. The graph showing the change of pH during the titration is called a *titration curve*.

EXAMPLE 2   Figure 17-1a shows the titration at 25 °C of 50.0 cm³ of an acid, either strong or weak, with strong base, and Fig. 17-1b the titration of 50.0 cm³ of a strong base and also a weak base with strong acid. All reagents are 0.100 M so that the end point always occurs on the addition of exactly 50.0 cm³ of titrant. The strong acid is HCl, the weak acid $HC_4H_7O_3$(β-hydroxybutyric acid), the strong base NaOH, and the weak base NH_3.

All curves display a very sharp rise or fall in pH at the end point. When HCl is titrated (Fig 17-1a) the pH rises slowly until very near the end point. The rise at the end point is much greater than for the $HC_4H_7O_3$ which starts at a

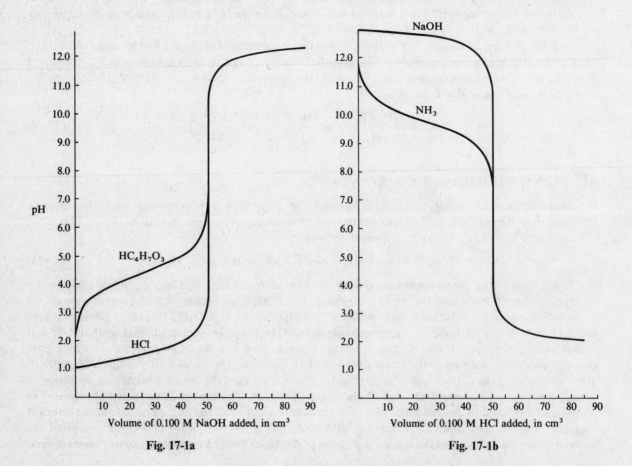

Fig. 17-1a   Fig. 17-1b

much higher pH, has an initial rapid rise, reverses its curvature, and eventually undergoes the very sharp rise typical of the end point. Beginning slightly beyond the end point the two curves are identical. The titration curves for NaOH and NH_3 (Fig 17-1b) are almost mirror images of those for HCl and $HC_4H_7O_3$ with decreasing pH. There is no practical reason for using a weak reagent as a titrant. If, for instance, NH_3 were used in place of NaOH in Fig. 17-1a the curve would level off beyond the end point at least 3 pH units lower, thus making it more difficult to discern the end point.

The points along a titration curve can be calculated by methods previously discussed in this chapter. Four parts of the curve can be distinguished. Consider the titration of acid with strong base.

1. Starting point. 0% neutralization.

    In the case of the strong acid, $[H^+]$ in the initial solution is simply the molar concentration of the acid. In the case of the weak acid, $[H^+]$ is calculated by the method used to compute the extent of ionization of any weak acid in terms of its ionization constant and molar concentration.
2. Approach to the end point. 5% to 95% neutralization.

    For the case of the strong acid, the neutralization reaction,

$$H^+ + OH^- \rightarrow H_2O$$

may be assumed to go to completion to the extent of the amount of base added. The amount of unreacted H^+ is then the difference between the initial amount of H^+ and the amount neutralized. To determine $[H^+]$ allowance is made for the dilution effect of increasing the total volume of the solution on addition of base.

For the case of the weak acid, the neutralization reaction may be written

$$HC_4H_7O_3 + OH^- \rightarrow C_4H_7O_3^- + H_2O$$

The amount of the hydroxybutyrate ion, $C_4H_7O_3^-$, is equal to the amount of base added. The amount of un-ionized acid, $HC_4H_7O_3$, is the difference between the initial amount and the amount neutralized. Then,

$$[H^+] = K_{a(HC_4H_7O_3)} \frac{[HC_4H_7O_3]}{[C_4H_7O_3^-]}$$

The above is Equation (17-10); all solutions in this region are buffers.
3. The end point. 100% neutralization.

    The pH at the end point is the same as for a solution of the salt containing the ions remaining at neutralization, NaCl or $NaC_4H_7O_3$. NaCl solutions are neutral and have a pH of 7.00. The other salt hydrolyzes, and the pH can be evaluated by solving the hydrolysis equilibria. The $HC_4H_7O_3$–NaOH titration has a pH greater than 7 at the end point because of the hydrolysis of the hydroxybutyrate ion.
4. Extension beyond the end point. Over 105% neutralization.

    With NaOH titrations, the excess OH^- beyond that needed for neutralization accumulates in the solution. The $[OH^-]$ is computed in terms of this excess and the total volume of solution. $[H^+]$ can then be computed from the K_w relationship.

$$[H^+] = \frac{K_w}{[OH^-]}$$

It makes no difference whether the acid titrated was weak or strong.

Points within 5% of the starting point or of the end point can be computed by the same equilibria, but some of the simplifying assumptions made above are no longer valid.

Polyprotic acids like H_3PO_4 may have two or more distinct end points, corresponding to neutralization of the first, second, and subsequent hydrogens. In such a case each end point would occur at a different pH.

The calculations for the titration curves of bases with strong acid (Fig. 17-1b) are done by similar methods.

The end point of a titration, the region of steepest rise in the titration curve, can be determined experimentally if an instrument is available for measuring the pH after each addition of base. A simpler means is to introduce into the solution a small amount of an indicator chosen so that its range lies within the steep vertical portion of the curve, ensuring a sharp color change at the end point.

Solved Problems

ACIDS AND BASES

17.1. Write the formulas for the conjugate bases of the following acids: (a) HCN, (b) HCO_3^-, (c) $N_2H_5^+$, (d) C_2H_5OH.

In each case the conjugate base is derived from the acid by the loss of a proton. In (d) the proton is lost from the highly electronegative oxygen rather than from carbon. (a) CN^-, (b) CO_3^{2-}, (c) N_2H_4, (d) $C_2H_5O^-$.

17.2. Write the formulas for the conjugate acids of the following bases: (a) $HC_2H_3O_2$, (b) HCO_3^-, (c) C_5H_5N, (d) $N_2H_5^+$.

In each case the conjugate acid is formed from the base by the addition of a proton. The proton is added to oxygen in (a) and (b), or to nitrogen in (c) and (d), all of which have unshared pairs available. In (a) it is added to the carbonyl ($-C=O$) oxygen.

(a) $H_2C_2H_3O_2^+$. This is a species which might form in liquid acetic acid upon addition of a strong acid.
(b) H_2CO_3. Note that HCO_3^- can serve as both an acid [Problem 17.1(b)] and a base.
(c) $C_5H_5NH^+$.
(d) $N_2H_6^{2+}$. Note that bases, like acids, can be polyfunctional. The second proton is accepted by N_2H_4, however, only with great difficulty.

17.3. Liquid NH_3, like water, is an amphiprotic solvent. Write the equation for the auto-ionization of NH_3.

$$2NH_3 \rightleftharpoons NH_4^+ + NH_2^-$$

17.4. Aniline, $C_6H_5NH_2$, is a weak organic base in aqueous solutions. Suggest a solvent in which aniline would become a strong base.

Obviously, a solvent is needed which has appreciably stronger acid properties than water. One such solvent, in which aniline is a strong base, is liquid acetic acid; many other acidic solvents would have the same effect.

17.5. NH_4ClO_4 and $HClO_4 \cdot H_2O$ both crystallize in the same orthorhombic structure, having unit-cell volumes of 0.395 and 0.370 nm^3, respectively. How do you account for the similarity in crystal structure and crystal dimension?

Both substances are ionic substances with lattice sites occupied by cations and anions. In "perchloric acid monohydrate," the cation is H_3O^+ and there are no H_2O *molecules* of hydration. The cations in the two crystals, H_3O^+ and NH_4^+, should occupy nearly equal amounts of space because they are isoelectronic (have the same number of electrons).

The data in this problem have been cited as one of the proofs for the existence of the hydronium ion, H_3O^+.

17.6. (a) Give an explanation for the decreasing acid strength in the series $HClO_4$, $HClO_3$, $HClO_2$. (b) What would be the relative basic strengths of ClO_4^-, ClO_3^-, ClO_2^-? (c) Keeping in mind the discussion of (a), how do you account for the fact that there is very little difference in acid strength in the series H_3PO_4, H_3PO_3, H_3PO_2?

(a) The Lewis structures are

Structures with double bonds to terminal oxygen atoms may also be written (see Problem 9.41). Since oxygen is more electronegative than chlorine, each terminal oxygen tends to withdraw electrons from the chlorine and in turn from the O—H bond, thus increasing the tendency for the proton to dissociate. In general, the greater the number of terminal oxygens in an oxy-acid, the stronger the acid.

(b) Members of a conjugate acid-base pair stand in complementary relationship to each other. The stronger an acid, the weaker its conjugate base. In (17-2), for example, the greater the strength of HA for a given B, the further the equilibrium distribution lies to the right; the further the equilibrium distribution lies to the right, the less effective is A^-, the base conjugate to HA, in acting as a base by accepting a proton. Thus the order of decreasing basic strength is ClO_2^-, ClO_3^-, ClO_4^-, the reverse of the order of decreasing strength of the conjugate acids.

(c) The hydrogens in these acids are not all bonded to oxygens. The Lewis structures are

(or alternative structures showing double bonds to terminal oxygens). The number of terminal oxygen atoms, one, is exactly the same in all three acids, so that, according to (a), no great differences in acidity are expected. Since the electronegativities of P and H are almost the same, there is no tendency for the phosphorus-bound H to ionize, nor to influence the ionization of the oxygen-bound H's

17.7. How do you account for the formation of $S_2O_3^{2-}$ from SO_3^{2-} and S in terms of Lewis acid theory?

A sulfur atom is electron-deficient and can thus be regarded as an acid. SO_3^{2-}, for which an octet structure can be written, is the base.

17.8. In combustion analysis of organic compounds, solid NaOH is used to absorb the CO_2 from the combustion gases. Account for this reaction in terms of Lewis acid theory.

The oxygen in the OH^- ion has three unshared pairs and is thus a Lewis base. To understand how CO_2 functions as an acid, note that the carbon in CO_2 is sp hybridized, but by reverting to sp^2 hybridization as in HCO_3^-, an orbital is made available to accept the base pair.

IONIZATION OF ACIDS AND BASES

17.9. At 25 °C, a 0.0100 M ammonia solution is 4.1% ionized. Calculate   (a) the concentration of the OH^- and NH_4^+ ions,   (b) the concentration of molecular ammonia,   (c) the ionization constant of aqueous ammonia,   (d) $[OH^-]$ after 0.0090 mol of NH_4Cl is added to 1 L of the above solution,   (e) [OH$^-$] of a solution prepared by dissolving 0.010 mol of NH$_3$ and 0.005 0 mol of HCl per liter.

$$NH_3 + H_2O \rightleftharpoons NH_4^+ + OH^-$$

The label on a solution refers to the stoichiometric or weight composition and does not indicate the concentration of any particular component of an ionic equilibrium. Thus the designation 0.010 0 M NH$_3$ means that the solution might have been made by dissolving 0.010 0 mol ammonia in enough water to make a liter of solution. It does not mean that the concentration of un-ionized ammonia in solution, [NH$_3$], is 0.010 0. Rather, the sum of the ionized and the un-ionized ammonia is equal to 0.010 0 mol per liter.

(a) $$[NH_4^+] = [OH^-] = (0.041)(0.010 0) = 0.000 41$$

(b) $$[NH_3] = 0.010 0 - 0.000 41 = 0.009 6$$

(c) $$K_b = \frac{[NH_4^+][OH^-]}{[NH_3]} = \frac{(0.000 41)(0.000 41)}{0.009 6} = 1.75 \times 10^{-5}$$

(d)   Since the base is so slightly ionized, we may assume that (1) the [NH$_4^+$] is completely derived from the NH$_4$Cl and that (2) the [NH$_3$] at equilibrium is the same as the stoichiometric molar concentration of the base. Then

$$K_b = \frac{[NH_4^+][OH^-]}{[NH_3]}$$

gives $$[OH^-] = \frac{K_b[NH_3]}{[NH_4^+]} = \frac{(1.75 \times 10^{-5})(0.010 0)}{0.009 0} = 1.94 \times 10^{-5}$$

The addition of NH$_4$Cl represses the ionization of NH$_3$, thus reducing greatly the [OH$^-$] of the solution; this phenomenon is called the *common-ion effect*.

Assumptions made to simplify the solution of a problem should always be verified after the problem has been solved. In this case we assumed that practically all the NH$_4^+$ came from the added NH$_4$Cl. In addition to this, there is a small amount of NH$_4^+$ resulting from the dissociation of the NH$_3$. From the chemical equation given in the statement of the problem, the amount of NH$_4^+$ coming from the NH$_3$ must equal the amount of OH$^-$, which we now know is 1.94×10^{-5} M. The correct total [NH$_4^+$] is then the sum of the contributions of NH$_4$Cl and NH$_3$, 0.009 0 + (1.94×10^{-5}). This sum is indeed equal to 0.009 0, well within our 10 percent allowance. We have thus justified our assumption.

The reader may wonder why we must make assumptions at all. This problem could have been solved by a more complete analysis as follows, without the assumption:

$$NH_3 + H_2O \rightleftharpoons NH_4^+ + OH^-$$

	NH$_3$	NH$_4^+$	OH$^-$
M at start from NH$_3$:	0.010 0	0	0
M at start from NH$_4^+$:	0	0.009 0	0
Change by reaction:	$-x$	$+x$	$+x$
M at equilibrium:	0.010 0 $- x$	0.009 0 $+ x$	x

$$[OH^-] = x = \frac{K_b[NH_3]}{[NH_4^+]} = \frac{(1.75 \times 10^{-5})(0.010 0 - x)}{0.009 0 + x}$$

This is a quadratic equation in x and can be solved by the usual methods. In the solution,

$$x = \frac{-b \pm \sqrt{b^2 - 4ac}}{2a}$$

the acceptable root requires the same sign for the square root as the sign of b. The value of the square root, however, is so close to the value of b, that the square root must be evaluated to four significant figures in order that x be known to even two figures. There are mathematical methods of solving the problem by approximations without having to evaluate the square root to four significant figures, but these mathematical approximations are essentially equivalent to the *chemical* approximations we had originally made.

In this case, failure to make the chemical approximations would have led to mathematical complications. Because of this, and because of the desirability of applying chemical intuition to problem analysis, we will make simplifying approximations in this book in advance, wherever they seem to make chemical sense. After solving the problem, we will always check the validity of our assumptions.

(e) Since HCl is a strong acid, the 0.005 0 mol of HCl will react completely with 0.005 0 mol of NH_3 to form 0.005 0 mol NH_4^+. Of the original 0.010 0 mol of NH_3, only half will remain as un-ionized ammonia.

$$[OH^-] = \frac{K_b[NH_3]}{[NH_4^+]} = \frac{(1.75 \times 10^{-5})(0.005\,0)}{0.005\,0} = 1.75 \times 10^{-5}$$

Check of assumption: The amount of NH_4^+ contributed by dissociation of NH_3 must be equal to the amount of OH^-, or 1.75×10^{-5} mol/L. This is indeed small compared with the 0.005 0 mol/L formed by neutralization of NH_3 with HCl.

17.10. Calculate the molar concentration of an acetic acid solution which is 2.0 percent ionized. K_a for $HC_2H_3O_2$ is 1.75×10^{-5} at 25 °C.

$$HC_2H_3O_2 \rightleftharpoons H^+ + C_2H_3O_2^-$$

Let x = molar concentration of acetic acid solution. Then

$$[H^+] = [C_2H_3O_2^-] = 0.020x \qquad \text{and} \qquad [HC_2H_3O_2] = x - 0.020x \approx x$$

the approximation being well within 10 percent.

$$\frac{[H^+][C_2H_3O_2^-]}{[HC_2H_3O_2]} = K_a \qquad \text{or} \qquad \frac{(0.020x)(0.020x)}{x} = 1.75 \times 10^{-5}$$

Solving, $(0.020)^2 x = 1.75 \times 10^{-5}$ or $x = 0.044$.

17.11. Calculate the percent ionization of a 1.00 M solution of hydrocyanic acid, HCN. K_a of HCN is 4.93×10^{-10}.

$$HCN \rightleftharpoons H^+ + CN^-$$

Since H^+ and CN^- are present in the solution only as a result of the ionization, their concentrations must be equal. (The contribution of the ionization of water to $[H^+]$ can be safely neglected.)

Let $x = [H^+] = [CN^-]$. Then $[HCN] = 1.00 - x$. Let us assume that x will be very small compared with 1.00, so that $[HCN] = 1.00$, within the allowed 10 percent error. Then

$$K_a = \frac{[H^+][CN^-]}{[HCN]} \qquad \text{or} \qquad 4.93 \times 10^{-10} = \frac{x^2}{1.00}$$

whence $x = 2.22 \times 10^{-5}$.

$$\text{Percent ionization} = \frac{\text{ionized HCN}}{\text{total HCN}} \times 100\% = \frac{2.22 \times 10^{-5} \text{ mol/L}}{1.00 \text{ mol/L}} \times 100\% = 0.002\,22\%$$

Check of assumption: $x(= 2.22 \times 10^{-5})$ is indeed very small compared with 1.00. (Or, as long as the percent ionization is less than 10 percent, the assumption is justified.)

17.12. The $[H^+]$ in a 0.020 M solution of benzoic acid is 1.1×10^{-3}. Compute K_a for the acid from this information.

$$HC_7H_5O_2 \rightleftharpoons H^+ + C_7H_5O_2^-$$

Since the hydrogen ion and benzoate ion come only from the ionization of the acid, their concentrations must be equal. (We have neglected the contribution of the ionization of water to $[H^+]$.)

$$[H^+] = [C_7H_5O_2^-] = 1.1 \times 10^{-3} \qquad [HC_7H_5O_2] = 0.020 - (1.1 \times 10^{-3}) = 0.019$$

$$K_a = \frac{[H^+][C_7H_5O_2^-]}{[HC_7H_5O_2]} = \frac{(1.1 \times 10^{-3})^2}{0.019} = 6.4 \times 10^{-5}$$

17.13. The ionization constant of formic acid, HCO_2H, is 1.77×10^{-4}. What is the percent ionization of a $0.001\,00$ M solution of this acid?

Let $x = [H^+] = [HCO_2^-]$; then $[HCO_2H] = 0.001\,00 - x$. Assume, as in Problem 17.11, that the percent ionization is less than 10 percent and that the formic acid concentration, $0.001\,00 - x$, may be approximated as $0.001\,00$. Then

$$K_a = \frac{[H^+][HCO_2^-]}{[HCO_2H]} = \frac{x^2}{0.001\,00} = 1.77 \times 10^{-4} \qquad \text{or} \qquad x = 4.2 \times 10^{-4}$$

In checking the assumption, we see that x is not negligible compared with $0.001\,00$. Therefore the assumption and the solution based on it must be rejected and the full quadratic form of the equation must be solved.

$$\frac{x^2}{0.001\,00 - x} = 1.77 \times 10^{-4}$$

Solving, $x = 3.4 \times 10^{-4}$. (We reject the negative root, -5.2×10^{-4}.)

$$\text{Percent ionization} = \frac{\text{ionized } HCO_2H}{\text{total } HCO_2H} \times 100\% = \frac{3.4 \times 10^{-4}}{0.001\,00} \times 100\% = 34\%$$

This exact solution (34 percent ionization) shows that the solution based on the original assumption was in error by about 20 percent.

17.14. What concentration of acetic acid is needed to give a $[H^+]$ of 3.5×10^{-4}? K_a is 1.75×10^{-5}.

Let $x =$ number of moles of acetic acid per liter.

$$[H^+] = [C_2H_3O_2^-] = 3.5 \times 10^{-4} \qquad [HC_2H_3O_2] = x - 3.5 \times 10^{-4}$$

$$\frac{[H^+][C_2H_3O_2^-]}{[HC_2H_3O_2]} = \frac{(3.5 \times 10^{-4})^2}{x - 3.5 \times 10^{-4}} = 1.75 \times 10^{-5} \qquad \text{or} \qquad x = 7.3 \times 10^{-3}$$

17.15. A 0.100 M solution of an acid (density $= 1.010$ g/cm^3) is 4.5 percent ionized. Compute the freezing point of the solution. The molecular weight of the acid is 300.

We must first determine the molality of the solution, i.e., the number of moles of acid dissolved in 1 kg of water.

$$\text{Mass of 1 L of solution} = (1\,000 \text{ cm}^3)(1.010 \text{ g/cm}^3) = 1\,010 \text{ g}$$

$$\text{Mass of solute in 1 L of solution} = (0.100 \text{ mol})(300 \text{ g/mol}) = 30 \text{ g}$$

$$\text{Mass of water in 1 L of solution} = 1\,010 \text{ g} - 30 \text{ g} = 980 \text{ g}$$

$$\text{Molality of solution} = \frac{0.100 \text{ mol acid}}{0.980 \text{ kg water}} = 0.102 \text{ mol/kg}$$

If the acid were not ionized at all, the freezing-point lowering would be

$$1.86 \times 0.102 = 0.190 \,°C$$

(see Chapter 14). Because of ionization, the total number of dissolved particles is greater than 0.102 mol per kg of solvent. The freezing-point depression is determined by the total number of dissolved particles, regardless of whether they are charged or uncharged.

Let $\alpha =$ fraction ionized. For every mole of acid added to the solution, there will be $(1 - \alpha)$ moles of un-ionized acid at equilibrium, α moles of H^+, and α moles of anion base conjugate to the acid, or a total of $(1 + \alpha)$ moles of dissolved particles. Hence the molality with respect to all dissolved particles is $(1 + \alpha)$ times the molality computed without regard to ionization.

$$\text{Freezing-point depression} = (1 + \alpha)(1.86)(0.102) = (1.045)(1.86)(0.102) = 0.198 \,°C$$

The freezing point of the solution is $-0.198 \,°C$.

17.16. A solution was made up to be 0.010 0 M in chloroacetic acid, $HC_2H_2O_2Cl$, and also 0.002 0 M in sodium chloroacetate, $NaC_2H_2O_2Cl$. K_a for chloroacetic acid is 1.40×10^{-3}. What is $[H^+]$ in the solution?

We could proceed as in part (d) of Problem 17.9 and assume that the concentration of chloroacetate ion can be approximated by the molar concentration of the sodium salt, 0.002 0, and that the extent of dissociation of the acid is small. Letting $[H^+] = x$,

$$[C_2H_2O_2Cl^-] = 0.002\,0 + x \approx 0.002\,0 \qquad [HC_2H_2O_2Cl] = 0.010\,0 - x \approx 0.010\,0$$

Then $\qquad x = [H^+] = K_a\dfrac{[HC_2H_2O_2Cl]}{[C_2H_2O_2Cl^-]} = (1.40 \times 10^{-3})\left(\dfrac{0.010\,0}{0.002\,0}\right) = 7.0 \times 10^{-3}$

Check: We see that the assumption led to a self-inconsistency. The value of x obtained, 7.0×10^{-3}, can be neglected in comparison with neither 0.002 0 nor 0.010 0.

We must start again, without making the simplifying assumptions.

$$x = [H^+] = (1.40 \times 10^{-3})\left(\frac{0.010\,0 - x}{0.002\,0 + x}\right)$$

The solution of the resulting quadratic equation gives the proper value for $[H^+]$, 2.4×10^{-3}.

Chloroacetic acid is apparently strong enough that the common-ion effect does not repress its ionization to a residual value small enough to be neglected. A hint of this result might have been taken from the relatively large value of K_a.

17.17. Calculate $[H^+]$ and $[C_2H_3O_2^-]$ in a solution that is 0.100 M in $HC_2H_3O_2$ and 0.050 M in HCl. K_a for $HC_2H_3O_2$ is 1.75×10^{-5}.

The HCl contributes so much more H^+ than the $HC_2H_3O_2$ that we can take $[H^+]$ as equal to the molar concentration of the HCl, 0.050; this is another example of the common-ion effect.

Then, if $[C_2H_3O_2^-] = x$, we have $[HC_2H_3O_2] = 0.100 - x \approx 0.100$ and

$$[C_2H_3O_2^-] = \frac{[HC_2H_3O_2]K_a}{[H^+]} = \frac{(0.100)(1.75 \times 10^{-5})}{0.050} = 3.5 \times 10^{-5}$$

Check of assumptions: (1) Contribution of acetic acid to $[H^+]$, x, is indeed small compared with 0.050; (2) x is indeed small compared with 0.100.

17.18. Calculate $[H^+]$, $[C_2H_3O_2^-]$, and $[CN^-]$ in a solution that is 0.100 M in $HC_2H_3O_2$ ($K_a = 1.75 \times 10^{-5}$) and 0.200 M in HCN ($K_a = 4.93 \times 10^{-10}$).

This problem is similar to the preceding one in that one of the acids, acetic, completely dominates the other in terms of contribution to the total $[H^+]$ of the solution. We base this assumption on the fact that K_a for $HC_2H_3O_2$ is much greater than for HCN; we will check the assumption after solving the problem. We will proceed by treating the acetic acid as if the HCN were not present.

Let $[H^+] = [C_2H_3O_2^-] = x$; then $[HC_2H_3O_2] = 0.100 - x \approx 0.100$ and

$$\frac{[H^+][C_2H_3O_2^-]}{[HC_2H_3O_2]} = \frac{x^2}{0.100} = 1.75 \times 10^{-5} \qquad \text{or} \qquad x = 1.32 \times 10^{-3}$$

Check of assumption: x is indeed small compared with 0.100.

Now we treat the HCN equilibrium established at a value of $[H^+]$ determined by the acetic acid, 1.32×10^{-3}. Let $[CN^-] = y$; then

$$[HCN] = 0.200 - y \approx 0.200$$

and $\qquad y = [CN^-] = \dfrac{K_a[HCN]}{[H^+]} = \dfrac{(4.93 \times 10^{-10})(0.200)}{1.32 \times 10^{-3}} = 7.5 \times 10^{-8}$

Check of assumptions: (1) y is indeed small compared with 0.200; (2) the amount of H^+ contributed by HCN ionization, equal to the amount of CN^- formed (7.5×10^{-8} mol/L), is indeed small compared with the amount of H^+ contributed by $HC_2H_3O_2$ (1.32×10^{-3} mol/L).

17.19. Calculate $[H^+]$ in a solution that is 0.100 M HCOOH ($K_a = 1.77 \times 10^{-4}$) and 0.100 M HOCN ($K_a = 3.3 \times 10^{-4}$).

This is a case in which two weak acids both contribute to $[H^+]$, neither contributing such a preponderant amount that the other's share can be neglected.

	HCOOH	$\rightleftharpoons$	H^+	+	HCO_2^-	HOCN	$\rightleftharpoons$	H^+	+	OCN^-
M (initial):	0.100		0		0	0.100		0		0
Change by reaction:	$-x$		$+x$		$+x$	$-y$		$+y$		$+y$
M (equilibrium):	$0.100 - x$		$x + y$		x	$0.100 - y$		$x + y$		y
Approx. M (equilibrium):	0.100		$x + y$		x	0.100		$x + y$		y

The final line in the above tabulation is based on the assumption that x and y are both small compared with 0.100.

$$\frac{x(x + y)}{0.100} = 1.77 \times 10^{-4} \qquad \frac{y(x + y)}{0.100} = 3.3 \times 10^{-4}$$

Dividing the HOCN equation by the HCOOH equation,

$$\frac{y}{x} = \frac{3.3}{1.77} = 1.86 \qquad \text{or} \qquad y = 1.86x$$

Subtracting the HCOOH equation from the HOCN equation,

$$\frac{y(x + y) - x(x + y)}{0.100} = 1.5 \times 10^{-4} \qquad \text{or} \qquad y^2 - x^2 = 1.5 \times 10^{-5}$$

Substitute $y = 1.86x$ into the last equation and solve to obtain $x = 2.5 \times 10^{-3}$. Then

$$y = 1.86x = 4.6 \times 10^{-3} \qquad \text{and} \qquad [H^+] = x + y = 7.1 \times 10^{-3}$$

Check of assumptions: The values of x and y are less than 10 percent of 0.100. (They are not much less than 10 percent. In this case we are just about at the limit of error we are allowing in this chapter.)

IONIZATION OF WATER

17.20. Calculate the $[H^+]$ and the $[OH^-]$ in 0.100 M $HC_2H_3O_2$ which is 1.31 percent ionized.

$$[H^+] = (0.0131)(0.100) = 1.31 \times 10^{-3}$$

$$[OH^-] = \frac{1.00 \times 10^{-14}}{[H^+]} = \frac{1.00 \times 10^{-14}}{1.31 \times 10^{-3}} = 7.6 \times 10^{-12}$$

Note that $[H^+]$ is computed as if the $HC_2H_3O_2$ were the only contributor, whereas $[OH^-]$ is based on the ionization of water. Obviously, if water ionizes to supply OH^-, it must supply an equal amount of H^+ at the same time. Implied in this solution is the assumption that water's contribution to $[H^+]$, 7.6×10^{-12} mol/L, is negligible compared with that of the $HC_2H_3O_2$. This assumption will be valid in all but the most dilute, or weakest, acid solutions. In comparing $[OH^-]$, however, water is the only source and it therefore cannot be overlooked.

17.21. Determine the $[OH^-]$ and the $[H^+]$ in 0.0100 M ammonia solution which is 4.1 percent ionized.

$$[OH^-] = (0.041)(0.0100) = 4.1 \times 10^{-4}$$

$$[H^+] = \frac{1.00 \times 10^{-14}}{[OH^-]} = \frac{1.00 \times 10^{-14}}{4.1 \times 10^{-4}} = 2.4 \times 10^{-11}$$

Here we have made the assumption that the contribution of water to $[OH^-]$ (equal to $[H^+]$, or 2.4×10^{-11}) is negligible compared with that of NH_3. K_w is used to compute $[H^+]$, since water is the only supplier of H^+. In general, $[H^+]$ for acidic solutions can usually be computed without regard to the water equilibrium; then K_w is used to compute $[OH^-]$. Conversely, $[OH^-]$ for basic solutions can usually be computed without regard to the water equilibrium; then K_w is used to compute $[H^+]$.

17.22. Express the following H^+ concentrations in terms of pH: (a) 1×10^{-3} mol/L,   (b) 5.4×10^{-9} mol/L.

(a) $\text{pH} = -\log [H^+] = -\log 10^{-3} = 3$

(b) $\text{pH} = -\log [H^+] = -\log (5.4 \times 10^{-9}) = -\log 5.4 + 9 = -0.73 + 9 = 8.27$

   Part (b) can be done on an electronic calculator by merely entering 5.4×10^{-9}, pressing LOG, and changing the sign of the result. The presentation here (and in subsequent problems) would be necessary if a table of logarithms had to be used. Students (almost all of whom have calculators) should study these examples anyway to gain a better understanding of how logarithms work.

17.23. Calculate the pH-values, assuming complete ionization, of   (a) 4.9×10^{-4} N acid,   (b) 0.0016 N base

(a)   Here $[H^+] = 4.9 \times 10^{-4}$ (see Chapter 12 for the definition of normality).

$$\text{pH} = -\log [H^+] = -\log (4.9 \times 10^{-4}) = -\log 4.9 + 4 = -0.69 + 4 = 3.31$$

(b) $[H^+] = \dfrac{10^{-14}}{[OH^-]} = \dfrac{10^{-14}}{1.6 \times 10^{-3}}$

$$\text{pH} = -\log [H^+] = -\log \frac{10^{-14}}{1.6 \times 10^{-3}} = -(-14 - \log 1.6 + 3) = 14 + 0.20 - 3 = 11.20$$

17.24. Change the following pH-values to $[H^+]$-values: (a) 4,   (b) 3.6.

(a) $[H^+] = 10^{-\text{pH}} = 10^{-4}$

(b) $[H^+] = 10^{-\text{pH}} = 10^{-3.6} = 10^{0.4-4} = 10^{0.4} \times 10^{-4}$

From a table of logarithms,

$$10^{0.4} = \text{antilog } 0.4 = 2.5$$

and so $[H^+] = 2.5 \times 10^{-4}$.

   Note that a negative number like -3.6 cannot be found in tables of logarithms, nor can -0.6. Only positive mantissas appear in the printed logarithm tables. The positive mantissa was achieved by adding and subtracting the next higher integer to the negative number: $-3.6 = 4.0 - 3.6 - 4.0 = 0.4 - 4.0$. Then 0.4 can be found to be the logarithm of 2.5.

17.25. What is the pH of   (a) 5.0×10^{-8} M HCl,   (b) 5.0×10^{-10} M HCl?

(a)   If we were to consider only the contribution of the HCl to the acidity of the solution, $[H^+]$ would be 5.0×10^{-8} and the pH would be greater than 7. This obviously cannot be true, because a solution of a pure acid, no matter how dilute, cannot be less acid than pure water alone. It is necessary in this problem to take into account something that we have omitted in all previous problems dealing with acid solutions, the contribution of water to the total acidity. A complete analysis of the water equilibrium is required.

$$\text{H}_2\text{O} \quad \rightleftharpoons \quad \text{H}^+ \quad + \quad \text{OH}^-$$

M from HCl:	5.0×10^{-8}	
Change by ionization of H_2O:	x	x
M at equilibrium:	$5.0 \times 10^{-8} + x$	x

$$K_w = [H^+][OH^-] = (5.0 \times 10^{-8} + x)x = 1.00 \times 10^{-14}$$

Solution of the quadratic equation gives the result: $x = 0.78 \times 10^{-7}$. Then $[H^+] = 5.0 \times 10^{-8} + x = 1.28 \times 10^{-7}$ and

$$pH = -\log (1.28 \times 10^{-7}) = 6.89$$

(b) Although the method of (a) could be used here, the problem can be simplified by noting that the HCl is so dilute as to make only a negligible contribution to $[H^+]$ as compared with the ionization of water. We may therefore write directly: $[H^+] = 1.00 \times 10^{-7}$ and $pH = 7.00$.

HYDROLYSIS

17.26. Calculate the extent of hydrolysis in a 0.0100 M solution of NH_4Cl. K_b for NH_3 is 1.75×10^{-5}.

$$NH_4^+ \rightleftharpoons NH_3 + H^+$$

$$K_a = \frac{[NH_3][H^+]}{[NH_4^+]} = \frac{K_w}{K_b} = \frac{1.00 \times 10^{-14}}{1.75 \times 10^{-5}} = 5.7 \times 10^{-10}$$

By the reaction equation, equal amounts of NH_3 and H^+ are formed. Let $x = [NH_3] = [H^+]$. Then

$$[NH_4^+] = 0.0100 - x \approx 0.0100$$

and

$$K_a = \frac{[NH_3][H^+]}{[NH_4^+]} \quad \text{gives} \quad 5.7 \times 10^{-10} = \frac{x^2}{0.0100}$$

Solving, $x = 2.4 \times 10^{-6}$. *Check of the approximation:* x is very small compared with 0.0100.

$$\text{Fraction hydrolyzed} = \frac{\text{amount hydrolyzed}}{\text{total amount}} = \frac{2.4 \times 10^{-6} \text{ mol/L}}{0.0100 \text{ mol/L}} = 2.4 \times 10^{-4} = 0.024\%$$

17.27. Calculate $[OH^-]$ in a 1.00 M solution of NaOCN. K_a for HOCN is 3.5×10^{-4}.

$$OCN^- + H_2O \rightleftharpoons HOCN + OH^-$$

$$K_b = \frac{[OH^-][HOCN]}{[OCN^-]} = \frac{K_w}{K_a} = \frac{1.00 \times 10^{-14}}{3.5 \times 10^{-4}} = 2.9 \times 10^{-11}$$

Since the source of OH^- and HOCN is the hydrolysis reaction, they must exist in equal concentrations; let $x = [OH^-] = [HOCN]$. Then

$$[OCN^-] = 1.00 - x \approx 1.00$$

and

$$K_b = 2.9 \times 10^{-11} = \frac{x^2}{1.00}$$

whence $x = [OH^-] = 5.4 \times 10^{-6}$.
   Check of approximation: x is very small compared with 1.00.

17.28. The acid ionization (hydrolysis) constant of Zn^{2+} is 3.3×10^{-10}. (a) Calculate the pH of a 0.0010 M solution of $ZnCl_2$. (b) What is the basic dissociation constant of $Zn(OH)^+$?

(a) $Zn^{2+} + H_2O \rightleftharpoons Zn(OH)^+ + H^+$ $K_a = \dfrac{[Zn(OH)^+][H^+]}{[Zn^{2+}]} = 3.3 \times 10^{-10}$

Let $x = [Zn(OH)^+] = [H^+]$. Then $[Zn^{2+}] = 0.0010 - x \approx 0.0010$ and

$$\frac{x^2}{0.0010} = 3.3 \times 10^{-10} \quad \text{or} \quad x = [H^+] = 5.7 \times 10^{-7}$$

$$pH = -\log (5.7 \times 10^{-7}) = -\log 10^{-6.2} = +6.2$$

Check of approximation: x is very small compared with 0.0010.

(b) Zn^{2+}, as an acid, is conjugate to the base $Zn(OH)^+$. For the basic dissociation,

$$Zn(OH)^+ \rightleftharpoons Zn^{2+} + OH^- \qquad K_b = \frac{K_w}{K_{a(Zn^{2+})}} = \frac{1.00 \times 10^{-14}}{3.3 \times 10^{-10}} = 3.0 \times 10^{-5}$$

17.29. Calculate the extent of hydrolysis and the pH of 0.0100 M $NH_4C_2H_3O_2$. K_a for $HC_2H_3O_2$ is 1.75×10^{-5} and K_b for NH_3 is 1.75×10^{-5}.

This is a case where both cation and anion hydrolyze.

$$\text{For } NH_4^+: \qquad K_a = \frac{K_w}{K_{b(NH_3)}} = \frac{1.00 \times 10^{-14}}{1.75 \times 10^{-5}} = 5.7 \times 10^{-10}$$

$$\text{For } C_2H_3O_2^-: \qquad K_b = \frac{K_w}{K_{a(HC_2H_3O_2)}} = \frac{1.00 \times 10^{-14}}{1.75 \times 10^{-5}} = 5.7 \times 10^{-10}$$

By coincidence, the hydrolysis constants for these two ions are identical. The production of H^+ by NH_4^+ hydrolysis must therefore exactly equal the production of OH^- by $C_2H_3O_2^-$ hydrolysis; and the H^+ and OH^- formed by hydrolysis neutralize each other to maintain the water equilibrium. The solution is thus neutral, $[H^+] = [OH^-] = 1.00 \times 10^{-7}$, and the pH is 7.00.

For NH_4^+ hydrolysis,

$$\frac{[NH_3][H^+]}{[NH_4^+]} = K_a = 5.7 \times 10^{-10}$$

Let $x = [NH_3]$. Then $0.0100 - x = [NH_4^+]$ and

$$\frac{x(1.00 \times 10^{-7})}{0.0100 - x} = 5.7 \times 10^{-10} \qquad \text{or} \qquad x = 5.7 \times 10^{-5}$$

$$\text{Percent } NH_4^+ \text{ hydrolyzed} = \frac{5.7 \times 10^{-5}}{0.0100} \times 100\% = 0.57\%$$

The percent hydrolysis of acetate ion must also be 0.57%, because the equilibrium constant for the hydrolysis is the same.

In comparing this result with Problem 17.26, note that the percent hydrolysis of NH_4^+ is greater in the presence of a hydrolyzing anion. The reason is that the mutual removal of some of the products of the two hydrolyses, H^+ and OH^-, by the water equilibrium reaction allows both hydrolyses to proceed to increasing extent.

17.30. Calculate the pH in a 0.100 M solution of NH_4OCN. K_b for NH_3 is 1.75×10^{-5} and K_a for HOCN is 3.5×10^{-4}.

As in Problem 17.29, both cation and anion hydrolyze. Since NH_3 is a weaker base than HOCN is an acid, however, NH_4^+ hydrolyzes more than OCN^-, and the pH of the solution is less than 7. In order to preserve electrical neutrality, there cannot be an appreciable difference between $[NH_4^+]$ and $[OCN^-]$. (The balancing of a slight difference could be accounted for by $[H^+]$ or $[OH^-]$.) Thus $[NH_3]$ must be practically equal to [HOCN], and we will indeed assume that they are equal.

Let $x = [NH_3] = [HOCN]$; then $0.100 - x = [NH_4^+] = [OCN^-]$.

$$\text{For } NH_4^+: \qquad K_a = \frac{[NH_3][H^+]}{[NH_4^+]} = \frac{K_w}{K_b} = \frac{1.00 \times 10^{-14}}{1.75 \times 10^{-5}} = 5.7 \times 10^{-10}$$

and
$$[H^+] = (5.7 \times 10^{-10})\left(\frac{0.100 - x}{x}\right) \qquad (1)$$

$$\text{For } OCN^-: \qquad K_b = \frac{[HOCN][OH^-]}{[OCN^-]} = \frac{K_w}{K_a} = \frac{1.00 \times 10^{-14}}{3.5 \times 10^{-4}} = 2.9 \times 10^{-11}$$

and
$$[OH^-] = (2.9 \times 10^{-11})\left(\frac{0.100 - x}{x}\right) \qquad (2)$$

Dividing (*1*) by (*2*),

$$\frac{[H^+]}{[OH^-]} = \frac{5.7 \times 10^{-10}}{2.9 \times 10^{-11}} = 19.7 \qquad\qquad (3)$$

Also, $[H^+]$ and $[OH^-]$ must satisfy the K_w relationship:

$$[H^+][OH^-] = K_w = 1.00 \times 10^{-14} \qquad\qquad (4)$$

Multiplying (*3*) by (*4*), we obtain

$$[H^+]^2 = 19.7 \times 10^{-14} \qquad [H^+] = 4.4 \times 10^{-7} \qquad pH = -\log[H^+] = 6.36$$

Check of assumption: Our assumption that $[NH_3] = [HOCN]$ is valid only if $[H^+]$ and $[OH^-]$ are much smaller than $[NH_3]$ and $[HOCN]$. We shall therefore solve for x, the $[NH_3]$ or $[HOCN]$. From (*1*),

$$x = \frac{(5.7 \times 10^{-10})(0.100 - x)}{[H^+]} = \frac{(5.7 \times 10^{-10})(0.100 - x)}{4.4 \times 10^{-7}} \approx 1.3 \times 10^{-4} \qquad\qquad (5)$$

Both $[H^+]$ and $[OH^-]$ are small compared with x.

It appears from the solution to this problem that the pH is independent of the concentration of NH_4OCN. Indeed this is true at sufficiently high concentrations. However x decreases with the initial concentration as shown in (*5*) so that at much lower concentrations the simplifying assumptions are no longer valid. (The problem can still be solved by explicitly invoking conservation of electrical charge but the solution becomes very complicated.)

POLYPROTIC ACIDS

17.31. Calculate the H^+ concentration of 0.10 M H_2S solution. K_1 and K_2 for H_2S are respectively 1.0×10^{-7} and 1.2×10^{-13}.

Most of the H^+ results from the primary ionization: $H_2S \rightleftharpoons H^+ + HS^-$.
Let $x = [H^+] = [HS^-]$. Then $[H_2S] = 0.10 - x \approx 0.10$.

$$K_1 = \frac{[H^+][HS^-]}{[H_2S]} \qquad \text{or} \qquad 1.0 \times 10^{-7} = \frac{x^2}{0.10} \qquad \text{or} \qquad x = 1.0 \times 10^{-4}$$

Check of assumptions: (1) x is indeed small compared with 0.10. (2) For the above value of $[H^+]$ and $[HS^-]$, the extent of the second dissociation is given by

$$[S^{2-}] = \frac{K_2[HS^-]}{[H^+]} = \frac{(1.2 \times 10^{-13})(1.0 \times 10^{-4})}{(1.0 \times 10^{-4})} = 1.2 \times 10^{-13}$$

The extent of the second dissociation is so small that it does not appreciably lower $[HS^-]$ or raise $[H^+]$ as calculated from the first dissociation only. This result, that the concentration of the conjugate base resulting from the second dissociation is numerically equal to K_2, is general whenever the extent of the second dissociation is less than 5 percent.

17.32. Calculate the concentration of $C_8H_4O_4^{2-}$ (*a*) in a 0.010 M solution of $H_2C_8H_4O_4$,   (*b*) in a solution which is 0.010 M with respect to $H_2C_8H_4O_4$ and 0.020 M with respect to HCl. The ionization constants for $H_2C_8H_4O_4$, phthalic acid, are

$$H_2C_8H_4O_4 \rightleftharpoons H^+ + HC_8H_4O_4^- \qquad K_1 = 1.3 \times 10^{-3}$$

$$HC_8H_4O_4^- \rightleftharpoons H^+ + C_8H_4O_4^{2-} \qquad K_2 = 3.9 \times 10^{-6}$$

(*a*)   If there were no second dissociation, the $[H^+]$ could be computed on the basis of the K_1 equation.

$$\frac{x^2}{0.010 - x} = 1.3 \times 10^{-3} \qquad \text{or} \qquad x = [H^+] = [HC_8H_4O_4^-] = 3.0 \times 10^{-3}$$

Note that it was necessary to solve the quadratic to obtain x. If we assume that the second dissociation does not appreciably affect $[H^+]$ or $[HC_8H_4O_4^-]$, then

$$[C_8H_4O_4^{2-}] = \frac{K_2[HC_8H_4O_4^-]}{[H^+]} = \frac{(3.9 \times 10^{-6})(3.0 \times 10^{-3})}{3.0 \times 10^{-3}} = 3.9 \times 10^{-6}$$

Check of assumption: The extent of the second dissociation relative to the first,

$$\frac{3.9 \times 10^{-6}}{3.0 \times 10^{-3}} = 1.3 \times 10^{-3} = 0.13\%$$

is sufficiently small to validate the assumption.

(b) The $[H^+]$ in solution may be assumed to be essentially that contributed by the HCl. Also, this large common-ion concentration represses the ionization of the phthalic acid, so that we assume that $[H_2C_8H_4O_4] = 0.010$. The most convenient equation to use is the K_1K_2 equation, since all the concentrations for this equation are known but one.

$$\frac{[H^+]^2[C_8H_4O_4^{2-}]}{[H_2C_8H_4O_4]} = K_1K_2 = (1.3 \times 10^{-3})(3.9 \times 10^{-6}) = 5.1 \times 10^{-9}$$

$$[C_8H_4O_4^{2-}] = \frac{(5.1 \times 10^{-9})[H_2C_8H_4O_4]}{[H^+]^2} = \frac{(5.1 \times 10^{-9})(0.010)}{(0.020)^2} = 1.3 \times 10^{-7}$$

Check of assumptions: Solving for the first dissociation,

$$[HC_8H_4O_4^-] = \frac{K_1[H_2C_8H_4O_4]}{[H^+]} = \frac{(1.3 \times 10^{-3})(0.010)}{0.020} = 6.5 \times 10^{-4}$$

The amount of H^+ contributed by this dissociation, 6.5×10^{-4} mol/L, is indeed less than 10 percent of the amount contributed by HCl (0.020 mol/L). The amount of H^+ contributed by the second dissociation is still less.

17.33. Calculate the extent of hydrolysis of 0.005 M K_2CrO_4. The ionization constants of H_2CrO_4 are $K_1 = 0.18$, $K_2 = 3.2 \times 10^{-7}$.

Just as in the ionization of polyprotic acids, so in the hydrolysis of their salts, the reaction proceeds in successive stages. The extent of the second stage is generally very small compared with the first. This is particularly true in this case, where H_2CrO_4 is quite a strong acid with respect to its first ionization. The equation of interest is

$$CrO_4^{2-} + H_2O \rightleftharpoons HCrO_4^- + OH^-$$

which indicates that the conjugate acid of the hydrolyzing CrO_4^{2-} is $HCrO_4^-$. As the ionization constant for $HCrO_4^-$ is K_2, the basic hydrolysis constant for the reaction is K_w/K_2.

$$\frac{[OH^-][HCrO_4^-]}{[CrO_4^{2-}]} = K_b = \frac{K_w}{K_2} = \frac{1.0 \times 10^{-14}}{3.2 \times 10^{-7}} = 3.1 \times 10^{-8}$$

Let $x = [OH^-] = [HCrO_4^-]$. Then $[CrO_4^{2-}] = 0.005 - x \approx 0.005$ and

$$\frac{x^2}{0.005} = 3.1 \times 10^{-8} \qquad \text{or} \qquad x = 1.2 \times 10^{-5}$$

$$\text{Fractional hydrolysis} = \frac{1.2 \times 10^{-5}}{0.005} = 2.4 \times 10^{-3} = 0.24\%$$

Check of assumption: x is indeed small compared with 0.005.

17.34. What is the pH of a 0.005 0 M solution of Na_2S? The ionization constants for H_2S are $K_1 = 1.0 \times 10^{-7}$ and $K_2 = 1.2 \times 10^{-13}$.

As in Problem 17.33, the first stage of hydrolysis, leading to HS^-, is predominant.

$$S^{2-} + H_2O \rightleftharpoons HS^- + OH^-$$

$$K_b = \frac{[HS^-][OH^-]}{[S^{2-}]} = \frac{K_w}{K_2} = \frac{1.00 \times 10^{-14}}{1.2 \times 10^{-13}} = 8.3 \times 10^{-2}$$

Because of the large value for K_b, it cannot be assumed that the equilibrium concentration of S^{2-} is approximately 0.005 0 mol/L, the stoichiometric concentration. In fact, the hydrolysis is so extensive that most of the S^{2-} is converted to HS^-.

Let $x = [S^{2-}]$, then $[HS^-] = [OH^-] = 0.005\,0 - x$.

$$\frac{(0.005\,0 - x)^2}{x} = 8.3 \times 10^{-2} \qquad \text{whence} \qquad x = 2.7 \times 10^{-4}$$

$$[OH^-] = 0.005\,0 - 2.7 \times 10^{-4} = 0.004\,7$$

$$pOH = -\log (4.7 \times 10^{-3}) = 2.33 \qquad pH = 14.00 - 2.33 = 11.67$$

Check of assumption: Consider the second stage of hydrolysis.

$$HS^- + H_2O \rightleftharpoons H_2S + OH^- \qquad K_b = \frac{K_w}{K_1} = \frac{1.00 \times 10^{-14}}{1.0 \times 10^{-7}} = 1.0 \times 10^{-7}$$

Solve for $[H_2S]$ by assuming the values of $[OH^-]$ and $[HS^-]$ already obtained.

$$[H_2S] = \frac{K_b[HS^-]}{[OH^-]} = \frac{(1.0 \times 10^{-7})(4.7 \times 10^{-3})}{4.7 \times 10^{-3}} = 1.0 \times 10^{-7}$$

The extent of the second hydrolysis compared with the first, $(1.6 \times 10^{-7})/(4.7 \times 10^{-3})$, is indeed small.

17.35. Calculate $[H^+]$, $[H_2PO_4^-]$, $[HPO_4^{2-}]$, and $[PO_4^{3-}]$ in 0.0100 M H_3PO_4. K_1, K_2, and K_3 are 7.52×10^{-3}, 6.23×10^{-8}, and 4.5×10^{-13}, respectively.

We begin by assuming that H^+ comes principally from the first stage of dissociation, and that the concentration of any anion formed by one stage of an ionization is not appreciably lowered by the succeeding stage of ionization.

$$H_3PO_4 \rightleftharpoons H^+ + H_2PO_4^- \qquad K_1 = 7.52 \times 10^{-3}$$

Let $[H^+] = [H_2PO_4^-] = x$. Then $[H_3PO_4] = 0.0100 - x$, and

$$\frac{x^2}{0.0100 - x} = 7.52 \times 10^{-3} \qquad \text{or} \qquad x = 0.005\,7$$

We next take the above value for $[H^+]$ and $[H_2PO_4^-]$ to solve for $[HPO_4^{2-}]$.

$$H_2PO_4^- \rightleftharpoons H^+ + HPO_4^{2-} \qquad K_2 = 6.23 \times 10^{-8}$$

$$[HPO_4^{2-}] = \frac{K_2[H_2PO_4^-]}{[H^+]} = \frac{(6.23 \times 10^{-8})(0.005\,7)}{0.005\,7} = 6.23 \times 10^{-8}$$

Check of assumption: The extent of the second dissociation compared with the first,

$$\frac{6.23 \times 10^{-8}}{5.7 \times 10^{-3}}$$

is indeed small.

Next we take the above values for $[H^+]$ and $[HPO_4^{2-}]$ to solve for $[PO_4^{3-}]$.

$$HPO_4^{2-} \rightleftharpoons H^+ + PO_4^{3-} \qquad K_3 = 4.5 \times 10^{-13}$$

$$[PO_4^{3-}] = \frac{K_3[HPO_4^{2-}]}{[H^+]} = \frac{(4.5 \times 10^{-13})(6.23 \times 10^{-8})}{5.7 \times 10^{-3}} = 4.9 \times 10^{-18}$$

Check of assumption: The depletion of HPO_4^- by the third stage, 4.9×10^{-18} mol/L, is an insignificant fraction of the amount present as a result of the second step, 6.23×10^{-8} mol/L.

17.36. What is the pH of 0.0100 M $NaHCO_3$? K_1 and K_2 for H_2CO_3 are 4.3×10^{-7} and 5.61×10^{-11}.

(H_2CO_3 in aqueous solution is in equilibrium with dissolved CO_2, the majority species. The value of K_1 given here is based on the total concentration of both neutral species. Since there is no effect on the stoichiometry or charge balance the problem can be worked as if all the neutral species were in the form H_2CO_3.) This has similarities to Problem 17.30 in that there is one reaction tending to make the solution acidic (the K_2 acid dissociation of HCO_3^-) and another reaction tending to make the solution basic (the hydrolysis of HCO_3^-).

$$HCO_3^- \rightleftharpoons H^+ + CO_3^{2-} \qquad\qquad K_2 = 5.61 \times 10^{-11} \qquad\qquad (1)$$

$$HCO_3^- + H_2O \rightleftharpoons OH^- + H_2CO_3 \qquad K_b = \frac{K_w}{K_1} = \frac{1.00 \times 10^{-14}}{4.3 \times 10^{-7}} = 2.3 \times 10^{-8} \qquad (2)$$

[Note that the hydrolysis constant for reaction (2) is related to K_1, because both hydrolysis and the K_1 equilibrium involve H_2CO_3 and HCO_3^-.] We see that the equilibrium constant for (2) is greater than that for (1); thus the pH is certain to exceed 7.

We assume that after self-neutralization both $[H^+]$ and $[OH^-]$ will be so small as to have no appreciable effect on the ionic charge balance. Therefore electrical neutrality can be preserved only by maintaining a fixed total anionic charge among the various carbonate species, since the cationic charge remains at 0.0100 M, the concentration of Na^+, regardless of the acid and base equilibria. In other words, for every negative charge removed by converting HCO_3^- to H_2CO_3, another negative charge must be created by converting HCO_3^- to CO_3^{2-}. This leads to the following conditions:

$$[H_2CO_3] = [CO_3^{2-}] = x \qquad\qquad [HCO_3^-] = 0.0100 - 2x \approx 0.0100$$

$$K_2 = \frac{[H^+][CO_3^{2-}]}{[HCO_3^-]} = \frac{[H^+]x}{0.0100} = 5.61 \times 10^{-11} \qquad\qquad (3)$$

$$K_b = \frac{[OH^-][H_2CO_3]}{[HCO_3^-]} = \frac{[OH^-]x}{0.0100} = 2.3 \times 10^{-8} \qquad\qquad (4)$$

Multiplying (3) and (4) and recalling that $[H^+][OH^-] = 1.00 \times 10^{-14}$, we obtain

$$\frac{(1.00 \times 10^{-14})x^2}{(0.0100)^2} = (5.61 \times 10^{-11})(2.3 \times 10^{-8}) \qquad \text{or} \qquad x = 1.14 \times 10^{-4}$$

Provisional check: $2x$ is indeed small compared with 0.0100.

We return now to (3).

$$[H^+] = \frac{(5.61 \times 10^{-11})[HCO_3^-]}{[CO_3^{2-}]} = \frac{(5.61 \times 10^{-11})(0.0100)}{1.14 \times 10^{-4}} = 4.9 \times 10^{-9}$$

$$pH = -\log[H^+] = -\log 4.9 + 9 = -0.69 + 9 = 8.31$$

Final check: Both $[H^+]$ and $[OH^-]$ are small compared with x, the shift in the concentrations of the other ions.

Alternatively, $[H^+]$ could have been computed from the K_1 equilibrium; the same result would have been obtained.

BUFFER SOLUTIONS, INDICATORS, AND TITRATION

17.37. A buffer solution of pH 8.50 is desired. (*a*) Starting with 0.0100 mol of KCN and the usual inorganic reagents of the laboratory, how would you prepare 1 L of the buffer solution? K_a for HCN is 4.93×10^{-10}. (*b*) By how much would the pH change after the addition of 5×10^{-5} mol $HClO_4$ to 100 cm^3 of the buffer? (*c*) By how much would the pH change after the addition of 5×10^{-5} mol NaOH to 100 cm^3 of the buffer? (*d*) By how much would the pH change after the addition of 5×10^{-5} mol NaOH to 100 cm^3 of pure water?

(*a*)   To find the desired $[H^+]$:

$$\log [H^+] = -pH = -8.50 = 0.50 - 9.00$$

$$[H^+] = \text{antilog } 0.50 \times \text{antilog } (-9.00) = 3.2 \times 10^{-9}$$

The buffer solution could be prepared by mixing CN^- (weak base) with HCN (weak acid) in the proper proportions so as to satisfy the ionization constant equilibrium for HCN.

$$HCN \rightleftharpoons H^+ + CN^- \qquad K_a = 4.93 \times 10^{-10} = \frac{[H^+][CN^-]}{[HCN]}$$

Hence
$$\frac{[CN^-]}{[HCN]} = \frac{K_a}{[H^+]} = \frac{4.93 \times 10^{-10}}{3.2 \times 10^{-9}} = 0.154 \tag{1}$$

This ratio of CN^- to HCN can be attained if some of the CN^- is neutralized with a strong acid, like HCl, to form an equivalent amount of HCN. The total cyanide available for both forms is 0.0100 mol. Let $x = [HCN]$; then $[CN^-] = 0.0100 - x$. Substituting in (*1*),

$$\frac{0.0100 - x}{x} = 0.154$$

from which $x = 0.0087$ and $0.0100 - x = 0.0013$. The buffer solution can be prepared by dissolving 0.0100 mol KCN and 0.0087 mol HCl in enough water to make up 1 L of solution.

(*b*)   100 cm^3 of the buffer contains

$$(0.0087 \text{ mol/L})(0.100 \text{ L}) = 8.7 \times 10^{-4} \text{ mol HCN}$$
and
$$(0.0013 \text{ mol/L})(0.100 \text{ L}) = 1.3 \times 10^{-4} \text{ mol CN}^-$$

The addition of 5×10^{-5} mol of strong acid will convert more CN^- to HCN. The resulting amount of HCN will be

$$(8.7 \times 10^{-4}) + (0.5 \times 10^{-4}) = 9.2 \times 10^{-4} \text{ mol}$$

and the resulting amount of CN^- will be

$$(1.3 \times 10^{-4}) - (0.5 \times 10^{-4}) = 0.8 \times 10^{-4} \text{ mol}$$

Only the ratio of the two concentrations is needed.

$$[H^+] = K_a \frac{[HCN]}{[CN^-]} = (4.93 \times 10^{-10}) \left(\frac{9.2}{0.8}\right) = 5.7 \times 10^{-9}$$

$$pH = -\log [H^+] = 9 - 0.75 = 8.25$$

The drop in pH caused by the addition of the acid is $8.50 - 8.25$, or 0.25 pH unit.

(*c*)   The addition of 5×10^{-5} mol of strong base will convert an equivalent amount of HCN to CN^-.

HCN: Resulting amount $= (8.7 \times 10^{-4}) - (0.5 \times 10^{-4}) = 8.2 \times 10^{-4}$ mol

CN^-:   Resulting amount $= (1.3 \times 10^{-4}) + (0.5 \times 10^{-4}) = 1.8 \times 10^{-4}$ mol

$$[H^+] = K_a \frac{[HCN]}{[CN^-]} = (4.93 \times 10^{-10}) \left(\frac{8.2}{1.8}\right) = 2.2 \times 10^{-9}$$

$$pH = -\log [H^+] = 9 - 0.35 = 8.65$$

The rise in pH caused by the addition of base is $8.65 - 8.50$, or 0.15 pH unit.

(*d*)
$$[OH^-] = \frac{5 \times 10^{-5} \text{ mol}}{0.100 \text{ L}} = 5 \times 10^{-4} \text{ M}$$

$$pOH = -\log [OH^-] = 4 - 0.70 = 3.30; \quad pH = 14.00 - pOH = 10.70$$

Pure water has a pH of 7.00. The rise caused by the addition of base is $10.70 - 7.00 = 3.70$ pH units, in sharp contrast to the small rise in the buffer solution.

17.38. If 0.00010 mol H_3PO_4 is added to a solution buffered at pH 7.00, what are the relative proportions of the four forms H_3PO_4, $H_2PO_4^-$, HPO_4^{2-}, PO_4^{3-}? K_1, K_2, and K_3 for phosphoric acid are respectively 7.52×10^{-3}, 6.23×10^{-8}, 4.5×10^{-13}.

Since the solution was previously well buffered, we can assume that the pH is not changed by addition of the phosphoric acid. Then, if $[H^+]$ is fixed, the ratio of two of the desired concentrations can be calculated from each of the ionization constant equations.

$$\frac{[H^+][H_2PO_4^-]}{[H_3PO_4]} = K_1 \qquad\qquad \frac{[H^+][HPO_4^{2-}]}{[H_2PO_4^-]} = K_2 \qquad\qquad \frac{[H^+][PO_4^{3-}]}{[HPO_4^{2-}]} = K_3$$

$$\frac{[H_3PO_4]}{[H_2PO_4^-]} = \frac{[H^+]}{K_1} \qquad\qquad \frac{[H_2PO_4^-]}{[HPO_4^{2-}]} = \frac{[H^+]}{K_2} \qquad\qquad \frac{[HPO_4^{2-}]}{[PO_4^{3-}]} = \frac{[H^+]}{K_3}$$

$$= \frac{1.00 \times 10^{-7}}{7.52 \times 10^{-3}} \qquad\qquad = \frac{1.00 \times 10^{-7}}{6.23 \times 10^{-8}} \qquad\qquad = \frac{1.00 \times 10^{-7}}{4.5 \times 10^{-13}}$$

$$= 1.33 \times 10^{-5} \qquad\qquad\quad = 1.61 \qquad\qquad\qquad\quad = 2.2 \times 10^5$$

Since the ratio $[H_3PO_4]/[H_2PO_4^-]$ is very small and the ratio $[HPO_4^{2-}]/[PO_4^{3-}]$ is very large, practically all of the material will exist as $H_2PO_4^-$ and HPO_4^{2-}. The sum of the amounts of these two ions will be practically equal to 0.000 10 mol; and if the total volume of solution is 1 L, the sum of these two ion concentrations will be 0.000 10 mol/L.

Let $x = [HPO_4^{2-}]$; then $[H_2PO_4^-] = 0.000\,10 - x$. Then

$$\frac{[H_2PO_4^-]}{[HPO_4^{2-}]} = 1.61 \qquad \text{gives} \qquad \frac{0.000\,10 - x}{x} = 1.61$$

Solving, $$x = [HPO_4^{2-}] = 3.8 \times 10^{-5}$$

$$[H_2PO_4^-] = 0.000\,10 - x = 6.2 \times 10^{-5}$$

$$[H_3PO_4] = (1.33 \times 10^{-5})[H_2PO_4^-] = (1.33 \times 10^{-5})(6.2 \times 10^{-5}) = 8.2 \times 10^{-10}$$

$$[PO_4^{3-}] = \frac{[HPO_4^{2-}]}{2.2 \times 10^5} = \frac{3.8 \times 10^{-5}}{2.2 \times 10^5} = 1.7 \times 10^{-10}$$

17.39. K_a for $HC_2H_3O_2$ is 1.75×10^{-5}. A 40.0-cm³ sample of 0.0100 M $HC_2H_3O_2$ is titrated with 0.0200 M NaOH. Calculate the pH after the addition of (a) 3.0 cm³, (b) 10.0 cm³, (c) 20.0 cm³, (d) 30.0 cm³ of the NaOH solution.

We can keep track of the changing amounts of the various species and the increasing volume by a scheme such as Table 17-2.

Table 17-2

	(a)	(b)	(c)	(d)
Amount of base added/L	0	0.003 0	0.010 0	0.020 0	0.030 0
Total volume/L	0.040 0	0.043 0	0.050 0	0.060 0	0.070 0
$n(HC_2H_3O_2)$, before neutralization	4.00×10^{-4}	4.00×10^{-4}	4.00×10^{-4}	4.00×10^{-4}	4.00×10^{-4}
$n(OH^-)$ added (Row 1 × 0.020 0 M)	0.0×10^{-4}	0.60×10^{-4}	2.00×10^{-4}	4.00×10^{-4}	6.00×10^{-4}
$n(C_2H_3O_2^-)$ formed	0.0×10^{-4}	0.60×10^{-4}	2.00×10^{-4}	4.00×10^{-4}	4.00×10^{-4}
$n(HC_2H_3O_2)$ remaining	4.00×10^{-4}	3.40×10^{-4}	2.00×10^{-4}	x	y
$n(OH^-)$ excess					2.00×10^{-4}

Note that the amount of acetic acid neutralized (amount of $C_2H_3O_2^-$) follows the amount of OH^- added, up to complete neutralization. Additional OH^-, having no more acid to neutralize, accumulates in the solution. Up to the end point, the amount of $HC_2H_3O_2$ remaining is obtained by simply subtracting the amount of $C_2H_3O_2^-$ from the initial amount of $HC_2H_3O_2$. At the end point and beyond, however, $[HC_2H_3O_2]$ cannot be set equal to zero but must be deduced from the hydrolysis equilibrium.

(a) and (b). Here absolute concentrations of conjugate acid and base are not needed; only their ratio is required.

	(a)	(b)
$[H^+] = \dfrac{K_a[HC_2H_3O_2]}{[C_2H_3O_2^-]}$	$\dfrac{(1.75 \times 10^{-5})(3.40)}{0.60}$ $= 9.9 \times 10^{-5}$	$\dfrac{(1.75 \times 10^{-5})(2.00)}{2.00}$ $= 1.75 \times 10^{-5}$
$pH = -\log [H^+]$	4.00	4.76

(c)
$$M(C_2H_3O_2^-) = \frac{4.00 \times 10^{-4} \text{ mol}}{0.0600 \text{ L}} = 6.7 \times 10^{-3} \text{ mol/L}$$

The solution at the end point is the same as 6.7×10^{-3} M $NaC_2H_3O_2$. Consider the hydrolysis of $NaC_2H_3O_2$:

$$C_2H_3O_2^- + H_2O \rightleftharpoons HC_2H_3O_2 + OH^-$$

Let $[HC_2H_3O_2] = [OH^-] = x$, $[C_2H_3O_2^-] = 6.7 \times 10^{-3} - x \approx 6.7 \times 10^{-3}$; then

$$\frac{[HC_2H_3O_2][OH^-]}{[C_2H_3O_2^-]} = \frac{x^2}{6.7 \times 10^{-3}} = K_b = \frac{1.00 \times 10^{-14}}{1.75 \times 10^{-5}} \qquad \text{or} \qquad x = 1.96 \times 10^{-6}$$

Check of assumption: x is indeed small compared with 6.7×10^{-3}.

$$pOH = -\log [OH^-] = -\log (1.9 \times 10^{-6}) = 6 - 0.29 = 5.71$$

$$pH = 14.00 - 5.71 = 8.29$$

(d) From the excess OH^- beyond that needed to neutralize all the acetic acid, we know that

$$M(OH^-) = \frac{2.0 \times 10^{-4} \text{ mol}}{0.070 \text{ L}} = 2.9 \times 10^{-3} \text{ mol/L}$$

Hence

$$pOH = -\log [OH^-] = -\log (2.9 \times 10^{-3}) = 2.54$$

$$pH = 14.00 - 2.54 = 11.46$$

17.40. Calculate a point on the titration curve for the addition of 2.0 cm³ of 0.0100 M NaOH to 50.0 cm³ of 0.0100 M chloroacetic acid, $HC_2H_2O_2Cl$. $K_a = 1.40 \times 10^{-3}$.

The simplifying assumption made in Problem 17.39 breaks down here. If the amount of chloroacetate ion formed were equivalent to the amount of NaOH added, we would have the following:

Amount OH^- added = $(0.0020 \text{ L})(0.010 \text{ mol/L}) = 2.0 \times 10^{-5}$ mol

Total volume = 0.0520 L

$$M(C_2H_2O_2Cl^-) = \frac{2.0 \times 10^{-5} \text{ mol}}{0.0520 \text{ L}} = 3.8 \times 10^{-4} \text{ mol/L} \qquad (1)$$

$$M(HC_2H_2O_2Cl) = \frac{(0.0500 \text{ L})(0.0100 \text{ mol/L})}{0.0520 \text{ L}} - 3.8 \times 10^{-4} \text{ mol/L} = 9.2 \times 10^{-3} \text{ mol/L} \qquad (2)$$

$$[H^+] = \frac{K_a[HC_2H_2O_2Cl]}{[C_2H_2O_2Cl^-]} = \frac{(1.40 \times 10^{-3})(9.2 \times 10^{-3})}{3.8 \times 10^{-4}} = 3.4 \times 10^{-2} \qquad (3)$$

This answer is obviously ridiculous. $[H^+]$ cannot possibly exceed the initial molar concentration of the acid. Apparently the amount of chloroacetate ion is greater than the equivalent amount of base added. This fact is related to the relatively strong acidity of the acid and to the appreciable ionization of the acid even before the titration begins. Mathematically, this is taken into account by an equation of electroneutrality, according to which there must be equal numbers of cationic and anionic charges in the solution.

$$[H^+] + [Na^+] = [C_2H_2O_2Cl^-] + [OH^-] \tag{4}$$

It is safe to drop the $[OH^-]$ term in (4) in this case, because it is so much smaller than $[C_2H_2O_2Cl^-]$. $[Na^+]$ is obtained from the amount of NaOH added and the total volume of solution.

$$[Na^+] = \frac{2.0 \times 10^{-5}}{0.0520} = 3.8 \times 10^{-4} \tag{5}$$

Chloroacetate ion can be computed from (4) and (5) with the neglect of $[OH^-]$.

$$[C_2H_2O_2Cl^-] = 3.8 \times 10^{-4} + [H^+] \tag{6}$$

The undissociated acid concentration is then the total molar concentration of acid (including its ion) minus $[C_2H_2O_2Cl^-]$.

$$[HC_2H_2O_2Cl] = \frac{(0.0100)(0.0500)}{0.0520} - (3.8 \times 10^{-4} + [H^+]) = 9.2 \times 10^{-3} - [H^+] \tag{7}$$

Note that (6) and (7) differ from (1) and (2) only in the inclusion of the $[H^+]$ terms.

Now we may return to the ionization equilibrium for the acid.

$$[H^+] = \frac{K_a[HC_2H_2O_2Cl]}{[C_2H_2O_2Cl^-]} = \frac{(1.40 \times 10^{-3})(9.2 \times 10^{-3} - [H^+])}{3.8 \times 10^{-4} + [H^+]}$$

The solution of the quadratic equation is $[H^+] = 2.8 \times 10^{-3}$, and pH $= -\log [H^+] = 2.55$.

The complication treated in this problem occurs whenever $[H^+]$ or $[OH^-]$ during partial neutralization cannot be neglected in comparison with the concentrations of other ions in solution. This is likely to be the case near the beginning of the titration of an acid which is only moderately weak.

17.41. An acid-base indicator has a K_a of 3.0×10^{-5}. The acid form of the indicator is red and the basic form is blue. (a) By how much must the pH change in order to change the indicator from 75% red to 75% blue? (b) For which of the titrations shown in Fig. 17-1a and 17-1b would this indicator be a suitable choice?

(a)
$$[H^+] = \frac{K_a[\text{acid}]}{[\text{base}]}$$

$$75\% \text{ red:} \qquad [H^+] = \frac{(3.0 \times 10^{-5})(75)}{25} = 9.0 \times 10^{-5} \qquad \text{pH} = 4.05$$

$$75\% \text{ blue:} \qquad [H^+] = \frac{(3.0 \times 10^{-5})(25)}{75} = 1.0 \times 10^{-5} \qquad \text{pH} = 5.00$$

The change in pH is $5.00 - 4.05 = 0.95$.

(b) The indicator changes its color in the pH range 4 to 5. In both titrations using HCl as titrant in Fig. 17-1b, the pH is falling rapidly in this range; thus the indicator would be suitable for both of them. In the titration of HCl with NaOH in Fig. 17-1a the pH is rising rapidly in this range and it could also be done with this indicator. This indicator, however, would not be suitable for the $HC_4H_7O_3$ titration which would change from red to blue long before the end point.

APPARENT EQUILIBRIUM CONSTANTS

17.42. One molecule of sucrose phosphate is synthesized biologically by the condensation of one molecule each of fructose and glucose phosphate.

$$\text{Fructose} + \text{glucose phosphate} \rightleftharpoons \text{sucrose phosphate}$$

K' for the reaction is 0.050 at a particular pH. To what volume should a solution containing 0.0500 mol each of fructose and glucose phophate be diluted so that there will be a 3.0 percent conversion to sucrose phosphate at equilibrium at the designated pH?

It is not necessary to consider the distribution of glucose phosphate and sucrose phosphate among their various states of ionization, since the K' expression appropriate for a particular pH is written in terms of the stoichiometric concentrations of the three principal reaction partners (excluding water and its ions), each being the sum of the concentrations of all the conjugate acid and basic forms.

The essential data can be represented in the following tabular form, where the extent of reaction is 3.0 percent of 0.0500 mol, or 0.0015 mol.

	Fructose +	glucose phosphate $\rightleftharpoons$	sucrose phosphate
n (initial):	0.0500	0.0500	0.0
Change in n by reaction:	-0.0015	-0.0015	$+0.0015$
n (equilibrium):	0.0485	0.0485	0.0015

Then, if the final volume is v liters,

$$K' = \frac{[\text{sucrose phosphate}]}{[\text{fructose}][\text{glucose phosphate}]} = \frac{0.0015/v}{(0.0485/v)^2} = 0.050$$

Solving, $v = 0.078$; the final volume should be 78 mL.

Supplementary Problems

Notes: 1. The numerical values tabulated for equilibrium constants differ in various books. The values selected for this book are internally consistent, and the computed answers are based on values given here. The temperature is assumed to be 25 °C unless there is a specific statement to the contrary.

2. Some of the problems at the end of the sections on Acids and Bases, Hydrolysis, Polyprotic Acids, and Buffers involve multiple equilibria. They may be omitted in a simplest treatment of ionic equilibrium.

ACIDS AND BASES

17.43. The self-ionization constant for pure formic acid, $K = [\text{HCOOH}_2^+][\text{HCOO}^-]$, has been estimated as 10^{-6} at room temperature. What percentage of formic acid molecules in pure formic acid, HCOOH, are converted to formate ion? The density of formic acid is 1.22 g/cm^3.

Ans.   0.004%

17.44. A certain reaction is catalyzed by acids, and the catalytic activity for 0.1 M solutions of the acids in water was found to decrease in the order HCl, HCOOH, HC$_2$H$_3$O$_2$. The same reaction takes place in anhydrous ammonia, but the three acids all have the same catalytic effect in 0.1 M solutions. Explain.

Ans.   The order of catalytic activity in water is the same as the order of acidity. In anhydrous ammonia all three acids are strong.

17.45. The amino acid glycine exists predominantly in the form $^+$NH$_3$CH$_2$COO$^-$. Write formulas for (a) the conjugate base and (b) the conjugate acid of glycine.

Ans.   (a) NH$_2$CH$_2$COO$^-$; (b) $^+$NH$_3$CH$_2$COOH

17.46. In the reaction of BeF_2 with $2F^-$ to form BeF_4^{2-}, which reactant is the Lewis acid and which is the Lewis base?

  Ans. BeF_2 is the acid and F^- is the base.

17.47. Calculate the ionization constant of formic acid, HCO_2H, which ionizes 4.2 percent in 0.10 M solution.

  Ans. 1.8×10^{-4}

17.48. A solution of acetic acid is 1.0 percent ionized. Determine the molar concentration of acetic acid and the $[H^+]$ of the solution. K_a of $HC_2H_3O_2$ is 1.75×10^{-5}.

  Ans.   0.17 M, 1.7×10^{-3}

17.49. The ionization constant of ammonia in water is 1.75×10^{-5}. Determine   (a) the degree of ionization and (b) the $[OH^-]$ of a 0.08 M solution of NH_3.

  Ans.   1.5 percent, 1.2×10^{-3}

17.50. Chloroacetic acid, a monoprotic acid, has a K_a of 1.40×10^{-3}. Compute the freezing point of a 0.10 M solution of this acid. Assume that the stoichiometric molar concentration and molality are the same in this case.

  Ans. $-0.21\ °C$

17.51. Determine $[OH^-]$ of a 0.050 M solution of ammonia to which has been added sufficient NH_4Cl to make the total $[NH_4^+]$ equal to 0.100. K_b of ammonia is 1.75×10^{-5}.

  Ans. 8.8×10^{-6}

17.52. Find the value of $[H^+]$ in a liter of solution in which are dissolved 0.080 mol $HC_2H_3O_2$ and 0.100 mol $NaC_2H_3O_2$. K_a for $HC_2H_3O_2$ is 1.75×10^{-5}.

  Ans. 1.4×10^{-5}

17.53. A 0.025 M solution of a monobasic acid had a freezing point of $-0.060\ °C$. What are K_a and pK_a for the acid?

  Ans. 3.0×10^{-3}, 2.52

17.54. Fluoroacetic acid has a K_a of 2.6×10^{-3}. What concentration of the acid is needed so that $[H^+]$ is 2.0×10^{-3}?

  Ans. 3.5×10^{-3} M

17.55. What is $[NH_4^+]$ in a solution that is 0.0200 M NH_3 and 0.0100 M KOH? K_b for NH_3 is 1.75×10^{-5}.

  Ans. 3.5×10^{-5}

17.56. What molar concentration of NH_3 provides a $[OH^-]$ of 1.5×10^{-3}? K_b for NH_3 is 1.75×10^{-5}.

  Ans.   0.13 M

17.57. What is $[HCOO^-]$ in a solution that is both 0.015 M HCOOH and 0.020 M HCl? K_a for HCOOH is 1.8×10^{-4}.

  Ans. 1.4×10^{-4}

17.58. What are $[H^+]$, $[C_3H_5O_3^-]$, and $[OC_6H_5^-]$ in a solution that is 0.030 M $HC_3H_5O_3$ and 0.100 M HOC_6H_5? K_a-values for $HC_3H_5O_3$ and HOC_6H_5 are 3.1×10^{-5} and 1.05×10^{-10}, respectively.

  Ans. $[H^+] = 9.6 \times 10^{-4}$, $[C_3H_5O_3^-] = 9.6 \times 10^{-4}$, $[OC_6H_5^-] = 1.1 \times 10^{-8}$

17.59. Find the value of $[OH^-]$ in a solution made by dissolving 0.0050 mol each of ammonia and pyridine in enough water to make 200 cm^3 of solution. K_b for ammonia and pyridine are 1.75×10^{-5} and 1.78×10^{-9}, respectively. What are the concentrations of ammonium and pyridinium ions?

Ans. $[OH^-] = [NH_4^+] = 6.5 \times 10^{-4}$, [pyridinium ion] $= 6.8 \times 10^{-8}$

17.60. Consider a solution of a monoprotic weak acid of acidity constant K_a. Calculate the minimum concentration, C, for which the percent ionization is less than 10 percent. Assume that activity coefficient corrections can be neglected.

Ans. $C = 90 K_a$

17.61. What is the percent ionization of 0.0065 M chloroacetic acid? K_a for this acid is 1.40×10^{-3}.

Ans. 37%

17.62. What concentration of dichloroacetic acid gives a $[H^+]$ of 8.5×10^{-3}? K_a for the acid is 3.32×10^{-2}.

Ans. 1.07×10^{-2} M

17.63. Calculate $[H^+]$ in a 0.200 M dichloroacetic acid solution that is also 0.100 M in sodium dichloroacetate. K_a for dichloroacetic acid is 3.32×10^{-2}.

Ans.   0.039

17.64. How much solid sodium dichloroacetate should be added to a liter of 0.100 M dichloroacetic acid to reduce $[H^+]$ to 0.030? K_a for dichloroacetic acid is 3.32×10^{-2}. Neglect the increase in volume of the solution on addition of the salt.

Ans.   0.047 mol

17.65. Calculate $[H^+]$ and $[C_2HO_2Cl_2^-]$ in solution that is 0.0100 M in HCl and 0.0100 M in $HC_2HO_2Cl_2$. K_a for $HC_2HO_2Cl_2$ (dichloroacetic acid) is 3.32×10^{-2}.

Ans.   0.0167, 0.0067

17.66. Calculate $[H^+]$, $[C_2H_3O_2^-]$, and $[C_7H_5O_2^-]$ in a solution that is 0.0200 M in $HC_2H_3O_2$ and 0.0100 M in $HC_7H_5O_2$. K_a-values for $HC_2H_3O_2$ and $HC_7H_5O_2$ are 1.75×10^{-5} and 6.46×10^{-5}, respectively.

Ans. 1.00×10^{-3}, 3.5×10^{-4}, 6.5×10^{-4}

17.67. Liquid ammonia ionizes to a slight extent. At $-50\,°C$, its ion product is $K_{NH_3} = [NH_4^+][NH_2^-] = 10^{-30}$. How many amide ions, NH_2^-, are present per mm^3 of pure liquid ammonia?

Ans.   600

IONIZATION OF WATER

17.68. Calculate the pH and pOH of the following solutions, assuming complete ionization: (a) 0.00345 N acid, (b) 0.000775 N acid,   (c) 0.00886 N base.

Ans.   (a) pH $= 2.46$, pOH $= 11.54$;   (b) 3.11, 10.89;   (c) 11.95, 2.05

17.69. Convert the following pH-values to $[H^+]$-values: (a) 4,   (b) 7,   (c) 2.50,   (d) 8.26.

Ans.   (a) 10^{-4};   (b) 10^{-7};   (c) 3.2×10^{-3};   (d) 5.5×10^{-9}

17.70. The $[H^+]$ of an HNO_3 solution is 1×10^{-3}, and the $[H^+]$ of an NaOH solution is 1×10^{-12}. Find the molar concentration and pH of each solution.

Ans. HNO_3: 0.001 M, pH = 3; NaOH: 0.01 M, pH = 12

17.71. Compute $[H^+]$ and $[OH^-]$ in a 0.0010 molar solution of a monobasic acid which is 4.2 percent ionized. What is the pH of the solution? What are K_a and pK_a for the acid?

Ans. $[H^+] = 4.2 \times 10^{-5}$, $[OH^-] = 2.4 \times 10^{-10}$, pH = 4.38, $K_a = 1.8 \times 10^{-6}$, $pK_a = 5.73$

17.72. Compute $[OH^-]$ and $[H^+]$ in a 0.10 N solution of a weak base which is 1.3 percent ionized. What is the pH of the solution?

Ans. $[OH^-] = 1.3 \times 10^{-3}$, $[H^+] = 7.7 \times 10^{-12}$, pH = 11.11

17.73. What is the pH of a solution containing 0.010 mol HCl per liter? Calculate the change in pH if 0.020 mol $NaC_2H_3O_2$ is added to a liter of this solution. K_a of $HC_2H_3O_2$ is 1.75×10^{-5}.

Ans.   Initial pH = 2.0, final pH = 4.76

17.74. The value of K_w at the physiological temperature 37 °C is 2.4×10^{-14}. What is the pH at the neutral point of water at this temperature, where there are equal numbers of H^+ and OH^-?

Ans.   6.81

17.75. What is the pH of 7.0×10^{-8} M acetic acid? What is the concentration of un-ionized acetic acid? K_a is 1.75×10^{-5} (*Hint*: Assume essentially complete ionization of the acetic acid in solving for $[H^+]$.)

Ans.   6.85, 5.6×10^{-10} M

17.76. Calculate $[OH^-]$ in a 0.0100 M solution of aniline, $C_6H_5NH_2$. K_b for the basic dissociation of aniline is 4.3×10^{-10}. What is the $[OH^-]$ in a 0.0100 M solution of aniline hydrochloride, which contains the ion $C_6H_5NH_3^+$?

Ans. 2.1×10^{-6}, 2.1×10^{-11}

17.77. Calculate the percent hydrolysis in a 0.0100 M solution of KCN. K_a for HCN is 4.93×10^{-10}.

Ans.   4.5%

17.78. The basic ionization constant for hydrazine, N_2H_4, is 9.6×10^{-7}. What would be the percent hydrolysis of 0.100 M N_2H_5Cl, a salt containing the acid ion conjugate to hydrazine base?

Ans.   0.032%

17.79. A 0.25 M solution of pyridinium chloride, $C_5H_6N^+Cl^-$, was found to have a pH of 2.93. What is K_b for the basic dissociation of pyridine, C_5H_5N?

Ans. 1.8×10^{-9}

17.80. K_a for the acid ionization of Fe^{3+} to $Fe(OH)^{2+}$ and H^+ is 6.5×10^{-3}. What is the maximum pH-value which could be used so that at least 95 percent of the total +III iron in a dilute solution is in the Fe^{3+} form?

Ans.   0.9

17.81. A 0.010 M solution of $PuO_2(NO_3)_2$ was found to have a pH of 3.80. What is the hydrolysis constant, K_a, for PuO_2^{2+}, and what is K_b for PuO_2OH^+?

Ans. $K_a = 2.5 \times 10^{-6}$, $K_b = 4.0 \times 10^{-9}$

17.82. Calculate the pH of 1.0×10^{-3} M sodium phenolate, $NaOC_6H_5$. K_a for HOC_6H_5 is 1.28×10^{-10}.

Ans.   10.39

17.83. Calculate $[H^+]$ and $[CN^-]$ in 0.0100 M NH_4CN. K_a for HCN is 4.93×10^{-10} and K_b for NH_3 is 1.75×10^{-5}.

Ans. 5.3×10^{-10}, 4.8×10^{-3}

POLYPROTIC ACIDS

17.84. Calculate the $[H^+]$ of a 0.050 M H_2S solution. K_1 of H_2S is 1.0×10^{-7}.

Ans. 7.1×10^{-5}

17.85. What is $[S^{2-}]$ in a 0.050 M H_2S solution? K_2 of H_2S is 1.2×10^{-13}.

Ans. 1.2×10^{-13}

17.86. What is $[S^-]$ in a solution that is 0.050 M H_2S and 0.0100 M HCl? Use data from Problems 17.84 and 17.85.

Ans. 6.0×10^{-18}

17.87. K_1 and K_2 for oxalic acid, $H_2C_2O_4$, are 5.9×10^{-2} and 6.4×10^{-5}. What is $[OH^-]$ in a 0.005 molar solution of $Na_2C_2O_4$?

Ans. 8.8×10^{-7}

17.88. Malonic acid is a dibasic acid having $K_1 = 1.49 \times 10^{-3}$ and $K_2 = 2.03 \times 10^{-6}$. Compute the concentration of the divalent malonate ion in   (a) 0.0010 M malonic acid,   (b) a solution that is 0.00010 M in malonic acid and 0.00040 M in HCl.

Ans.   (a) 2.0×10^{-6};   (b) 3.2×10^{-7}

17.89. Compute the pH of a 0.010 M solution of H_3PO_4. K_1 and K_2 for H_3PO_4 are respectively 7.52×10^{-3} and 6.23×10^{-8}.

Ans.   2.24

17.90. What is $[H^+]$ in a 0.0060 M H_2SO_4 solution? The first ionization of H_2SO_4 is complete and the second ionization has a K_2 of 1.20×10^{-2}. What is $[SO_4^{2-}]$ in the same solution?

Ans. $[H^+] = 9.4 \times 10^{-3}$, $[SO_4^{2-}] = 3.4 \times 10^{-3}$

17.91. Ethylenediamine, $NH_2C_2H_4NH_2$, is a base that can add one or two protons. The successive pK_b-values for the reaction of the neutral base and of the monocation with water are 3.288 and 6.436, respectively. In a 0.0100 M solution of ethylenediamine, what are the concentrations of the singly charged cation and of the doubly charged cation?

Ans. 2.03×10^{-3} mol/L, 3.66×10^{-7} mol/L

17.92. pK_1 and pK_2 for pyrophosphoric acid are 0.85 and 1.49, respectively. Neglecting the third and fourth dissociations of this tetraprotic acid, what would be the concentration of the divalent anion in a 0.050 M solution of the acid?

Ans. 1.5×10^{-2} M

17.93. What is $[CO_3^{2-}]$ in a 0.001 0 M Na_2CO_3 solution after the hydrolysis reactions have come to equilibrium? K_1 and K_2 for H_2CO_3 are 4.30×10^{-7} and 5.61×10^{-11}, respectively.

   Ans. 6.6×10^{-4}

17.94. Calculate the pH of 0.050 M NaH_2PO_4 and of 0.002 0 M Na_3PO_4. K_1, K_2, and K_3 of H_3PO_4 are 7.52×10^{-3}, 6.23×10^{-8}, and 4.5×10^{-13}, respectively.

   Ans.   4.7, 11.27

17.95. Citric acid is a polyprotic acid with pK_1, pK_2, and pK_3 equal to 3.15, 4.77, and 6.39, respectively. Calculate the concentrations of H^+, the monovalent anion, the divalent anion, and the trivalent anion in 0.010 0 M citric acid.

   Ans. 2.3×10^{-3} M, 2.3×10^{-3} M, 1.7×10^{-5} M, 2.9×10^{-9} M

17.96. The amino acid glycine, NH_2CH_2COOH, is basic because of its NH_2— group and acidic because of its —COOH group. By a process equivalent to base dissociation, glycine can acquire an additional proton to form $^+NH_3CH_2COOH$. The resulting cation may be considered to be a diprotic acid, since one proton from the —COOH group and one proton from the $^+NH_3$— group may be lost. The pK_a values for these processes are 2.35 and 9.78, respectively. In a 0.010 0 M solution of neutral glycine, what is the pH and what percent of the glycine is in the cationic form at equilibrium?

   Ans.   6.14, 0.016%

BUFFER SOLUTIONS, INDICATORS, AND TITRATIONS

17.97. A buffer solution was prepared by dissolving 0.020 0 mol propionic acid and 0.015 mol sodium propionate in enough water to make 1 L of solution. (*a*) What is the pH of the buffer? (*b*) What would be the pH change if 1.0×10^{-5} mol HCl were added to 10 cm^3 of the buffer. (*c*) What would be the pH change if 1.0×10^{-5} mol NaOH were added to 10 cm^3 of the buffer? K_a for propionic acid is 1.34×10^{-5}.

   Ans.   (*a*) 4.75; (*b*) −0.05; (*c*) +0.05

17.98. The base imidazole has a K_b of 1.11×10^{-7}. (*a*) In what amounts should 0.020 0 M HCl and 0.020 0 M imidazole be mixed to make 100 cm^3 of a buffer at pH 7.00? (*b*) If the resulting buffer is diluted to 1 L, what is the pH of the diluted buffer?

   Ans.   (*a*) 34 cm^3 acid, 66 cm^3 base; (*b*) 7.00

17.99. In the titration of HCl with NaOH represented in Fig. 17-1*a*, calculate the pH after the addition of a total of 20.0, 30.0, and 60.0 cm^3 of NaOH.

   Ans.   1.37, 1.60, 11.96

17.100. In the titration of β-hydroxybutyric acid, $HC_4H_7O_3$, with NaOH represented in Fig. 17-1*a*, calculate the pH after the addition of a total of 20.0, 30.0, and 70.0 cm^3 of NaOH. pK_a for $HC_4H_7O_3$ is 4.70.

   Ans.   4.52, 4.88, 12.22

17.101. In the titration of NH_3 with HCl represented in Fig. 17-1*b*, calculate the total volume of the HCl solution needed to bring the pH to 10.00 and to 8.00. K_b for NH_3 is 1.75×10^{-5}.

   Ans.   7.4 cm^3, 47.3 cm^3

17.102. The dye bromcresol green has a pK_a-value of 4.95. For which of the four titrations shown in Fig. 17-1 would bromcresol green be a suitable end-point indicator?

Ans. HCl vs. NaOH, NaOH vs. HCl, NH_3 vs. HCl

17.103. Bromphenol blue is an indicator with a K_a-value of 5.84×10^{-5}. What percentage of this indicator is in its basic form at a pH of 4.84?

Ans. 80%

17.104. Calculate the pH and $[NH_3]$ at the end point in the titration of NH_3 with HCl, at the concentrations indicated for Fig 17-1*b*. K_b for NH_3 is 1.75×10^{-5}.

Ans. 5.27, 5.4×10^{-6}

17.105. If 0.001 0 mol of citric acid is dissolved in 1 L of a solution buffered at pH 5.00 (without changing the volume), what will be the equilibrium concentrations of citric acid, its monovalent anion, the divalent anion, and the trivalent anion? Use pK-values from Problem 17.95.

Ans. 5.0×10^{-6} M, 3.6×10^{-4} M, 6.1×10^{-4} M, 2.5×10^{-5} M

17.106. If 0.000 500 mol $NaHCO_3$ is added to a large volume of a solution buffered at pH 8.00, how much material will exist in each of the three forms H_2CO_3, HCO_3^-, and CO_3^{2-}? K_1 and K_2 of H_2CO_3 are, respectively, 4.30×10^{-7} and 5.61×10^{-11}.

Ans. 1.13×10^{-5} mol, 4.86×10^{-4} mol, 2.73×10^{-6} mol

17.107. A buffer solution of pH 6.70 can be prepared by employing solutions of NaH_2PO_4 and Na_2HPO_4. If 0.005 0 mol NaH_2PO_4 is weighed out, how much Na_2HPO_4 must be used to make 1 L of the solution? Take K-values from Problem 17.94.   *Ans.* 0.001 6 mol

17.108. Now much NaOH must be added to 1 L of 0.010 M H_3BO_3 to make a buffer solution of pH 10.10? H_3BO_3 is a monoprotic acid with $K_a = 5.8 \times 10^{-10}$.

Ans. 0.008 8 mol

17.109. A buffer solution was prepared by dissolving 0.050 mol formic acid and 0.060 mol sodium formate in enough water to make 1 L of solution. K_a for formic acid is 1.77×10^{-4}. (*a*) Calculate the pH of the solution. (*b*) If this solution were diluted to ten times its volume, what would be the pH? (*c*) If the solution in (*b*) were diluted to ten times *its* volume, what would be the pH?

Ans. (*a*) 3.83; (*b*) 3.85; (*c*) 4.00

APPARENT EQUILIBRIUM CONSTANTS

17.110. The equilibrium between fumaric acid, $H_2C_4H_2O_4$, and malic acid, $H_2C_4H_4O_5$, can be made to occur in water solution in the presence of the enzyme fumarase. K' for the reaction

$$H_2C_4H_2O_4 + H_2O \rightleftharpoons H_2C_4H_4O_5$$

in aqueous solution is 3.5 at a specified acidity. How much pure fumaric acid should be weighed out to make 1 L of a solution at the standard acidity which is to contain 0.20 mol malic acid after enzymatic equilibration?

Ans. 0.26 mol

17.111. The following equilibria of amino acids have been studied in dilute aqueous solution.

(*1*)   Glutamic acid + pyruvic acid $\rightleftharpoons$ α-ketoglutaric acid + alanine

(*2*)   Glutamic acid + oxalacetic acid $\rightleftharpoons$ α-ketoglutaric acid + aspartic acid

(*3*)   Alanine + oxalacetic acid $\rightleftharpoons$ pyruvic acid + aspartic acid

One molecule of each substance appears in each of these balanced equations. K' for reaction (*1*) is 2.07, and for reaction (*2*) is 5.44, at a specified acidity. What is K' for reaction (*3*) at the same acidity? [*Hint*: The equilibrium of reaction (*3*) in the presence of a small amount of glutamic acid is the same as in the absence of glutamic acid.]

Ans.   2.63

Chapter 18

Complex Ions; Precipitates

COORDINATION COMPLEXES

This section deals with the equilibria involving metal ions, their complexing ligands, and the resulting coordination complexes (see Chapter 9). Some of these complexes are so stable that salts can be prepared from them that show no appreciable amounts of the separate constituents. An example is the ferricyanide ion, $Fe(CN)_6^{3-}$, solutions of which are quite different in properties from those of Fe^{3+} or of CN^-. More commonly, the complex is not so stable and in solution is partly dissociated into its components. In such a case there is an equilibrium constant which regulates the allowable simultaneous values of the concentrations of the various species. An example is $FeBr^{2+}$, which may be formed or dissociated easily, depending on slight modifications of experimental conditions.

$$Fe^{3+} + Br^- \rightleftharpoons FeBr^{2+} \qquad K_s = \frac{[FeBr^{2+}]}{[Fe^{3+}][Br^-]}$$

The constant K_s is called a *stability constant*; the larger its value, the stabler the complex.

In some cases several ligands may be complexed, one at a time. A separate equilibrium equation may be written for each *successive* ligand addition; e.g.,

$$Cd^{2+} + CN^- \rightleftharpoons CdCN^+ \qquad K_1 = \frac{[CdCN^+]}{[Cd^{2+}][CN^-]}$$

$$CdCN^+ + CN^- \rightleftharpoons Cd(CN)_2 \qquad K_2 = \frac{[Cd(CN)_2]}{[CdCN^+][CN^-]}$$

Analogous equations can be written for the addition of a third and fourth cyanide, with constants K_3 and K_4. In addition to the stepwise formation equilibria, we can write a single overall equation for the formation of a complex containing several ligands from the free cation and ligands.

$$Cd^{2+} + 4CN^- \rightleftharpoons Cd(CN)_4^{2-} \qquad K_s = \frac{[Cd(CN)_4^{2-}]}{[Cd^{2+}][CN^-]^4}$$

It can easily be shown that in this case $K_s = K_1 K_2 K_3 K_4$.

Equilibria between a complex ion and its components are sometimes written in reverse.

$$Cd(CN)_4^{2-} \rightleftharpoons Cd^{2+} + 4CN^- \qquad K_d = \frac{[Cd^{2+}][CN^-]^4}{[Cd(CN)_4^{2-}]}$$

The *dissociation constant*, K_d, is the reciprocal of the overall stability constant, K_s.

Other symbols commonly used in texts and tables of data are

Synonymous with K_s: K_f or K_{form} (formation constant)

Synonymous with K_d: K_{diss} or K_{inst} (instability constant)

SOLUBILITY PRODUCT

Consider the equilibrium between solid AgCl and its dissolved ions in a saturated solution:

$$AgCl \ (solid) \rightleftharpoons Ag^+ + Cl^- \ (in \ solution)$$

The expression for the equilibrium constant is

$$K_{sp} = [Ag^+][Cl^-]$$

where the convention of excluding terms for solids has been honored. This particular case of the equilibrium constant for the solution of ionic substances is called the *solubility product* (K_{sp}). Solubility products for some other saturated solutions of a slightly soluble salt are

$$K_{sp} \text{ of } BaCO_3 = [Ba^{2+}][CO_3^{2-}]$$

$$K_{sp} \text{ of } CaF_2 = [Ca^{2+}][F^-]^2$$

$$K_{sp} \text{ of } Bi_2S_3 = [Bi^{3+}]^2 [S^{2-}]^3$$

In each of these expressions, the exponents are derived from the balanced equation in the usual manner. As in previous chapters, the temperature will be taken as 25 °C unless a statement is made to the contrary.

APPLICATIONS OF SOLUBILITY PRODUCT TO PRECIPITATION

Precipitation

The solubility product enables us to explain and predict the degree of completeness of precipitation reactions. Whenever the product of the appropriate powers of the concentrations of any two ions in a solution *exceeds* the value of the corresponding solubility product, the cation-anion combinations will be precipitated until the product of the concentrations of these two ions remaining in solution (raised to their respective powers) again attains the value of the solubility product.

EXAMPLE 1   When some NaF is added to a saturated solution of CaF_2, the $[F^-]$ is greatly increased and the product of the ion concentrations, $[Ca^{2+}][F^-]^2$, may temporarily exceed the value of the solubility product. To restore equilibrium, some Ca^{2+} unites with an equivalent quantity of F^- to form solid CaF_2, until $[Ca^{2+}][F^-]^2$ for the ions remaining in solution again attains the value of the solubility product. Note that in this case the final value of $[F^-]$ is much greater than twice that of $[Ca^{2+}]$, since NaF yields a large contribution to the total $[F^-]$.

The ion concentrations that appear in the expression for the solubility product refer only to the simple ions *in solution*, and do not include the material in the precipitate. Additional equilibria may exist between the simple ions and complexes *in solution*, in case soluble complexes form; these equilibria would be governed by their own stability constants.

Solution of precipitates

The product of the concentrations of any two ions (raised to the appropriate powers) in a solution is called the *ion product*, Q, analogous to Q, the *reaction quotient* of Chapter 16. Whenever Q is *less than* the value of the corresponding solubility product, the solution is not saturated. If some of the corresponding solid salt is in contact with the solution, more of the salt will dissolve.

EXAMPLE 2   Suppose HCl (which supplies H^+) is added to a saturated solution of $Mg(OH)_2$ in equilibrium with some undissolved solute. The H^+ removes nearly all the OH^- in solution to form H_2O. This greatly decreases the $[OH^-]$, and more $Mg(OH)_2$ dissolves so that the ion concentration product can again attain the value of K_{sp} for $Mg(OH)_2$.

Prevention of precipitation

To prevent the precipitation of a slightly soluble salt, some substance must be added which will keep the concentration of one of the ions so low that the solubility product of the salt is not attained.

EXAMPLE 3 H_2S will not precipitate FeS from a strongly acid (HCl) solution of Fe^{2+}. The large $[H^+]$ furnished by the hydrochloric acid represses the ionization of H_2S (common-ion effect) and thus reduces the $[S^{2-}]$ to so low a value that the solubility product of FeS is not reached.

Solved Problems

COORDINATION COMPLEXES

18.1. One liter of solution was prepared containing 0.001 0 mol of silver in the $+I$ oxidation state and 1.00 mol of NH_3. What is the concentration of free Ag^+ in the solution at equilibrium? K_d for $Ag(NH_3)_2^+$ is 6.0×10^{-8}.

Most of the silver, approximately 0.001 0 mol, will be in the form of the complex ion $Ag(NH_3)_2^+$. The concentration of free NH_3 at equilibrium is practically unchanged from 1.00 mol/L, since only 0.002 00 mol of NH_3 would be used up to form 0.001 0 mol of complex.

$$Ag(NH_3)_2^+ \rightleftharpoons Ag^+ + 2NH_3$$

$$K_d = \frac{[Ag^+][NH_3]^2}{[Ag(NH_3)_2^+]} \qquad \text{or} \qquad 6.0 \times 10^{-8} = \frac{[Ag^+](1.00)^2}{0.001\,0}$$

Solving $[Ag^+] = 6.0 \times 10^{-11}$; i.e., an equilibrium concentration of 6.0×10^{-11} mol/L.

18.2. K_1 for the complexation of one NH_3 with Ag^+ is 2.0×10^3. (a) With reference to Problem 18.1, what is the concentration of $Ag(NH_3)^+$? (b) What is K_2 for this system?

(a) K_1 refers to the following:

$$Ag^+ + NH_3 \rightleftharpoons Ag(NH_3)^+ \qquad K_1 = \frac{[Ag(NH_3)^+]}{[Ag^+][NH_3]}$$

Then, from Problem 18.1,

$$[Ag(NH_3)^+] = K_1([Ag^+][NH_3]) = (2.0 \times 10^3)(6.0 \times 10^{-11})(1.00) = 1.2 \times 10^{-7}$$

This problem is actually a check on an assumption made in Problem 18.1; namely, that practically all the dissolved silver was in the complex $Ag(NH_3)_2^+$. If $[Ag(NH_3)^+]$ had turned out to be greater than about 1×10^{-4}, the assumption would have been shown to be incorrect.

(b) K_1, K_2, and K_d are interrelated.

$$K_2 = \frac{[Ag(NH_3)_2^+]}{[Ag(NH_3)^+][NH_3]} = \frac{[Ag(NH_3)_2^+]/[Ag^+][NH_3]^2}{[Ag(NH_3)^+]/[Ag^+][NH_3]} = \frac{1/K_d}{K_1}$$

$$= \frac{1}{K_1 K_d} = \frac{1}{(2.0 \times 10^3)(6.0 \times 10^{-8})} = 8.3 \times 10^3$$

18.3. How much NH_3 should be added to a solution of 0.001 0 M $Cu(NO_3)_2$ in order to reduce $[Cu^{2+}]$ to 10^{-13}? K_d for $Cu(NH_3)_4^{2+}$ is 4.35×10^{-13}. Neglect the amount of copper in complexes containing fewer than 4 ammonias per copper.

$$Cu(NH_3)_4^{2+} \rightleftharpoons Cu^{2+} + 4NH_3 \qquad K_d = \frac{[Cu^{2+}][NH_3]^4}{[Cu(NH_3)_4^{2+}]} = 4.35 \times 10^{-13}$$

Since the sum of the concentrations of copper in the complex and in the free ionic state must equal 0.001 0 mol/L, and since the amount of the free ion is very small, the concentration of the complex is taken to be 0.001 0 mol/L.

Let $x = [NH_3]$. Then

$$\frac{(10^{-13})(x^4)}{0.001\,0} = 4.35 \times 10^{-13} \qquad \text{or} \qquad x^4 = 4.35 \times 10^{-3} \qquad \text{or} \qquad x = 0.26$$

The concentration of NH_3 at equilibrium is 0.26 mol/L. The amount of NH_3 used up in forming 0.0010 mol/L of complex is 0.0040 mol/L, an amount negligible compared with the amount remaining at equilibrium. Hence the amount of NH_3 to be added is 0.26 mol/L.

18.4. A solution was prepared in which, prior to complexation, Cd(II) was 0.00025 M and I^- was 0.0100 M. K_1 and K_2 for complexation of Cd^{2+} with I^- are 190 and 44, respectively. What percentages of Cd(II) at equilibrium are Cd^{2+}, $Cd\,I^+$, and $Cd\,I_2$?

Let $[Cd^{2+}] = x$, $[Cd\,I^+] = y$, and $[Cd\,I_2] = z$. Assume that $[I^-]$ remains 0.0100 within the precision of this calculation. (At most only 0.0005 could be complexed.)

$$Cd^{2+} + I^- \rightleftharpoons Cd\,I^+ \qquad\qquad Cd\,I^+ + I^- \rightleftharpoons Cd\,I_2$$

$$x \quad 0.0100 \quad y \qquad\qquad\qquad y \quad 0.0100 \quad z$$

$$K_1 = \frac{y}{0.0100x} = 190 \qquad y = 1.90x \qquad K_2 = \frac{z}{0.0100y} = 44 \qquad z = 0.44y$$

$$z = (0.44)(1.90x) = 0.84x$$

$$x + y + z = 0.00025 = x + 1.90x + 0.84x = 3.74x$$

$$x = 6.7 \times 10^{-5}\ M \qquad y = 12.7 \times 10^{-5}\ M \qquad z = 5.6 \times 10^{-5}\ M$$

$$[Cd^{2+}] = \frac{6.7 \times 10^{-5}\ M}{2.5 \times 10^{-4}\ M} \times 100\% = 27\% \qquad [Cd\,I^+] = \frac{12.7 \times 10^{-5}\ M}{2.5 \times 10^{-4}\ M} \times 100\% = 51\%$$

$$[Cd\,I_2] = \frac{5.6 \times 10^{-5}\ M}{2.5 \times 10^{-4}\ M} \times 100\% = 22\%$$

SOLUBILITY PRODUCT AND PRECIPITATION

18.5. The solubility of $PbSO_4$ in water is 0.038 g/L. Calculate the solubility product of $PbSO_4$.

$$PbSO_4\ (solid) \rightleftharpoons Pb^{2+} + SO_4^{2-}\ (in\ solution)$$

The concentrations of the ions must be expressed in moles per liter. To convert 0.038 g/L to moles of ions per liter, divide by the formula weight of $PbSO_4$ (303).

$$0.038\ g/L = \frac{0.038\ g/L}{303\ g/mol} = 1.25 \times 10^{-4}\ mol/L$$

Since 1.25×10^{-4} mol of dissolved $PbSO_4$ yields 1.25×10^{-4} mol each of Pb^{2+} and SO_4^{2-},

$$K_{sp} = [Pb^{2+}][SO_4^{2-}] = (1.25 \times 10^{-4})(1.25 \times 10^{-4}) = 1.6 \times 10^{-8}$$

This method may be applied to any fairly insoluble salt whose ions do not hydrolyze appreciably or form soluble complexes. Sulfides, carbonates, and phosphates, and salts of many of the transition metals like iron, must be treated by taking into account hydrolysis and, in some cases, complexation. Some examples will be given in the problems that follow.

18.6. The solubility of Ag_2CrO_4 in water is 0.022 g/L. Determine the solubility product.

$$Ag_2CrO_4 \rightleftharpoons 2Ag^+ + CrO_4^{2-}$$

To convert 0.022 g/L to mol/L of the ions, divide by the formula weight of Ag_2CrO_4 (332).

$$0.022\ g/L = \frac{0.022\ g/L}{332\ g/mol} = 6.6 \times 10^{-5}\ mol/L$$

Since 1 mol of dissolved Ag_2CrO_4 yields 2 mol Ag^+ and 1 mol CrO_4^{2-},

$$[Ag^+] = 2(6.6 \times 10^{-5}) = 1.3 \times 10^{-4} \qquad [CrO_4^{2-}] = 6.6 \times 10^{-5}$$

and $$K_{sp} = [Ag^+]^2[CrO_4^{2-}] = (1.3 \times 10^{-4})^2(6.6 \times 10^{-5}) = 1.1 \times 10^{-12}$$

18.7. The solubility product of $Pb(IO_3)_2$ is 2.5×10^{-13}. What is the solubility of $Pb(IO_3)_2$ (a) in mol/L and (b) in g/L?

(a) Let the solubility of $Pb(IO_3)_2$ be x mol/L. Then $[Pb^{2+}] = x$ and $[IO_3^-] = 2x$.

$$[Pb^{2+}][IO_3^-]^2 = K_{sp} \qquad gives \qquad x(2x)^2 = 2.5 \times 10^{-13}$$

Then $4x^3 = 2.5 \times 10^{-13}$, $x^3 = 62 \times 10^{-15}$, and $x = 4.0 \times 10^{-5}$.

(b) $$Solubility = (4.0 \times 10^{-5} \text{ mol/L})(557 \text{ g/mol}) = 0.022 \text{ g/L}$$

18.8. The $[Ag^+]$ of a solution is 4×10^{-3}. Calculate the $[Cl^-]$ that must be exceeded before AgCl can precipitate. The solubility product of AgCl at $25\,°C$ is 1.8×10^{-10}.

$$[Ag^+][Cl^-] = K_{sp}$$

$$(4 \times 10^{-3})[Cl^-] = 1.8 \times 10^{-10}$$

$$[Cl^-] = 5 \times 10^{-8}$$

Hence a $[Cl^-]$ of 5×10^{-8} must be exceeded before AgCl precipitates. This problem differs from the previous ones in that the two ions forming the precipitate are furnished to the solution independently. This represents a typical analytical situation, in which some soluble chloride is added to precipitate silver ion present in a solution.

18.9. Calculate the solubility of CaF_2 in 0.015 M NaF solution. K_{sp} of CaF_2 is 3.9×10^{-11}.

Because of the large concentration of the common ion, F^-, the solubility will be quite low, and the additional F^- from the dissolved CaF_2 will be assumed negligible compared to that from the NaF. Hence,

$$[Ca^{2+}][F^-]^2 = K_{sp} = 3.9 \times 10^{-11} = [Ca^{2+}](0.015)^2$$

$$[Ca^{2+}] = \frac{3.9 \times 10^{-11}}{(0.015)^2} = 1.7 \times 10^{-7}$$

The solubility will be 1.7×10^{-7} mol/L.

The $[F^-]$ from the CaF_2 will be $2 \times 1.7 \times 10^{-7} = 3.4 \times 10^{-7}$ M, which can be ignored, as assumed, compared to 0.015 M.

18.10. Calculate the concentration of Ag^+, CrO_4^{2-}, NO_3^-, and K^+ after 30 cm^3 of 0.010 M $AgNO_3$ is mixed with 20 cm^3 of 0.010 M K_2CrO_4 and equilibrium is attained. K_{sp} for Ag_2CrO_4 is 1.1×10^{-12}.

If precipitation did not occur the following concentrations would be found, allowing for the dilution upon mixing:

$$[CrO_4^{2-}] = (\tfrac{20}{50})(0.010) = 0.0040 \text{ M}; \quad [K^+] = 2[CrO_4^{2-}] = 0.0080 \text{ M}$$

$$[Ag^+] = [NO_3^-] = (\tfrac{30}{50})(0.010) = 0.0060 \text{ M}$$

Since K^+ and NO_3^- do not react, the above values of $[K^+]$ and $[NO_3^-]$ will prevail after equilibration. To determine whether a precipitate will form calculate the *ion product*, Q, and compare its value to K_{sp}.

$$Q = [Ag^+]^2[CrO_4^{2-}] = (0.0060)^2(0.004) = 1.4 \times 10^{-7} \quad K_{sp} = 1.1 \times 10^{-12}$$

Since $Q \gg K_{sp}$, a precipitate will form.

Only 0.003 mol/L of CrO_4^{2-} is required to precipitate all the Ag^+. The excess CrO_4^{2-}, 0.001 0 mol/L, ensures that very little Ag^+ will remain in solution. The $[CrO_4^{2-}]$ equivalent to the remaining $[Ag^+]$ will also be very small compared to the 0.001 0 mol/L excess. Hence, after Ag_2CrO_4(s) precipitates:

$$K_{sp} = 1.1 \times 10^{-12} = [Ag^+]^2[CrO_4^{2-}] = [Ag^+]^2[0.001\,0]$$

$$[Ag^+]^2 = 1.1 \times 10^{-9}; \quad [Ag^+] = 3.3 \times 10^{-5}$$

The above value is small enough to justify the assumptions made. The final solution contains 3.3×10^{-5} M Ag^+ and 0.001 0 M CrO_4^{2-}.

18.11. Calculate the solubility of AgCN in a buffer solution of pH 3.00. K_{sp} for AgCN is 6.0×10^{-17} and K_a for HCN is 4.93×10^{-10}.

In this solution the silver that dissolves remains as Ag^+, but the cyanide that dissolves is converted mostly to HCN on account of the fixed acidity of the buffer. [The complex $Ag(CN)_2^-$ forms appreciably only at higher cyanide ion concentrations.] First calculate the ratio of [HCN] to $[CN^-]$ at this pH.

$$\frac{[H^+][CN^-]}{[HCN]} = K_a \quad \text{or} \quad \frac{[HCN]}{[CN^-]} = \frac{[H^+]}{K_a} = \frac{1.0 \times 10^{-3}}{4.93 \times 10^{-10}} = 2.0 \times 10^6$$

The two equilibria can be combined to give an overall K for the dissolution process:

$$AgCN \rightleftharpoons Ag^+ + CN^- \qquad K_{sp}$$

Subtract: $\qquad \dfrac{HCN \rightleftharpoons H^+ + CN^- \qquad K_a}{AgCN + H^+ \rightleftharpoons Ag^+ + HCN \qquad K = \dfrac{K_{sp}}{K_a} = \dfrac{6.0 \times 10^{-17}}{4.93 \times 10^{-10}} = 1.22 \times 10^{-7}}$

Let the solubility of AgCN be x mol/L; then

$$x = [Ag^+] \text{ at equilibrium} \qquad x = [CN^-] + [HCN]$$

Very little error is made by neglecting $[CN^-]$ in comparison to [HCN] (1 part in 2 million) and equating [HCN] to x.

$$K = 1.22 \times 10^{-7} = \frac{[Ag^+][HCN]}{[H^+]} = \frac{x^2}{1.00 \times 10^{-3}} \qquad x = 1.1 \times 10^{-5}$$

18.12. Calculate the NH_4^+ ion concentration (derived from NH_4Cl) needed to prevent $Mg(OH)_2$ from precipitating in 1 L of solution which contains 0.010 mol of ammonia and 0.001 0 mol of Mg^{2+}. The ionization constant of ammonia is 1.75×10^{-5}. The solubility product of $Mg(OH)_2$ is 7.1×10^{-12}.

First, find the maximum $[OH^-]$ that can be present in the solution without precipitation of $Mg(OH)_2$.

$$[Mg^{2+}][OH^-]^2 = 7.1 \times 10^{-12}$$

$$[OH^-] = \sqrt{\frac{7.1 \times 10^{-12}}{[Mg^{2+}]}} = \sqrt{\frac{7.1 \times 10^{-12}}{0.001\,0}} = \sqrt{71 \times 10^{-10}} = 8 \times 10^{-5}$$

Now find the $[NH_4^+]$ (derived from NH_4Cl) needed to repress the ionization of NH_3 so that the $[OH^-]$ will not exceed 8×10^{-5}.

$$\frac{[NH_4^+][OH^-]}{[NH_3]} = 1.75 \times 10^{-5} \quad \text{or} \quad \frac{[NH_4^+](8 \times 10^{-5})}{0.010} = 1.75 \times 10^{-5}$$

whence $[NH_4^+] = 2.2 \times 10^{-3}$. (Since 0.010 M ammonia is only slightly ionized, especially in the presence of excess NH_4^+, the $[NH_3]$ may be considered to be 0.010.)

18.13. Given that 2×10^{-4} mol each of Mn^{2+} and Cu^{2+} was contained in 1 L of a 0.003 M $HClO_4$ solution, and this solution was saturated with H_2S. Determine whether or not each of these ions,

Mn^{2+} and Cu^{2+}, will precipitate as the sulfide. The solubility of H_2S, 0.10 mol/L, is assumed to be independent of the presence of other materials in the solution. K_{sp} of MnS is 3×10^{-14}; of CuS, 8×10^{-37}. K_1 and K_2 for H_2S are 1.0×10^{-7} and 1.2×10^{-13}, respectively.

$[H_2S] = 0.10$, since the solution is saturated with H_2S; $[H^+] = 0.003$, since the H_2S contributes negligible H^+ compared with $HClO_4$. Calculate $[S^{2-}]$ from the combined ionization constants.

$$\frac{[H^+]^2[S^{2-}]}{[H_2S]} = K_1 K_2 = (1.0 \times 10^{-7})(1.2 \times 10^{-13}) = 1.2 \times 10^{-20}$$

$$[S^{2-}] = (1.2 \times 10^{-20})\frac{[H_2S]}{[H^+]^2} = (1.2 \times 10^{-20})\frac{0.10}{(0.003)^2} = 1.3 \times 10^{-16}$$

Now form the ion product, Q, in each case and compare it to K_{sp}.

For MnS: $Q = [Mn^{2+}][S^{2-}] = (2 \times 10^{-4})(1.3 \times 10^{-16}) = 2.6 \times 10^{-20} < 3 \times 10^{-14} = K_{sp}$

For CuS: $Q = [Cu^{2+}][S^{2-}] = (2 \times 10^{-4})(1.3 \times 10^{-16}) = 2.6 \times 10^{-20} > 8 \times 10^{-37} = K_{sp}$

Thus MnS remains in solution while CuS precipitates. This example illustrates how two metal ions can be separated by controlling the acidity of the solution while adding a reagent.

Actually, as the precipitation of CuS occurs, some additional H^+ will be added from the H_2S, and the equilibrium concentration of S^{2-} will decrease, but not enough to change the outcome predicted above. This feature will be treated in greater detail in the following problem.

18.14. In Problem 18.13, how much Cu^{2+} escapes precipitation?

Most of the Cu^{2+} will precipitate, and an increment of $2 \times 2 \times 10^{-4}$ mol/L of H^+ will be added to the solution from the H_2S, bringing the total $[H^+]$ to 0.003 4 M. Making this correction

$$[S^{2-}] = (1.2 \times 10^{-20})\frac{0.10}{(0.003\,4)^2} = 1.0 \times 10^{-16}$$

$$[Cu^{2+}] = \frac{K_{sp}}{[S^{2-}]} = \frac{8 \times 10^{-37}}{1.0 \times 18^{-16}} = 8 \times 10^{-21}$$

Thus, 8×10^{-21} mol Cu^{2+} remains in solution. On a percentage basis the amount of Cu^{2+} remaining unprecipitated is

$$\frac{8 \times 10^{-21}}{2 \times 10^{-4}} \times 100 = 4 \times 10^{-15}\%$$

18.15. If the solution in Problem 18.13 is made neutral by lowering the $[H^+]$ to 10^{-7}, will MnS precipitate?

$$[S^{2-}] = (1.2 \times 10^{-20})\frac{0.10}{(10^{-7})^2} = 1.2 \times 10^{-7}$$

$$Q = [Mn^{2+}][S^{2-}] = (2 \times 10^{-4})(1.2 \times 10^{-7}) = 2.4 \times 10^{-11}$$

But 2.4×10^{-11} is greater than the K_{sp} of MnS (3×10^{-14}). Hence MnS will precipitate.

Lowering the $[H^+]$ to 10^{-7} increases the ionization of H_2S to such an extent that sufficient S^{2-} is furnished for the solubility product of MnS to be exceeded.

18.16. How much NH_3 must be added to a 0.004 M Ag^+ solution to prevent the precipitation of AgCl when $[Cl^-]$ reaches 0.001? K_{sp} for AgCl is 1.8×10^{-10} and K_d for $Ag(NH_3)_2^+$ is 6.0×10^{-8}.

Just as acids may be used to lower the concentration of anions in solution, so complexing agents may be used in some cases to lower the concentration of cations. In this problem, the addition of NH_3 converts most of the silver to the complex ion, $Ag(NH_3)_2^+$. The upper limit for the uncomplexed $[Ag^+]$ without formation of a precipitate can be calculated from the solubility product.

$$[Ag^+][Cl^-] = 1.8 \times 10^{-10} \quad \text{or} \quad [Ag^+] = \frac{1.8 \times 10^{-10}}{[Cl^-]} = \frac{1.8 \times 10^{-10}}{0.001} = 1.8 \times 10^{-7}$$

Enough NH_3 must be added to keep the $[Ag^+]$ below 1.8×10^{-7}. The concentration of $Ag(NH_3)_2^+$ at this limit would then be $0.004 - (1.8 \times 10^{-7})$, or practically 0.004.

$$\frac{[Ag^+][NH_3]^2}{[Ag(NH_3)_2^+]} = K_d \quad \text{or} \quad [NH_3]^2 = \frac{K_d[Ag(NH_3)_2^+]}{[Ag^+]} = \frac{(6.0 \times 10^{-8})(0.004)}{1.8 \times 10^{-7}} = 1.33 \times 10^{-3}$$

and $[NH_3] = 0.036$. The amount of NH_3 that must be added is equal to the sum of the amount of free NH_3 remaining in the solution and the amount of NH_3 used up in forming 0.004 mol/L of the complex ion, $Ag(NH_3)_2^+$. This sum is $0.036 + 2(0.004) = 0.044$ mol NH_3 to be added per liter.

18.17. What is the solubility of AgSCN in 0.0030 M NH_3? K_{sp} for AgSCN is 1.1×10^{-12}, and K_d for $Ag(NH_3)_2^+$ is 6.0×10^{-8}.

We may assume that practically all the dissolved silver will exist as the complex ion, $Ag(NH_3)_2^+$. Then, if the solubility of AgSCN is x mol/L, $x = [SCN^-] = [Ag(NH_3)_2^+]$. The concentration of uncomplexed Ag^+ can be estimated from K_d, assuming for the sake of simplification that $[NH_3]$ is essentially unchanged.

$$\frac{[Ag^+][NH_3]^2}{[Ag(NH_3)_2^+]} = K_d \quad \text{or} \quad [Ag^+] = \frac{K_d[Ag(NH_3)_2^+]}{[NH_3]^2} = \frac{(6.0 \times 10^{-8})x}{(0.003)^2} = 6.7 \times 10^{-3}x$$

This result validates our first assumption: The ratio of uncomplexed to complexed silver in solution is only 6.7×10^{-3}. The two equilibria can be combined to give an overall K for the dissolution process:

$$\begin{array}{lll} \text{AgSCN} & \rightleftharpoons Ag^+ + SCN^- & K_{sp} \\ \text{Subtract:} \quad Ag(NH_3)_2^+ & \rightleftharpoons Ag^+ + 2NH_3 & K_d \\ \hline \text{AgSCN} + 2NH_3 \rightleftharpoons Ag(NH_3)_2^+ + SCN^- & K = \dfrac{K_{sp}}{K_d} = \dfrac{1.1 \times 10^{-12}}{6.0 \times 10^{-8}} = 1.8 \times 10^{-5} \end{array}$$

$$K = 1.8 \times 10^{-5} = \frac{[Ag(NH_3)_2^+][SCN^-]}{[NH_3]^2} = \frac{x^2}{(0.0030)^2} \qquad x = 1.3 \times 10^{-5}$$

The validity of the second assumption is confirmed by the answer. If 1.3×10^{-5} mol of complex is formed per liter, the amount of NH_3 used up for complex formation is $2(1.3 \times 10^{-5}) = 2.6 \times 10^{-5}$ mol/L. The concentration of the remaining free NH_3 in solution is practically unchanged from its initial value, 0.0030 mol/L.

18.18. Calculate the simultaneous solubility of CaF_2 and SrF_2. K_{sp} for these two salts are 3.9×10^{-11} and 2.9×10^{-9}, respectively.

The two solubilities are not independent of each other because there is a common ion, F^-. We will first assume that most of the F^- in the saturated solution is contributed by the SrF_2, since its K_{sp} is so much larger than that of CaF_2. We can then proceed to solve for the solubility of SrF_2 as if the CaF_2 were not present. If the solubility of SrF_2 is x mol/L, $x = [Sr^{2+}]$, $2x = [F^-]$. Then

$$4x^3 = K_{sp} = 2.9 \times 10^{-9} \quad \text{or} \quad x = 9 \times 10^{-4}$$

The CaF_2 solubility will have to adapt to the concentration of F^- set by the SrF_2 solubility.

$$[Ca^{2+}] = \frac{K_{sp}}{[F^-]^2} = \frac{3.9 \times 10^{-11}}{(2 \times 9 \times 10^{-4})^2} = 1.2 \times 10^{-5}$$

i.e., the solubility of CaF_2 is 1.2×10^{-5} mol/L.

Check of assumption: The amount of F^- contributed by the solubility of CaF_2 is twice the concentration of Ca^{2+}, or 2.4×10^{-5} mol/L. This is indeed small compared with the amount contributed by SrF_2, $2 \times 9 \times 10^{-4} = 1.8 \times 10^{-3}$ mol/L.

A more general solution requiring no assumptions is as follows: Let $x = [Ca^{2+}]$; $x[F^-]^2 = 3.9 \times 10^{-11}$; $[Sr^{2+}][F^-]^2 = 2.9 \times 10^{-9}$.

$$\text{Dividing:} \quad \frac{[Sr^{2+}][F^-]^2}{x[F^-]^2} = \frac{2.9 \times 10^{-9}}{3.9 \times 10^{-11}} = 74.4 \qquad [Sr^{2+}] = 74.4x$$

$$[F^-] = 2([Ca^{2+}] + [Sr^{2+}]) = 2(x + 74.4x) = 2 \times 75.4x = 151x$$

Then substituting in the K_{sp} for CaF_2:

$$(x)(151x)^2 = 3.9 \times 10^{-11} \qquad x = 1.2 \times 10^{-5} \qquad \text{(the solubility of } CaF_2\text{)}$$

$$74.4x = 8.9 \times 10^{-4} \qquad \text{(the solubility of } SrF_2\text{)}$$

18.19. Calculate the simultaneous solubility of AgSCN and AgBr. The solubility products for these two salts are 1.1×10^{-12} and 5.0×10^{-13}.

Since the solubilities are not very different the second approach of Problem 18.18 is *required*. Let $x = [Br^-]$; then take the ratio of the K_{sp}'s.

$$\frac{[SCN^-]}{[Br^-]} = \frac{1.1 \times 10^{-12}}{5.0 \times 10^{-13}} = \frac{[SCN^-]}{x} \qquad [SCN^-] = 2.2x$$

$$[Ag^+] = [SCN^-] + [Br^-] = 2.2x + x = 3.2x$$

Then substituting in K_{sp} for AgBr

$$[Ag^+][Br^-] = (3.2x)(x) = 5.0 \times 10^{-13}; \, 3.2x^2 = 5.0 \times 10^{-13}$$

$$x = 4.0 \times 10^{-7} \qquad \text{(solubility of AgBr)}$$

$$2.2x = 8.8 \times 10^{-7} \qquad \text{(solubility of AgSCN)}$$

18.20. Calculate the solubility of MnS in pure water. K_{sp} is 3×10^{-14}. K_1 and K_2 for H_2S are 1.0×10^{-7} and 1.2×10^{-13}

This problem differs from the analogous problems dealing with chromates, oxalates, halides, sulfates, and iodates. The difference lies in the extensive hydrolysis of the sulfide ion.

$$S^{2-} + H_2O \rightleftharpoons HS^- + OH^- \qquad K_b = \frac{K_w}{K_2} = \frac{10^{-14}}{1.2 \times 10^{-13}} = 0.083$$

If x mol/L is the solubility of MnS, we cannot simply equate x to $[S^{2-}]$. Instead, $x = [S^{2-}] + [HS^-] + [H_2S]$. To simplify, we first assume that the first stage of hydrolysis is almost complete and that the second stage proceeds to only a slight extent. In other words, $x = [Mn^{2+}] = [HS^-] = [OH^-]$.

$$[S^{2-}] = \frac{[HS^-][OH^-]}{K_b} = \frac{x^2}{0.083}$$

At equilibrium,

$$[Mn^{2+}][S^{2-}] = \frac{x(x)^2}{0.083} = K_{sp} = 3 \times 10^{-14} \qquad \text{or} \qquad x = 1.4 \times 10^{-5}$$

Check of assumptions:

1.
$$[S^{2-}] = \frac{x^2}{0.083} = \frac{(1.4 \times 10^{-5})^2}{0.083} = 2.4 \times 10^{-9}$$

$[S^{2-}]$ is indeed negligible compared with $[HS^-]$.

2. $$[H_2S] = \frac{[H^+][HS^-]}{K_1} = \frac{K_w[HS^-]}{[OH^-]K_1} = \frac{10^{-14}x}{10^{-7}x} = 10^{-7}$$

$[H_2S]$ is also small compared with $[HS^-]$.

The above approximations would not be valid for sulfides like CuS which are much more insoluble than MnS. First, the water dissociation would begin to play an important role in determining $[OH^-]$. Second, the second stage of hydrolysis, producing $[H_2S]$, would not be negligible compared with the first. Even for MnS, an additional complication arises because of the complexation of Mn^{2+} with OH^-. The full treatment of sulfide solubilities is a complicated problem because of the multiple equilibria which must be considered.

If hydrolysis had not been considered, the answer would have been simply $\sqrt{3 \times 10^{-14}} = 1.7 \times 10^{-7}$, underestimating the solubility by a factor of 80! ($1.4 \times 10^{-5}/1.7 \times 10^{-7} \cong 80$.) Thus hydrolysis greatly increases the solubility of sulfides.

18.21. The following solutions were mixed: 500 cm³ of 0.0100 M $AgNO_3$ and 500 cm³ of a solution that was both 0.0100 M in NaCl and 0.0100 M in NaBr. K_{sp} for AgCl and for AgBr are 1.8×10^{-10} and 5.0×10^{-13}. Calculate $[Ag^+]$, $[Cl^-]$, and $[Br^-]$ in the equilibrium solution.

If there were no precipitation, the diluting effect of mixing would make

$$[Ag^+] = [Cl^-] = [Br^-] = \tfrac{1}{2}(0.0100) = 0.0050$$

AgBr is the more insoluble salt and would take precedence in the precipitation process. To find whether AgCl also precipitates, we may assume that it does not. In this case, only Ag^+ and Br^- would be removed by precipitation, and the concentrations of these two ions in solution would remain equal to each other.

$$[Ag^+][Br^-] = [Ag^+]^2 = K_{sp} = 5.0 \times 10^{-13}$$

or $$[Ag^+] = [Br^-] = 7.1 \times 10^{-7}$$

We now examine the ion product for AgCl.

$$[Ag^+][Cl^-] = (7.1 \times 10^{-7})(5.0 \times 10^{-3}) = 3.5 \times 10^{-9}$$

Since this ion product exceeds K_{sp} for AgCl, at least some AgCl must also precipitate. In other words, our first assumption was wrong.

Since both halides precipitate, both solubility product requirements must be met simultaneously.

$$[Ag^+][Cl^-] = 1.8 \times 10^{-10} \tag{1}$$

$$[Ag^+][Br^-] = 5.0 \times 10^{-13} \tag{2}$$

The third equation needed to define the three unknowns is an equation expressing the balancing of positive and negative charges in solution: $[Na^+] + [Ag^+] = [Cl^-] + [Br^-] + [NO_3^-]$.

$$0.0100 + [Ag^+] = [Cl^-] + [Br^-] + 0.0050$$

or $$[Cl^-] + [Br^-] - [Ag^+] = 0.0050 \tag{3}$$

Dividing (1) by (2), $[Cl^-]/[Br^-] = 360$, and we see that Br^- plays a negligible role in the total anion concentration of the solution. Also, $[Ag^+]$ must be negligible in (3) because of the insolubility of the two silver salts. We thus assume in (3) that $[Cl^-] = 0.0050$. From (1),

$$[Ag^+] = \frac{1.8 \times 10^{-10}}{[Cl^-]} = \frac{1.8 \times 10^{-10}}{0.0050} = 3.6 \times 10^{-8}$$

From (2),

$$[Br^-] = \frac{5.0 \times 10^{-13}}{[Ag^+]} = \frac{5.0 \times 10^{-13}}{3.6 \times 10^{-8}} = 1.4 \times 10^{-5}$$

Check of assumptions: Both $[Ag^+]$ and $[Br^-]$ are negligible compared with 0.0050.

Note that in general, in the presence of both precipitates the ratio of the anion concentrations must be the same as the corresponding ratio of the K_{sp}'s (as can be confirmed by the above results). Also observe that the addition of a few extra drops of $AgNO_3$ (not enough to decrease $[Cl^-]$ substantially) will produce more $AgCl(s)$ but will not change the above answers.

18.22. How much Ag^+ would remain in solution after mixing equal volumes of 0.080 M $AgNO_3$ and 0.080 M HOCN? K_{sp} for AgOCN is 2.3×10^{-7}. K_a for HOCN is 3.5×10^{-4}.

The overall reaction is

$$Ag^+ + HOCN \rightleftharpoons AgOCN + H^+ \tag{1}$$

Assume this goes nearly to completion, then write the K for the reverse reaction by combining K_{sp} and K_a.

$$AgOCN \rightleftharpoons Ag^+ + OCN^- \qquad K_{sp}$$

Subtract:
$$\underline{HOCN \rightleftharpoons H^+ + OCN^- \qquad K_a}$$

$$AgOCN + H^+ \rightleftharpoons Ag^+ + HOCN \qquad K = \frac{K_{sp}}{K_a}$$

Because of the dilution upon mixing the initial $[Ag^+]$ was 0.040 M, which would yield 0.040 M H^+ and leave no appreciable excess HOCN upon completion of reaction (1). Hence if $[Ag^+] = x$, then $[HOCN] = x$, and $[H^+] = 0.040 - x$.

$$K = \frac{[Ag^+][HOCN]}{[H^+]} = \frac{x^2}{0.040 - x} = \frac{2.3 \times 10^{-7}}{3.5 \times 10^{-4}} = 6.6 \times 10^{-4}$$

Solving the quadratic, $x = 4.8 \times 10^{-3} = $ remaining $[Ag^+]$.

Thus, although stoichiometrically equivalent amounts of reagents were mixed about one-eighth $(4.8 \times 10^{-3}/0.040)$ of the silver failed to precipitate. The H^+ generated in the precipitation reaction suppressed the ionization of the HOCN with the result that an insufficient $[OCN^-]$ was available.

Supplementary Problems

COMPLEX IONS

18.23. A 0.001 0 mol sample of solid NaCl was added to 1 L of 0.010 M $Hg(NO_3)_2$. Calculate the $[Cl^-]$ equilibrated with the newly formed $HgCl^+$. K_1 for $HgCl^+$ formation is 5.5×10^6. Neglect the K_2 equilibrium.

Ans. 2×10^{-8}

18.24. What is the $[Cd^{2+}]$ in 1 L of solution prepared by dissolving 0.001 mol $Cd(NO_3)_2$ and 1.5 mol NH_3? K_d for the dissociation of $Cd(NH_3)_4^{2+}$ into Cd^{2+} and $4NH_3$ is 3.6×10^{-8}. Neglect the amount of cadmium in complexes containing fewer than 4 ammonia groups.

Ans. 7.1×10^{-12}

18.25. Silver ion forms $Ag(CN)_2^-$ in the presence of excess CN^-. How much KCN should be added to 1 L of a 0.000 5 M Ag^+ solution in order to reduce $[Ag^+]$ to 1.0×10^{-19}? K_d for the complete dissociation of $Ag(CN)_2^-$ is 3.3×10^{-21}.

Ans.   0.005 mol

18.26. A recent investigation of the complexation of SCN^- with Fe^{3+} led to values of 130, 16, and 1.0 for K_1, K_2, and K_3, respectively. What is the overall formation constant of $Fe(SCN)_3$ from its component ions, and what is the dissociation constant of $Fe(SCN)_3$ into its simplest ions on the basis of these data?

Ans. $K_s = 2.1 \times 10^3$, $K_d = 4.8 \times 10^{-4}$

18.27. Sr^{2+} forms a very unstable complex with NO_3^-. A solution that was nominally 0.00100 M $Sr(ClO_4)_2$ and 0.050 M KNO_3 was found to have only 75 percent of its strontium in the uncomplexed Sr^{2+} form, the balance being $Sr(NO_3)^+$. What is K_1 for complexation?

Ans.   6.7

18.28. A solution made up to be 0.0100 M $Co(NO_3)_2$ and 0.0200 M N_2H_4 at a total ionic strength of 1 was found to have an equilibrium $[Co^{2+}]$ of 6.2 × 10⁻³. Assuming that the only complex formed was $Co(N_2H_4)^{2+}$, what is the apparent K_1 for complex formation at this ionic strength?

Ans.   38

18.29. Equal volumes of 0.0010 M $Fe(ClO_4)_3$ and 0.10 M KSCN were mixed. Using the data in Problem 18.26, find the equilibrium percentages of the iron existing as Fe^{3+}, $FeSCN^{2+}$, $Fe(SCN)_2^+$, and $Fe(SCN)_3$.

Ans.   8%, 50%, 40%, 2%

18.30. What is the concentration of free Cd^{2+} in 0.005 M $CdCl_2$? K_1 for chloride complexation of Cd^{2+} is 100; K_2 need not be considered.

Ans.   2.8 × 10⁻³ mol/L

SOLUBILITY PRODUCT AND PRECIPITATION

18.31. Calculate the solubility products of the following compounds. The solubilities are given in moles per liter. (a) $BaSO_4$, 1.05 × 10⁻⁵ mol/L; (b) TlBr, 1.9 × 10⁻³ mol/L; (c) $Mg(OH)_2$, 1.21 × 10⁻⁴ mol/L; (d) $Ag_2C_2O_4$, 1.15 × 10⁻⁴ mol/L; (e) $La(IO_3)_3$, 7.8 × 10⁻⁴ mol/L.

Ans.   (a) 1.1 × 10⁻¹⁰; (b) 3.6 × 10⁻⁶; (c) 7.1 × 10⁻¹²; (d) 6.1 × 10⁻¹²; (e) 1.0 × 10⁻¹¹

18.32. Calculate the solubility products of the following salts. Solubilities are given in grams per liter. (a) CaC_2O_4, 0.0055 g/L; (b) $BaCrO_4$, 0.0037 g/L; (c) CaF_2, 0.017 g/L.

Ans.   (a) 1.8 × 10⁻⁹; (b) 2.1 × 10⁻¹⁰; (c) 4.1 × 10⁻¹¹

18.33. The solubility product of SrF_2 at 25 °C is 2.9 × 10⁻⁹. (a) Determine the solubility of SrF_2 at 25 °C, in mol/L and in mg/cm³. (b) What are $[Sr^{2+}]$ and $[F^-]$ (in mol/L) in a saturated solution of SrF_2?

Ans.   (a) 9 × 10⁻⁴ mol/L, 0.11 mg/cm³; (b) $[Sr^{2+}]$ = 9 × 10⁻⁴, $[F^-]$ = 1.8 × 10⁻³

18.34. What $[SO_4^{2-}]$ must be exceeded to produce a $RaSO_4$ precipitate in 500 cm³ of a solution containing 0.00010 mol of Ra^{2+}? K_{sp} of $RaSO_4$ is 4 × 10⁻¹¹.

Ans.   2 × 10⁻⁷

18.35. A solution has a Mg^{2+} concentration of 0.001 mol/L. Will $Mg(OH)_2$ precipitate if the OH^- concentration of the solution is (a) 10⁻⁵ mol/L, (b) 10⁻³ mol/L? K_{sp} of $Mg(OH)_2$ is 7.1 × 10⁻¹².

Ans.   (a) no; (b) yes

18.36. Radioactive tracers provide a convenient means of measuring the small concentrations encountered in determining K_{sp} values. Exactly 20.0 cm³ of 0.0100 M $AgNO_3$ solution containing radioactive silver, with an activity of 29 610 counts/min per cm³, was mixed with 100 cm³ of 0.0100 M KIO_3, and the mixture diluted to exactly 400 cm³. After equilibrium was reached, a portion of solution was filtered to remove any solids and was found to have an activity of 47.4 counts/min per cm³. Calculate K_{sp} of $AgIO_3$.

Ans.   3.2 × 10⁻⁸

18.37. An old procedure for the determination of sulfur in gasoline involves the precipitation of barium sulfate as the final chemical step, and specifies that no more than 1 μg of sulfur be permitted to remain behind in solution. If the precipitation is made from 400 cm^3 what must the concentration of excess barium ion be? K_{sp} for BaSO$_4$ is 1.1×10^{-10}.

Ans.   [Ba^{2+}] must be at least 1.4×10^{-3} M

18.38. A solution contains 0.0100 mol/L each of Cd^{2+} and Mg^{2+}. (*a*) To what pH should it be raised to precipitate the maximum amount of one metal (identify which it is) as the hydroxide without precipitating the other? (*b*) What fraction of the precipitated metal will still remain in solution? (*c*) If the pH were 0.50 unit less what fraction would remain in solution? K_{sp} values are 7.1×10^{-12} for Mg(OH)$_2$ and 4.5×10^{-15} for Cd(OH)$_2$.

Ans.   (*a*) 9.43, to precipitate Cd(OH)$_2$;   (*b*) 6.3×10^{-4};   (*c*) 6.2×10^{-3}

18.39. After solid SrCO$_3$ was equilibrated with a pH 8.60 buffer, the solution was found to have [Sr^{2+}] = 1.6×10^{-4}. What is the solubility product for SrCO$_3$? K_2 for carbonic acid is 5.61×10^{-11}.

Ans. 5.6×10^{-10}

18.40. Calculate the solubility at 25 °C of CaCO$_3$ in a closed vessel containing a solution of pH 8.60. K_{sp} for CaCO$_3$ is 7.55×10^{-9}. K_2 for carbonic acid is 5.61×10^{-11}.

Ans. 5.9×10^{-4} mol/L

18.41. How much AgBr could dissolve in 1 L of 0.40 M NH$_3$? K_{sp} for AgBr is 5.0×10^{-13}, and K_d for Ag(NH$_3$)$_2^+$ is 6.0×10^{-8}.

Ans. 1.2×10^{-3} mol

18.42. Solution A was made by mixing equal volumes of 0.0010 M Cd^{2+} and 0.0072 M OH$^-$ solutions as neutral salt and strong base, respectively. Solution B was made by mixing equal volumes of 0.0010 M Cd^{2+} and a standard KI solution. What was the concentration of the standard KI solution if the final [Cd^{2+}] in solutions A and B was the same? K_{sp} for Cd(OH)$_2$ is 4.5×10^{-15}, and K_s for formation of CdI$_4^{2-}$ from its simple ions is 4×10^5. Neglect the amount of cadmium in all iodide complexes other than CdI$_4^{2-}$.

Ans.   2.3 M

18.43. Ag$_2$SO$_4$ and SrSO$_4$ are both shaken up with pure water. K_{sp}-values for these two salts are 1.5×10^{-5} and 3.2×10^{-7}, respectively. Evaluate [Ag$^+$] and [Sr^{2+}] in the resulting saturated solution.

Ans. 3.1×10^{-5}, 2.1×10^{-5}

18.44. Calculate [F$^-$] in a solution saturated with respect to both MgF$_2$ and SrF$_2$ (made by dissolving both MgF$_2$ and SrF$_2$ in H$_2$O until an excess of both solids remains). K_{sp}-values for the two salts are 6.6×10^{-9} and 2.9×10^{-9}, respectively.

Ans. 2.7×10^{-3}

18.45. Equal volumes of 0.0200 M AgNO$_3$ and 0.0200 M HCN were mixed. Calculate [Ag$^+$] at equilibrium. K_{sp} for AgCN is 6.0×10^{-17} and K_a for HCN is 4.93×10^{-10}.

Ans. 3.5×10^{-5}

18.46. Equal volumes of 0.0100 M Sr(NO$_3$)$_2$ and 0.0100 M NaHSO$_4$ were mixed. Calculate [Sr^{2+}] and [H$^+$] at equilibrium. K_{sp} for SrSO$_4$ is 3.2×10^{-7} and K_a for HSO$_4^-$ (the same as K_2 for H$_2$SO$_4$) is 1.2×10^{-2}. Take into account the amount of H$^+$ needed to balance the charge of the SO$_4^{2-}$ remaining in the solution.

Ans. 6.7×10^{-4}, 4.8×10^{-3}

18.47. Excess solid $Ag_2C_2O_4$ is shaken with   (a) 0.0010 M HNO_3,   (b) 0.00030 M HNO_3. What is the equilibrium value of $[Ag^+]$ in the resulting solution? K_{sp} for $Ag_2C_2O_4$ is 6×10^{-12}; K_2 for $H_2C_2O_4$ is 6.4×10^{-5}. K_1 is so large that the concentration of free oxalic acid is of no importance in this problem.

Ans.   (a) 5.4×10^{-4};   (b) 3.5×10^{-4}

18.48. How much solid $Na_2S_2O_3$ should be added to 1 L of water so that 0.00050 mol $Cd(OH)_2$ could just barely dissolve? K_{sp} for $Cd(OH)_2$ is 4.5×10^{-15}. K_1 and K_2 for $S_2O_3^{2-}$ complexation with Cd^{2+} are 8.3×10^3 and 2.5×10^2, respectively. [As part of the problem, determine whether CdS_2O_3 or $Cd(S_2O_3)_2^{2-}$ is the predominant species in solution.]

Ans.   0.23 mol

In the problems below, use the following physical constants for H_2S:

$$\text{Solubility} = 0.10 \text{ mol/L} \qquad K_1 = 1.0 \times 10^{-7} \qquad K_2 = 1.2 \times 10^{-13}$$

18.49. What is the maximum possible $[Ag^+]$ in a saturated H_2S solution from which precipitation has not occurred? Solubility product of Ag_2S is 6.7×10^{-50}.

Ans. 7.5×10^{-19}

18.50. Determine the $[S^{2-}]$ in a saturated H_2S solution to which enough HCl has been added to produce a $[H^+]$ of 2×10^{-4}.

Ans. 3×10^{-14}

18.51. Will FeS precipitate in a saturated H_2S solution if the solution contains 0.01 mol/L of Fe^{2+} and (a) 0.2 mol/L of H^+?   (b) 0.001 mol/L of H^+? K_{sp} of FeS is 8×10^{-19}.

Ans.   (a) no;   (b) yes

18.52. Given that 0.0010 mol each of Cd^{2+} and Fe^{2+} is contained in 1 L of 0.020 M HCl and that this solution is saturated with H_2S. K_{sp} of CdS is 1.4×10^{-29}, of FeS, 8×10^{-19}.   (a) Determine whether or not each of these ions will precipitate as the sulfide.   (b) How much Cd^{2+} remains in solution at equilibrium?

Ans.   (a) only CdS precipitates;   (b) 4.7×10^{-12} mol

18.53. In an attempted determination of the solubility product of Tl_2S, the solubility of this compound in pure CO_2-free water was determined as 3.6×10^{-6} mol/L. What is the computed K_{sp}? Assume that the dissolved sulfide hydrolyzes practically completely to HS^-, and that the further hydrolysis to H_2S can be neglected.

Ans. 8×10^{-21}

18.54. Calculate the solubility of FeS in pure water. $K_{sp} = 8 \times 10^{-19}$. (*Hint*: The second stage of hydrolysis, producing H_2S, cannot be neglected.)

Ans. 4×10^{-7} mol/L

Chapter 19

Electrochemistry

In this chapter we deal with two aspects of the connection between chemistry and electricity: *electrolysis*, or the decomposition of matter accompanying the passage of electricity through it, and *galvanic cell action*, or the role of a chemical reaction as an electric generator.

ELECTRICAL UNITS

The *coulomb* (C) is the SI unit of electrical charge. From the point of view of fundamental particles, the *elementary* unit is the charge of one proton, which is equal in magnitude to the charge of one electron. No chemical particle is known whose charge is not an integral multiple of this elementary charge, the value of which is 1.602×10^{-19} C.

Electric current is the rate of flow of charge. The SI unit, the *ampere* (A), is a flow rate of one coulomb per second (1 A = 1 C/s).

The *electrical potential difference* between two points in a circuit causes the transfer of charge from one point to the other. The *volt* (V) is the SI unit of electrical potential difference. When a charge of 1 C moves through a potential difference of 1 V it gains 1 J of energy:

$$\text{Energy in joules} = (\text{charge in coulombs}) \times (\text{potential difference in volts})$$
$$= (\text{current in amperes}) \times (\text{time in seconds}) \times (\text{potential difference in volts}) \quad (19\text{-}1)$$

The *watt* (W) is the SI unit of *power* (electrical and other); one watt is developed when 1 J of work is performed in a time of 1 s. From (*19-1*),

$$\text{Power in watts} = (\text{current in amperes}) \times (\text{potential difference in volts}) \quad (19\text{-}2)$$

In electrochemistry we are normally concerned with direct current (DC). However the above discussion of units applies equally well to alternating current (AC), the conventional type of power in the home and laboratory.

FARADAY'S LAWS OF ELECTROLYSIS

1. The mass of any substance liberated or deposited at an electrode is proportional to the electrical charge (i.e., the number of coulombs) that has passed through the electrolyte.
2. The masses of different substances liberated or deposited by the same quantity of electricity (i.e., by the same number of coulombs) are proportional to the equivalent weights (Chapter 12) of the various substances.

These laws, found empirically by Faraday over half a century prior to the discovery of the electron, can now be shown to be simple consequences of the electrical nature of matter. In any electrolysis, a reduction must occur at the cathode to remove electrons flowing into the electrode and an oxidation must occur at the anode to supply the electrons that leave the electrolytic cell at this electrode. By the principle of continuity of current, electrons must be discharged at the cathode at exactly the same rate at which they are supplied to the anode. By definition of the equivalent weight for oxidation-reduction reactions (that fraction of the formula weight associated with the transfer of one electron), the number of *gram-equivalents* of electrode reaction must be proportional to the amount of charge transported into or out of the electrolytic cell and must, indeed, be equal to the number of *moles of electrons* transported in the circuit. The *Faraday constant* ($\mathscr{F}$) is equal to the charge of one mole of electrons:

$$\mathscr{F} = (1.602 \times 10^{-19} \text{ C/electron})(6.022 \times 10^{23} \text{ electrons/mol}) = 9.65 \times 10^4 \text{ C/mol}$$

The usual symbol, $n(e^-)$, may be used to refer to the number of moles of electronic charge, i.e., the number of gram-equivalents.

The equivalent weight needed for electrolytic calculations can be found by writing the balanced half-reactions for the electrode processes. Thus in the electrolytic reduction of Cu^{2+}, the cathode reaction is

$$Cu^{2+} + 2e^- \rightarrow Cu$$

The equivalent weight of copper is $\frac{1}{2}$ the atomic weight. If a solution of Cu^+ were electrolyzed, the equivalent weight would be the atomic weight of copper, because only one electron is captured per copper atom formed:

$$Cu^+ + e^- \rightarrow Cu$$

Special information about the electrode reactions is often needed in order to calculate the equivalent weight for electrolysis, just as in ordinary oxidation-reduction reactions. If a solution containing Fe^{3+} is electrolyzed at low voltages, the electrode reaction for the iron might be

$$Fe^{3+} + e^- \rightarrow Fe^{2+}$$

and the equivalent weight of iron would equal the atomic weight. At higher voltages, the reaction might be

$$Fe^{3+} + 3e^- \rightarrow Fe$$

and the equivalent weight of iron would be one-third of the atomic weight.

GALVANIC CELLS

Many oxidation-reduction reactions may be carried out in such a way as to generate electricity. In principle, this may always be done for spontaneous, aqueous, oxidation-reduction reactions. Such an arrangement for the production of an electric current is called a *galvanic* or *electrochemical cell*. The experimental requirements are

1. The oxidizing and reducing agents are not in physical contact with each other but are contained in separate compartments, called *half-cells*. Each half-cell contains a solution and a metallic conductor (electrode).

2. The reducing or oxidizing agent in a half-cell may be either the electrode itself, a solid substance deposited on the electrode, a gas which bubbles around the electrode, or a solute in the solution which bathes the electrode. Just as in electrolysis, the electrode at which reduction occurs is termed the *cathode* and that at which oxidation occurs is termed the *anode*.

3. The solutions of the two half-cells are connected in some way that allows ions to move between them. Among the possible arrangements to accomplish this are   (*a*) careful layering of the less dense solution over the denser one;   (*b*) separation of the two solutions by a porous substance, such as fritted glass, unglazed porcelain, or a fiber permeated with some electrolyte solution;   (*c*) insertion of a connecting electrolyte solution (a *salt bridge*) between the two solutions.

The potential developed across the two electrodes causes an electric current to flow and the half-cell reactions to proceed if the electrodes are connected to each other by an outside conducting circuit.

STANDARD HALF-CELL POTENTIALS

The reaction occurring in each half-cell may be represented by an ion-electron partial equation of the type described in Chapter 11. The whole-cell operation involves a flow of electrons in the external circuit. The electrons generated in the oxidation half-reaction enter the anode, travel through the external circuit to the cathode, and are consumed at the cathode by the reduction half-reaction. From the electrical principle of equal current at all points in a nonbranching circuit, the number of electrons generated in the oxidation must exactly balance the number of electrons consumed in the reduction. This demands the same rule for combining two half-reactions into a balanced whole equation that was used in Chapter 11.

In the half-cell of the reducing agent, the oxidation product accumulates during operation of the cell. The reducing agent together with its oxidation product, known as a *couple*, are thus found in the same compartment during cell operation. Similarly, the other half-cell contains a couple consisting of the oxidizing agent and its reduction product. An arbitrary couple, consisting of both product and reactant of an oxidation-reduction half-reaction, may sometimes be the reducing part of a galvanic cell and

Table 19-1.   Standard Electrode Potentials at 25 °C

Reaction	$E°/V$
$F_2 + 2e^- \rightarrow 2F^-$	2.87
$S_2O_8^{2-} + 2e^- \rightarrow 2SO_4^{2-}$	1.96
$Co^{3+} + e^- \rightarrow Co^{2+}$	1.92
$H_2O_2 + 2H^+ + 2e^- \rightarrow 2H_2O$	1.763
$Ce^{4+} + e^- \rightarrow Ce^{3+}$ (in 1 M HClO$_4$)	1.70
$MnO_4^- + 8H^+ + 5e^- \rightarrow Mn^{2+} + 4H_2O$	1.51
$Cl_2 + 2e^- \rightarrow 2Cl^-$	1.358
$Tl^{3+} + 2e^- \rightarrow Tl^+$	1.25
$MnO_2 + 4H^+ + 2e^- \rightarrow Mn^{2+} + 2H_2O$	1.23
$O_2 + 4H^+ + 4e^- \rightarrow 2H_2O$	1.229
$Br_2 + 2e^- \rightarrow 2Br^-$	1.065
$AuCl_4^- + 3e^- \rightarrow Au + 4Cl^-$	1.002
$Pd^{2+} + 2e^- \rightarrow Pd$	0.915
$Ag^+ + e^- \rightarrow Ag$	0.799 1
$Fe^{3+} + e^- \rightarrow Fe^{2+}$	0.771
$O_2 + 2H^+ + 2e^- \rightarrow H_2O_2$	0.695
$I_2(s) + 2e^- \rightarrow 2I^-$	0.535
$Cu^+ + e^- \rightarrow Cu$	0.520
$Fe(CN)_6^{3-} + e^- \rightarrow Fe(CN)_6^{4-}$	0.361
$Co(dip)_3^{3+} + e^- \rightarrow Co(dip)_3^{2+}$	0.34
$Cu^{2+} + 2e^- \rightarrow Cu$	0.34
$Ge^{2+} + 2e^- \rightarrow Ge$	0.247
$PdI_4^{2-} + 2e^- \rightarrow Pd + 4I^-$	0.18
$Sn^{4+} + 2e^- \rightarrow Sn^{2+}$	0.15
$Ag(S_2O_3)_2^{3-} + e^- \rightarrow Ag + 2S_2O_3^{2-}$	0.017
$2H^+ + 2e^- \rightarrow H_2$	0.000 0
$Ge^{4+} + 2e^- \rightarrow Ge^{2+}$	0.00
$Pb^{2+} + 2e^- \rightarrow Pb$	−0.126
$Sn^{2+} + 2e^- \rightarrow Sn$	−0.14
$Ni^{2+} + 2e^- \rightarrow Ni$	−0.257
$Tl^+ + e^- \rightarrow Tl$	−0.336
$Cd^{2+} + 2e^- \rightarrow Cd$	−0.403
$Fe^{2+} + 2e^- \rightarrow Fe$	−0.44
$Zn^{2+} + 2e^- \rightarrow Zn$	−0.762 6
$Na^+ + e^- \rightarrow Na$	−2.713
$Li^+ + e^- \rightarrow Li$	−3.040

sometimes the oxidizing part, depending on what the other couple is. The (Fe^{3+}/Fe^{2+}) couple, for example, takes an oxidizing role when paired against the strongly reducing (Zn^{2+}/Zn) couple, and a reducing role when paired against the strongly oxidizing (Ce^{4+}/Ce^{3+}) couple. (The oxidized form is listed first.)

Each couple has an intrinsic ability to consume electrons. This ability can be assigned a numerical value, called the *electrode potential*. Electrode potentials may be represented by symbols such as $E(Fe^{3+}/Fe^{2+})$ or $E(Zn^{2+}/Zn)$. When two couples are combined into a whole cell, the couple with the higher electrode potential provides the oxidizing agent, absorbs electrons at its electrode, and is positive. The other couple provides the reducing agent, generates electrons at its electrode, and is negative. The driving force for the flow of current is the *algebraic difference* between the two electrode potentials and is equal numerically to the voltage output of the cell under the limiting operating condition of zero current.

The numerical value of an electrode potential depends on the nature of the particular chemicals, on the temperature, and on the concentrations of the various members of the couple. For purposes of reference, half-cell potentials are tabulated for standard states of all the chemicals, defined as 1 atm pressure for each gas,* the pure substance for each liquid or solid, and 1 mol/L for every nongaseous solute appearing in the balanced half-cell reaction. Such reference potentials are called *standard electrode potentials* and are designated by the symbol $E°$. The same symbol is used for the standard potential of the whole cell, the value that could be realized in measurement at the standard states of all reactants and products. A partial listing of standard electrode potentials at 25 °C is given in Table 19-1. Since only *differences* between two electrode potentials can be measured by the cell voltage, there is an arbitrary zero point on this scale, assigned to the potential of the standard hydrogen electrode, (H^+/H_2).

A couple with a large positive $E°$, like (F_2/F^-), is strongly oxidizing, captures electrons from the electrode, and is experimentally *positive* with respect to a standard hydrogen electrode. Conversely, a couple with a large negative $E°$, like (Li^+/Li), is strongly reducing; that is, it undergoes oxidation and thereby transfers electrons to the electrode. Such a standard half-cell is experimentally *negative* with respect to a standard hydrogen electrode, since the external circuit receives electrons from the half-cell.

COMBINATIONS OF COUPLES

There is another way of combining two half-cell reactions, in which the electrons do not cancel. This cannot correspond to a whole cell, where the electrons always cancel. The situation is a hypothetical one often used to calculate an unknown half-cell potential on the basis of two known half-cell potentials. In such a case the electron number cannot be left out of consideration. As discussed in the next section, if n is the number of electrons in the half-reaction, $nE°$ is proportional to the free energy associated with it. As in Chapter 16, when one adds two reactions, one may also add their free energies to get the free enegy of the overall reaction. The rule for this case is as follows:

If two reduction half-reactions are added or subtracted to give a third reduction half-reaction, the two $nE°$ products are added or subtracted correspondingly to give the $nE°$-value for the resulting half-reaction.

EXAMPLE 1

	$E°$	n	$nE°$
$Fe^{2+} + 2e^- \rightarrow Fe$	-0.44 V	2	-0.88 V
$Fe^{3+} + e^- \rightarrow Fe^{2+}$	0.77	1	0.77
SUM: $Fe^{3+} + 3e^- \rightarrow Fe$		3	-0.11 V

Since $nE°$ for the resulting half-reaction is -0.11 V, and n is 3, $E°$ must be $-0.11/3 = -0.04$ V.

---

* The necessity for converting to 1 bar is recognized but had not yet been dealt with in international listings by the end of 1988. Changes will be extremely small.

This rule makes it possible to reduce the length of compiled tables, since many half-reactions can be computed even if they are not tabulated.

FREE ENERGY, NONSTANDARD POTENTIALS, AND THE DIRECTION OF OXIDATION-REDUCTION REACTIONS

The discussion following (16-4) indicated that the *decrease* in free energy of a system can be equated to the maximum amount of work the system can perform at constant temperature and pressure in forms other than expansive or contractive work. At this point, we can make use of that principle by noting from (19-1) that the electrical work performed by a galvanic cell is equal to the voltage times the electrical charge transferred at either electrode. For the passage of n moles of electrons, the charge transferred is $n\mathscr{F}$. Thus, the electrical work in joules equals $n\mathscr{F}E$. The *maximum* amount of work performed by a galvanic cell is the value of $n\mathscr{F}E$ under conditions where the electrode processes are reversible. Reversibility may be approximated for many electrodes if the cell potential is measured at very low current; such measurements indeed are the basis of the values used in the construction of tables like Table 19-1. Since the electrical work is the only form of work other than expansive or contractive work performed in the typical galvanic cell, the free-energy principle may be stated as

$$\Delta G = -n\mathscr{F}E \qquad (19\text{-}3)$$

The dependence of a cell potential on the concentrations of the reactants and products may be derived from the known dependence of ΔG upon concentration, (16-6).

$$E = -\frac{\Delta G}{n\mathscr{F}} = -\frac{\Delta G^\circ + RT \ln Q}{n\mathscr{F}} \qquad (19\text{-}4)$$

where, as in Chapter 16, $R = 8.314$ J/K when ΔG is expressed in joules. When all reactants and products are in their standard states, $Q = 1$ (see Chapter 16) and $E = E^\circ$; thus, $\Delta G^\circ = -n\mathscr{F}E^\circ$. Substituting back into (19-4), we obtain the *Nernst equation*:

$$E = E^\circ - \frac{RT}{n\mathscr{F}} \ln Q \qquad (19\text{-}5)$$

When the constants are combined and the natural logarithm converted to a common logarithm, the expression for the potential (in volts) at 25 °C becomes

$$E = E^\circ - \frac{0.059\,2}{n} \log Q \qquad (19\text{-}6)$$

In (19-6), n has been made dimensionless. For a half-reaction, n is the number of electrons in the half-equation; for a whole-cell reaction, n is the number of electrons in *one* of the multiplied half-equations before canceling the electrons. The Nernst equation is closely related to the laws of chemical equilibrium. Le Chatelier's principle applies to the potential of a cell in the same sense as it applies to the yield of an equilibrium process. Since Q has product concentrations in the numerator and reactant concentrations in the denominator, an increased concentration of product reduces the potential and an increased concentration of reactant raises the potential.

The same type of equation may be used to describe the concentration dependence of the potential of a single half-cell, i.e., of an electrode potential. In this case Q, the reaction quotient, contains terms in the numerator corresponding to the products of the balanced half-cell reaction written as a reduction, terms in the denominator corresponding to the reactants, but no terms for the electron. For example:

$$E(\text{Fe}^{3+}/\text{Fe}^{2+}) = E^\circ(\text{Fe}^{3+}/\text{Fe}^{2+}) - 0.059\,2 \log \frac{[\text{Fe}^{2+}]}{[\text{Fe}^{3+}]}$$

$$E(\text{MnO}_4^-/\text{Mn}^{2+}) = E^\circ(\text{MnO}_4^-/\text{Mn}^{2+}) - \frac{0.059\,2}{5} \log \frac{[\text{Mn}^{2+}]}{[\text{MnO}_4^-][\text{H}^+]^8}$$

Spontaneous oxidation-reduction reactions

If the potential of a whole cell is positive, the free-energy change is negative, by (*19-3*). Then the corresponding oxidation-reduction reaction is spontaneous as written, and a galvanic cell would operate spontaneously, with electrons provided to the external circuit at the half-cell where oxidation takes place. If the potential is negative, the free-energy change is positive, and the corresponding reaction does not take place spontaneously. These statements about the direction of spontaneous reaction are valid whether the reaction is set up as a galvanic cell or as an ordinary process where the reactants and products are all mixed in the same vessel, since the free-energy change for a reaction depends on the concentrations but not on the way in which the reaction is carried out. In particular, if both reduced and oxidized members of the two couples are mixed, all substances being at standard states (unit relative concentrations), any reducing agent can reduce an oxidizing agent occurring higher in the table of standard electrode potentials. The same rule of relative position may be applied for general values of the concentrations, with E for each half-cell, as computed from the Nernst equation, replacing $E°$. Usually the qualitative prediction based on $E°$-values is not changed even for moderate deviations from standard states if the two $E°$-values are separated by at least several tenths of a volt.

It should be noted that predictions under this rule indicate what reactions *might* occur but say nothing about the rate at which they *do* occur.

Electrode reactions in electrolysis

A nonspontaneous oxidation-reduction reaction, for which the calculated cell potential is negative, may be induced by electrolysis, i.e., by using an external electrical potential to force electrons into the couple undergoing reduction and to extract electrons from the couple undergoing oxidation. The *minimum* external potential required for electrolysis has the magnitude of the computed cell potential of the reaction. (The actual electrolyzing potential exceeds this minimum because of the irreversibility of electrode processes occurring at nonzero rates.)

Under nearly reversible conditions, we have the following rule:

Of all possible reductions that might occur electrolytically at a cathode, the one for which the corresponding electrode potential is algebraically greatest is the most likely. Conversely, the most probable anodic oxidation is that for which the corresponding electrode (reduction) potential is algebraically the least.

In applying the rule, the student should keep in mind that: (*a*) a solute molecule or ion may undergo oxidation or reduction; (*b*) the anode electrode may itself undergo oxidation; (*c*) the solvent may undergo oxidation or reduction.

EXAMPLE 2   Let us illustrate possibility (*c*) for water at 25 °C.
Reduction to molecular hydrogen is found in Table 19-1:

$$2H^+ + 2e^- \rightarrow H_2 \qquad E° = 0.0000 \text{ V}$$

Assume that H_2 gas is allowed to accumulate to a partial pressure of 1 atm. In neutral solutions, where $[H^+] = 10^{-7}$ and the H_2 pressure retains its unit value,

$$E = E° - \frac{0.0592}{2} \log \frac{P(H_2)}{[H^+]^2} = 0.0000 - 0.0296 \log \frac{1}{(10^{-7})^2}$$

$$= 0.0000 - 0.0296 \times 14 = -0.414 \text{ V}$$

Water is thus more difficult to reduce, but hydrogen easier to oxidize, in neutral solutions than in acids.
For the *oxidation* of water to molecular oxygen, the most appropriate entry in Table 19-1 is

$$O_2 + 4H^+ + 4e^- \rightarrow 2H_2O \qquad E° = 1.229 \text{ V}$$

The standard states to which this $E°$-value refers are 1 atm for oxygen gas and 1 mol/L for H^+. We can calculate E for the above half-cell for neutral solutions, in which $[H^+] = 10^{-7}$, by using the Nernst equation. Assuming that the oxygen remains at its standard state, $P(O_2) = 1$ atm,

$$E = E° - \frac{0.059\,2}{4} \log \frac{1}{[H^+]^4 P(O_2)} = 1.229 - 0.014\,8 \log \frac{1}{(10^{-7})^4}$$

$$= 1.229 - 0.014\,8 \times 28 = 0.815 \text{ V}$$

It is thus harder for oxygen to be reduced, but easier for water to be oxidized, in neutral than in acid solutions.

USE OF $E°'$-VALUES

The above examples show that practical values for electrode potentials may be very different from $E°$-values for reactions in almost-neutral solutions involving H^+ or OH^-. Since most reactions in cellular biology occur in solutions near pH 7, it has become customary for biochemists to define $E°'$ as an electrode potential at which all reactants but H^+ and OH^- are at unit concentration and $[H^+]$ is 10^{-7}. The values calculated in the previous section, -0.414 V and 0.815 V, may be considered the $E°'$ values for the H^+/H_2 and the O_2/H_2O couples, respectively. Then, for any reaction carried out in a buffer solution at pH 7, a modified form of the Nernst equation may be given for the electrode potential at 25 °C:

$$\text{At pH 7:} \qquad E = E°' - \frac{0.059\,2}{n} \log \frac{[\text{reduced}]}{[\text{oxidized}]} \qquad (19\text{-}7)$$

In (19-7), [reduced]/[oxidized] is similar to Q but does not include $[H^+]$ or $[OH^-]$ terms.

Since in practice the concentrations of the principal oxidized and reduced species are often of the same order of magnitude, E-values do not differ much from $E°'$-values. Another advantage of using an $E°'$-value that incorporates as a standard state the $[H^+]$ at a pH of 7 is that we need not worry about the various states of ionization of the oxidized or reduced species. The relative distribution of ionization states is fixed at constant pH, and it has become customary in biochemical applications to define [reduced] or [oxidized] as the relative stoichiometric concentration, i.e., the numerical value of the sum of the molar concentrations of all the ionization states of the reductant or oxidant.

Solved Problems

ELECTRICAL UNITS

19.1. A lamp draws a current of 2.0 A. Find the charge in coulombs used by the lamp in 30 s.

Charge in coulombs = (current in amperes) × (time in seconds) = (2.0 A) (30 s) = 60 C

19.2. Compute the time required to pass 36 000 C through an electroplating bath using a current of 5 A.

$$\text{Time in seconds} = \frac{\text{charge in coulombs}}{\text{current in amperes}} = \frac{36\,000 \text{ C}}{5 \text{ A}} = 7\,200 \text{ s} = 2 \text{ h}$$

19.3. A dynamo delivers 15 A at 120 V. (a) Compute the power in kilowatts supplied by the dynamo. (b) How much electric energy, in kilowatt-hours, is supplied by the dynamo in 2 h? (c) What is the cost of this energy at 6¢ per kilowatt-hour?

(a)   Power = (15 A)(120 V) = 1 800 W = 1.8 kW

(b)   Energy = (1.8 kW)(2 h) = 3.6 kW·h

(c)   Cost = (3.6 kW·h)(6¢/kW·h) = 21.6¢

19.4. A resistance heater was wound around a 50-g metallic cylinder. A current of 0.65 A was passed through the heater for 24 s while the measured voltage drop across the heater was 5.4 V. The temperature of the cylinder was 22.5 °C before the heating period and 29.8 °C at the end. If heat losses to the environment can be neglected, what is the specific heat of the cylinder metal in J/g · K?

By (19-1),

$$\text{Energy input} = (0.65 \text{ A}) (5.4 \text{ V}) (24 \text{ s}) = 84 \text{ J}$$

But, also,

$$\text{Energy input} = (\text{mass}) \times (\text{specific heat}) \times (\text{temperature rise})$$

Therefore,

$$84 \text{ J} = (50 \text{ g})(\text{specific heat})[(29.8 - 22.5) \text{ K}]$$

from which specific heat = 0.23 J/g · K.

19.5. How many electrons per second pass through a cross section of a copper wire carrying 10^{-16} A?

Since 1 A = 1 C/s,

$$\text{Rate} = \frac{10^{-16} \text{ C/s}}{1.6 \times 10^{-19} \text{ C/electron}} = 600 \text{ electrons/s}$$

FARADAY'S LAWS OF ELECTROLYSIS

19.6. Exactly 0.2 mol of electrons is passed through three electrolytic cells in series. One contains silver ion, one zinc ion, and one ferric ion. Assume that the only cathode reaction in each cell is the reduction of the ion to the metal. How many grams of each metal will be deposited?

One mol of electrons liberates 1 g-eq of an element. Equivalent weights of Ag^+, Zn^{2+}, and Fe^{3+} are

$$\frac{107.9}{1} = 107.9 \qquad \frac{65.39}{2} = 32.69 \qquad \frac{55.85}{3} = 18.62$$

respectively.

$$\text{Ag deposited} = (0.2 \text{ mol } e^-)(107.9 \text{ g/mol } e^-) = 21.58 \text{ g}$$

$$\text{Zn deposited} = (0.2 \text{ mol } e^-)(32.69 \text{ g/mol } e^-) = 6.54 \text{ g}$$

$$\text{Fe deposited} = (0.2 \text{ mol } e^-)(18.62 \text{ g/mol } e^-) = 3.72 \text{ g}$$

19.7. A current of 5.00 A flowing for exactly 30 min deposits 3.048 g of zinc at the cathode. Calculate the equivalent weight of zinc from this information.

$$\text{Number of coulombs used} = (5.00 \text{ A})[(30 \times 60)\text{s}] = 9.00 \times 10^3 \text{ C}$$

$$n(e^-) \text{ used} = \frac{9.00 \times 10^3 \text{ C}}{9.65 \times 10^4 \text{ C/mol } e^-} = 0.093\,3 \text{ mol } e^-$$

$$\text{Equivalent weight} = \text{mass deposited by 1 mol } e^- = \frac{3.048 \text{ g}}{0.093\,3 \text{ mol } e^-} = 32.7 \text{ g/g-eq}$$

19.8. A certain current liberates 0.504 g of hydrogen in 2 h. How many grams of oxygen and of copper (from Cu^{2+} solution) can be liberated by the same current flowing for the same time?

Masses of different substances liberated by the same number of coulombs are proportional to their equivalent weights. Equivalent weight of hydrogen is 1.008; of oxygen, 8.00; of copper, 31.8.

$$\text{Number of g-eq of hydrogen in } 0.504 \text{ g} = \frac{0.504 \text{ g}}{1.008 \text{ g/g-eq}} = \tfrac{1}{2} \text{ g-eq}$$

Then $\tfrac{1}{2}$ g-eq each of oxygen and copper can be liberated.

$$\text{Mass of oxygen liberated} = (\tfrac{1}{2} \text{ g-eq})(8.00 \text{ g/g-eq}) = 4.00 \text{ g} \qquad \text{(see below)}$$

$$\text{Mass of copper liberated} = (\tfrac{1}{2} \text{ g-eq})(31.8 \text{ g/g-eq}) = 15.9 \text{ g}$$

The equivalent weight of any substance in an electrolysis is determined from the balanced electrode reaction. For the liberation of oxygen from water the anode reaction is

$$2H_2O \rightarrow O_2 + 4H^+ + 4e^-$$

The molecular weight of O_2 is 32.00. The equivalent weight is the molecular weight divided by the number of electrons which must flow to produce one molecule, or $32.00/4 = 8.00$.

19.9.   The same quantity of electricity that liberated 2.158 g of silver was passed through a solution of a gold salt and 1.314 g of gold was deposited. The equivalent weight of silver is 107.9. Calculate the equivalent weight of gold. What is the oxidation state of gold in this gold salt?

$$\text{Number of g-eq of Ag in } 2.158 \text{ g} = \frac{2.158 \text{ g}}{107.9 \text{ g/g-eq}} = 0.020\,00 \text{ g-eq Ag}$$

Then 1.314 g Au must represent 0.020 00 g-eq, and so

$$\text{Equivalent weight of Au} = \frac{1.314 \text{ g}}{0.020\,00 \text{ g-eq}} = 65.70 \text{ g/g-eq}$$

$$\text{Oxidation State} = \text{number of electrons needed to form one gold atom by reduction}$$

$$= \frac{\text{atomic weight of Au}}{\text{equivalent weight of Au}} = \frac{197.0}{65.7} = 3 \text{ or III}$$

19.10.   How long would it take to deposit 100 g Al from an electrolytic cell containing Al_2O_3 at a current of 125 A? Assume that Al formation is the only cathode reaction.

$$\text{Equivalent weight of Al} = \tfrac{1}{3}(\text{atomic weight}) = \tfrac{1}{3}(27.0) = 9.0 \text{ g Al/mol } e^-$$

$$n(e^-) = \frac{100 \text{ g Al}}{9.0 \text{ g Al/mol } e^-} = 11.1 \text{ mol } e^-$$

$$\text{Time} = \frac{\text{charge}}{\text{current}} = \frac{(11.1 \text{ mol } e^-)(9.65 \times 10^4 \text{ C/mol } e^-)}{125 \text{ A}} = 8.6 \times 10^3 \text{ s} = 2.4 \text{ h}$$

19.11.   A current of 15.0 A is employed to plate nickel in a $NiSO_4$ bath. Both Ni and H_2 are formed at the cathode. The current efficiency with respect to formation of Ni is 60 percent.   (a) How many grams of nickel are plated on the cathode per hour?   (b) What is the thickness of the plating if the cathode consists of a sheet of metal 4.0 cm square which is coated on both faces? The density of nickel is 8.9 g/cm^3.   (c) What volume of H_2 (S.T.P.) is formed per hour?

(a) $\text{Total number of coulombs used} = (15.0 \text{ A})(3\,600 \text{ s}) = 5.40 \times 10^4 \text{ C}$

$$n(e^-) \text{ used} = \frac{5.40 \times 10^4 \text{ C}}{9.65 \times 10^4 \text{ C/mol } e^-} = 0.560 \text{ mol } e^-$$

$$\text{Number of g-eq Ni deposited} = (0.60)(0.560 \text{ mol } e^-)(1 \text{ g-eq Ni/mol } e^-) = 0.336 \text{ g-eq Ni}$$

Equivalent weight of Ni $= \frac{1}{2}$(atomic weight) $= \frac{1}{2}(58.70) = 29.4$ g/g-eq

Mass of Ni deposited $= (0.336$ g-eq$)(29.4$ g/g-eq$) = 9.9$ g

(b)   Area of two faces $= 2(4.0$ cm$)(4.0$ cm$) = 32$ cm^2

Volume of 9.9 g Ni $= \frac{\text{mass}}{\text{density}} = \frac{9.9 \text{ g}}{8.9 \text{ g/cm}^3} = 1.11$ cm^3

Thickness of plating $= \frac{\text{volume}}{\text{area}} = \frac{1.11 \text{ cm}^3}{32 \text{ cm}^2} = 0.035$ cm

(c)   Number of g-eq H$_2$ liberated $= (0.40)(0.560 \, \mathscr{F})(1$ g-eq H$_2/\mathscr{F}) = 0.224$ g-eq H$_2$

Volume of 1 g-eq ($\frac{1}{2}$ mol) H$_2 = \frac{1}{2}(22.4$ L$) = 11.2$ L H$_2$

Volume of H$_2$ liberated $= (0.224$ g-eq$)(11.2$ L/g-eq$) = 2.51$ L H$_2$

Note the use of the symbol $\mathscr{F}$, the Faraday unit, which is synonymous with the unit, mol e^-

19.12. How many coulombs must be supplied to a cell for the electrolytic production of 245 g of NaClO$_4$ from NaClO$_3$? Because of side reactions, the anode efficiency for the desired reaction is 60 percent.

First it is necessary to know the equivalent weight of NaClO$_4$ for this reaction. The balanced anode reaction equation is

$$\text{ClO}_3^- + \text{H}_2\text{O} \rightarrow \text{ClO}_4^- + 2\text{H}^+ + 2e^-$$

Equivalent weight of NaClO$_4 = \frac{\text{formula weight}}{\text{number of electrons transferred}} = \frac{122.4}{2} = 61.2$

Number of g-eq NaClO$_4 = \frac{245 \text{ g}}{61.2 \text{ g/g-eq}} = 4.00$ g-eq NaClO$_4$

$n(e^-)$ required $= \frac{4.00 \text{ g-eq}}{0.60 \text{ g-eq anode product/mol } e^-} = 6.7$ mol e^-

Number of coulombs required $= (6.7$ mol $e^-)(9.6 \times 10^4$ C/mol $e^-) = 6.4 \times 10^5$ C

GALVANIC CELLS AND ELECTRODE PROCESSES

19.13. What is the standard potential of a cell that uses the (Zn^{2+}/Zn) and (Ag$^+$/Ag) couples? Which couple is negative? Write the equation for the cell reaction occurring at unit relative concentrations (the standard states).

The standard electrode potentials for (Zn^{2+}/Zn) and (Ag$^+$/Ag) are, from Table 19-1, -0.763 V and $+0.799$ V, respectively. The standard potential of the cell is the difference between these two numbers, $0.799 - (-0.763) = 1.562$ V. The silver electrode potential is higher, and thus silver ion is the oxidizing agent. The zinc couple provides the reducing agent and is the negative electrode. The equation is

$$\text{Zn} + 2\text{Ag}^+ \rightarrow \text{Zn}^{2+} + 2\text{Ag}$$

19.14. Can Fe^{3+} oxidize Br$^-$ to Br$_2$ at unit relative concentrations?

From Table 19-1, the (Fe^{3+}/Fe^{2+}) couple has a lower standard electrode potential, 0.771 V, than the (Br$_2$/Br$^-$) couple, 1.065 V. Therefore Fe^{2+} can reduce Br$_2$, but Br$^-$ cannot reduce Fe^{3+}; i.e., Fe^{3+} cannot oxidize Br$^-$. On the other hand, I$^-$, occurring at a much lower standard electrode potential, 0.535 V, is easily oxidized by Fe^{3+} to I$_2$.

19.15. What is the standard electrode potential for (MnO_4^-/MnO_2) in acid solution?

The reduction half-reaction for this couple is

$$MnO_4^- + 4H^+ + 3e^- \rightarrow MnO_2 + 2H_2O$$

which can be written as the difference of two half-reactions whose electrode potentials are listed in Table 19-1. $nE°$-values may be correspondingly subtracted.

		n	$E°$	$nE°$
$MnO_4^- + 8H^+$	$+ 5e^- \rightarrow Mn^{2+} + 4H_2O$	5	1.51 V	7.55 V
$MnO_2 + 4H^+$	$+ 2e^- \rightarrow Mn^{2+} + 2H_2O$	2	1.23	2.46
Difference: $MnO_4^- - MnO_2 + 4H^+ + 3e^- \rightarrow 2H_2O$		3		5.09 V

Rearranging, $MnO_4^- + 4H^+ + 3e^- \rightarrow MnO_2 + 2H_2O$, the desired reaction, in which $n = 3$.

$$E° \text{ for the desired reaction} = \frac{5.09}{3} = 1.70 \text{ V}$$

19.16. Predict the stabilities at 25 °C of aqueous solutions of the uncomplexed intermediate oxidation states of (a) thallium and (b) copper.

(a) The question is whether the intermediate state, Tl^+, spontaneously decomposes into the lower and higher states, Tl and Tl^3. The supposed disproportionation reaction,

$$3Tl^+ \rightarrow 2Tl + Tl^{3+}$$

could be written in the ion-electron method as

	$E°$ (red) for the couple (Table 19-1)	
$2 \times (Tl^+ + e^- \rightarrow Tl)$	-0.336 V	(1)
$Tl^+ \rightarrow Tl^{3+} + 2e^-$	1.25 V	(2)

In (1), the (Tl^+/Tl) couple functions as oxidizing agent; in (2), the (Tl^{3+}/Tl^+) couple functions as reducing agent. The reaction would occur at unit concentrations if $E°$ for the reducing couple were less than $E°$ for the oxidizing couple. Since 1.25 V is greater than -0.336 V, the reaction cannot occur as written. We conclude that Tl^+ does not spontaneously decompose into Tl and Tl^{3+}. On the contrary, the reverse reaction is a spontaneous one:

$$2Tl + Tl^{3+} \rightarrow 3Tl^+$$

This means that Tl(III) salts are unstable in solution in the presence of metallic Tl.

(b) The supposed disproportionation of Cu^+ would take the following form:

$$2Cu^+ \rightarrow Cu + Cu^{2+}$$

The ion-electron partial equations are

	$E°$ (red) for the couple (Table 19-1)
$Cu^+ + e^- \rightarrow Cu$	0.520 V
$Cu^+ \rightarrow Cu^{2+} + e^-$	0.160 V (computed)

This process could occur if $E°$ for the supposed reducing couple, (Cu^{2+}/Cu^+), were less than $E°$ for the oxidizing couple, (Cu^+/Cu). Indeed, $+0.16$ V (computed by the method of Problem 19.15) is less than 0.52 V. Therefore Cu^+ is unstable to disproportionation in solution. Compounds of Cu(I) can exist only as extremely insoluble substances or as such stable complexes that only a very small concentration of free Cu^+ can exist in solution.

19.17. (a) What is the potential of the cell containing the (Zn^{2+}/Zn) and (Cu^{2+}/Cu) couples if the Zn^{2+} and Cu^{2+} concentrations are 0.1 M and 10^{-9} M, respectively, at 25 °C?   (b) What is ΔG for the reduction of 1 mol of Cu^{2+} by Zn at the indicated concentrations of the ions, and what is $\Delta G°$ for the reaction, both at 25 °C?

(a)   The cell reaction is $Zn + Cu^{2+} \rightarrow Zn^{2+} + Cu$, with an n-value of 2.

$$E = E° - \frac{0.059\,2 \log Q}{n}$$

$E°$, the standard cell potential, is equal to the difference between the standard electrode potentials,

$$0.34 - (-0.76) = 1.10 \text{ V}$$

Q, the concentration function, does not include terms for the solid metals, because the metals are in their standard states.

$$E = 1.10 - \frac{0.059\,2}{2} \log \frac{[Zn^{2+}]}{[Cu^{2+}]} = 1.10 - 0.029\,6 \log \frac{10^{-1}}{10^{-9}}$$

$$= 1.10 - (0.029\,6)(8) = 0.86 \text{ V}$$

(b)   From (19-3),

$$\Delta G = -n\mathscr{F}E = -(2 \text{ mol } e^-)(9.65 \times 10^4 \text{ C/mol } e^-)(0.86 \text{ V}) = -166 \times 10^3 \text{ C} \cdot \text{V} = -166 \text{ kJ}$$

$$\Delta G° = -n\mathscr{F}E° = -(2 \times 9.65 \times 10^4 \times 1.10) \text{ J} = -212 \text{ kJ}$$

19.18. By how much is the oxidizing power of the (MnO_4^-/Mn^{2+}) couple decreased if the H^+ concentration is decreased from 1 M to 10^{-4} M at 25 °C?

The half-cell reaction for the reduction is

$$MnO_4^- + 8H^+ + 5e^- \rightarrow Mn^{2+} + 4H_2O$$

with an n-value of 5. Assume that only the H^+ concentration deviates from 1 mol/L.

$$E - E° = -\frac{0.059\,2}{5} \log \frac{[Mn^{2+}]}{[MnO_4^-][H^+]^8} = -0.011\,8 \log \frac{1}{(1)(10^{-4})^8}$$

$$= -(0.011\,8)(32) = -0.38 \text{ V}$$

The couple has moved down the table 0.38 V (to a position of less oxidizing power) from its standard value.

19.19. In the continued electrolysis of each of the following solutions at pH 7.0 and 25 °C, predict the main product at each electrode if there are no (irreversible) electrode polarization effects: (a) 1 M $NiSO_4$ with palladium electrodes;   (b) 1 M $NiBr_2$ with inert electrodes;   (c) 1 M Na_2SO_4 with Cu electrodes.

(a)   *Cathode Reaction.*

The following two possible reduction processes may be considered:

		$E°$
(1)	$Ni^{2+} + 2e^- \rightarrow Ni$	-0.25 V
(2)	$2H^+ + 2e^- \rightarrow H_2$	0.00

By the rule that the most probable cathode process is that for which the corresponding electrode potential is algebraically greatest, the hydrogen couple is favored at unit concentrations. Allowing for the effect of the pH 7.0 buffer, however, E for (2) is lowered to -0.41 V, as calculated in Example 2. The reduction of nickel then becomes the favored process.

Anode Reaction.

Three possible oxidation processes may be considered, the reverses of the following reduction half-reactions:

$$E^\circ$$

$$(3) \quad O_2 + 4H^+ + 4e^- \rightarrow 2H_2O \qquad 1.23 \text{ V}$$

$$(4) \qquad\qquad Pd^{2+} + 2e^- \rightarrow Pd \qquad 0.915$$

$$(5) \qquad\quad S_2O_8^{2-} + 2e^- \rightarrow 2SO_4^{2-} \qquad 1.96$$

The standard potentials are reasonable values to take in considering (4) and (5). Although the initial concentrations of Pd^{2+} and $S_2O_8^{2-}$ are zero, they would increase during prolonged electrolysis if these species were the principal products. In the case of (3), however, the buffering of the solution prevents the buildup of $[H^+]$, and it would be more appropriate to take the E-value calculated for pH 7.0 in Example 2, 0.82 V. It is apparent that of the three possible anode reactions, (3) has the smallest E-value and would thus occur most readily.

In conclusion, the electrode processes to be expected are

Anode: $\quad 2H_2O \rightarrow O_2 + 4H^+ + 4e^-$

Cathode: $\quad Ni^{2+} + 2e^- \rightarrow Ni$

Overall: $\quad 2H_2O + 2Ni^{2+} \rightarrow O_2 + 4H^+ + 2Ni$

(b) *Cathode Reaction.*

As in (a), Ni reduction would occur.

Anode Reaction.

The expression "inert electrode" is often used to indicate that we may neglect reaction of the electrode itself, either by virtue of the intrinsically high value of its electrode potential or because of polarization effects related to the preparation of the electrode surface. The remaining possible anode reactions are the reverses of the following:

$$E^\circ$$

$$(6) \quad Br_2 + 2e^- \qquad\quad \rightarrow 2Br^- \qquad 1.065 \text{ V}$$

$$(3) \quad O_2 + 4H^+ + 4e^- \rightarrow 2H_2O \qquad 1.23$$

When the E-value for (3) is computed as 0.82 V for pH 7.0, as in (a), oxygen evolution takes precedence. [In practice, "overvoltage" or polarization is more difficult to avoid in the case of reactions involving gases (O_2) as compared with liquids and dissolved solutes, so that electrolysis of $NiBr_2$ at most electrodes would probably lead to Br_2 formation.]

(c) *Cathode Reaction.*

The new couple to be considered is the sodium couple, the reduction reaction for which is

$$E^\circ$$

$$(7) \quad Na^+ + e^- \rightarrow Na \qquad -2.71 \text{ V}$$

This E-value is much lower than that of (2), the evolution of H_2 at pH 7.0, -0.41 V. Therefore, hydrogen evolution will occur at the cathode.

Anode Reaction.

In addition to (3) and (5), the reaction of the Cu anode must be considered, the reverse of which is

$$E^\circ$$

$$(8) \quad Cu^{2+} + 2e^- \rightarrow Cu \qquad 0.34 \text{ V}$$

Process (8) has the lowest E-value, and copper dissolution would thus take precedence over oxygen evolution.

19.20. Knowing that K_{sp} for AgCl is 1.8×10^{-10}, calculate E for a silver–silver chloride electrode immersed in 1 M KCl at 25 °C.

The electrode process is a special case of the (Ag$^+$/Ag) couple, except that silver in the $+$I state collects as solid AgCl on the electrode itself. Even solid AgCl, however, has some Ag$^+$ in equilibrium with it in solution. This [Ag$^+$] can be computed from the K_{sp} equation:

$$[Ag^+] = \frac{K_{sp}}{[Cl^-]} = \frac{1.8 \times 10^{-10}}{1} = 1.8 \times 10^{-10}$$

This value for [Ag$^+$] can be inserted into the Nernst equation for the (Ag$^+$/Ag) half-reaction.

$$Ag^+ + e^- \rightarrow Ag \qquad E^\circ = 0.799 \text{ V}$$

$$E = E^\circ - \frac{0.059\,2}{1} \log \frac{1}{[Ag^+]} = 0.799 - 0.059\,2 \log \frac{1}{1.8 \times 10^{-10}} = 0.799 - 0.577 = 0.222 \text{ V}$$

19.21. From data in Table 19-1, calculate the overall stability constant, K_s, of $Ag(S_2O_3)_2^{3-}$ at 25 °C.

There are two entries in the table for couples connecting the zero and $+$I oxidation states of silver.

$$E^\circ$$

$$(1) \quad Ag^+ + e^- \qquad \rightarrow Ag \qquad\qquad 0.799 \text{ V}$$

$$(2) \quad Ag(S_2O_3)_2^{3-} + e^- \rightarrow Ag + 2S_2O_3^{2-} \qquad 0.017$$

Process (1) refers to the couple in which Ag$^+$ is at unit concentration; the Ag$^+$ concentration to which the E° for (2) refers is that value which satisfies the complex ion equilibrium when the other species are at unit concentration.

$$[Ag^+] = \frac{[Ag(S_2O_3)_2^{3-}]}{K_s[S_2O_3^{2-}]^2} = \frac{1}{K_s}$$

In other words, the standard conditions for couple (2) may be thought of as a nonstandard condition, [Ag$^+$] $= 1/K_s$, for couple (1). Then the Nernst equation for (1) gives

$$E = E^\circ - 0.059\,2 \log \frac{1}{[Ag^+]} \qquad \text{or} \qquad 0.017 = 0.799 - 0.059\,2 \log K_s$$

from which

$$\log K_s = \frac{0.799 - 0.017}{0.059\,2} = 13.21 \qquad \text{and} \qquad K_s = 1.6 \times 10^{13}$$

19.22. (a) At equimolar concentrations of Fe^{2+} and Fe^{3+}, what must [Ag$^+$] be so that the voltage of the galvanic cell made from the (Ag$^+$/Ag) and (Fe^{3+}/Fe^{2+}) electrodes equals zero?

$$Fe^{2+} + Ag^+ \rightleftharpoons Fe^{3+} + Ag$$

(b) Determine the equilibrium constant at 25 °C for the reaction.

(a)   For the reaction as written,

$$E^\circ = E^\circ(Ag^+/Ag) - E^\circ(Fe^{3+}/Fe^{2+}) = 0.799 - 0.771 = 0.028 \text{ V}$$

From the Nernst equation,

$$E = E^\circ - 0.059\,2 \log \frac{[Fe^{3+}]}{[Fe^{2+}][Ag^+]}$$

$$0 = 0.028 - 0.059\,2 \log \frac{1}{[Ag^+]} = 0.028 + 0.059\,2 \log [Ag^+]$$

$$\log [Ag^+] = -\frac{0.028}{0.059\,2} = -0.47$$

$$[Ag^+] = 0.34$$

(b) To find the equilibrium constant, we must combine the relationship between K and $\Delta G°$, (16-8), with the relationship between $\Delta G°$ and $E°$ obtained from (19-3), $\Delta G° = -n\mathscr{F}E°$.

$$\log K = -\frac{\Delta G°}{2.303RT} = \frac{n\mathscr{F}E°}{2.303RT} = \frac{nE°}{0.059\,2}$$

Note that the same combination of constants, of value 0.059 2 at 25 °C, occurs here as in the Nernst equation.

$$\log K = \frac{0.028}{0.059\,2} = 0.47 \quad\text{whence}\quad K = \frac{[Fe^{3+}]}{[Fe^{2+}][Ag^+]} = 3.0$$

Part (a) above could have been solved alternatively by using the equilibrium constant, noting that $[Fe^{2+}] = [Fe^{3+}]$ and solving for $[Ag^+]$:

$$[Ag^+] = \frac{[Fe^{3+}]}{[Fe^{2+}] \times 3.0} = \frac{1}{3.0} = 0.33$$

The two methods must be equivalent because the voltage of a galvanic cell becomes zero when the two couples are at equilibrium with each other.

19.23. An excess of liquid mercury was added to a 1.00×10^{-3} M acidified solution of Fe^{3+}. It was found that only 5.4 percent of the iron remained as Fe^{3+} at equilibrium at 25 °C. Calculate $E°(Hg_2^{2+}/Hg)$, assuming that the only reaction that occurred was

$$2Hg + 2Fe^{3+} \rightarrow Hg_2^{2+} + 2Fe^{2+}$$

First we calculate the equilibrium constant for the reaction. At equilibrium,

$$[Fe^{3+}] = (0.054)(1.00 \times 10^{-3}) = 5.4 \times 10^{-5}$$

$$[Fe^{2+}] = (1 - 0.054)(1.00 \times 10^{-3}) = 9.46 \times 10^{-4}$$

$$[Hg_2^{2+}] = \tfrac{1}{2}[Fe^{2+}] = 4.73 \times 10^{-4}$$

Liquid Hg is in excess and is in its standard state.

$$K = \frac{[Hg_2^{2+}][Fe^{2+}]^2}{[Fe^{3+}]^2} = \frac{(4.73 \times 10^{-4})(9.46 \times 10^{-4})^2}{(5.4 \times 10^{-5})^2} = 0.145$$

The standard potential of the cell corresponding to the reaction may be computed from the relation found in Problem 19.22(b).

$$E° = \frac{0.059\,2}{n} \log K = \frac{(0.059\,2)(-0.839)}{2} = -0.025 \text{ V}$$

For the reaction as written,

$$E° = E°(Fe^{3+}/Fe^{2+}) - E°(Hg_2^{2+}/Hg)$$

or $$E°(Hg_2^{2+}/Hg) = E°(Fe^{3+}/Fe^{2+}) - E° = 0.771 - (-0.025) = 0.796 \text{ V}$$

19.24. $E°'$-values for the (pyruvate/lactate) and (NAD^+/NADH) couples at pH 7 and 25 °C are -0.19 V and -0.32 V, respectively. In the enzymatic reduction of pyruvate to lactate (2 electrons per molecule) by NADH (2 electrons per molecule), what is $E°'$ for the reaction and what is $\Delta G°'$ at pH 7 per mole of pyruvate reduced?

Except for H_2O and its ions, the reaction may be written

$$\text{Pyruvate} + \text{NADH} \rightleftharpoons \text{lactate} + \text{NAD}^+$$

Then $$E°' = E°'(\text{pyruvate/lactate}) - E°'(\text{NAD}^+/\text{NADH}) = -0.19 - (-0.32) = 0.13 \text{ V}$$

and $$\Delta G°' = -n\mathscr{F}E°' = -(2 \text{ mol } e^-)(9.65 \times 10^4 \text{ C/mol } e^-)(0.13 \text{ V})$$

$$= -2.5 \times 10^4 \text{ J} = -25 \text{ kJ}$$

Supplementary Problems

ELECTRICAL UNITS

19.25. How many coulombs per hour pass through an electroplating bath which uses a current of 5 A?

   Ans. 1.8×10^4 C/h

19.26. Compute the cost at 5¢ per kilowatt-hour of operating for 8 hours an electric motor which takes 15 A at 110 V.

   Ans.   66¢

19.27. A tank containing 0.2 m³ of water was used as a constant-temperature bath. How long would it take to heat the bath from 20 °C to 25 °C with a 250-W immersion heater? Neglect the heat capacity of the tank frame and any heat losses to the air.

   Ans.   4.6 h

19.28. The specific heat of a liquid was measured by placing 100 g of the liquid in a calorimeter. The liquid was heated by an electrical immersion coil. The heat capacity of the calorimeter together with the coil was previously determined to be 31.4 J/K. With the 100-g sample in place in the calorimeter, a current of 0.500 A was passed through the immersion coil for exactly 3 min. The voltage across the terminals of the coil was measured to be 1.50 V. The temperature of the sample rose by 0.800 °C. Find the specific heat capacity of the liquid.

   Ans.   1.38 kJ/kg·K

19.29. The heat of solution of NH_4NO_3 in water was determined by measuring the amount of electrical work needed to compensate for the cooling which would otherwise occur when the salt dissolves. After the NH_4NO_3 was added to the water, electric energy was provided by passage of a current through a resistance coil until the temperature of the solution reached the value it had prior to the addition of the salt. In a typical experiment, 4.4 g NH_4NO_3 was added to 200 g water. A current of 0.75 A was provided through the heater coil, and the voltage across the terminals was 6.0 V. The current was applied for exactly 5.2 min. Calculate ΔH for the solution of 1 mol NH_4NO_3 in enough water to give the same concentration as was attained in the above experiment.

   Ans.   25.5 kJ

FARADAY'S LAWS

   In the electrolysis problems, assume that the electrode efficiency is 100 percent for the principal electrode reaction, unless a statement is made in the problem to the contrary.

19.30. What current is required to pass 1 mol of electrons per hour through an electrodeposition cell? How many grams of aluminum and of cadmium will be liberated by 1 mol of electrons?

   Ans.   26.8 A, 8.99 g Al, 56.2 g Cd

19.31. What mass of aluminum is deposited electrolytically in 30 min by a current of 40 A?

   Ans.   6.7 g

19.32. How many amperes are required to deposit on the cathode 5.00 g of gold per hour from a solution containing a salt of gold in the +III oxidation state?

Ans.   2.04 A

19.33. How many hours will it take to produce 100 lb of electrolytic chlorine from NaCl in a cell that carries 1 000 A? The anode efficiency for the chlorine reaction is 85 percent.

Ans.   40.4 h

19.34. A given quantity of electricity passes through two separate electrolytic cells containing solutions of $AgNO_3$ and $SnCl_2$, respectively. If 2.00 g of silver is deposited in one cell, how many grams of tin are deposited in the other cell?

Ans.   1.10 g Sn

19.35. An electrolytic cell contains a solution of $CuSO_4$ and an anode of impure copper. How many kilograms of copper will be refined (deposited on the cathode) by 150 A maintained for 12.0 h?

Ans.   2.13 kg Cu

19.36. How many hours are required for a current of 3.0 A to decompose electrolytically 18 g water?

Ans.   18 h

19.37. Hydrogen peroxide can be prepared by the successive reactions

$$2NH_4HSO_4 \rightarrow H_2 + (NH_4)_2S_2O_8$$

$$(NH_4)_2S_2O_8 + 2H_2O \rightarrow 2NH_4HSO_4 + H_2O_2$$

The first reaction is an electrolytic reaction, and the second a steam distillation. What current would have to be used in the first reaction to produce enough intermediate to yield 100 g of pure H_2O_2 per hour? Assume 50 percent anode current efficiency.

Ans.   315 A

19.38. The electrodes in a lead storage battery are made of Pb and PbO_2. The overall reaction during discharge is

$$Pb + PbO_2 + 2H_2SO_4 \rightarrow 2PbSO_4 + 2H_2O$$

(a) What is the minimum amount (mass in pounds) of lead (counting the lead in both free and combined forms) in a battery if the battery is designed to deliver 100 A · h? Assume a 25 percent "coefficient of use"; this is the percent of the Pb and PbO_2 in the battery case that actually is available for the electrode reactions.   (b) If the average voltage of a storage battery is 2.00 V under zero load, what is the approximate free-energy change for the reaction as written above?

Ans.   (a) 6.8 lb lead;   (b) −386 kJ

19.39. Neglecting electrode polarization effects, predict the principal product at each electrode in the continued electrolysis at 25 °C of each of the following: (a) 1 M $Fe_2(SO_4)_3$ with inert electrodes in 0.1 M H_2SO_4,   (b) 1 M LiCl with silver electrodes,   (c) 1 M $FeSO_4$ with inert electrodes at pH 7.0,   (d) molten NaF with inert electrodes.

Ans.   (a) Fe^{2+} and O_2;   (b) H_2 and AgCl;   (c) H_2 and Fe^{3+};   (d) Na and F_2

19.40. A galvanic cell was operated under almost ideally reversible conditions at a current of 10^{-16} A.   (a) At this current, how long would it take to deliver 1 mol of electrons?   (b) How many electrons would be delivered by the cell to a pulsed measuring circuit in 10 ms of operation?

Ans.   (a) 3×10^{13} years;   (b) 6

GALVANIC CELLS AND OXIDATION-REDUCTION

All problems refer to 25 °C.

19.41. (a) What is the standard potential of the cell made up of the (Cd^{2+}/Cd) and (Cu^{2+}/Cu) couples?   (b) Which couple is positive?

  Ans.   (a) 0.74 V;   (b) (Cu^{2+}/Cu)

19.42. What is the standard potential of the cell containing the (Sn^{2+}/Sn) and (Br_2/Br^-) couples?

  Ans.   1.21 V

19.43. Why are Co^{3+} salts unstable in water?

  Ans. Co^{3+} can oxidize H_2O, the principal products being Co^{2+} and O_2.

19.44. If H_2O_2 is mixed with Fe^{2+}, which reaction is more likely, the oxidation of Fe^{2+} to Fe^{3+}, or the reduction of Fe^{2+} to Fe? Write the reaction for each possibility and compute the standard potential of the equivalent electrochemical cell.

  Ans.   More likely: $H_2O_2 + 2H^+ + 2Fe^{2+} \rightarrow 2H_2O + 2Fe^{3+}$; $E° = 0.99$ V. Less likely: $H_2O_2 + Fe^{2+} \rightarrow$ $Fe + O_2 + 2H^+$. The reverse of this reaction occurs with a standard potential of 1.14 V.

19.45. What substance can be used to oxidize fluorides to fluorine?

  Ans.   Fluorides may be oxidized electrolytically, but not chemically by any substance listed in Table 19-1.

19.46. Are Fe^{2+} solutions stable in the air? Why can such solutions be preserved by the presence of iron nails?

  Ans. O_2 oxidizes Fe^{2+} to Fe^{3+}, but Fe reduces Fe^{3+} to Fe^{2+}.

19.47. What is the standard potential of the (Tl^{3+}/Tl) electrode?

  Ans.   0.72 V

19.48. Which of the following intermediate oxidation states is stable with respect to disproportionation in oxygen-free noncomplexing media: germanium(II), tin(II)?

  Ans.   tin(II)

19.49. Would H_2O_2 behave as oxidant or reductant with respect to each of the following couples at standard concentrations: (a) (I_2/I^-),   (b) $(S_2O_8^{2-}/SO_4^{2-})$,   (c) (Fe^{3+}/Fe^{2+})?

  Ans.   (a) oxidant.   (b) reductant.   (c) both; in fact, iron salts in either $+II$ or $+III$ state catalyze the self-oxidation-reduction of H_2O_2.

19.50. What is the potential of a cell containing two hydrogen electrodes, the negative one in contact with 10^{-8} molar H^+ and the positive one in contact with 0.025 molar H^+?

  Ans.   0.379 V

19.51. Compute the potential of the (Ag^+/Ag) couple with respect to (Cu^{2+}/Cu) if the concentrations of Ag^+ and Cu^{2+} are 4.2×10^{-6} M and 1.3×10^{-3} M, respectively. What is the value of ΔG for the reduction of 1 mol of Cu^{2+} by Ag at the indicated ion concentrations?

  Ans.   0.23 V, -44 kJ

19.52. Copper can reduce zinc ions if the resultant copper ions can be kept at a sufficiently low concentration by the formation of an insoluble salt. What is the maximum concentration of Cu^{2+} in solution if this reaction is to occur, when Zn^{2+} is 1 molar?

Ans. 6×10^{-38} M

19.53. Evaluate the equilibrium constant for the reaction

$$Fe(CN)_6^{4-} + Co(dip)_3^{3+} \rightleftharpoons Fe(CN)_6^{3-} + Co(dip)_3^{2+}$$

Ans. 0.5

19.54. When a rod of metallic lead was added to a 0.0100 M solution of $Co(en)_3^{3+}$, it was found that 68 percent of the cobalt complex was reduced to $Co(en)_3^{2+}$ by the lead. (*a*) Find the value of K for

$$Pb + 2Co(en)_3^{3+} \rightleftharpoons Pb^{2+} + 2Co(en)_3^{2+}$$

(*b*) What is the value of $E°(Co(en)_3^{3+}/Co(en)_3^{2+})$?

Ans. (*a*) 0.0154;   (*b*) −0.180 V

19.55. A (Tl^+/Tl) couple was prepared by saturating 0.1 M KBr with TlBr and allowing the Tl^+ from the relatively insoluble bromide to equilibrate. This couple was observed to have a potential of −0.443 V with respect to a (Pb^{2+}/Pb) couple in which Pb^{2+} was 0.1 molar. What is the solubility product of TlBr?

Ans. 3.7×10^{-6}

19.56. K_d for complete dissociation of $Ag(NH_3)_2^+$ into Ag^+ and NH_3 is 6.0×10^{-8}. Calculate $E°$ for the following half-reaction by reference to Table 19-1:

$$Ag(NH_3)_2^+ + e^- \rightarrow Ag + 2NH_3$$

Ans. 0.372 V

19.57. Calculate K_s for formation of PdI_4^{2-} from Pd^{2+} and I^-.

Ans. 10^{25}

19.58. Reference tables give the following entry:

$$HO_2^- + H_2O + 2e^- \rightarrow 3OH^- \qquad E° = 0.88 \text{ V}$$

Combining this information with relevant entries in Table 19-1, find K_1 for the acid dissociation of H_2O_2.

Ans. 10^{-12}

19.59. Calculate $\Delta G^{°\prime}$ for the oxidation of 1 mol of flavin (2 electrons per molecule) by molecular oxygen (1 atm) at pH 7. The $E^{°\prime}$-value for the flavin couple is −0.22 V. (*Hint*: Be sure to use $E^{°\prime}$ rather than $E°$ for the reduction of O_2 to H_2O.)

Ans. −201 kJ

Chapter 20

Rates of Reactions

The discussion of chemical equilibrium or of the direction of spontaneous processes is often accompanied by the statement that thermodynamics does not explain the *rate* at which spontaneous processes occur or chemical equilibrium is achieved. The branch of chemistry that treats the rates of reactions is called *chemical kinetics*. There are two main objectives of chemical kinetics: the systematization of data dealing with the dependence of the rates of reactions on controllable variables, and the inference of the molecular mechanism by which reactions occur from the observed rates. This chapter deals largely with systematics.

RATE CONSTANTS AND THE ORDER OF REACTIONS

Reactions that occur completely within a single phase of matter, particularly a liquid or gaseous phase, are called *homogeneous reactions*. Reactions that take place at least in part at the interface between two phases, such as solid and liquid or solid and gas, are called *heterogeneous reactions*. The discussion and problems in this chapter will concern homogeneous reactions, unless a specific statement to the contrary is made.

According to the *law of mass action*, the rate of a chemical reaction at a given temperature, expressed as the amount reacting per unit volume per unit time, depends only on the concentrations of the various substances influencing the rate (and not, for example, on the size of the reaction vessel). The substances that influence the rate are usually one or more of the reactants, occasionally one of the products, and sometimes a catalyst that does not appear in the balanced overall chemical equation. The dependence of the rate on the concentrations can be expressed in many cases as a direct proportionality, in which the concentrations appear to the zero, first, or second power. The power to which the concentration of a substance appears in the rate expression is called the *order of the reaction* with respect to that substance. Some examples follow.

(*1*)   Rate $= k_1[A]$ First-order in A

(*2*)   Rate $= k_2[A][B]$ First-order in A and first-order in B

(*3*)   Rate $= k_3[A]^2$ Second-order in A

(*4*)   Rate $= k_4[A]^2[B]$ Second-order in A and first-order in B

(*5*)   Rate $= k_5$ Zero-order

Note: In all expressions involving rates, [X] denotes the concentration of X (not the relative concentration) and carries the units of concentration (e.g., mol/L).

The order cannot be deduced from the chemical equation but must be ascertained by experiment. The overall order of a reaction is the sum of the orders with respect to the various substances. In the above examples, the values of the overall order are 1, 2, 2, 3, and 0, respectively.

The proportionality factors, k, called the *rate constants* or the *specific rates*, are constant at a given temperature but may depend on the temperature. There are dimensions to k, and the units should be expressed when k-values are tabulated. The rate itself is defined as the change of concentration of a reactant or product per unit time. If A is a reactant and C a product, the rate might be expressed as

$$\text{Rate} = -\frac{\Delta[A]}{\Delta t} \quad \text{or as} \quad \text{Rate} = \frac{\Delta[C]}{\Delta t}$$

where $\Delta[X]$ is the *change* in the concentration of X and Δt is the time interval over which the change is measured. (If the rate changes rapidly, Δt should be small. The law of mass action applies only at very small values of Δt, in which case the derivative notation of calculus,

$$-\frac{d[A]}{dt} \quad \text{or} \quad \frac{d[C]}{dt}$$

replaces the ratio expression.) The minus sign is used to express the rate in terms of a reactant concentration (which decreases during the reaction) and a plus sign for a product concentration (which increases during the reaction), so that the rate is always a positive quantity.

EXAMPLE 1   For some reactions, there is no ambiguity in defining the rate. For example, in

$$CH_3OH + HCOOH \rightarrow HCOOCH_3 + H_2O$$

the rate could be expressed as

$$-\frac{\Delta[CH_3OH]}{\Delta t} \quad \text{or} \quad -\frac{\Delta[HCOOH]}{\Delta t} \quad \text{or} \quad \frac{\Delta[HCOOCH_3]}{\Delta t}$$

since these three ratios are all equal when concentrations are in molar units. In the following reaction, however,

$$N_2 + 3H_2 \rightarrow 2NH_3$$

the coefficients for the different substances are not the same; the concentration of H_2 decreases three times as fast as that of N_2 in the same molar units, and NH_3 is formed twice as fast as N_2 is used up. Any one of the three ratios

$$-\frac{\Delta[N_2]}{\Delta t} \qquad -\frac{\Delta[H_2]}{\Delta t} \qquad \frac{\Delta[NH_3]}{\Delta t}$$

could be used to specify the rate, but the choice should be specified because the numerical value of k (though not the order) depends on the choice.

First-order reactions

In the particular case of a first-order reaction, where the rate is proportional to the concentration of the reactant [A], as in (*1*) above, the methods of integral calculus give

$$[A] = [A]_0 e^{-kt} \tag{20-1}$$

(See Appendix C for the definition of e, the base of natural logarithms.) $[A]_0$ is the value of [A] at the beginning of the experiment, when $t = 0$. The logarithmic form can be derived from the above.

$$k = -\frac{2.303 \log \dfrac{[A]}{[A]_0}}{t} \tag{20-2}$$

The amount of time, $t_{1/2}$, for a first-order reaction to achieve 50 percent completion, is given by (*20-2*) as

$$t_{1/2} = -\frac{2.303 \log \frac{1}{2}}{k} = \frac{0.693}{k} \tag{20-3}$$

Note that $t_{1/2}$, called the *half-life*, is independent of the initial concentration of A; this independence of $[A]_0$ is characteristic only of first-order reactions.

Other rate laws

Fractional orders also occur (for example, $\frac{1}{2}$- or $\frac{3}{2}$-order, in which the rate is proportional to $[A]^{1/2}$ or $[A]^{3/2}$, respectively). Some reaction rates may not be expressible in the proportional form altogether. An example of a complex rate expression is the following:

$$(6) \quad \text{Rate} = \frac{k_1[A]^2}{1 + k_2[A]}$$

The rate of a heterogeneous reaction may be proportional to the interfacial area of contact between the phases, as well as to concentrations of reactants within a particular phase, as in the case of many reactions catalyzed at surfaces.

ENERGY OF ACTIVATION

The temperature dependence of a reaction rate can be represented by the *Arrhenius equation*:

$$k = Ae^{-E_a/RT} \tag{20-4}$$

The *pre-exponential factor*, A, is also called the *frequency factor*, and E_a is called the *energy of activation*; the units of E_a (and of RT) are J/mol or cal/mol. Both A and E_a may be considered to be constant, at least over a narrow temperature range. From (*20-4*) the rate constants at two different temperatures are related by

$$\log \frac{k_2}{k_1} = \frac{E_a}{2.303R} \left(\frac{1}{T_1} - \frac{1}{T_2} \right) \tag{20-5}$$

MECHANISM OF REACTIONS

It has been said that a chemical equation accurately portrays the nature of the starting and final materials in a chemical reaction, but that the arrow conveniently covers our ignorance of what happens in between. How many successive steps are there in the overall process? What are the spatial and energetic requirements for the interactions in each step? What is the rate of each step? Although rate measurements usually describe only the overall reaction, the measurement of rates under different conditions often contributes to the understanding of the mechanism.

Molecularity

The molecularity specifies the number of molecules interacting in an individual mechanistic step. A *unimolecular* reaction is a step in which a single molecule spontaneously undergoes a reaction; a *bimolecular* reaction requires the coming together of two molecules; a *termolecular* reaction is one in which three molecules interact in a single step. There are no known single steps of higher molecularity than termolecular.

A unimolecular reaction is first-order in the species that undergoes the spontaneous rearrangement or decomposition. This is true because there is an intrinsic probability that a molecule will undergo the reaction in a unit time interval, and the overall rate per unit volume is the product of this probability per unit time and the total number of molecules per unit volume (which in turn is proportional to the concentration).

A bimolecular reaction rate is proportional to the frequency of collisions between two molecules of the reacting species. It is known from kinetic theory that the frequency of collisions between two like molecules, A, is proportional to $[A]^2$, and the frequency of collisions between an A and a B molecule is proportional to $[A][B]$. If the species whose molecules collide are starting materials in limited concentrations, the reaction is second-order, following the rate equation of either type (*3*) or type (*2*).

A termolecular reaction rate is proportional to the frequency of three-body collisions, which in turn is proportional to $[A]^3$, $[A]^2[B]$ or $[A][B][C]$, depending on whether the three colliding molecules represent one, two, or three different species. If these species are starting materials, the reaction is third-order.

Although the order of a single mechanistic step can be predicted from the molecularity, the molecularity of a step, or steps, cannot be predicted from the order of the overall reaction. There are a number of complications which make it impossible to conclude automatically that a first-order reaction is unimolecular, that a second-order reaction is bimolecular, or that a third-order reaction is termolecular.

In many cases, the reaction is a sequence of steps, and the overall rate may be governed by the slowest step. Experimental conditions might interchange the relative speeds of different steps, and the order would appear to change. Another complication arises because the slowest, or rate-limiting, step may involve the reaction of an unstable intermediate species, and it is necessary to express the concentration of this species in terms of the reactants before the overall order of the reaction can be deciphered. The solved problems will give examples of some of these complexities.

Energetics

Most molecular collisions do not result in reaction. Even if the requisite number of molecules came together, only those possessing enough energy, usually far in excess of the average energy, can undergo the violent distortions of bond lengths and angles necessary for the rearrangements that lead to chemical reaction. The energy of activation, E_a, is a measure of the excess energy needed for molecules to react, and the exponential term in the Arrhenius equation, (20-4), is of the order of magnitude of the fraction of molecules possessing this amount of excess energy.

The *activated state* has been defined as that distorted combination of reacting molecules having the minimum amount of excess energy so that the combination, or *complex*, could as easily rearrange to form the products as revert to the reactants. The energy of the activated state may be likened to the potential energy of a mountain pass which a climber must cross in order to go from one side of the mountain to another. For a set of reversible reactions, the same activated state must be crossed for the reactions in the two directions. Thus, for reactions carried out at constant pressure,

$$(1 \text{ mol})[E_a(\text{forward}) - E_a(\text{reverse})] = \Delta H \tag{20-6}$$

The factor (1 mol) has been inserted in (20-6) so that the enthalpy change of reaction will, as usual, have the units of energy. These relationships are shown graphically in Fig. 20-1 below.

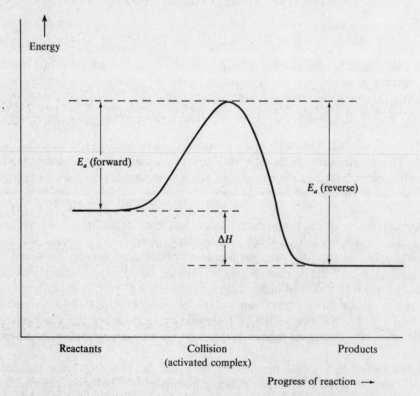

Fig. 20-1   **Energy relationships in a reaction (forward reaction is exothermic).**

Solved Problems

RATE CONSTANTS AND THE ORDER OF REACTION

20.1. In a catalytic experiment involving the Haber process, $N_2 + 3H_2 \rightarrow 2NH_3$, the rate of reaction was measured as

$$\frac{\Delta[NH_3]}{\Delta t} = 2.0 \times 10^{-4} \, \text{mol} \cdot \text{L}^{-1} \cdot \text{s}^{-1}$$

If there were no side reactions, what was the rate of reaction expressed in terms of (a) N_2, (b) H_2?

(a) From the coefficients in the balanced equation, $\Delta n(N_2) = -\frac{1}{2}\Delta n(NH_3)$. Therefore,

$$-\frac{\Delta[N_2]}{\Delta t} = \frac{1}{2}\frac{\Delta[NH_3]}{\Delta t} = 1.0 \times 10^{-4} \, \text{mol} \cdot \text{L}^{-1} \cdot \text{s}^{-1}$$

(b) Similarly,

$$-\frac{\Delta[H_2]}{\Delta t} = \frac{3}{2}\frac{\Delta[NH_3]}{\Delta t} = 3.0 \times 10^{-4} \, \text{mol} \cdot \text{L}^{-1} \cdot \text{s}^{-1}$$

20.2. What are the units of the rate constant, k, for (a) a zero-order reaction, (b) a first-order reaction, (c) a second-order reaction, (d) a third-order reaction, (e) a half-order reaction, when concentrations are expressed in molar units and time in seconds?

In each case, we write out the full rate equation and find the units of k that will satisfy the equation.

(a)
$$-\frac{\Delta[A]}{\Delta t} = k$$

$$\text{Units of } k = \text{units of } \frac{[A]}{t} = \frac{\text{mol/L}}{\text{s}} = \text{mol} \cdot \text{L}^{-1} \cdot \text{s}^{-1}$$

Note that the units of $\Delta[A]$, the change in concentration, are the same as the units of $[A]$ itself; similarly for Δt.

(b)
$$-\frac{\Delta[A]}{\Delta t} = k[A] \qquad \text{or} \qquad k = -\frac{1}{[A]}\frac{\Delta[A]}{\Delta t}$$

$$\text{Units of } k = \frac{1}{\text{mol/L}} \times \frac{\text{mol/L}}{\text{s}} = \text{s}^{-1}$$

First-order reactions are the only reactions for which k has the same numerical value, regardless of the units used for expressing the concentrations of the reactants or products.

(c)
$$-\frac{\Delta[A]}{\Delta t} = k[A]^2 \qquad\qquad -\frac{\Delta[A]}{\Delta t} = k[A][B]$$

$$k = \frac{1}{[A]^2}\frac{\Delta[A]}{\Delta t} \qquad\qquad k = -\frac{1}{[A][B]}\frac{\Delta[A]}{\Delta t}$$

$$\text{Units of } k = \frac{1}{(\text{mol/L})^2} \times \frac{\text{mol/L}}{\text{s}} = \text{L} \cdot \text{mol}^{-1} \cdot \text{s}^{-1}$$

Note that the units of k depend on the *total* order of the reaction, not on the way the total order is composed of the orders with respect to different reactants.

(d)
$$-\frac{\Delta[A]}{\Delta t} = k[A]^3 \quad \text{or} \quad k = -\frac{1}{[A]^3}\frac{\Delta[A]}{\Delta t}$$

$$\text{Units of } k = \frac{1}{(\text{mol/L})^3} \times \frac{\text{mol/L}}{\text{s}} = L^2 \cdot mol^{-2} \cdot s^{-1}$$

(e)
$$-\frac{\Delta[A]}{\Delta t} = k[A]^{1/2} \quad \text{or} \quad k = -\frac{1}{[A]^{1/2}}\frac{\Delta[A]}{\Delta t}$$

$$\text{Units of } k = \frac{1}{(\text{mol/L})^{1/2}} \times \frac{\text{mol/L}}{\text{s}} = mol^{1/2} \cdot L^{-1/2} \cdot s^{-1}$$

20.3. In a certain polluted atmosphere containing O_3 at a steady-state concentration of 2.0×10^{-8} mol/L, the hourly production of O_3 by all sources was estimated as 7.2×10^{-15} mol/L. If the only mechanism for destruction of O_3 is the second-order reaction $2O_3 \rightarrow 3O_2$, calculate the rate constant for the destruction reaction, defined by the rate law for $-\Delta[O_3]/\Delta t$.

At the steady state, the rate of destruction of O_3 must equal the rate of its generation, 7.2×10^{-15} mol $\cdot L^{-1} \cdot h^{-1}$. From the second-order rate law,

$$-\frac{\Delta[O_3]}{\Delta t} = k[O_3]^2$$

$$k = -\frac{1}{[O_3]^2}\frac{\Delta[O_3]}{\Delta t} = \frac{1}{(2.0 \times 10^{-8} \text{ mol/L})^2}\frac{7.2 \times 10^{-15} \text{ mol} \cdot L^{-1} \cdot h^{-1}}{3.6 \times 10^3 \text{ s} \cdot h^{-1}} = 5 \times 10^{-3} \text{ L} \cdot mol^{-1} \cdot s^{-1}$$

20.4. A viral preparation was inactivated in a chemical bath. The inactivation process was found to be first-order in virus concentration, and at the beginning of the experiment 2.0 percent of the virus was found to be inactivated per minute. Evaluate k for the inactivation process.

From the first-order rate law,

$$-\frac{\Delta[A]}{\Delta t} = k[A], \quad \text{or} \quad k = -\frac{\Delta[A]}{[A]}\frac{1}{\Delta t}$$

It is seen that only the *fractional* change in concentration, $-\Delta[A]/[A]$, is needed; namely, 0.020 when $\Delta t = 1$ min $= 60$ s. This form of the equation may be used for the *initial* rate when the value of [A] is not changing appreciably; that condition is certainly met when only 2 percent is inactivated in the first minute.

$$k = \frac{0.020}{60 \text{ s}} = 3.3 \times 10^{-4} \text{ s}^{-1}$$

20.5. For the process described in Problem 20.4, how much time would be required for the virus to become   (a) 50 percent inactivated,   (b) 75 percent inactivated?

The method used in Problem 20.4 cannot be used here because [A] changes appreciably over the course of the reaction. Equation (20-1), (20-2), or (20-3) is appropriate here.

(a)   The time for 50 percent reaction is the half-life, and (20-3) gives

$$t_{1/2} = \frac{0.693}{k} = \frac{0.693}{3.3 \times 10^{-4} \text{ s}^{-1}} = 2.1 \times 10^3 \text{ s} = 35 \text{ min}$$

(b) Equation (20-2) may be used. If 75 percent of the virus is inactivated, the fraction remaining, $[A]/[A]_0$, is 0.25.

$$t = -\frac{2.303 \log \dfrac{[A]}{[A]_0}}{k} = -\frac{2.303 \log 0.25}{3.3 \times 10^{-4}\,s^{-1}} = 4.2 \times 10^3\,s = 70\,min$$

An alternate solution is to apply the half-life concept twice. Since it takes 35 min for half of the virus to become inactivated regardless of the initial concentration, the time for the virus to be reduced from 50 percent to 25 percent full strength will be another half-life. The total time for reduction to $\frac{1}{4}$ strength is thus two half-lives, or 70 min.

Similarly, the total time required to reduce the initial activity to $\frac{1}{8}$ is three half-lives; to $\frac{1}{16}$, four half-lives; and so on. This method can be used only for first-order reactions.

20.6. If in the fermentation of sugar in an enzymatic solution that is initially 0.12 M the concentration of the sugar is reduced to 0.06 M in 10 h and to 0.03 M in 20 h, what is the order of the reaction and what is the rate constant?

This problem is analogous to Problem 20.5. Since doubling the time doubles the fractional reduction of the reactant concentration, the reaction must be first-order. Alternatively, the reduction of the sugar concentration from 0.06 M to 0.03 M may be thought of as a new experiment with an initial concentration of 0.06 M. Since the same half-life (10 h) was observed in both experiments, the reaction must be first-order because only in a first-order reaction is the half-life independent of the initial concentration. The rate constant may be evaluated from the half-life via (20-3).

$$k = \frac{0.693}{t_{1/2}} = \frac{0.693}{10\,h} = 6.9 \times 10^{-2}\,h^{-1}$$

$$= \frac{6.9 \times 10^{-2}\,h^{-1}}{3.6 \times 10^3\,s \cdot h^{-1}} = 1.9 \times 10^{-5}\,s^{-1}$$

20.7. A reaction between substances A and B is represented stoichiometrically by $A + B \rightarrow C$. Observations on the rate of this reaction are obtained in three separate experiments as follows:

	Initial Concentrations		Duration of Experiment, $\Delta t/h$	Final Concentration, $[A]_f/M$
	$[A]_0/M$	$[B]_0/M$		
(1)	0.1000	1.0	0.50	0.0975
(2)	0.1000	2.0	0.50	0.0900
(3)	0.0500	1.0	2.00	0.0450

What is the order with respect to each reactant, and what is the value of the rate constant?

First, let us tabulate the initial rate of the reaction for each experiment, noting that $\Delta[A] = [A]_f - [A]_0$ is in each case small enough in magnitude to allow the rate to be expressed in terms of changes over the entire time of the experiment.

	Initial Concentrations				Initial Rates $-\dfrac{\Delta[A]}{\Delta t} \Big/ M \cdot h^{-1}$
	$[A]_0/M$	$[B]_0/M$	$\Delta[A]/M$	$\Delta t/h$	
(1)	0.1000	1.0	−0.0025	0.50	0.0050
(2)	0.1000	2.0	−0.0100	0.50	0.0200
(3)	0.0500	1.0	−0.0050	2.00	0.0025

In comparing experiments (*1*) and (*2*), we note that [A] is the same in both, but [B] is twice as great in (*2*). Since the rate in (*2*) is 4 times that in (*1*), the reaction must be second-order in B.

In comparing experiments (*1*) and (*3*), we note that [B] is the same in both, but [A] is twice as great in (*1*). Since the rate in (*1*) is twice that in (*3*), the reaction must be first-order in A.

The rate equation may be written accordingly as

$$-\frac{\Delta[A]}{\Delta t} = k[A][B]^2$$

and k may be evaluated from any of the experiments; we take (*1*) as an example, using average values of [A] and [B].

$$k = \frac{-\dfrac{\Delta[A]}{\Delta t}}{[A][B]^2} = \frac{0.0050 \text{ M} \cdot \text{h}^{-1}}{(0.099 \text{ M})(1.0 \text{ M})^2} = 0.051 \text{ L}^2 \cdot \text{mol}^{-2} \cdot \text{h}^{-1}$$

or $1.4 \times 10^{-5} \text{ L}^2 \cdot \text{mol}^{-2} \cdot \text{s}^{-1}$. The reader should confirm that the same result will be obtained by using data from either of the other two experiments.

20.8. The esterification of acetic anhydride by ethyl alcohol can be represented by the following balanced equation:

$$\underset{\text{A}}{(CH_3CO)_2O} + \underset{\text{B}}{C_2H_5OH} \rightarrow CH_3COOC_2H_5 + CH_3COOH$$

When the reaction is carried out in dilute hexane solution, the rate may be represented by $k[A][B]$. When ethyl alcohol (B) is the solvent, the rate may be represented by $k[A]$. (The values of k are not the same in the two cases.) Explain the difference in the apparent order of the reaction.

When a solvent is also a reactant, its concentration is so large compared with the extent of reaction that it does not change. (Compare the convention we had used, in studying the thermodynamics of aqueous solutions, of viewing water as always being in its standard state in all dilute solutions.) Thus, the dependence of the rate on the concentration of ethyl alcohol cannot be determined unless ethyl alcohol becomes a solute in some other solvent, so that its concentration can be varied.

20.9. *trans*-1,2-Dideuterocyclopropane (A) undergoes a first-order decomposition. The observed rate constant at a certain temperature, measured in terms of the disappearance of A, was $1.52 \times 10^{-4} \text{ s}^{-1}$. Analysis of the reaction products showed that the reaction followed two parallel paths, one leading to dideuteropropene (B) and the other to *cis*-1,2-dideuterocyclopropane (C). B was found to constitute 11.2 percent of the reaction product, independently of the extent of reaction. What is the order of the reaction for each of the paths, and what is the value of the rate constant for the formation of each of the products?

Since the percentage distribution of B and C is always the same, B and C must be formed at a fixed ratio of rates. Since A disappears by a first-order process, B and C must each be formed by a first-order process. We can then write

$$\frac{\Delta[B]}{\Delta t} = k_B[A] \qquad \frac{\Delta[C]}{\Delta t} = k_C[A]$$

We now want an equation to represent the rate of formation of all products—that is, B and C combined—which should equal the rate of disappearance of A.

$$\frac{\Delta([B] + [C])}{\Delta t} = \frac{\Delta[B]}{\Delta t} + \frac{\Delta[C]}{\Delta t} = (k_B + k_C)[A] = k_A[A]$$

or
$$k_A = k_B + k_C$$

In other words, the rate constant for the disappearance of A is equal to the sum of the rate constants for the formation of B and of C. Also, since B is formed at a rate equal to 11.2 percent of the rate of disappearance of A,

$$\frac{\Delta[B]}{\Delta t} = k_B[A] = (0.112)\left(-\frac{\Delta[A]}{\Delta t}\right) = (0.112)k_A[A]$$

or
$$k_B = (0.112)k_A = (0.112)(1.52 \times 10^{-4}\,\text{s}^{-1}) = 1.7 \times 10^{-5}\,\text{s}^{-1}$$

Then
$$k_C = k_A - k_B = 1.52 \times 10^{-4} - 1.7 \times 10^{-5} = 1.35 \times 10^{-4}\,\text{s}^{-1}$$

20.10. The complexation of Fe^{2+} with the chelating agent dipyridyl (abbreviated dip) has been studied kinetically in both the forward and reverse directions. For the complexation reaction

$$Fe^{2+} + 3\,\text{dip} \rightarrow Fe(\text{dip})_3^{2+}$$

the rate of formation of the complex at 25 °C is given by

$$\text{Rate} = (1.45 \times 10^{13}\,\text{L}^3 \cdot \text{mol}^{-3} \cdot \text{s}^{-1})[Fe^{2+}][\text{dip}]^3$$

and for the reverse of the above reaction, the rate of disappearance of the complex is

$$(1.22 \times 10^{-4}\,\text{s}^{-1})[Fe(\text{dip})_3^{2+}]$$

What is K_s, the stability constant for the complex?

Not all reactions can be studied conveniently in both directions. When this is possible, we know that at dynamic equilibrium the rate of formation of the complex must equal the rate of decomposition, since concentrations of the various species stay fixed.

$$\text{Rate of forward reaction} = \text{rate of reverse reaction}$$

$$(1.45 \times 10^{13}\,\text{L}^3 \cdot \text{mol}^{-3} \cdot \text{s}^{-1})[Fe^{2+}][\text{dip}]^3 = (1.22 \times 10^{-4}\,\text{s}^{-1})[Fe(\text{dip})_3^{2+}]$$

Solving,

$$K_s = \frac{[Fe(\text{dip})_3^{2+}]}{[Fe^{2+}][\text{dip}]^3} = \frac{1.45 \times 10^{13}}{1.22 \times 10^{-4}} = 1.19 \times 10^{17}$$

(In the equilibrium constant equation, as opposed to rate equations, [X] is dimensionless, being the concentration relative to the standard state of 1 mol/L.)

ENERGY OF ACTIVATION AND MECHANISM OF REACTIONS

20.11. The decomposition of N_2O into N_2 and O in the presence of gaseous argon follows second-order kinetics, with

$$k = (5.0 \times 10^{11}\,\text{L} \cdot \text{mol}^{-1} \cdot \text{s}^{-1})e^{-29\,000\,\text{K}/T}$$

What is the energy of activation of this reaction?

Comparing the equation for k in this case with (20-4), we note that the exponent of e is $-E_a/RT$.

$$\frac{E_a}{RT} = \frac{29\,000\,\text{K}}{T}$$

$$E_a = (29\,000\,\text{K})R = (29\,000\,\text{K})(8.314\,\text{J} \cdot \text{K}^{-1} \cdot \text{mol}^{-1}) = 241\,\text{kJ} \cdot \text{mol}^{-1}$$

20.12. The first-order rate constant for the hydrolysis of CH_3Cl in H_2O has a value of 3.32×10^{-10} s^{-1} at 25 °C and 3.13×10^{-9} s^{-1} at 40 °C. What is the value of the energy of activation?

Equation (20-5) may be solved for E_a.

$$E_a = 2.303R\left(\frac{T_1 T_2}{T_2 - T_1}\right)\left(\log \frac{k_2}{k_1}\right)$$

$$= (2.303)(8.314 \text{ J} \cdot \text{K}^{-1} \cdot \text{mol}^{-1})\left(\frac{298 \times 313}{313 - 298} \text{ K}\right)\left(\log \frac{3.13 \times 10^{-9}}{3.32 \times 10^{-10}}\right)$$

$$= (119 \text{ kJ} \cdot \text{mol}^{-1})(\log 9.4) = 116 \text{ kJ} \cdot \text{mol}^{-1}$$

20.13. A second-order reaction whose rate constant at 800 °C was found to be 5.0×10^{-3} L·mol^{-1}·s^{-1} has an activation energy of 45 kJ·mol^{-1}. What is the value of the rate constant at 875 °C?

Equation (20-5) may be solved for k_2, the rate constant at the higher temperature.

$$\log \frac{k_2}{k_1} = \frac{E_a(T_2 - T_1)}{2.303RT_1 T_2}$$

$$= \frac{(4.5 \times 10^4 \text{ J} \cdot \text{mol}^{-1})[(875 - 800) \text{ K}]}{(2.303)(8.314 \text{ J} \cdot \text{K}^{-1} \cdot \text{mol}^{-1})(1073 \text{ K})(1148 \text{ K})} = 0.1431$$

$$\frac{k_2}{k_1} = 1.39$$

$$k_2 = (1.39)(5.0 \times 10^{-3} \text{ L} \cdot \text{mol}^{-1} \cdot \text{s}^{-1}) = 7.0 \times 10^{-3} \text{ L} \cdot \text{mol}^{-1} \cdot \text{s}^{-1}$$

20.14. The *trans* → *cis* isomerization of 1,2-dichloroethylene proceeds with an energy of activation of 55.3 kcal·mol^{-1}. ΔH associated with the reaction is 1.0 kcal. What do you predict is the value of E_a for the reverse *cis* → *trans* isomerization?

From (20-6),

$$E_a(\text{reverse}) = E_a(\text{forward}) - \frac{\Delta H}{1 \text{ mol}} = 55.3 - 1.0 = 54.3 \text{ kcal} \cdot \text{mol}^{-1}$$

20.15. A gaseous molecule A can undergo a unimolecular decomposition into C if it acquires a critical amount of energy. Such a suitably energized molecule of A, designated as A*, can be formed by a collision between two ordinary A molecules. Competing with the unimolecular decomposition of A* into C is the bimolecular deactivation of A* by collision with an ordinary A molecule. (a) Write a balanced equation and the rate law for each of the above steps.   (b) Making the assumption that A* disappears by all processes at the same rate at which it is formed, what would be the rate law for the formation of C, in terms of [A] and constants of individual steps only? (c) What limiting order in A would the formation of C exhibit at low pressures of A, and what limiting order at high pressures?

(a)   (1)   Activation: $A + A \rightarrow A^* + A$ $\dfrac{\Delta[A^*]}{\Delta t} = k_1[A]^2$

                  (2)   Deactivation: $A^* + A \rightarrow A + A$ $-\dfrac{\Delta[A^*]}{\Delta t} = k_2[A^*][A]$

                  (3)   Reaction: $A^* \rightarrow C$ $-\dfrac{\Delta[A^*]}{\Delta t} = k_3[A^*]$

(b) Note that A* appears in each of the three separate steps. The net change in [A*] can be evaluated by summing over all three steps.

$$\left(\frac{\Delta[A^*]}{\Delta t}\right)_{net} = k_1[A]^2 - k_2[A^*][A] - k_3[A^*]$$

From the assumption that the net rate of change of [A*] is zero, the right-hand side can be equated to zero.

$$k_1[A]^2 - k_2[A^*][A] - k_3[A^*] = 0 \qquad \text{or} \qquad [A^*] = \frac{k_1[A]^2}{k_3 + k_2[A]}$$

Inserting this value into the rate law for step (3), and recognizing that $-\Delta[A^*] = \Delta[C]$ for this step,

$$(4) \qquad \frac{\Delta[C]}{\Delta t} = \frac{k_3 k_1 [A]^2}{k_3 + k_2[A]}$$

Thus, the formation of C follows a complex kinetic formulation, not represented by a simple order.

(c) At very low pressures (i.e., small [A]), the second term in the denominator on the right side of (4) becomes negligible in comparison with the first.

$$\text{Low-pressure limit:} \qquad \frac{\Delta[C]}{\Delta t} = \frac{k_3 k_1 [A]^2}{k_3} = k_1[A]^2$$

The order appears to be second, with k_1 the second-order rate constant.

At very high pressures, on the other hand, the first term of the denominator becomes negligible in comparison with the second.

$$\text{High-pressure limit:} \qquad \frac{\Delta[C]}{\Delta t} = \frac{k_3 k_1 [A]^2}{k_2[A]} = \frac{k_3 k_1}{k_2}[A]$$

Now the order appears to be first, with $k_3 k_1/k_2$ as the apparent rate constant.

This problem illustrates the concept of the rate-limiting step. In the familiar example of a bucket brigade, in which a water bucket passes sequentially from one person to his neighbor until it finally reaches the reservoir that is to be filled, the overall rate of transfer of water cannot be greater than the rate of the slowest step. Here the activation step (1) becomes the slowest step when [A] is low and the reaction step (3) becomes the slowest step when [A] is high. Step (1), depending on the square of [A], is more sensitive to pressure than step (3).

20.16. For the hydrolysis of methyl formate, $HCOOCH_3$, in acid solutions,

$$\text{Rate} = k[HCOOCH_3][H^+]$$

The balanced equation is $HCOOCH_3 + H_2O \rightarrow HCOOH + CH_3OH$. Why does $[H^+]$ appear in the rate law, when it does not appear in the balanced equation?

H^+ is a catalyst for the reaction. It is actually a reactant in an early intermediate stage of the reaction and is then released back into the solution at a later stage.

20.17. The conversion of the D-optical isomer of gaseous

$$\begin{array}{c} \text{I} \\ | \\ C_2H_5-CH \\ | \\ CH_3 \end{array}$$

into the L-isomer in the presence of iodine vapor follows the law rate $= kP(A)P(I_2)^{1/2}$, where A represents the D-isomer. (Partial pressures are a legitimate form for expression of concentrations in rate laws.) Suggest a mechanism that could account for the fractional order.

I_2 can undergo a slight dissociation into I atoms. This system easily comes to equilibrium.

$$I_2 \rightleftharpoons 2I \qquad K_p = \frac{P(I)^2}{P(I_2)}$$

The partial pressure of iodine atoms can be evaluated in terms of this equilibrium as

$$P(I) = K_p^{1/2} \times P(I_2)^{1/2}$$

If an intermediate stage of the reaction involves the addition of an iodine atom to A, followed by a subsequent loss of the iodine atom that was initially a part of the A molecule, and if the addition of I to A is bimolecular (with a rate constant k_2) and rate-determining, then:

Observed rate = rate of I addition

$$= k_2 P(A)P(I) = k_2 P(A) \times K_p^{1/2} \times P(I_2)^{1/2} = (k_2 K_p^{1/2})P(A)P(I_2)^{1/2}$$

The numerical value of the term in parentheses, $(k_2 K_p^{1/2})$, is the numerical value of the apparent rate constant for the overall $\frac{3}{2}$-order reaction.

The above mechanism is plausible and is consistent with the observations. We cannot be sure from the kinetic data alone, however, that there is not some other mechanism which is also consistent with the observations. Other types of experiment are needed to confirm a mechanism based on rate data. However, a proposed mechanism that yields a rate law other than the observed rate law can surely be rejected.

Supplementary Problems

RATE CONSTANTS AND THE ORDER OF REACTION

20.18. For the reaction

$$3\,BrO^- \rightarrow BrO_3^- + 2\,Br^-$$

in alkaline aqueous solution, the value of the second-order (in BrO^-) rate constant at 80 °C in the rate law for $-\Delta[BrO^-]/\Delta t$ was found to be 0.056 $L \cdot mol^{-1} \cdot s^{-1}$. What is the rate constant when the rate law is written for   (a) $\Delta[BrO_3^-]/\Delta t$,   (b) $\Delta[Br^-]/\Delta t$?

Ans.   (a) 0.018 7 $L \cdot mol^{-1} \cdot s^{-1}$;   (b) 0.037 $L \cdot mol^{-1} \cdot s^{-1}$

20.19. The hydrolysis of methyl acetate in alkaline solution,

$$CH_3COOCH_3 + OH^- \rightarrow CH_3COO^- + CH_3OH$$

followed rate = $k[CH_3OOCH_3][OH^-]$, with k equal to 0.137 $L \cdot mol^{-1} \cdot s^{-1}$ at 25 °C. A reaction mixture was prepared to have initial concentrations of methyl acetate and OH^- of 0.050 M each. How long would it take for 5.0 percent of the methyl acetate to be hydrolyzed at 25 °C?

Ans.   7.3 s

20.20. A first-order reaction in aqueous solution was too fast to be detected by a procedure that could have followed a reaction having a half-life of at least 2.0 ns. What is the minimum value of k for this reaction?

Ans. $3.5 \times 10^8 \text{ s}^{-1}$

20.21. Gaseous cyclobutene isomerizes to butadiene in a first-order process which has a k-value at 153 °C of $3.3 \times 10^{-4} \text{ s}^{-1}$. How many minutes would it take for the isomerization to proceed 40 percent to completion at this temperature?

Ans.   26 min

20.22. The approach to the following equilibrium was observed kinetically from both directions:

$$PtCl_4^{2-} + H_2O \rightleftharpoons Pt(H_2O)Cl_3^- + Cl^-$$

At 25 °C, it was found that at 0.3 ionic strength:

$$-\frac{\Delta[PtCl_4^{2-}]}{\Delta t} = (3.9 \times 10^{-5}\ s^{-1})[PtCl_4^{2-}] - (2.1 \times 10^{-3}\ L \cdot mol^{-1} \cdot s^{-1})[Pt(H_2O)Cl_3^-][Cl^-]$$

What is the value of K_4' for the complexation of the fourth Cl^- by Pt(II) (the apparent equilibrium constant for the reverse of the reaction as written above) at 0.3 ionic strength?

Ans.   54

20.23. The following reaction was studied at 25 °C in benzene solution containing 0.1 M pyridine:

$$CH_3OH + (C_6H_5)_3CCl \rightarrow CH_3OC(C_6H_5)_3 + HCl$$
$$\quad A \qquad\qquad B \qquad\qquad\qquad C$$

The following sets of data were obtained in three separate experiments.

	\multicolumn{3}{c}{Initial Concentrations}		
	$[A]_0/M$	$[B]_0/M$	$[C]_0/M$	$\Delta t/min$	Final $[C]/M$
(1)	0.1000	0.0500	0.0000	25	0.0033
(2)	0.1000	0.1000	0.0000	15.0	0.0039
(3)	0.2000	0.1000	0.0000	7.5	0.0077

What rate law is consistent with the above data and what is the best average value for the rate constant, expressed in seconds and molar concentration units?

Ans.   rate = $k[A^2][B]$, $k = 4.6 \times 10^{-3}\ L^2 \cdot mol^{-2} \cdot s^{-1}$

20.24. Bicyclohexane was found to undergo two parallel first-order rearrangements. At 730 K, the first-order rate constant for the formation of cyclohexene was measured as $1.26 \times 10^{-4}\ s^{-1}$, and for the formation of methylcyclopentene the rate constant was $3.8 \times 10^{-5}\ s^{-1}$. What was the percentage distribution of the rearrangement products?

Ans.   77% cyclohexene, 23% methylcyclopentene

ENERGY OF ACTIVATION AND MECHANISM OF REACTIONS

20.25. The rate constant for the first-order decomposition of ethylene oxide into CH_4 and CO may be described by the equation

$$\log k(\text{in } s^{-1}) = 14.34 - \frac{1.25 \times 10^4\ K}{T}$$

(a) What is the energy of activation for this reaction?   (b) What is the value of k at 670 K?

Ans.   (a) 239 kJ·mol^{-1};   (b) $4.8 \times 10^{-5}\ s^{-1}$

20.26. The first-order gaseous decomposition of N_2O_4 into NO_2 has a k-value of $4.5 \times 10^3\ s^{-1}$ at 1 °C and an energy of activation of 58 kJ·mol^{-1}. At what temperature would k be $1.00 \times 10^4\ s^{-1}$?

Ans.   10 °C

20.27. Biochemists often define Q_{10} for a reaction as the ratio of the rate constant at 37 °C to the rate constant at 27 °C. What must be the energy of activation for a reaction that has a Q_{10} of 2.5?

Ans.   71 kJ·mol^{-1}

20.28. In gaseous reactions important for the understanding of the upper atmosphere, H_2O and O react bimolecularly to form two OH radicals. ΔH for this reaction is 72 kJ at 500 K and E_a is 77 kJ·mol^{-1}. Estimate E_a for the bimolecular recombination of two OH radicals to form H_2O and O.

Ans.   5 kJ·mol^{-1}

20.29. H_2 and I_2 react bimolecularly in the gas phase to form HI, and HI in turn decomposes bimolecularly into H_2 and I_2. The energies of activation for these two reactions were observed to be 163 kJ·mol^{-1} and 184 kJ·mol^{-1}, respectively, over the same temperature ranges near 100 °C. What do you predict from these data for ΔH of the gaseous reaction $H_2 + I_2 \rightleftharpoons 2HI$ at 100 °C?

Ans.   −21 kJ

20.30. Predict the form of the rate law for the reaction $2A + B \rightarrow$ products, if the first step is the reversible dimerization of A, followed by reaction of A_2 with B in a bimolecular step. Assume that the equilibrium concentration of A_2 is very small compared with [A].

Ans.   rate $= k[A]^2[B]$

20.31. The following reaction was observed in aqueous solution:

$$2Cu^{2+} + 6CN^- \rightarrow 2Cu(CN)_2^- + (CN)_2$$

and the rate was found to be of the form $k[Cu^{2+}]^2[CN^-]^6$. If the first step is the rapid development of the complexation equilibrium to form the relatively unstable (with respect to reversal of the complexation step) $Cu(CN)_3^-$, what rate-limiting step could account for the observed kinetic data?

Ans.   bimolecular decomposition: $2Cu(CN)_3^- \rightarrow 2Cu(CN)_2^- + (CN)_2$

20.32. The hydrolysis of $(i\text{-}C_3H_7O)_2POF$ was studied at different acidities. The apparent first-order rate constant, k, at a particular temperature was found to depend on pH but not on the nature or concentration of the buffer used to regulate the pH. k was fairly constant from pH 4 to pH 7, but rose from this constant value with decreasing pH below 4 or with increasing pH above 7. What is the nature of the phenomenon responsible for this behavior?

Ans.   The reaction is catalyzed by both H^+ and OH^-.

20.33. It has been found that the rates of reaction of a ketone in mildly basic solution are identical for the following three reactions: (*a*) reaction with Br_2 leading to the substitution of an H on the ketone by a Br,   (*b*) conversion of the D-isomer of the ketone into an equimolar mixture of the D- and L-isomers,   (*c*) isotopic exchange of a hydrogen atom on the carbon next to the C=O group of the ketone by a deuterium atom in the solvent. The rate of each of these reactions is equal to $k[\text{ketone}][OH^-]$ and is independent of $[Br_2]$. What can be concluded about the mechanism from these observations?

Ans.   The rate-determining step for all three reactions must be the preliminary reaction of the ketone with OH^-, probably leading to the conjugate base of the ketone. The conjugate base subsequently reacts very rapidly with   (*a*) Br_2,   (*b*) some acid in the medium, or   (*c*) the deuterated solvent.

Chapter 21

Nuclear Processes

In ordinary chemical reactions, the atoms of the reactant molecules regroup themselves to form the product molecules. In such reactions, the outer electrons of the atoms undergo rearrangements in being transferred wholly or in part from one atom to another. The atomic nuclei, on the other hand, change their positions with respect to each other but are themselves unchanged.

There are reactions in which the nuclei themselves are broken down, and in which the product materials do not contain the same elements as the reactants. The chemistry of such nuclear reactions is called nuclear chemistry. Spontaneous disintegration of individual nuclei is usually accompanied by the emission of highly penetrating radiation and is known as radioactivity. Other nuclear changes result from the interaction of a neutron or high-energy photon with a nucleus, or the impact of some high-velocity charged particle upon a target nucleus. The results of such nuclear bombardment are highly dependent on the energy.

FUNDAMENTAL PARTICLES

For the purposes of this chapter, the discussion of the fundamental particles occurring in nuclear reactions will be limited to the *nucleons* (the *proton* and *neutron*), the (negative) *electron*, and the *positive electron* (or *positron*). In Table 21-1 the masses of these particles are given in atomic mass units (Chapter 2) and their charges are expressed as multiples of the elementary charge $e = 1.602 \times 10^{-19}$ C.

Table 21-1

Particle	Symbol	Mass/u	Charge/e
Proton	p	1.007 276 5	+1
Neutron	n	1.008 664 9	0
Electron	e^-, β^-	0.000 548 6	−1
Positron	e^+, β^+	0.000 548 6	+1

BINDING ENERGIES

The mass of an atom, in general, is not equal to the sum of the masses of its component protons, neutrons, and electrons. If we could imagine a reaction in which free protons, neutrons, and electrons combine to form an atom, we would find that for all nuclides except ^{1}H the mass of the atom is slightly less than the mass of the component parts and also that a tremendous amount of energy is released when the reaction occurs. The loss in mass is exactly equivalent to the released energy, according to the Einstein equation,

$$E = mc^2$$

Energy = (change in mass) × (velocity of light)^2

This energy-equivalent of the calculated loss of mass is called the *binding energy* of the atom or nucleus. When m is expressed in kilograms and c in meters per second, E is in joules. A more convenient unit of energy for nuclear reactions is the MeV (see Chapter 8). The Einstein equation gives

$$\text{Energy} = \frac{(\text{change in mass in u})(1.661 \times 10^{-27} \text{ kg/u})(2.998 \times 10^8 \text{ m/s})^2}{1.602 \times 10^{-13} \text{ J/MeV}}$$

$$= (932 \text{ MeV}) \times (\text{change in mass in u})$$

NUCLEAR EQUATIONS

The rules for balancing nuclear equations are different from the rules for balancing ordinary chemical equations.

1. Each particle is assigned a superscript equal to its mass number or number of nucleons, A, and a subscript equal to its atomic number or nuclear charge, Z.
2. A free proton is the nucleus of the hydrogen atom, and is therefore assigned the notation ^1_1H.
3. A free neutron is assigned zero atomic number because it has no charge. The mass number of a neutron is 1. The full notation for a neutron is ^1_0n.
4. An electron (e^- or β^-) is assigned the mass number zero and the atomic number -1; hence the full notation $^{\,0}_{-1}e$.
5. A positron (e^+ or β^+) is assigned the mass number zero and the atomic number $+1$; hence the full notation $^{\,0}_{+1}e$.
6. An *alpha particle* (α-particle) is a helium nucleus, and is therefore represented by the full notation ^4_2He (or $^4_2\alpha$).
7. A *gamma ray* (γ) is a photon (Chapter 8); it has zero mass number and zero charge, and is notated $^0_0\gamma$.
8. In a balanced equation the sum of the subscripts (atomic numbers), written or implied, must be the same on the two sides of the equation. The sum of the superscripts (mass numbers), written or implied, must also be the same on the two sides of the equation. Thus the equation for the first step in the radioactive decay of ^{226}Ra is

$$^{226}_{88}\text{Ra} \rightarrow\, ^{222}_{86}\text{Rn} + {}^4_2\text{He}$$

Many nuclear processes may be indicated by a shorthand notation, in which a light bombarding particle and a light product particle are represented by symbols in parentheses between the symbols for the initial target nucleus and the final product nucleus. The symbols n, p, d, α, β^-, β^+, γ are used to represent neutron, proton, deuteron (^2_1H), alpha, electron, positron, and gamma rays, respectively. The atomic numbers are commonly omitted because the symbol for any element implies its atomic number. Examples of the corresponding long- and shorthand notations for several reactions follow.

$$^{14}_{7}\text{N} + {}^1_1\text{H} \;\rightarrow\, ^{11}_{6}\text{C} + {}^4_2\text{He} \qquad ^{14}\text{N}(p, \alpha)^{11}\text{C}$$

$$^{27}_{13}\text{Al} + {}^1_0n \;\rightarrow\, ^{27}_{12}\text{Mg} + {}^1_1\text{H} \qquad ^{27}\text{Al}(n, p)^{27}\text{Mg}$$

$$^{55}_{25}\text{Mn} + {}^2_1\text{H} \rightarrow\, ^{55}_{26}\text{Fe} + 2{}^1_0n \qquad ^{55}\text{Mn}(d, 2n)^{55}\text{Fe}$$

Just as an ordinary *chemical* equation is a shortened version of the complete thermochemical equation which expresses both energy and mass balance, so each nuclear equation has associated with it a term, either written or implied, expressing energy balance. The symbol Q is usually used to designate the net energy *released* when all reactant and product particles of matter are at zero velocity. Q is the energy equivalent of the mass decrease accompanying the reaction, and is usually expressed in MeV.

RADIOCHEMISTRY

Special properties of radioactive nuclides make them useful tracers for following complex processes. *Radiochemistry* is that branch of chemistry which involves the applications of radioactivity to chemical problems, as well as the chemical processing of radioactive substances.

A radioactive nuclide is spontaneously converted to another nuclide by one of the following processes, in which there is an overall decrease in mass.

1. *Alpha decay:*   An α-particle is emitted and the daughter nucleus has an atomic number, Z, two units less, and a mass number, A, four units less than the parent's. Thus

$$^{226}_{88}\text{Ra} \rightarrow ^{222}_{86}\text{Rn} + ^{4}_{2}\alpha$$

2. *β^- decay:*   An electron is emitted and the daughter has a Z-value one unit greater than the parent's, with no change in A. Thus

$$^{31}_{14}\text{Si} \rightarrow ^{31}_{15}\text{P} + ^{0}_{-1}\beta$$

3. *β^+-decay:*   A positron is emitted and the daughter has a Z-value one unit less than the parent's, with no change in A. Thus

$$^{40}_{21}\text{Sc} \rightarrow ^{40}_{20}\text{Ca} + ^{0}_{+1}\beta$$

The emitted positron is itself unstable in the neighborhood of electrons and is normally consumed, after being slowed down by collisions, in the following annihilation reaction:

$$^{0}_{+1}\beta + ^{0}_{-1}\beta \rightarrow ^{0}_{0}\gamma + ^{0}_{0}\gamma$$

4. *K-electron capture:*   By capturing an orbital electron within its own atom, a nucleus can reduce its Z-value by one unit without a change in A. Thus

$$^{7}_{4}\text{Be} + ^{0}_{-1}e(K\text{-capture}) \rightarrow ^{7}_{3}\text{Li}$$

A radioactive nucleus decays by a first-order process, so that (*20-1*), (*20-2*), and (*20-3*) apply. The stability of a nucleus with respect to spontaneous decay may be indicated by its first-order rate constant, k, or by the half-life, $t_{1/2}$.

Radioactivity is measured by observing the high-energy particles produced directly or indirectly as a result of the disintegration process. A convenient unit of radioactivity is the *curie*, defined by

$$1 \text{ Ci} = 3.700\,0 \times 10^{10} \text{ disintegrations per second}$$

The activity of a sample, expressed in curies, depends both on the number of atoms of the radioactive nuclide (i.e., the mass of the sample) and on the half-life (or disintegration rate constant). See Problem 21.14. The subunits *millicurie* (mCi), *microcurie* (μCi), etc., are also used.

Solved Problems

21.1.   How many protons, neutrons, and electrons are there in each of the following atoms: (*a*) ^{3}He, (*b*) ^{12}C,   (*c*) ^{206}Pb?

   (*a*)   From the atomic weight table (last page of this book), we see that the atomic number of He is 2; therefore the nucleus must contain 2 protons. Since the mass number of this isotope is 3, the sum of the numbers of protons and neutrons is 3; therefore there is 1 neutron. The number of electrons in the atom is the same as the atomic number, 2.

   (*b*)   The atomic number of carbon is 6; hence the nucleus must contain 6 protons. The number of neutrons is $12 - 6 = 6$. The number of electrons is the same as the atomic number, 6.

(c) The atomic number of lead is 82; hence there are 82 protons in the nucleus. The number of neutrons is $206 - 82 = 124$. There are 82 electrons.

21.2. Complete the following nuclear equations:

(a) $^{14}_{7}\text{N} + ^{4}_{2}\text{He} \rightarrow ^{17}_{8}\text{O} + \cdots$ (d) $^{30}_{15}\text{P} \rightarrow ^{30}_{14}\text{S} + \cdots$

(b) $^{9}_{4}\text{Be} + ^{4}_{2}\text{He} \rightarrow ^{12}_{6}\text{C} + \cdots$ (e) $^{3}_{1}\text{H} \rightarrow ^{3}_{2}\text{He} + \cdots$

(c) $^{9}_{4}\text{Be}(p, \alpha) \ldots$ (f) $^{43}_{20}\text{Ca}(\alpha, \ldots)^{46}_{21}\text{Sc}$

(a) The sum of the subscripts on the left is $7 + 2 = 9$. The subscript of the first product on the right is 8. Hence the second product on the right must have a subscript (nuclear charge) of 1.
      The sum of the superscripts on the left is $14 + 4 = 18$. The superscript of the first product on the right is 17. Hence the second product on the right must have a superscript (mass number) of 1.
      The particle with nuclear charge 1 and mass number 1 is the proton, $^{1}_{1}\text{H}$.

(b) The nuclear charge of the second product particle (its subscript) is $(4 + 2) - 6 = 0$. The mass number of the particle (its superscript) is $(9 + 4) - 12 = 1$. Hence the particle must be the neutron, $^{1}_{0}n$.

(c) The reactants, $^{9}_{4}\text{Be}$ and $^{1}_{1}\text{H}$, have a combined nuclear charge of 5 and mass number of 10. In addition to the α-particle, a product will be formed of charge $5 - 2 = 3$, and mass $10 - 4 = 6$. This is $^{6}_{3}\text{Li}$, since lithium is the element of atomic number 3.

(d) The nuclear charge of the second particle is $15 - 14 = +1$. The mass number is $30 - 30 = 0$. Hence the particle must be the positron, $_{+1}^{0}e$.

(e) The nuclear charge of the second particle is $1 - 2 = -1$. Its mass number is $3 - 3 = 0$. Hence the particle must be an electron, $_{-1}^{0}e$.

(f) The reactants, $^{43}_{20}\text{Ca}$ and $^{4}_{2}\text{He}$, have a combined nuclear charge of 22 and mass number of 47. The ejected product will have a charge $22 - 21 = 1$, and mass $47 - 46 = 1$. This is a proton and should be represented within the parentheses by p.

21.3. What is the total binding energy of ^{12}C and what is the average binding energy per nucleon?

Although "binding energy" is a term referring to the nucleus, it is more convenient to use the mass of the whole atom (nuclide) in calculations, since these are the masses that are given in tables. If $M(X)$ is the atomic mass of nuclide X,

$$M(\text{nucleus}) = M(X) - ZM(e^-) \tag{1}$$

The nucleus of X consists of Z protons and $A - Z$ neutrons. Hence its binding energy is given by

$$\text{BE} = \{ZM(p) + (A - Z)M(n)\} - M(\text{nucleus}) \tag{2}$$

Applying (1) both to the nucleus of X and to the proton, which is a $^{1}_{1}\text{H}$ nucleus, and substituting in (2),

$$\text{BE} = \{Z[M(^{1}_{1}\text{H}) - M(e^-)] + (A - Z)M(n)\} - [M(X) - ZM(e^-)]$$
$$= \{ZM(^{1}_{1}\text{H}) + (A - Z)M(n)\} - M(X)$$

In other words, nuclear masses can be replaced by atomic (nuclidic) masses in calculating the binding energy. Whole atom masses can, in fact, be used for mass-difference calculations in all nuclear reaction types discussed in this chapter, except for β^+ processes where there is a resulting annihilation of two electron masses (one β^+ and one β^-).

The data needed for ^{12}C can be obtained from Tables 2-1 and 21-1.

Mass of 6 ^{1}H atoms = 6×1.00783	= 6.04698
Mass of 6 neutrons = 6×1.00867	= $\underline{6.05202}$
Total mass of component particles	= 12.09900
Mass of ^{12}C	= $\underline{12.00000}$
Loss in mass on formation of ^{12}C	= 0.09900u
Binding energy = $(932 \text{ MeV})(0.0990)$	= 92.3 MeV

Since there are 12 nucleons (protons and neutrons), the average binding energy per nucleon is (92.3 MeV)/12, or 7.69 MeV.

21.4. Evaluate Q for the $^7\text{Li}(p, n)^7\text{Be}$ reaction.

The change in mass for the reaction must be computed.

Reactants		Products	
^7_3Li	7.016 00	$^1_0 n$	1.008 66
^1_1H	1.007 83	^7_4Be	7.016 93
	8.023 83		8.025 59

Increase of mass = 8.025 59 − 8.023 83 = 0.001 76u

A corresponding net amount of energy must be consumed, equal to (932)(0.001 76) MeV, or 1.64 MeV; thus $Q = -1.64$ MeV. This energy is supplied as kinetic energy of the bombarding proton and is *part* of the acceleration requirement for the proton supplied by the particle accelerator.

21.5. The Q-value for the $^3\text{He}\,(n, p)$ reaction is 0.76 MeV. What is the nuclidic mass of ^3He?

The reaction is

$$^3_2\text{He} + {}^1_0 n \to {}^1_1\text{H} + {}^3_1\text{H}$$

The mass loss must be 0.76/932 = 0.000 82 u.

The mass balance can be calculated on the basis of whole atoms.

Reactants		Products	
^3He	x	^1H	1.007 83
n	1.008 66	^3H	3.016 05
	$x + 1.008\,66$		4.023 88

Then $(x + 1.008\,66) - 4.023\,88 = 0.000\,82$ or $x = 3.016\,04$ u.

21.6. Calculate the maximum kinetic energy of the β^- emitted in the radioactive decay of ^6He. Assume that the β^- has its maximum energy when no other emission accompanies the process.

The process referred to is

$$^6_2\text{He} \to {}^6_3\text{Li} + {}^0_{-1}\beta$$

In computing the mass change during this process, only the whole atomic masses of ^6He and ^6Li need be considered.

$$\text{Mass of } {}^6\text{He} = 6.018\ 89$$

$$\text{Mass of } {}^6\text{Li} \ = \underline{6.015\ 12}$$

$$\text{Loss in mass} = 0.003\ 77 \text{ u}$$

$$\text{Energy equivalent} = (932)(0.003\ 77) \text{ MeV} = 3.51 \text{ MeV}$$

The maximum kinetic energy of the β^- particle is 3.51 MeV.

21.7. ^{13}N decays by β^+ emission. The maximum kinetic energy of the β^+ is 1.20 MeV. What is the nuclidic mass of ^{13}N?

The reaction is

$$^{13}_{7}\text{N} \to {}^{13}_{6}\text{C} + {}^0_{+1}\beta$$

This is the type of process, mentioned in Problem 21.3, in which a simple difference of whole atom masses is not the desired quantity.

$$\text{Mass difference} = [M(\text{nucleus}) \text{ for } {}^{13}\text{N}] - [M(\text{nucleus}) \text{ for } {}^{13}\text{C}] - M(e)$$
$$= [M({}^{13}\text{N}) - 7M(e)] - [M({}^{13}\text{C}) - 6M(e)] - M(e)$$
$$= M({}^{13}\text{N}) - M({}^{13}\text{C}) - 2M(e) = M({}^{13}\text{N}) - 13.003\,35 - 2(0.000\,55)$$
$$= M({}^{13}\text{N}) - 13.004\,45$$

This expression must equal the mass equivalent of the maximum kinetic energy of the β^+.

$$\frac{1.20 \text{ MeV}}{932 \text{ MeV/u}} = 0.001\,29 \text{ u}$$

Then

$$0.001\,29 = M({}^{13}\text{N}) - 13.004\,45 \qquad \text{or} \qquad M({}^{13}\text{N}) = 13.005\,74 \text{ u}$$

21.8. Consider the two nuclides of mass number 7, ${}^7\text{Li}$ and ${}^7\text{Be}$. Which of the two is stabler? How does the unstable nuclide decay into the stable one?

Table 2-1 shows that ${}^7\text{Be}$ has a larger mass than ${}^7\text{Li}$. Thus ${}^7_4\text{Be}$ can decay spontaneously into ${}^7_3\text{Li}$, but not vice versa. There are two types of decay process in which Z is decreased by one unit without a change in mass number A: β^+ emission and K-capture. These two processes have different mass balance requirements.

Assume that the process is β^+ emission.

$$ {}^7_4\text{Be} \rightarrow {}^{\,0}_{+1}\beta + {}^7_3\text{Li} $$

It was shown in Problem 21.7 (third line of mass difference equation) that a positron emission occurs (i.e., Q is positive and the reaction is spontaneous) only if the *nuclidic* mass of the parent species exceeds the *nuclidic* mass of the daughter by at least twice the rest mass of the electron, $2(0.000\,55) = 0.001\,10$ u. In the present case the actual mass difference between parent and daughter nuclides is $7.016\,93 - 7.016\,00 = 0.000\,93$ u. We thus see that positron emission in this case is impossible. By elimination, we conclude that ${}^7\text{Be}$ undergoes K-capture.

Note that we have predicted only that ${}^7\text{Be}$ *should* decay by K-capture into ${}^7\text{Li}$. We have said nothing about the rate of such a process. *Measurements* show the half-life of the process to be 53 days.

21.9. An isotopic species of lithium hydride, ${}^6\text{Li}{}^2\text{H}$, is a potential nuclear fuel, on the basis of the following reaction:

$$ {}^6_3\text{Li} + {}^2_1\text{H} \rightarrow 2{}^4_2\text{He} $$

Calculate the expected power production, in megawatts, associated with the consumption of 1.00 g of ${}^6\text{Li}{}^2\text{H}$ per day. Assume 100 percent efficiency in the process.

The change in mass for the reaction is first computed.

$$\text{Mass of } {}^6_3\text{Li} = 6.015\,12$$
$$\text{Mass of } {}^2_1\text{H} = \underline{2.014\,10}$$
$$\text{Total mass of reactants} = 8.029\,22$$
$$\text{Mass of products} = 2(4.002\,60) = \underline{8.005\,20}$$
$$\text{Loss in mass} = 0.024\,02 \text{ u}$$

$$\text{Energy per atomic event} = (0.024\,02 \text{ u})(932 \times 10^6 \text{ eV/u})(1.602 \times 10^{-19} \text{ J/eV})$$
$$= 3.59 \times 10^{-12} \text{ J}$$

$$\text{Energy per mol LiH} = (3.59 \times 10^{-12} \text{ J})(6.02 \times 10^{23} \text{ mol}^{-1}) = 2.16 \times 10^{12} \text{ J/mol}$$

$$\text{Power production per g LiH} = \frac{(2.16 \times 10^{12} \text{ J/mol})/(8.03 \text{ g/mol})}{(24 \text{ h})(3.6 \times 10^3 \text{ s/h})}$$

$$= 3.11 \times 10^6 \text{ W/g} = 3.11 \text{ MW/g}$$

21.10. ^{18}F is found to undergo 90 percent radioactive decay in 366 min. What is its computed half-life from this observation?

The rate constant for the decay can be found from (20-2). 90 percent decay corresponds to 10 percent, or 0.10, survival. In dealing with radioactive decay the total population of radioactive element is used in place of its concentration. So in place of the concentration ratio $[A]/[A]_0$, put the ratio of the numbers of atoms N/N_0, or moles, or masses, of radioactive element. The mass of radioactive element encountered in the laboratory is exceedingly small; a typical sample can be measured only by its activity. Since its activity is proportional to its population the observed ratio of activities A/A_0 can be used in place of the number ratio N/N_0.

$$k = -\frac{2.303 \log \dfrac{N}{N_0}}{t} = -\frac{2.303 \log 0.10}{366 \text{ min}} = 6.29 \times 10^{-3} \text{ min}^{-1}$$

Then the half-life can be computed from (20-3).

$$t_{1/2} = \frac{0.693}{k} = \frac{0.693}{6.29 \times 10^{-3} \text{ min}^{-1}} = 110 \text{ min}$$

21.11. Estimate the age of an Egyptian mummy from which a piece of the linen wrapping was analyzed and found to have a ^{14}C activity of 8.1 counts (disintegrations) per minute, per gram of carbon.

The half-life of ^{14}C is 5 730 years. It is generally assumed that over at least the past 30 000 years the ^{14}C content of atmospheric carbon (as CO_2) has been roughly constant. Living plants, which obtained their carbon from the air by photosynthesis, thus have had over this period a constant activity, whose value has been found to be 15.3 counts per minute, per gram of carbon.

After the flax was harvested and made into linen there was no longer any absorption of ^{14}C from the atmosphere and the activity decayed. Use (20-3) to find k, then (20-2) to find the time, t, for the activity to decay to its present level.

$$k = \frac{0.693}{t_{1/2}} = \frac{0.693}{5\,730 \text{ y}} = 1.209 \times 10^{-4} \text{ y}^{-1}$$

$$t = \frac{-2.303}{k} \log \left(\frac{A}{A_0}\right) = \frac{-2.303}{1.209 \times 10^{-4} \text{ y}^{-1}} \log \left(\frac{8.1}{15.3}\right) = \frac{(-2.303)(-0.276)}{1.209 \times 10^{-4} \text{ y}^{-1}} = 5\,260 \text{ y}$$

21.12. A sample of uraninite, a uranium-containing mineral, was found on analysis to contain 0.214 g of lead for every gram of uranium. Assuming that the lead all resulted from the radioactive disintegration of the uranium since the geological formation of the uraninite and that all isotopes of uranium other than ^{238}U can be neglected, estimate the date when the mineral was formed in the earth's crust. The half-life of ^{238}U is 4.5×10^9 years.

The radioactive decay of ^{238}U leads, after 14 steps, to the stable lead isotope, ^{206}Pb. The first of these steps, the α-decay of ^{238}U with a 4.5×10^9 y half-life, is intrinsically more than 10^4 times as slow as any of the subsequent steps. As a result, the time required for the first step accounts for essentially all the time required for the entire 14-step sequence.

In a sample containing 1 g U, there is

$$\frac{0.214 \text{ g Pb}}{206 \text{ g/mol}} = 1.04 \times 10^{-3} \text{ mol Pb}$$

and

$$\frac{1.000 \text{ g U}}{238 \text{ g/mol}} = 4.20 \times 10^{-3} \text{ mol U}$$

If each atom of lead in the mineral today is the daughter of a uranium atom that existed at the time of the formation of the mineral, then the original number of moles of uranium in the sample would have been

$$(1.04 + 4.20) \times 10^{-3} = 5.24 \times 10^{-3}$$

Then the fraction remaining is

$$\frac{N}{N_0} = \frac{4.20 \times 10^{-3}}{5.24 \times 10^{-3}} = 0.802$$

Letting t be the elapsed time from the formation of the mineral in the earth's crust to the present time, we have

$$k = \frac{0.693}{t_{1/2}} = \frac{0.693}{4.5 \times 10^9 \text{ y}} = 1.54 \times 10^{-10} \text{ y}^{-1}$$

$$t = -\frac{2.303 \log \dfrac{N}{N_0}}{k} = -\frac{-2.303 \log 0.802}{1.54 \times 10^{-10} \text{ y}^{-1}} = 1.4 \times 10^9 \text{ y}$$

21.13. A sample of $^{14}CO_2$ was to be mixed with ordinary CO_2 for a biological tracer experiment. In order that 10 cm^3 (S.T.P.) of the diluted gas should have 10^4 disintegrations per minute, how many microcuries of radioactive carbon are needed to prepare 60 L of the diluted gas?

$$\text{Total activity} = \frac{10^4 \text{ dis/min}}{10 \text{ cm}^3} \times \frac{(60 \text{ L})(10^3 \text{ cm}^3/\text{L})}{60 \text{ s/min}}$$

$$= (10^6 \text{ dis/s})\left(\frac{1 \text{ Ci}}{3.7 \times 10^{10} \text{ dis/s}}\right)\left(\frac{10^6 \text{ } \mu\text{Ci}}{1 \text{ Ci}}\right) = 27 \text{ } \mu\text{Ci}$$

To find the *mass* of ^{14}C needed to provide the 27 μCi, the procedure of Problem 21.14 would be followed.

21.14. The half-life of ^{40}K is 1.25×10^9 years. What mass of this nuclide has an activity of 1 μ Ci?

Let us first calculate the rate constant and express it in s^{-1}.

$$k = \frac{0.693}{t_{1/2}} = \frac{0.693}{(1.25 \times 10^9 \text{ y})(365 \text{ d/y})(24 \text{ h/d})(3.6 \times 10^3 \text{ s/h})} = 1.76 \times 10^{-17} \text{ s}^{-1}$$

The disintegration rate is an instantaneous rate measured under conditions of essential constancy of the concentration (i.e., the population) of ^{40}K atoms. The form of the rate equation is that used in Chapter 20, with the numerical value of the rate taken from the definition of the curie.

$$\text{Rate} = -\frac{\Delta N}{\Delta t} = kN = (3.70 \times 10^{10} \text{ dis} \cdot \text{s}^{-1} \cdot \text{Ci}^{-1})(10^{-6} \text{ Ci} \cdot \mu\text{Ci}^{-1})$$

$$= 3.70 \times 10^4 \text{ dis} \cdot \text{s}^{-1} \cdot \mu\text{Ci}^{-1}$$

$$N = \frac{\text{rate}}{k} = \frac{3.70 \times 10^4 \text{ atoms} \cdot \text{s}^{-1} \cdot \mu\text{Ci}^{-1}}{1.76 \times 10^{-17} \text{ s}^{-1}} = 2.10 \times 10^{21} \text{ atoms} \cdot \mu\text{Ci}^{-1}$$

and the corresponding mass is

$$\frac{(2.10 \times 10^{21} \text{ atoms} \cdot \mu\text{Ci}^{-1})(40 \text{ g } ^{40}K/\text{mol})}{6.0 \times 10^{23} \text{ atoms/mol}} = 0.140 \text{ g } ^{40}K/\mu\text{Ci}$$

21.15. ^{227}Ac has a half-life of 21.8 years with respect to radioactive decay. The decay follows two parallel paths, one leading to ^{227}Th and one leading to ^{223}Fr. The percentage yields of these two daughter nuclides are 1.4 percent and 98.6 percent, respectively. What is the rate constant, in y^{-1}, for each of the separate paths?

The rate constant for the decay of Ac, k_{Ac}, can be computed from the half-life.

$$k_{Ac} = \frac{0.693}{t_{1/2}} = \frac{0.693}{21.8 \text{ y}} = 3.18 \times 10^{-2} \text{ y}^{-1}$$

As in Problem 20.9, the overall rate constant for a set of parallel first-order reactions is equal to the sum of the separate rate constants,

$$k_{Ac} = k_{Th} + k_{Fr}$$

and the fractional yield of either process is equal to the ratio of the rate constant for that process to the overall rate constant.

$$k_{Th} = (\text{fractional yield of Th}) \times k_{Ac} = (0.014)(3.18 \times 10^{-2} \, y^{-1}) = 4.5 \times 10^{-4} \, y^{-1}$$

$$k_{Fr} = (\text{fractional yield of Fr}) \times k_{Ac} = (0.986)(3.18 \times 10^{-2} \, y^{-1}) = 3.14 \times 10^{-2} \, y^{-1}$$

Supplementary Problems

21.16. Determine the number of (a) nuclear protons, (b) nuclear neutrons, (c) electrons, in each of the following atoms: (1) ^{70}Ge, (2) ^{72}Ge, (3) ^{9}Be, (4) ^{235}U.

Ans. (1): (a) 32; (b) 38; (c) 32   (3): (a) 4; (b) 5; (c) 4
      (2): (a) 32; (b) 40; (c) 32   (4): (a) 92; (b) 143; (c) 92

21.17. Write the complete nuclear symbols for natural fluorine and natural arsenic. Each has only one stable isotope.

Ans. $^{19}_{9}F$, $^{75}_{33}As$

21.18. By natural radioactivity, ^{238}U emits an α-particle. The heavy residual nucleus is called UX_1. UX_1 in turn emits a β^--particle. The heavy residual nucleus from this radioactive process is called UX_2. Determine the atomic numbers and mass numbers of (a) UX_1 and (b) UX_2.

Ans. (a) 90, 234; (b) 91, 234

21.19. By radioactivity, $^{239}_{93}Np$ emits a β^--particle. The residual heavy nucleus is also radioactive and gives rise to ^{235}U by its radioactive process. What small particle is emitted simultaneously with the formation of ^{235}U?

Ans. α-particle

21.20. Complete the following equations.

(a) $^{23}_{11}Na + ^{4}_{2}He \rightarrow ^{26}_{12}Mg + ?$ (c) $^{106}Ag \rightarrow ^{106}Cd + ?$

(b) $^{64}_{29}Cu \rightarrow ^{0}_{+1}\beta + ?$ (d) $^{10}_{5}B + ^{4}_{2}He \rightarrow ^{13}_{7}N + ?$

Ans. (a) $^{1}_{1}H$; (b) $^{64}_{28}Ni$; (c) $^{0}_{-1}\beta$; (d) $^{1}_{0}n$

21.21. Complete the notations for the following nuclear processes.
(a) $^{24}Mg(d, \alpha)$?   (c) $^{40}Ar(\alpha, p)$?   (e) $^{130}Te(d, 2n)$?   (g) $^{59}Co(n, \alpha)$?
(b) $^{26}Mg(d, p)$?   (d) $^{12}C(d, n)$?   (f) $^{55}Mn(n, \gamma)$?

Ans. (a) ^{22}Na; (b) ^{27}Mg; (c) ^{43}K; (d) ^{13}N; (e) ^{130}I; (f) ^{56}Mn; (g) ^{56}Mn

21.22. If a nuclide of an element in Group I (alkali metals) of the periodic table undergoes radioactive decay by emitting positrons, what is the valence expected for the resulting element?

Ans. zero

21.23. An alkaline earth element is radioactive. It and its daughter elements decay by emitting 3 alpha particles in succession. In what group should the resulting element be found?

Ans. Group IV (C group)

21.24. If an atom of ^{235}U, after absorption of a slow neutron, undergoes fission to form an atom of ^{139}Xe and an atom of ^{94}Sr, what other particles are produced, and how many?

Ans.   3 neutrons

21.25. Which is the more unstable of each of the following pairs, and in each case what type of process could the unstable nucleus undergo: (a) ^{16}C, ^{16}N;   (b) ^{18}F, ^{18}Ne?

Ans.   (a) ^{16}C, β^--decay;   (b) ^{18}Ne, both β^+-decay and K-electron capture are possibilities on the basis of data furnished here.

21.26. One of the stablest nuclei is ^{55}Mn. Its nuclidic mass is 54.938 u. Determine its total binding energy and average binding energy per nucleon.

Ans.   483 MeV, 8.78 MeV per nucleon

21.27. How much energy is released during each of the following fusion reactions?

$$(a) \ {}^1_1H + {}^7_3Li \rightarrow 2{}^4_2He \qquad (b) \ {}^3_1H + {}^2_1H \rightarrow {}^4_2He + {}^1_0n$$

Ans.   (a) 17.4 MeV;   (b) 17.6 MeV

21.28. ^{14}C is believed to be made in the upper atmosphere by an (n, p) process on ^{14}N. What is Q for this reaction?

Ans.   0.62 MeV

21.29. In the reaction ^{32}S$(n, \gamma)^{33}$S with slow neutrons, the γ is produced with an energy of 8.65 MeV. What is the nuclidic mass of ^{33}S?

Ans.   32.971 46 u

21.30. If a β^+ and a β^- annihilate each other and their rest masses are converted into two γ-rays of equal energy, what is the energy in MeV of each γ?

Ans.   0.51 MeV

21.31. ΔE for the combustion of a mole of ethylene in oxygen is -1.4×10^3 kJ. What would be the loss in mass (expressed in u) accompanying the oxidation of one molecule of ethylene?

Ans. 1.6×10^{-8} u. (This value is so small compared to the molecular mass that the change in mass, as in all chemical reactions, is ordinarily not taken into account.)

21.32. The sun's energy is believed to come from a series of nuclear reactions, the overall result of which is the transformation of four hydrogen atoms into one helium atom. How much energy is released in the formation of one helium atom? (Include the annihilation energy of the two positrons formed in the nuclear reactions with two electrons.)

Ans.   26.8 MeV

21.33. It is proposed to use the nuclear fusion reaction

$$2{}^2_1H \rightarrow {}^3_1H + {}^1_1H + \text{energy}$$

to produce industrial electric power. If the output is to be 50 MW and the energy of the above reaction is used with 30 percent efficiency, how many grams of deuterium fuel will be needed per day?

Ans.   149 g/day

21.34. A pure radiochemical preparation was observed to disintegrate at the rate of 4 280 counts per minute at 1:35 p.m. At 4:55 p.m. of the same day, the disintegration rate of the sample was only 1 070 counts per minute. The disintegration rate is proportional to the number of radioactive atoms in the sample. What is the half-life of the material?

Ans.   100 min

21.35. An atomic battery for pocket watches has been developed which uses the beta particles from ^{147}Pm as the primary energy source. The half-life of ^{147}Pm is 2.62 years. How long would it take for the rate of beta emission in the battery to be reduced to 10 percent of its initial value.

Ans.   8.7 years

21.36. A set of piston rings weighing 120 g was irradiated with neutrons in a nuclear reactor, converting some of the cobalt present in the steel to ^{60}Co. Irradiation continued until the total ^{60}Co activity was 360 mCi. The rings were inserted in an automobile engine which was operated for 24 h under average conditions, following which 0.27 μCi of ^{60}Co activity was found in the oil filter. Calculate the rate of wear of the piston rings in mg/y based on the assumption that all the eroded metal was captured in the oil filter.

Ans.   33 mg/y

21.37. A charcoal sample taken from a fire pit in an archeologist's excavation of a rock shelter was believed to have been formed when early occupants of the shelter burned wood for cooking. A 100-mg sample of pure carbon from the charcoal was found in 1979 to have a disintegration rate of 0.25 count per minute. In what millennium did the tree grow from which the archeological sample was taken? Use the data from Problem 21.11.

Ans.   15 millennia ago

21.38. All naturally occurring rubidium ores contain ^{87}Sr, resulting from the beta decay of ^{87}Rb. In naturally occurring rubidium, 278 of every 1 000 rubidium atoms are ^{87}Rb. A mineral containing 0.85 % rubidium was analyzed and found to contain 0.008 9 % strontium. Assuming that all of the strontium originated by radioactive decay of ^{87}Rb, estimate the age of the mineral. ^{87}Rb has a half-life of 4.9×10^{10} years.

Ans. 2.6×10^9 years

21.39. Transuranium elements were originally believed not to occur in nature because of their relatively short half-lives. Then ^{244}Pu was reported in a natural ore. The half-life of ^{244}Pu is 8.08×10^7 years. If this element is stabler than any of its radioactive predecessors and thus has not been produced in this ore in significant amounts since the ore was deposited, what fraction of the original ^{244}Pu content would still be present if the ore is assumed to be 5×10^9 years old.

Ans. 10^{-19}

21.40. Prior to the use of nuclear weapons, the specific activity of ^{14}C in soluble ocean carbonates was found to be 16 disintegrations per minute per gram of carbon. The amount of carbon in these carbonates has been estimated as 4.5×10^{16} kg. How many megacuries of ^{14}C did the ocean carbonates contain?

Ans.   320 MCi

21.41. If the limit of a particular detection system is 0.002 disintegration per second for a 1 g sample, what would be the maximum half-life that this system could detect on a 1-g sample of a nuclide whose mass number is around 200?

Ans. 3×10^{16} years

21.42. The activity of 30 μg of ^{247}Cm is 2.8 nCi. Calculate the disintegration rate constant and the half-life of ^{247}Cm.

Ans. 1.42×10^{-15} s^{-1}, 1.55×10^7 years

21.43. How much heat would be developed per hour from a 1-Ci ^{14}C source if all the energy of the β^--decay were imprisoned?

Ans.   3 J/h

Appendix A

Exponents

A. The following is a partial list of powers of 10.

$10^0 = 1$

$10^1 = 10$

$10^2 = 10 \times 10 = 100$

$10^3 = 10 \times 10 \times 10 = 1\,000$

$10^4 = 10 \times 10 \times 10 \times 10 = 10\,000$

$10^5 = 10 \times 10 \times 10 \times 10 \times 10 = 100\,000$

$10^6 = 10 \times 10 \times 10 \times 10 \times 10 \times 10 = 1\,000\,000$

$10^{-1} = \dfrac{1}{10} = 0.1$

$10^{-2} = \dfrac{1}{10^2} = \dfrac{1}{100} = 0.01$

$10^{-3} = \dfrac{1}{10^3} = \dfrac{1}{1000} = 0.001$

$10^{-4} = \dfrac{1}{10^4} = \dfrac{1}{10\,000} = 0.000\,1$

In the expression 10^5, the *base* is 10 and the *exponent* is 5.

B. In multiplication, exponents of like bases are added.

(1) $a^3 \times a^5 = a^{3+5} = a^8$
(2) $10^2 \times 10^3 = 10^{2+3} = 10^5$
(3) $10 \times 10 = 10^{1+1} = 10^2$

(4) $10^7 \times 10^{-3} = 10^{7-3} = 10^4$
(5) $(4 \times 10^4)(2 \times 10^{-6}) = 8 \times 10^{4-6} = 8 \times 10^{-2}$
(6) $(2 \times 10^5)(3 \times 10^{-2}) = 6 \times 10^{5-2} = 6 \times 10^3$

C. In division, exponents of like bases are subtracted.

(1) $\dfrac{a^5}{a^3} = a^{5-3} = a^2$

(2) $\dfrac{10^2}{10^5} = 10^{2-5} = 10^{-3}$

(3) $\dfrac{8 \times 10^2}{2 \times 10^{-6}} = \dfrac{8}{2} \times 10^{2+6} = 4 \times 10^8$

(4) $\dfrac{5.6 \times 10^{-2}}{1.6 \times 10^4} = \dfrac{5.6}{1.6} \times 10^{-2-4} = 3.5 \times 10^{-6}$

D. Any number may be expressed as an integral power of 10, or as the product of two numbers one of which is an integral power of 10 (e.g. $300 = 3 \times 10^2$).

(1) $22\,400 = 2.24 \times 10^4$
(2) $7\,200\,000 = 7.2 \times 10^6$
(3) $454 = 4.54 \times 10^2$
(4) $0.454 = 4.54 \times 10^{-1}$

(5) $0.045\,4 = 4.54 \times 10^{-2}$
(6) $0.000\,06 = 6 \times 10^{-5}$
(7) $0.003\,06 = 3.06 \times 10^{-3}$
(8) $0.000\,000\,5 = 5 \times 10^{-7}$

Moving the decimal point one place to the right is equivalent to multiplying a number by 10; moving the decimal point two places to the right is equivalent to multiplying by 100, and so on. Whenever the decimal point is moved to the right by *n* places, compensation can be achieved by *dividing* at the same time by 10^n; the value of the number remains unchanged. Thus

$$0.032\,5 = \frac{3.25}{10^2} = 3.25 \times 10^{-2}$$

Moving the decimal point one place to the left is equivalent to dividing by 10. Whenever the decimal point is moved to the left n places, compensation can be achieved by *multiplying* at the same time by 10^n; the value of the number remains unchanged. For example,

$$7\,296 = 72.96 \times 10^2 = 7.296 \times 10^3$$

E.   An expression with an exponent of zero is equal to 1.

  (1) $a^0 = 1$ **(2)** $10^0 = 1$ **(3)** $(3 \times 10)^0 = 1$ **(4)** $7 \times 10^0 = 7$ **(5)** $8.2 \times 10^0 = 8.2$

F.   A factor may be transferred from the numerator to the denominator of a fraction, or vice versa, by changing the sign of the exponent.

  (1) $10^{-4} = \dfrac{1}{10^4}$ **(2)** $5 \times 10^{-3} = \dfrac{5}{10^3}$ **(3)** $\dfrac{7}{10^{-2}} = 7 \times 10^2$ **(4)** $-5a^{-2} = -\dfrac{5}{a^2}$

G.   The meaning of the fractional exponent is illustrated by the following.

  (1) $10^{2/3} = \sqrt[3]{10^2}$ **(2)** $10^{3/2} = \sqrt{10^3}$ **(3)** $10^{1/2} = \sqrt{10}$ **(4)** $4^{3/2} = \sqrt{4^3} = \sqrt{64} = 8$

H.   **(1)** $(10^3)^2 = 10^{3 \times 2} = 10^6$ **(2)** $(10^{-2})^3 = 10^{-2 \times 3} = 10^{-6}$ **(3)** $(a^3)^{-2} = a^{-6}$

I.   To extract the square root of a power of 10, divide the exponent by 2. If the exponent is an odd number it should be increased or decreased by 1, and the coefficient adjusted accordingly. To extract the cube root of a power of 10, adjust so that the exponent is divisible by 3; then divide the exponent by 3. The coefficients are treated independently.

  (1) $\sqrt{90\,000} = \sqrt{9 \times 10^4} = \sqrt{9} \times \sqrt{10^4} = 3 \times 10^2$ or 300
  (2) $\sqrt{3.6 \times 10^3} = \sqrt{36 \times 10^2} = \sqrt{36} \times \sqrt{10^2} = 6 \times 10^1$ or 60
  (3) $\sqrt{4.9 \times 10^{-5}} = \sqrt{49 \times 10^{-6}} = \sqrt{49} \times \sqrt{10^{-6}} = 7 \times 10^{-3}$ or 0.007
  (4) $\sqrt[3]{8 \times 10^9} = \sqrt[3]{8} \times \sqrt[3]{10^9} = 2 \times 10^3$ or 2000
  (5) $\sqrt[3]{1.25 \times 10^5} = \sqrt[3]{125 \times 10^3} = \sqrt[3]{125} \times \sqrt[3]{10^3} = 5 \times 10$ or 50

J.   Multiplication and division of numbers expressed as powers of ten.

  (1) $8\,000 \times 2\,500 = (8 \times 10^3)(2.5 \times 10^3) = 20 \times 10^6 = 2 \times 10^7$ or $20\,000\,000$

  (2) $\dfrac{48\,000\,000}{1\,200} = \dfrac{48 \times 10^6}{12 \times 10^2} = 4 \times 10^{6-2} = 4 \times 10^4$ or $40\,000$

  (3) $\dfrac{0.007\,8}{120} = \dfrac{7.8 \times 10^{-3}}{1.2 \times 10^2} = 6.5 \times 10^{-5}$ or $0.000\,065$

  (4) $(4 \times 10^{-3})(5 \times 10^4)^2 = (4 \times 10^{-3})(5^2 \times 10^8) = 4 \times 5^2 \times 10^{-3+8} = 100 \times 10^5 = 1 \times 10^7$

  (5) $\dfrac{(6\,000\,000)(0.000\,04)^4}{(800)^2(0.000\,2)^3} = \dfrac{(6 \times 10^6)(4 \times 10^{-5})^4}{(8 \times 10^2)^2(2 \times 10^{-4})^3} = \dfrac{6 \times 4^4}{8^2 \times 2^3} \times \dfrac{10^6 \times 10^{-20}}{10^4 \times 10^{-12}}$

$$= \dfrac{6 \times 256}{64 \times 8} \times \dfrac{10^{6-20}}{10^{4-12}} = 3 \times \dfrac{10^{-14}}{10^{-8}} = 3 \times 10^{-6}$$

(6) $(\sqrt{4.0 \times 10^{-6}})(\sqrt{8.1 \times 10^3})(\sqrt{0.001\,6}) = (\sqrt{4.0 \times 10^{-6}})(\sqrt{81 \times 10^2})(\sqrt{16 \times 10^{-4}})$

$$= (2 \times 10^{-3})(9 \times 10^1)(4 \times 10^{-2})$$

$$= 72 \times 10^{-4} = 7.2 \times 10^{-3} \quad \text{or} \quad 0.007\,2$$

(7) $(\sqrt[3]{6.4 \times 10^{-2}})(\sqrt[3]{27\,000})(\sqrt[3]{2.16 \times 10^{-4}})$

$$= (\sqrt[3]{64 \times 10^{-3}})(\sqrt[3]{27 \times 10^3})(\sqrt[3]{27 \times 10^3})(\sqrt[3]{216 \times 10^{-6}})$$

$$= (4 \times 10^{-1})(3 \times 10^1)(6 \times 10^{-2})$$

$$= 72 \times 10^{-2} \text{ or } 0.72$$

Appendix B

Significant Figures

INTRODUCTION

The numerical value of every observed measurement is an approximation. No physical measurement, such as mass, length, time, volume, velocity, is ever absolutely correct. The accuracy (reliability) of every measurement is limited by the reliability of the measuring instrument, which is never absolutely reliable.

Consider that the length of an object is recorded as 15.7 cm. By convention, this means that the length was measured to the *nearest* tenth of a centimeter and that its exact value lies between 15.65 and 15.75 cm. If this measurement were exact to the nearest hundredth of a centimeter, it would have been recorded as 15.70 cm. The value 15.7 cm represents *three significant figures* (1, 5, 7), while 15.70 cm represents *four significant figures* (1, 5, 7, 0). A significant figure is one which is known to be reasonably reliable.

Similarly, a recorded mass of 3.406 2 g, observed with an analytical balance, means that the mass of the object was determined to the nearest tenth of a milligram and represents five significant figures (3, 4, 0, 6, 2), the last figure (2) being reasonably correct and guaranteeing the certainty of the preceding four figures.

A 50-cm^3 buret has markings 0.1 cm^3 apart, and the hundredths of a cubic centimeter are estimated. A recorded volume of 41.83 cm^3 represents four significant figures. The last figure (3), being estimated, may be in error by one or two digits in either direction. The preceding three figures (4, 1, 8) are completely certain.

In elementary measurements in chemistry and physics, the last digit is an estimated figure and is considered as a significant figure.

ZEROS

A recorded volume of 28 cm^3 represents two significant figures (2, 8). If this same volume were written as 0.028 L, it would still contain only two significant figures. Zeros appearing as the first figures of a number are not significant, since they merely locate the decimal point. However, the values 0.028 0 L and 0.280 L represent three significant figures (2, 8, and the last zero); the value 1.028 L represents four significant figures (1, 0, 2, 8); and the value 1.028 0 L represents five significant figures (1, 0, 2, 8, 0). Similarly, the value 19.00 for the atomic weight of fluorine contains four significant figures.

The statement that a body of ore weighs 9 800 lb does not indicate definitely the accuracy of the weighing. The last two zeros may have been used merely to locate the decimal point. If it was weighed to the nearest hundred pounds, the weight contains only two significant figures and may be written exponentially as 9.8×10^3 lb. If weighed to the nearest ten pounds it may be written as 9.80×10^3 lb, which indicates that the value is accurate to three significant figures. Since the zero in this case is not needed to locate the decimal point, it must be a significant figure. If the object was weighed to the nearest pound, the weight could be written as 9.800×10^3 lb (four significant figures). Likewise, the statement that the velocity of light is 186 000 mi/s is accurate to three significant figures, since this value is accurate only to the nearest thousand miles per second; to avoid confusion, it may be written as 1.86×10^5 mi/s. (It is customary to place the decimal point after the first significant figure.)

EXACT NUMBERS

Some numerical values are exact to as many significant figures as necessary, by definition. Included in this category are the numerical equivalents of prefixes used in unit definition. For example, 1 cm = 0.01 m by definition, and the units conversion factor, 1.0×10^{-2} m/cm, is exact to an infinite number of significant figures.

Other numerical values are exact by definition. For example, the atomic weight scale was established by fixing the mass of one atom of ^{12}C as 12.0000 u. As many more zeros could be added as desired. Other examples include the definition of the inch (1 in = 2.5400 cm) and the calorie (1 cal = 4.18400 J).

ROUNDING OFF

A number is rounded off to the desired number of significant figures by dropping one or more digits to the right. When the first digit dropped is less than 5, the last digit retained should remain unchanged; when it is greater than 5, 1 is added to the last digit retained. When it is exactly 5, 1 is added to the last digit retained if that digit is odd. Thus the quantity 51.75 g may be rounded off to 51.8 g; 51.65 g to 51.6 g; 51.85 g to 51.8 g. When more than one digit is to be dropped, rounding off should be done in a block, not one digit at a time.

ADDITION AND SUBTRACTION

The answer should be rounded off after adding or subtracting, so as to retain digits only as far as the first column containing estimated figures. (Remember that the last significant figure is estimated.)

Examples   Add the following quantities expressed in grams.

(1)	25.340	(2)	58.0	(3)	4.20	(4)	415.5
	5.465		0.0038		1.6523		3.64
	0.322		0.00001		0.015		0.238
	31.127 g (*Ans.*)		58.00381		5.8673		419.378
			= 58.0 g (*Ans.*)		= 5.87 g (*Ans.*)		= 419.4 g (*Ans.*)

An alternative procedure is to round off the individual numbers before performing the arithmetic operation, retaining only as many columns to the right of the decimal as would give a digit in every item to be added or subtracted. Examples **(2)**, **(3)**, and **(4)** above would be done as follows:

(2)	58.0	(3)	4.20	(4)	415.5
	0.0		1.65		3.6
	0.0		0.02		0.2
	58.0 g		5.87 g		419.3 g

Note that the answer to **(4)** differs by one in the last place from the previous answer. The last place, however, is known to have some uncertainty in it.

MULTIPLICATION AND DIVISION

The answer should be rounded off to contain only as many significant figures as are contained in the least exact factor. For example, when multiplying 7.485×8.61, or when dividing $0.1642 \div 1.52$, the answer should be given in three significant figures.

This rule is an approximation to a more exact statement that the fractional or percentage error of a product or quotient cannot be any less than the fractional or percentage error of any one factor. For this reason, numbers whose first significant figure is 1 (or occasionally 2) must contain an additional significant figure to have a given fractional error in comparison with a number beginning with 8 or 9.

Consider the division

$$\frac{9.84}{9.3} = 1.06$$

By the approximate rule, the answer should be 1.1 (two significant figures). However, a difference of 1 in the last place of 9.3 (9.3 ± 0.1) results in an error of about 1 percent, while a difference of 1 in the last place of 1.1 (1.1 ± 0.1) yields an error of roughly 10 percent. Thus the answer 1.1 is of much lower percentage accuracy than 9.3. Hence in this case the answer should be 1.06, since a difference of 1 in the last place of the least exact factor used in the calculation (9.3) yields a percentage of error about the same (about 1 percent) as a difference of 1 in the last place of 1.06 (1.06 ± 0.01). Similarly, $0.92 \times 1.13 = 1.04$.

In nearly all practical chemical calculations, a precision of only two to four significant figures is required. Therefore the student need not perform multiplications and divisions manually. Even if an electronic calculator is not available, an inexpensive 10-in slide rule is accurate to three significant figures, and a table of 4-place logarithms is accurate to four significant figures.

Appendix C

Logarithms

INTRODUCTION

Common Logarithms

The logarithm (log, in abbreviated form) of a positive number is the exponent, or power, of a given base that is required to produce that number. For example, since $1000 = 10^3$, $100 = 10^2$, $10 = 10^1$, $1 = 10^0$, then the logarithms of $1\,000$, 100, 10, 1, to be base 10 are respectively 3, 2, 1, 0. It is obvious that $10^{1.5377}$ will give some number greater than 10 (which is 10^1) but smaller than 100 (10^2). Actually, $10^{1.5377} = 34.49$; hence log $34.49 = 1.537\,7$.

The system of logarithms whose base is 10 (called the *common* or *Briggsian* system) had been widely used for making numerical computations (multiplication and division) before the advent of electronic calculators and computers. The techniques will be presented later in this appendix and numerical tables are provided in Appendix D. Students who have calculators will not need these techniques, but should learn them for the sake of improving their understanding of logarithms and as a backup method of calculation.

Natural Logarithms

The logarithmic function that occurs commonly in physics and chemistry as part of the solution to certain differential equations has as its base not the number 10 but the transcendental number $e = 2.718\,28\ldots$. To differentiate between the common and the *natural* or *Napierian* logarithms, a more explicit notation could be used: $\log_{10} N = x$ and $\log_e N = y$, where $10^x = N$ and $e^y = N$. In this book, and in many chemistry and physics books, the notation log N is used to indicate the logarithm to the base 10, and ln N to indicate the natural logarithm to the base e.

Although it is possible to construct logarithm tables for the base e, or for any other base for that matter, there is a simple relationship between any two base systems such that a number found by the use of the common logarithmic table can easily be converted to the logarithm to any other base. Specifically, for the interconversion of common and natural logarithms, it can be shown that

$$\ln N = 2.303 \log N \tag{1}$$

Other Properties of Logarithms

Negative numbers do not have logarithms.

The logarithm of a number greater than 1 is positive; of a number less than 1 it is negative. The logarithm of 0 is infinitely negative.

Given the value of the logarithm of a number to a certain base, the process of finding the number is called taking the antilogarithm. Taking the antilogarithm is synonymous with raising the base to a power equal to the logarithm.

Logarithms themselves have no units. Also in principle one cannot take the logarithm of a number with units, but in fact one often does so when using relationships for which the units have been designated. For example, $pH = -\log[H^+]$ provided that $[H^+]$ is in units of mol/L.

OBTAINING LOGS ON AN ELECTRONIC CALCULATOR

Enter the number, either in direct or exponential notation. Be sure it has a positive sign. Press LOG for the common log or LN for the natural log and the result will be displayed. To find the antilog of a value, enter it, adjust its sign to plus or minus as required, and press first INV (inverse), then LOG or LN, depending whether the value is a common log or natural log. Some calculators have the functions 10^x and e^x in place of INV LOG and INV LN, respectively.

BASIC PRINCIPLES OF LOGARITHMS

Since logarithms are exponents, all properties of exponents are also properties of logarithms.

A. The logarithm of the product of two numbers is the sum of their logarithms.

$$\log ab = \log a + \log b \qquad \log (5\,280 \times 48) = \log 5\,280 + \log 48$$

B. The logarithm of the quotient of two numbers is the logarithm of the numerator minus the logarithm of the denominator.

$$\log \frac{a}{b} = \log a - \log b \qquad \log \frac{536}{24.5} = \log 536 - \log 24.5$$

C. The log of the nth power of a number is n times the log of the number.

$$\log a^n = n \log a \qquad \log (4.28)^3 = 3 \log 4.28$$

This rule can be used to derive (*1*). If

$$x = \log N$$

that is, if

$$10^x = N$$

then from Basic Principle of Logarithms **C** above, applied now to the base e,

$$\ln N = \ln 10^x = x \ln 10 = (\log N)(\ln 10)$$

The numerical value of $\ln 10$ is $2.302\,585\ldots$; so to four significant figures,

$$\ln N = 2.303 \log N$$

D. The log of the nth root of a number is the log of the number, divided by n.

$$\log \sqrt[n]{a} = \frac{1}{n} \log a \qquad \log \sqrt{32} = \frac{1}{2} \log 32 \qquad \log \sqrt[3]{792} = \frac{1}{3} \log 792$$

This is a special case of **C**, since the nth root of a number is the number raised to the $(1/n)$th power.

USING COMMON LOGARITHMS FOR CALCULATION

Definition of Terms

The number before the decimal point is the *characteristic* of the log, and the decimal fraction part is the *mantissa* of the log. In the example, $\log 34.49 = 1.537\,7$, the characteristic is 1 and the mantissa is .537 7.

The mantissa of the log of a number is found in tables, printed without the decimal point. Each mantissa in the tables is understood to have a decimal point preceding it, and the mantissa is always considered positive.

The characteristic is determined by inspection from the number itself according to the following rules:

(1) For a number greater than 1, the characteristic is positive and is one *less* than the number of digits before the decimal point. For example:

Number	5297	348	900	34.8	60	4.764	3
Characteristic	3	2	2	1	1	0	0

(2) For a positive number less than 1, the characteristic is negative and its magnitude is one *more* than the number of zeros immediately following the decimal point. The negative sign of the characteristic is written in one of three ways:　(*a*) above the characteristic, as $\bar{1}, \bar{2}$, and so on;　(*b*) as 9. -10, 8, -10, and so on;　(*c*) in front of the number in the usual way, as $-1, -2$, and so on. Thus the characteristic of the logarithm of 0.348 5 is $\bar{1}$, 9, -10, or -1; of the logarithm of 0.051 3, it is $\bar{2}$, 8. -10, or -2.

Using the Common Logarithm Tables

Suppose it is required to find the complete log of the number 728. In the table of logarithms in Appendix D glance down the N column to 72, then horizontally to the right to column 8 and note the entry 8621 which is the required mantissa. Since the characteristic is 2, log 728 = 2.862 1. (This means that $728 = 10^{2.8621}$).

The mantissa for log 72.8, for log 7.28, for log 0.728, for log 0.072 8, etc., is .862 1, but the characteristics differ. Thus:

$$\begin{aligned}
\log 728 &= 2.862\,1 & \log 0.728 &= \bar{1}.862\,1 & \text{or} \quad 9.862\,1 - 10 & \quad \text{or} \quad -1 + .862\,1 \\
\log 72.8 &= 1.862\,1 & \log 0.072\,8 &= \bar{2}.8621 & \text{or} \quad 8.862\,1 - 10 & \quad \text{or} \quad -2 + .862\,1 \\
\log 7.28 &= 0.862\,1 & \log 0.007\,28 &= \bar{3}.8621 & \text{or} \quad 7.862\,1 - 10 & \quad \text{or} \quad -3 + .862\,1
\end{aligned}$$

Note that if a negative characteristic is written by the third method given above, -3 for example, a $+$ sign must be written before the mantissa because the mantissa is always positive.

To find log 46.38: Glance down the N column to 46, then horizontally to column 3 and note the mantissa 6 656. Moving farther to the right along the same line, the figure 7 is found under column 8 of Proportional Parts. The required mantissa is .665 6 + .000 7 = .666 3. Since the characteristic is 1, log 46.38 = 1.666 3.

The mantissa for log 463 8, for log 463.8, for log 46.38, etc., is .666 3, but the characteristics differ, as in the example above.

Exercises.　Find the logarithms of the following numbers.

(1) 454	(6) 0.621	*Ans.*
(2) 5280	(7) 0.946 3	
(3) 96 500	(8) 0.035 3	
(4) 30.48	(9) 0.002 2	
(5) 1.057	(10) 0.000 264 5	

Ans.
(1) 2.657 1　　(6) $\bar{1}.793\,1$　or　$9.793\,1 - 10$　or　$-1 + .793\,1$
(2) 3.722 6　　(7) $\bar{1}.976\,0$　or　$9.976\,0 - 10$　or　$-1 + .976\,0$
(3) 4.984 5　　(8) $\bar{2}.547\,8$　or　$8.547\,8 - 10$　or　$-2 + .547\,8$
(4) 1.484 0　　(9) $\bar{3}.342\,4$　or　$7.342\,4 - 10$　or　$-3 + .342\,4$
(5) 0.024 1　　(10) $\bar{4}.422\,4$　or　$6.422\,4 - 10$　or　$-4 + .422\,4$

Sometimes the log of a number must be used in an algebraic equation, such as $y = 7.5 \log x$, or in graphs. If x is greater than 1, log x is positive and there is no special problem. If x is between zero and 1, however, log x is negative. This negative log, according to the above rules, is written as the sum of a positive mantissa and a negative characteristic. For algebraic manipulations it is preferable to treat log x as a single number with a definite sign, either positive or negative. For such a purpose, $\bar{2}.748\,6$ would be written as $-1.251\,4$, obtained by adding -2 and $+.748\,6$ algebraically.

Exercises.　Write the logarithms of the following numbers as quantities suitable for substitution in an algebraic equation.

(1) 0.275　(2) 0.000 394　(3) 0.014 9　　*Ans.* (1) $-0.560\,7$　(2) $-3.404\,5$　(3) $-1.826\,8$

Using Tables of Antilogarithms

The antilogarithm is the number corresponding to a given logarithm. "The antilog of 3" means "the number whose log is 3"; that number is obviously 1 000. In general, antilog $N = 10^N$.

Suppose it is required to find the antilog of 2.674 7, i.e. the number whose log is 2.674 7. The characteristic is 2 and the mantissa is .674 7. Using the table of Antilogarithms in Appendix D, locate 67 in the first column, then move horizontally to column 4 and note the digits 4 721. Moving farther to the right along the same line, the entry 8 is found under column 7 of Proportional Parts. Adding 8 to 4 721 gives 4 729. Since the characteristic is 2, there are three digits to the left of the decimal point. Hence 472.9 is the required number.

It should be understood that the antilog of 1.674 7 is 47.29; the antilog of 0.674 7 is 4.729; the antilog of 9.674 7 − 10 is 0.472 9; etc. On the other hand, the antilog of −1.674 7 must be rewritten as antilog of $\overline{2}.325\,3$, or 8.325 3 − 10, or −2 + .325 3, before the tables may be used, because only positive mantissas are found in the table.

Exercises.   Find the numbers corresponding to the following logarithms.

(1) 3.156 8	(7) 0.000 8	*Ans.* (1) 1 435	(7) 1.002
(2) 1.693 4	(8) 9.750 7 − 10 or $\overline{1}.750\,7$ or −1 + .750 7	(2) 49.37	(8) 0.563 2
(3) 5.693 4	(9) 8.003 4 − 10 or $\overline{2}.003\,4$ or −2 + .003 4	(3) 493 700	(9) 0.010 08
(4) 2.500 0	(10) 7.200 6 − 10 or $\overline{3}.200\,6$ or −3 + .200 6	(4) 316.2	(10) 0.001 587
(5) 2.043 6	(11) −0.243 6	(5) 110.6	(11) 0.570 7
(6) 0.914 2	(12) −3.762 9	(6) 8.208	(12) 0.000 172 6

Logarithms of Numbers Expressed in Scientific Notation

$$\log(4.50 \times 10^7) = \log 4.50 + \log 10^7 = 7 + \log 4.50 = 7 + 0.653\,2 = 7.653\,2$$
$$\log(4.50 \times 10^{-4}) = \log 4.50 + \log 10^{-4} = -4 + \log 4.50 = -4 + 0.653\,2 = \overline{4}.653\,2 = 6.653\,2 - 10$$

In general, if a number is expressed as a product of two factors, the first being a number between 1 and 10, and the second being an integral power of ten, the logarithm has as its mantissa the logarithm of the first factor and as its characteristic the exponent of ten in the second factor.

Exercises.   Find the logarithms of the following numbers.

(1) 3.75×10^2	(3) 6.6×10^{-27}	(5) 60.3×10^{-8}
(2) 6.02×10^{23}	(4) 0.75×10^4	(6) 2.09×10^{-15}

Ans. (1) 2.574 0	(3) $\overline{27}.819\,5 = 3.819\,5 - 30$ $= -27 + .819\,5$	(5) $\overline{7}.780\,3 = 3.780\,3 - 10 = -7 + .780\,3$
(2) 23.779 6	(4) 3.875 1	(6) $\overline{15}.320\,1 = 5.320\,1 - 20 = -15 + .320\,1$

Conversely, an antilogarithm can be expressed directly in scientific notation. The power of 10 is the characteristic; the coefficient of the power of 10 is the antilog of the mantissa, with the decimal point in the antilog following the first digit.

$$\text{antilog } 3.842\,0 = (\text{antilog } .842\,0) \times (\text{antilog } 3) = 6.95 \times 10^3$$
$$\text{antilog } \overline{3}.842\,0 = (\text{antilog } .842\,0) \times [\text{antilog}(-3)] = 6.95 \times 10^{-3}$$
$$\text{antilog } -3.842\,0 = \text{antilog } \overline{4}.158\,0 = (\text{antilog } .158\,0) \times [\text{antilog}(-4)] = 1.439 \times 10^4$$

Exercises.   Write the antilogarithms of the following in scientific notation.

(1) 10.476 9	(3) 5.040 3	(5) 7.621 6
(2) $\overline{19}.204\,6$	(4) 4.140 2 − 20	(6) −8.276 3

Ans. (1) 2.998×10^{10}	(3) 1.097×10^5	(5) 4.184×10^7
(2) 1.602×10^{-19}	(4) 1.381×10^{-16}	(6) 5.292×10^{-9}

Illustrations of the Use of Logarithms

1.   Find the value of $487 \times 2.45 \times 0.038\,7$.

$$\text{Let}\quad x = 487 \times 2.45 \times 0.038\,7.$$
$$\log x = \log 487 + \log 2.45 + \log 0.038\,7$$
$$= 1.664\,4$$
$$x = \text{antilog } 1.664\,4 = 46.17\quad\text{or}\quad 46.2$$
$$\text{(to three significant figures)}$$

log 487	=	2.687 5
log 2.45	=	0.389 2
log 0.038 7	=	8.587 7 − 10 (add)
log x	=	11.664 4 − 10
or		1.664 4

2. Find $x = \dfrac{136.3}{65.38}$. $\log 136.3 = 2.134\,5$

$\log x = \log 136.3 - \log 65.38 = 0.319\,1$ $\log 65.38 = \underline{1.815\,4}$ (subtract)

$x = \text{antilog } 0.319\,1 = 2.085$ $\log x = 0.319\,1$

3. Find $x = \dfrac{1}{22.4}$. $\log 1 = 0 = 10.000\,0 - 10$

$\log x = \log 1 - \log 22.4 = 8.649\,8 - 10$ $\log 22.4 = \underline{1.350\,2}$ (subtract)

$x = \text{antilog } 8.649\,8 - 10 = 0.044\,65$ or $0.044\,6$ $\log x = 8.649\,8 - 10$

(three significant figures)

Adding or subtracting $10,000\,0 - 10$, $20,000\,0 - 20$, etc., to or from any logarithm does not change its value.

4.   Find $x = \dfrac{17.5 \times 1.92}{0.283 \times 0.031\,4}$.

$\log x = (\log 17.5 + \log 1.92) - \qquad (\log 0.283 + \log 0.031\,4)$

$\log 17.5 = 1.243\,0 \qquad\qquad\qquad \log 0.283 \ = \ 9.451\,8 - 10$

$\log 1.92 = \underline{0.283\,3}$ (add)$\qquad\qquad \log 0.031\,4 = \underline{8.496\,9 - 10}$ (add)

$\qquad\qquad 1.526\,3$ or $11.526\,3 - 10 \qquad\qquad 17.948\,7 - 20$ or $7.948\,7 - 10$

$\log 17.5 + \log 1.92 \quad = 11.526\,3 - 10$

$\log 0.283 + \log 0.031\,4 = \ 7.948\,7 - 10$ (subtract)

$\log x = \ 3.577\,6$

$x = \text{antilog } 3.577\,6 = 378\,1$ or 3.78×10^3

5.   Find $x = (6.138)^3$.

$\log x = 3(\log 6.138) = 3(0.788\,1) = 2.364\,3$

$x = \text{antilog } 2.364\,3 = 231.4$

6.   Find $x = \sqrt{7\,514}$ or $(7\,514)^{1/2}$.

$\log x = \frac{1}{2}(\log 7\,514) = \frac{1}{2}(3.875\,8) = 1.937\,9$

$x = \text{antilog } 1.937\,9 = 86.68$

7.   Find $x = \sqrt[3]{0.059\,2}$ or $(0.059\,2)^{1/3}$.

$\log x = \frac{1}{3}\log 0.059\,2 = \frac{1}{3}(8.772\,3 - 10) = \frac{1}{3}(28.772\,3 - 30) = 9.590\,8 - 10$

$x = \text{antilog } 9.590\,8 - 10 = 0.389\,8$ or 0.390 (three significant figures)

8.   Find $x = \sqrt{(152)^3}$.

$\log x = \frac{1}{2}(3 \log 152) = \frac{1}{2}(3 \times 2.181\,8) = \frac{1}{2}(6.545\,4) = 3.272\,7$

$x = \text{antilog } 3.272\,7 = 1\,874$ or 1.87×10^3 (three significant figures)

9.   Find $(6.8 \times 10^{-4})^3$ or $(6.8)^3 \times 10^{-12}$.

$\log(6.8)^3 = 3(\log 6.8) = 3(0.832\,5) = 2.497\,5$

$(6.8)^3 = \text{antilog } 2.497\,5 = 314.4$ or 3.1×10^2 (two significant figures)

Then $(6.8 \times 10^{-4})^3 = 3.1 \times 10^2 \times 10^{-12} = 3.1 \times 10^{-10}$

10.   Find $\sqrt{8.31 \times 10^{-11}}$ or $\sqrt{83.1 \times 10^{-12}}$ or $\sqrt{83.1} \times 10^{-6}$.

$\log\sqrt{83.1} = \frac{1}{2}(\log 83.1) = \frac{1}{2}(1.919\,6) = 0.959\,8$

$\sqrt{83.1} = \text{antilog } 0.959\,8 = 9.116$ or 9.12

Then $\sqrt{8.31 \times 10^{-11}} = 9.12 \times 10^{-6}$

11. Find $x = 97^{1.665}$ $\log 1.665 = .221\,4$

$\log x = 1.665(\log 97)$ $\log 1.987 = .298\,3$

      $= 1.665(1.986\,8) = 3.309$ $\log(1.665 \times 1.987) = .519\,7$

   $x = \text{antilog } 3.309 = 2.04 \times 10^3$ $1.665 \times 1.987 = 3.309$

Note that the logarithm of a logarithm must be taken if logarithm tables are used to perform this type of calculation.

12. Find $x = \ln 28.25$.

       $x = 2.303 \log 28.25 = 2.303 \times 1.451\,0$ $\log 2.303 \;\; = .362\,3$

                                        $\log 1.451\,0 = .161\,7$ (add)

   $\log x = \log 2.303 + \log 1.451\,0 = .524\,0$ $\log x \;\;\;\;\;\; = .524\,0$

       $x = \text{antilog } 0.524\,0 = 3.342$

13. Find $x = 57.9 \ln(3.25 \times 10^4)$. $\log 57.9 = 1.762\,7$

                                           $\log 2.303 = .362\,3$

       $x = 57.9 \times 2.303 \log(3.25 \times 10^4)$ $\log 4.512 = .654\,4$ (add)

       $= 57.9 \times 2.303 \times 4.512$ $2.779\,4$

  $\log x = \log 57.9 + \log 2.303 + \log 4.512$

       $= 2.779\,4$

       $x = \text{antilog } 2.779\,4 = 602$

Appendix D

Four-Place Common Logarithms

N	0	1	2	3	4	5	6	7	8	9	Proportional Parts 1	2	3	4	5	6	7	8	9
10	0000	0043	0086	0128	0170	0212	0253	0294	0334	0374	4	8	12	17	21	25	29	33	37
11	0414	0453	0492	0531	0569	0607	0645	0682	0719	0755	4	8	11	15	19	23	26	30	34
12	0792	0828	0864	0899	0934	0969	1004	1038	1072	1106	3	7	10	14	17	21	24	28	31
13	1139	1173	1206	1239	1271	1303	1335	1367	1399	1430	3	6	10	13	16	19	23	26	29
14	1461	1492	1523	1553	1584	1614	1644	1673	1703	1732	3	6	9	12	15	18	21	24	27
15	1761	1790	1818	1847	1875	1903	1931	1959	1987	2014	3	6	8	11	14	17	20	22	25
16	2041	2068	2095	2122	2148	2175	2201	2227	2253	2279	3	5	8	11	13	16	18	21	24
17	2304	2330	2355	2380	2405	2430	2455	2480	2504	2529	2	5	7	10	12	15	17	20	22
18	2553	2577	2601	2625	2648	2672	2695	2718	2742	2765	2	5	7	9	12	14	16	19	21
19	2788	2810	2833	2856	2878	2900	2923	2945	2967	2989	2	4	7	9	11	13	16	18	20
20	3010	3032	3054	3075	3096	3118	3139	3160	3181	3201	2	4	6	8	11	13	15	17	19
21	3222	3243	3263	3284	3304	3324	3345	3365	3385	3404	2	4	6	8	10	12	14	16	18
22	3424	3444	3464	3483	3502	3522	3541	3560	3579	3598	2	4	6	8	10	12	14	15	17
23	3617	3636	3655	3674	3692	3711	3729	3747	3766	3784	2	4	6	7	9	11	13	15	17
24	3802	3820	3838	3856	3874	3892	3909	3927	3945	3962	2	4	5	7	9	11	12	14	16
25	3979	3997	4014	4031	4048	4065	4082	4099	4116	4133	2	3	5	7	9	10	12	14	15
26	4150	4166	4183	4200	4216	4232	4249	4265	4281	4298	2	3	5	7	8	10	11	13	15
27	4314	4330	4346	4362	4378	4393	4409	4425	4440	4456	2	3	5	6	8	9	11	13	14
28	4472	4487	4502	4518	4533	4548	4564	4579	4594	4609	2	3	5	6	8	9	11	12	14
29	4624	4639	4654	4669	4683	4698	4713	4728	4742	4757	1	3	4	6	7	9	10	12	13
30	4771	4786	4800	4814	4829	4843	4857	4871	4886	4900	1	3	4	6	7	9	10	11	13
31	4914	4928	4942	4955	4969	4983	4997	5011	5024	5038	1	3	4	6	7	8	10	11	12
32	5051	5065	5079	5092	5105	5119	5132	5145	5159	5172	1	3	4	5	7	8	9	11	12
33	5185	5198	5211	5224	5237	5250	5263	5276	5289	5302	1	3	4	5	6	8	9	10	12
34	5315	5328	5340	5353	5366	5378	5391	5403	5416	5428	1	3	4	5	6	8	9	10	11
35	5441	5453	5465	5478	5490	5502	5514	5527	5539	5551	1	2	4	5	6	7	9	10	11
36	5563	5575	5587	5599	5611	5623	5635	5647	5658	5670	1	2	4	5	6	7	8	10	11
37	5682	5694	5705	5717	5729	5740	5752	5763	5775	5786	1	2	3	5	6	7	8	9	10
38	5798	5809	5821	5832	5843	5855	5866	5877	5888	5899	1	2	3	5	6	7	8	9	10
39	5911	5922	5933	5944	5955	5966	5977	5988	5999	6010	1	2	3	4	5	7	8	9	10
40	6021	6031	6042	6053	6064	6075	6085	6096	6107	6117	1	2	3	4	5	6	8	9	10
41	6128	6138	6149	6160	6170	6180	6191	6201	6212	6222	1	2	3	4	5	6	7	8	9
42	6232	6243	6253	6263	6274	6284	6294	6304	6314	6325	1	2	3	4	5	6	7	8	9
43	6335	6345	6355	6365	6375	6385	6395	6405	6415	6425	1	2	3	4	5	6	7	8	9
44	6435	6444	6454	6464	6474	6484	6493	6503	6513	6522	1	2	3	4	5	6	7	8	9
45	6532	6542	6551	6561	6571	6580	6590	6599	6609	6618	1	2	3	4	5	6	7	8	9
46	6628	6637	6646	6656	6665	6675	6684	6693	6702	6712	1	2	3	4	5	6	7	7	8
47	6721	6730	6739	6749	6758	6767	6776	6785	6794	6803	1	2	3	4	5	5	6	7	8
48	6812	6821	6830	6839	6848	6857	6866	6875	6884	6893	1	2	3	4	4	5	6	7	8
49	6902	6911	6920	6928	6937	6946	6955	6964	6972	6981	1	2	3	4	4	5	6	7	8
50	6990	6998	7007	7016	7024	7033	7042	7050	7059	7067	1	2	3	3	4	5	6	7	8
51	7076	7084	7093	7101	7110	7118	7126	7135	7143	7152	1	2	3	3	4	5	6	7	8
52	7160	7168	7177	7185	7193	7202	7210	7218	7226	7235	1	2	2	3	4	5	6	7	7
53	7243	7251	7259	7267	7275	7284	7292	7300	7308	7316	1	2	2	3	4	5	6	6	7
54	7324	7332	7340	7348	7356	7364	7372	7380	7388	7396	1	2	2	3	4	5	6	6	7
N	0	1	2	3	4	5	6	7	8	9	1	2	3	4	5	6	7	8	9

N	0	1	2	3	4	5	6	7	8	9	Proportional Parts								
											1	2	3	4	5	6	7	8	9
55	7404	7412	7419	7427	7435	7443	7451	7459	7466	7474	1	2	2	3	4	5	5	6	7
56	7482	7490	7497	7505	7513	7520	7528	7536	7543	7551	1	2	2	3	4	5	5	6	7
57	7559	7566	7574	7582	7589	7597	7604	7612	7619	7627	1	2	2	3	4	5	5	6	7
58	7634	7642	7649	7657	7664	7672	7679	7686	7694	7701	1	1	2	3	4	4	5	6	7
59	7709	7716	7723	7731	7738	7745	7752	7760	7767	7774	1	1	2	3	4	4	5	6	7
60	7782	7789	7796	7803	7810	7818	7825	7832	7839	7846	1	1	2	3	4	4	5	6	6
61	7853	7860	7868	7875	7882	7889	7896	7903	7910	7917	1	1	2	3	4	4	5	6	6
62	7924	7931	7938	7945	7952	7959	7966	7973	7980	7987	1	1	2	3	3	4	5	6	6
63	7993	8000	8007	8014	8021	8028	8035	8041	8048	8055	1	1	2	3	3	4	5	5	6
64	8062	8069	8075	8082	8089	8096	8102	8109	8116	8122	1	1	2	3	3	4	5	5	6
65	8129	8136	8142	8149	8156	8162	8169	8176	8182	8189	1	1	2	3	3	4	5	5	6
66	8195	8202	8209	8215	8222	8228	8235	8241	8248	8254	1	1	2	3	3	4	5	5	6
67	8261	8267	8274	8280	8287	8293	8299	8306	8312	8319	1	1	2	3	3	4	5	5	6
68	8325	8331	8338	8344	8351	8357	8363	8370	8376	8382	1	1	2	3	3	4	4	5	6
69	8388	8395	8401	8407	8414	8420	8426	8432	8439	8445	1	1	2	2	3	4	4	5	6
70	8451	8457	8463	8470	8476	8482	8488	8494	8500	8506	1	1	2	2	3	4	4	5	6
71	8513	8519	8525	8531	8537	8543	8549	8555	8561	8567	1	1	2	2	3	4	4	5	5
72	8573	8579	8585	8591	8597	8603	8609	8615	8621	8627	1	1	2	2	3	4	4	5	5
73	8633	8639	8645	8651	8657	8663	8669	8675	8681	8686	1	1	2	2	3	4	4	5	5
74	8692	8698	8704	8710	8716	8722	8727	8733	8739	8745	1	1	2	2	3	4	4	5	5
75	8751	8756	8762	8768	8774	8779	8785	8791	8797	8802	1	1	2	2	3	3	4	5	5
76	8808	8814	8820	8825	8831	8837	8842	8848	8854	8859	1	1	2	2	3	3	4	5	5
77	8865	8871	8876	8882	8887	8893	8899	8904	8910	8915	1	1	2	2	3	3	4	4	5
78	8921	8927	8932	8938	8943	8949	8954	8960	8965	8971	1	1	2	2	3	3	4	4	5
79	8976	8982	8987	8993	8998	9004	9009	9015	9020	9025	1	1	2	2	3	3	4	4	5
80	9031	9036	9042	9047	9053	9058	9063	9069	9074	9079	1	1	2	2	3	3	4	4	5
81	9085	9090	9096	9101	9106	9112	9117	9122	9128	9133	1	1	2	2	3	3	4	4	5
82	9138	9143	9149	9154	9159	9165	9170	9175	9180	9186	1	1	2	2	3	3	4	4	5
83	9191	9196	9201	9206	9212	9217	9222	9227	9232	9238	1	1	2	2	3	3	4	4	5
84	9243	9248	9253	9258	9263	9269	9274	9279	9284	9289	1	1	2	2	3	3	4	4	5
85	9294	9299	9304	9309	9315	9320	9325	9330	9335	9340	1	1	2	2	3	3	4	4	5
86	9345	9350	9355	9360	9365	9370	9375	9380	9385	9390	1	1	2	2	3	3	4	4	5
87	9395	9400	9405	9410	9415	9420	9425	9430	9435	9440	0	1	1	2	2	3	3	4	4
88	9445	9450	9455	9460	9465	9469	9474	9479	9484	9489	0	1	1	2	2	3	3	4	4
89	9494	9499	9504	9509	9513	9518	9523	9528	9533	9538	0	1	1	2	2	3	3	4	4
90	9542	9547	9552	9557	9562	9566	9571	9576	9581	9586	0	1	1	2	2	3	3	4	4
91	9590	9595	9600	9605	9609	9614	9619	9624	9628	9633	0	1	1	2	2	3	3	4	4
92	9638	9643	9647	9652	9657	9661	9666	9671	9675	9680	0	1	1	2	2	3	3	4	4
93	9685	9689	9694	9699	9703	9708	9713	9717	9722	9727	0	1	1	2	2	3	3	4	4
94	9731	9736	9741	9745	9750	9754	9759	9763	9768	9773	0	1	1	2	2	3	3	4	4
95	9777	9782	9786	9791	9795	9800	9805	9809	9814	9818	0	1	1	2	2	3	3	4	4
96	9823	9827	9832	9836	9841	9845	9850	9854	9859	9863	0	1	1	2	2	3	3	4	4
97	9868	9872	9877	9881	9886	9890	9894	9899	9903	9908	0	1	1	2	2	3	3	4	4
98	9912	9917	9921	9926	9930	9934	9939	9943	9948	9952	0	1	1	2	2	3	3	4	4
99	9956	9961	9965	9969	9974	9978	9983	9987	9991	9996	0	1	1	2	2	3	3	3	4
N	0	1	2	3	4	5	6	7	8	9	1	2	3	4	5	6	7	8	9

Antilogarithms

	0	1	2	3	4	5	6	7	8	9	Proportional Parts								
											1	2	3	4	5	6	7	8	9
.00	1000	1002	1005	1007	1009	1012	1014	1016	1019	1021	0	0	1	1	1	1	2	2	2
.01	1023	1026	1028	1030	1033	1035	1038	1040	1042	1045	0	0	1	1	1	1	2	2	2
.02	1047	1050	1052	1054	1057	1059	1062	1064	1067	1069	0	0	1	1	1	1	2	2	2
.03	1072	1074	1076	1079	1081	1084	1086	1089	1091	1094	0	0	1	1	1	1	2	2	2
.04	1096	1099	1102	1104	1107	1109	1112	1114	1117	1119	0	1	1	1	1	2	2	2	2
.05	1122	1125	1127	1130	1132	1135	1138	1140	1143	1146	0	1	1	1	1	2	2	2	2
.06	1148	1151	1153	1156	1159	1161	1164	1167	1169	1172	0	1	1	1	1	2	2	2	2
.07	1175	1178	1180	1183	1186	1189	1191	1194	1197	1199	0	1	1	1	1	2	2	2	2
.08	1202	1205	1208	1211	1213	1216	1219	1222	1225	1227	0	1	1	1	1	2	2	2	3
.09	1230	1233	1236	1239	1242	1245	1247	1250	1253	1256	0	1	1	1	1	2	2	2	3
.10	1259	1262	1265	1268	1271	1274	1276	1279	1282	1285	0	1	1	1	1	2	2	2	3
.11	1288	1291	1294	1297	1300	1303	1306	1309	1312	1315	0	1	1	1	2	2	2	2	3
.12	1318	1321	1324	1327	1330	1334	1337	1340	1343	1346	0	1	1	1	2	2	2	3	3
.13	1349	1352	1355	1358	1361	1365	1368	1371	1374	1377	0	1	1	1	2	2	2	3	3
.14	1380	1384	1387	1390	1393	1396	1400	1403	1406	1409	0	1	1	1	2	2	2	3	3
.15	1413	1416	1419	1422	1426	1429	1432	1435	1439	1442	0	1	1	1	2	2	2	3	3
.16	1445	1449	1452	1455	1459	1462	1466	1469	1472	1476	0	1	1	1	2	2	2	3	3
.17	1479	1483	1486	1489	1493	1496	1500	1503	1507	1510	0	1	1	1	2	2	2	3	3
.18	1514	1517	1521	1524	1528	1531	1535	1538	1542	1545	0	1	1	1	2	2	2	3	3
.19	1549	1552	1556	1560	1563	1567	1570	1574	1578	1581	0	1	1	1	2	2	3	3	3
.20	1585	1589	1592	1596	1600	1603	1607	1611	1614	1618	0	1	1	1	2	2	3	3	3
.21	1622	1626	1629	1633	1637	1641	1644	1648	1652	1656	0	1	1	2	2	2	3	3	3
.22	1660	1663	1667	1671	1675	1679	1683	1687	1690	1694	0	1	1	2	2	2	3	3	3
.23	1698	1702	1706	1710	1714	1718	1722	1726	1730	1734	0	1	1	2	2	2	3	3	4
.24	1738	1742	1746	1750	1754	1758	1762	1766	1770	1774	0	1	1	2	2	2	3	3	4
.25	1778	1782	1786	1791	1795	1799	1803	1807	1811	1816	0	1	1	2	2	2	3	3	4
.26	1820	1824	1828	1832	1837	1841	1845	1849	1854	1858	0	1	1	2	2	3	3	3	4
.27	1862	1866	1871	1875	1879	1884	1888	1892	1897	1901	0	1	1	2	2	3	3	3	4
.28	1905	1910	1914	1919	1923	1928	1932	1936	1941	1945	0	1	1	2	2	3	3	4	4
.29	1950	1954	1959	1963	1968	1972	1977	1982	1986	1991	0	1	1	2	2	3	3	4	4
.30	1995	2000	2004	2009	2014	2018	2023	2028	2032	2037	0	1	1	2	2	3	3	4	4
.31	2042	2046	2051	2056	2061	2065	2070	2075	2080	2084	0	1	1	2	2	3	3	4	4
.32	2089	2094	2099	2104	2109	2113	2118	2123	2128	2133	0	1	1	2	2	3	3	4	4
.33	2138	2143	2148	2153	2158	2163	2168	2173	2178	2183	0	1	1	2	2	3	3	4	4
.34	2188	2193	2198	2203	2208	2213	2218	2223	2228	2234	1	1	2	2	3	3	4	4	5
.35	2239	2244	2249	2254	2259	2265	2270	2275	2280	2286	1	1	2	2	3	3	4	4	5
.36	2291	2296	2301	2307	2312	2317	2323	2328	2333	2339	1	1	2	2	3	3	4	4	5
.37	2344	2350	2355	2360	2366	2371	2377	2382	2388	2393	1	1	2	2	3	3	4	4	5
.38	2399	2404	2410	2415	2421	2427	2432	2438	2443	2449	1	1	2	2	3	3	4	4	5
.39	2455	2460	2466	2472	2477	2483	2489	2495	2500	2506	1	1	2	2	3	3	4	5	5
.40	2512	2518	2523	2529	2535	2541	2547	2553	2559	2564	1	1	2	2	3	4	4	5	5
.41	2570	2576	2582	2588	2594	2600	2606	2612	2618	2624	1	1	2	2	3	4	4	5	5
.42	2630	2636	2642	2649	2655	2661	2667	2673	2679	2685	1	1	2	2	3	4	4	5	6
.43	2692	2698	2704	2710	2716	2723	2729	2735	2742	2748	1	1	2	3	3	4	4	5	6
.44	2754	2761	2767	2773	2780	2786	2793	2799	2805	2812	1	1	2	3	3	4	4	5	6
.45	2818	2825	2831	2838	2844	2851	2858	2864	2871	2877	1	1	2	3	3	4	5	5	6
.46	2884	2891	2897	2904	2911	2917	2924	2931	2938	2944	1	1	2	3	3	4	5	5	6
.47	2951	2958	2965	2972	2979	2985	2992	2999	3006	3013	1	1	2	3	3	4	5	5	6
.48	3020	3027	3034	3041	3048	3055	3062	3069	3076	3083	1	1	2	3	4	4	5	6	6
.49	3090	3097	3105	3112	3119	3126	3133	3141	3148	3155	1	1	2	3	4	4	5	6	6
	0	1	2	3	4	5	6	7	8	9	1	2	3	4	5	6	7	8	9

Antilogarithms

	0	1	2	3	4	5	6	7	8	9	1	2	3	4	5	6	7	8	9
											Proportional Parts								
.50	3162	3170	3177	3184	3192	3199	3206	3214	3221	3228	1	1	2	3	4	4	5	6	7
.51	3236	3243	3251	3258	3266	3273	3281	3289	3296	3304	1	2	2	3	4	5	5	6	7
.52	3311	3319	3327	3334	3342	3350	3357	3365	3373	3381	1	2	2	3	4	5	5	6	7
.53	3388	3396	3404	3412	3420	3428	3436	3443	3451	3459	1	2	2	3	4	5	6	6	7
.54	3467	3475	3483	3491	3499	3508	3516	3524	3532	3540	1	2	2	3	4	5	6	6	7
.55	3548	3556	3565	3573	3581	3589	3597	3606	3614	3622	1	2	2	3	4	5	6	7	7
.56	3631	3639	3648	3656	3664	3673	3681	3690	3698	3707	1	2	3	3	4	5	6	7	8
.57	3715	3724	3733	3741	3750	3758	3767	3776	3784	3793	1	2	3	3	4	5	6	7	8
.58	3802	3811	3819	3828	3837	3846	3855	3864	3873	3882	1	2	3	4	4	5	6	7	8
.59	3890	3899	3908	3917	3926	3936	3945	3954	3963	3972	1	2	3	4	5	5	6	7	8
.60	3981	3990	3999	4009	4018	4027	4036	4046	4055	4064	1	2	3	4	5	6	6	7	8
.61	4074	4083	4093	4102	4111	4121	4130	4140	4150	4159	1	2	3	4	5	6	7	8	9
.62	4169	4178	4188	4198	4207	4217	4227	4236	4246	4256	1	2	3	4	5	6	7	8	9
.63	4266	4276	4285	4295	4305	4315	4325	4335	4345	4355	1	2	3	4	5	6	7	8	9
.64	4365	4375	4385	4395	4406	4416	4426	4436	4446	4457	1	2	3	4	5	6	7	8	9
.65	4467	4477	4487	4498	4508	4519	4529	4539	4550	4560	1	2	3	4	5	6	7	8	9
.66	4571	4581	4592	4603	4613	4624	4634	4645	4656	4667	1	2	3	4	5	6	7	9	10
.67	4677	4688	4699	4710	4721	4732	4742	4753	4764	4775	1	2	3	4	5	7	8	9	10
.68	4786	4797	4808	4819	4831	4842	4853	4864	4875	4887	1	2	3	4	6	7	8	9	10
.69	4898	4909	4920	4932	4943	4955	4966	4977	4989	5000	1	2	3	5	6	7	8	9	10
.70	5012	5023	5035	5047	5058	5070	5082	5093	5105	5117	1	2	4	5	6	7	8	9	11
.71	5129	5140	5152	5164	5176	5188	5200	5212	5224	5236	1	2	4	5	6	7	8	10	11
.72	5248	5260	5272	5284	5297	5309	5321	5333	5346	5358	1	2	4	5	6	7	9	10	11
.73	5370	5383	5395	5408	5420	5433	5445	5458	5470	5483	1	3	4	5	6	8	9	10	11
.74	5495	5508	5521	5534	5546	5559	5572	5585	5598	5610	1	3	4	5	6	8	9	10	12
.75	5623	5636	5649	5662	5675	5689	5702	5715	5728	5741	1	3	4	5	7	8	9	10	12
.76	5754	5768	5781	5794	5808	5821	5834	5848	5861	5875	1	3	4	5	7	8	9	11	12
.77	5888	5902	5916	5929	5943	5957	5970	5984	5998	6012	1	3	4	5	7	8	10	11	12
.78	6026	6039	6053	6067	6081	6095	6109	6124	6138	6152	1	3	4	6	7	8	10	11	13
.79	6166	6180	6194	6209	6223	6237	6252	6266	6281	6295	1	3	4	6	7	9	10	11	13
.80	6310	6324	6339	6353	6368	6383	6397	6412	6427	6442	1	3	4	6	7	9	10	12	13
.81	6457	6471	6486	6501	6516	6531	6546	6561	6577	6592	2	3	5	6	8	9	11	12	14
.82	6607	6622	6637	6653	6668	6683	6699	6714	6730	6745	2	3	5	6	8	9	11	12	14
.83	6761	6776	6792	6808	6823	6839	6855	6871	6887	6902	2	3	5	6	8	9	11	13	14
.84	6918	6934	6950	6966	6982	6998	7015	7031	7047	7063	2	3	5	6	8	10	11	13	15
.85	7079	7096	7112	7129	7145	7161	7178	7194	7211	7228	2	3	5	7	8	10	12	13	15
.86	7244	7261	7278	7295	7311	7328	7345	7362	7379	7396	2	3	5	7	8	10	12	13	15
.87	7413	7430	7447	7464	7482	7499	7516	7534	7551	7568	2	3	5	7	9	10	12	14	16
.88	7586	7603	7621	7638	7656	7674	7691	7709	7727	7745	2	4	5	7	9	11	12	14	16
.89	7762	7780	7798	7816	7834	7852	7870	7889	7907	7925	2	4	5	7	9	11	13	14	16
.90	7943	7962	7980	7998	8017	8035	8054	8072	8091	8110	2	4	6	7	9	11	13	15	17
.91	8128	8147	8166	8185	8204	8222	8241	8260	8279	8299	2	4	6	8	9	11	13	15	17
.92	8318	8337	8356	8375	8395	8414	8433	8453	8472	8492	2	4	6	8	10	12	14	15	17
.93	8511	8531	8551	8570	8590	8610	8630	8650	8670	8690	2	4	6	8	10	12	14	16	18
.94	8710	8730	8750	8770	8790	8810	8831	8851	8872	8892	2	4	6	8	10	12	14	16	18
.95	8913	8933	8954	8974	8995	9016	9036	9057	9078	9099	2	4	6	8	10	12	15	17	19
.96	9120	9141	9162	9183	9204	9226	9247	9268	9290	9311	2	4	6	8	11	13	15	17	19
.97	9333	9354	9376	9397	9419	9441	9462	9484	9506	9528	2	4	7	9	11	13	15	17	20
.98	9550	9572	9594	9616	9638	9661	9683	9705	9727	9750	2	4	7	9	11	13	16	18	20
.99	9772	9795	9817	9840	9863	9886	9908	9931	9954	9977	2	5	7	9	11	14	16	18	20
	0	1	2	3	4	5	6	7	8	9	1	2	3	4	5	6	7	8	9

Index

Table of Atomic Weights (1986)

Scaled to the relative atomic mass $A_r(^{12}C) = 12$

Name	Symbol	Atomic Number	Atomic Weight	Name	Symbol	Atomic Number	Atomic Weight
Actinium	Ac	89	227.027 8	Hydrogen	H	1	1.007 94
Aluminum	Al	13	26.981 539	Indium	In	49	114.82
Americium	Am	95	(243)	Iodine	I	53	126.904 47
Antimony	Sb	51	121.75	Iridium	Ir	77	192.22
Argon	Ar	18	39.948	Iron	Fe	26	55.847
Arsenic	As	33	74.921 59	Krypton	Kr	36	83.80
Astatine	At	85	(210)	Lanthanum	La	57	138.905 5
Barium	Ba	56	137.327	Lawrencium	Lr	103	(260)
Berkelium	Bk	97	(247)	Lead	Pb	82	207.2
Beryllium	Be	4	9.012 182	Lithium	Li	3	6.941
Bismuth	Bi	83	208.980 37	Lutetium	Lu	71	174.967
Boron	B	5	10.811	Magnesium	Mg	12	24.305 0
Bromine	Br	35	79.904	Manganese	Mn	25	54.938 05
Cadmium	Cd	48	112.411	Mendelevium	Md	101	(258)
Calcium	Ca	20	40.078	Mercury	Hg	80	200.59
Californium	Cf	98	(251)	Molybdenum	Mo	42	95.94
Carbon	C	6	12.011	Neodymium	Nd	60	144.24
Cerium	Ce	58	140.115	Neon	Ne	10	20.179 7
Cesium	Cs	55	132.905 43	Neptunium	Np	93	237.048 2
Chlorine	Cl	17	35.452 7	Nickel	Ni	28	58.69
Chromium	Cr	24	51.996 1	Niobium	Nb	41	92.906 38
Cobalt	Co	27	58.933 20	Nitrogen	N	7	14.006 74
Copper	Cu	29	63.546	Nobelium	No	102	(259)
Curium	Cm	96	(247)	Osmium	Os	76	190.2
Dysprosium	Dy	66	162.50	Oxygen	O	8	15.999 4
Einsteinium	Es	99	(252)	Palladium	Pd	46	106.42
Erbium	Er	68	167.26	Phosphorus	P	15	30.973 762
Europium	Eu	63	151.965	Platinum	Pt	78	195.08
Fermium	Fm	100	(257)	Plutonium	Pu	94	(244)
Fluorine	F	9	18.998 403 2	Polonium	Po	84	(209)
Francium	Fr	87	(223)	Potassium	K	19	39.098 3
Gadolinium	Gd	64	157.25	Praseodymium	Pr	59	140.907 65
Gallium	Ga	31	69.723	Promethium	Pm	61	(145)
Germanium	Ge	32	72.61	Protactinium	Pa	91	231.035 9
Gold	Au	79	196.966 54	Radium	Ra	88	226.025 4
Hafnium	Hf	72	178.49	Radon	Rn	86	(222)
Helium	He	2	4.002 602	Rhenium	Re	75	186.207
Holmium	Ho	67	164.930 32	Rhodium	Rh	45	102.905 50

TABLE OF ATOMIC WEIGHTS

Name	Symbol	Atomic Number	Atomic Weight	Name	Symbol	Atomic Number	Atomic Weight
Rubidium	Rb	37	85.4678	Thulium	Tm	69	168.934 21
Ruthenium	Ru	44	101.07	Tin	Sn	50	118.710
Samarium	Sm	62	150.36	Titanium	Ti	22	47.88
Scandium	Sc	21	44.955 910	Tungsten	W	74	183.85
Selenium	Se	34	78.96	(Wolfram)			
Silicon	Si	14	28.0855	Unnilhexium	Unh	106	(263)
Silver	Ag	47	107.8682	Unnilpentium	Unp	105	(262)
Sodium	Na	11	22.989 768	Unnilquadium	Unq	104	(261)
Strontium	Sr	38	87.62	Unnilseptium	Uns	107	(262)
Sulfur	S	16	32.066	Uranium	U	92	238.0289
Tantalum	Ta	73	180.9479	Vanadium	V	23	50.941 5
Technetium	Tc	43	(98)	Xenon	Xe	54	131.29
Tellurium	Te	52	127.60	Ytterbium	Yb	70	173.04
Terbium	Tb	65	158.925 34	Yttrium	Y	39	88.905 85
Thallium	Tl	81	204.3833	Zinc	Zn	30	65.39
Thorium	Th	90	232.0381	Zirconium	Zr	40	91.224

A value in parentheses for a radioactive element is the atomic mass number of the isotope of that element of longest known half-life.